Handbook of Lipids

Volume I

Handbook of Lipids
Volume I

Edited by **Kyle Walker**

R Callisto Reference

New York

Published by Callisto Reference,
106 Park Avenue, Suite 200,
New York, NY 10016, USA
www.callistoreference.com

Handbook of Lipids: Volume I
Edited by Kyle Walker

International Standard Book Number: 978-1-63239-402-6 (Hardback)

Printed in the United States of America.

Contents

Preface

The term 'lipids', refers to a group of naturally occurring organic compounds which includes fats, sterols and vitamins among numerous others. Some of their characteristics are solubility in nonpolar organic solvents like acetone, ether, benzene and chloroform, their inability to get dissolved in hydrogen dioxide among numerous others. Lipids perform important biological functions such as serving the function of structural components of cell membranes, storing energy, signaling among others. These compounds also find applications in nanotechnology, cosmetics and food industries. Lipids can essentially be categorized into hydrophobic or amphiphilic small molecules. Some of their varieties are:

(1) Fatty Acids are a type of lipids are all moderate to long chain fatty acids and are essentially esters.
(2) Soaps and Detergents are salts and carboxylic acids which are made up of longer chains which contain more than eight carbons. Both hydrophobic (alkyl) as well as hydrophilic (CO_2) regions are present in the same molecule.
(3) Fats and Oils are tri-esters of fatty acids which contain 1,2,3-trihydroxypropane (glycerol). These types of lipids are present in both animal and plants, and happen to be a major lipid group in our diet.
(4) Waxes are essentially esters of fatty acids. These contain long chains of a hydroxyl group known as monohydric alcohols. They also contain hydrocarbons.
(5) Phospholipids are the main constituents of cell membranes.

Besides these, there are many more types of lipids like Terpenes, Steroids, Lipid Soluble, and Vitamins Biosynthetic Pathways.

Hopefully, this text will be used by students, and will continue to be of value to them throughout their subsequent professional careers. However, I wish to clarify here that this does not mean that this book has been written only for students. In fact, I hope that this book will cater to research scholars, biologists, manufacturing engineers and chemists alike. I convey our heartfelt thanks to all the contributors and to the team at the publishing house for their encouragement and excellent technical assistance as and when required.

Editor

Nuclear Receptors in Nonalcoholic Fatty Liver Disease

Jorge A. López-Velázquez, Luis D. Carrillo-Córdova, Norberto C. Chávez-Tapia, Misael Uribe, and Nahum Méndez-Sánchez

Liver Research Unit, Medica Sur Clinic & Foundation, Puente de Piedra 150, Colonia Toriello Guerra, 14050 Tlalpan, Mexico City, Mexico

Correspondence should be addressed to Nahum Méndez-Sánchez, nmendez@medicasur.org.mx

Academic Editor: Piero Portincasa

Nuclear receptors comprise a superfamily of ligand-activated transcription factors that are involved in important aspects of hepatic physiology and pathophysiology. There are about 48 nuclear receptors in the human. These nuclear receptors are regulators of many hepatic processes including hepatic lipid and glucose metabolism, bile acid homeostasis, drug detoxification, inflammation, regeneration, fibrosis, and tumor formation. Some of these receptors are sensitive to the levels of molecules that control lipid metabolism including fatty acids, oxysterols, and lipophilic molecules. These receptors direct such molecules to the transcriptional networks and may play roles in the pathogenesis and treatment of nonalcoholic fatty liver disease. Understanding the mechanisms underlying the involvement of nuclear receptors in the pathogenesis of nonalcoholic fatty liver disease may offer targets for the development of new treatments for this liver disease.

1. Introduction

Liver diseases are a serious problem throughout the world. In Mexico, since 2000, cirrhosis and other chronic liver diseases have become among the main causes of mortality [1]. The incidence and prevalence of liver diseases are increasing along with changes in lifestyle and population aging, and these diseases were responsible for 20,941 deaths in 2007 [2].

In Mexico, the incidence of metabolic syndrome is also increasing. The metabolic syndrome has recently been associated with nonalcoholic fatty liver disease (NAFLD), and about 90% of patients with NAFLD have more than one feature of the metabolic syndrome [3]. The severity of NAFLD is one factor contributing to the development of nonalcoholic steatohepatitis (NASH), cirrhosis, and hepatocellular carcinoma [4, 5]. The growing obesity epidemic requires a better understanding of the genetic networks and signal transduction pathways that regulate the pathogenesis of these conditions. A clear definition of the mechanisms responsible for metabolic control may provide new knowledge for the development of new drugs, with novel mechanisms of action, for the treatment of chronic liver diseases.

The ability of individual nuclear receptors (NRs) to regulate multiple genetic networks in different tissues and their own ligands may represent a new class of potential drugs targets. To elucidate the challenges involved in developing such drugs, this paper focuses on the role of hepatic NRs in lipid metabolism and the possible effects on the physiopathology of NAFLD.

2. Nonalcoholic Fatty Liver Disease

NAFLD is defined by the accumulation of triglycerides in the form of droplets (micro- and macrovesicles) within hepatocytes [6]. The mechanism involves impaired insulin regulation, which affects fat and glucose metabolism (intermediary metabolism) in the liver, skeletal muscle, and adipose tissue, a condition known as insulin resistance. Insulin resistance increases free fatty acids and hepatic *de novo* lipogenesis, causes dysfunction in fatty acid oxidation, and alters very-low-density lipoprotein (VLDL) triglyceride export [7].

NAFLD is associated with insulin resistance, obesity, and a lifestyle characterized by physical inactivity and an unlimited supply of high-fat foods. However, more recent studies

TABLE 1: Nuclear receptors in hepatic lipid metabolism.

RXR partner	Ligands	Official name	Role in hepatic lipid metabolism
LXRα	Oxysterols (22(R)-hydroxycholesterol, 24(S)-hydroxycholesterol, 24(S),25-epoxycholesterol, 27-hydroxycholesterol) and fatty acids	NR1H3	(i) Increases fatty acid synthesis, TG level, HDL level, cholesterol secretion (ii) Upregulation of SREBP-c $\left\{\begin{array}{l}\text{FAS}\\\text{ACC}\\\text{SCD1}\end{array}\right.$ (iii) Upregulation of ChREBP, Angptl3 (iv) Downregulation of ApoA-V
PPARα	Fatty acids, fibrates, statins, eicosanoids, and leukotrienes	NR1C1	(i) Promotes fatty acid oxidation (by lipoprotein lipase activation) (ii) Improves insulin resistance (iii) Suppression: acyl CoA oxidase (ACO-OX), acyl CoA synthase (ACS), enoyl-CoA hydratase, malic enzyme, HMG-CoA synthase, mitochondrial enzymes, APOA1 and APOCIII
FXR	Bile acids, pregnadiene, and fexaramine	NR1H4	(i) Induces lipoprotein metabolism genes/clearance represses hepatic genes involved in the synthesis of TG (ii) Induces human PPARα (iii) Increases hepatic expression of receptors VLDL (iv) Reduces: hepatic lipogenesis and plasma triglyceride and cholesterol levels (v) Decreases expression of proteins apoC-III and Angptl3 (inhibitors of LPL)
PXR	Pregnanes, progesterone, and glucocorticoids, LCA, xenobiotics/drugs, rifampicin	NR1I2	(i) Induces lipogenesis by increasing expression of the fatty acid translocase CD36, SCD-1, and long-chain free fatty acid elongase (ii) Suppression of several genes involved in fatty acid β-oxidation (PPARα, thiolase, carnitine palmitoyltransferase 1a (Cpt1a), and mitochondrial 3-hydroxy-3-methylglutaryl CoA synthase 2 (Hmgcs2))
CAR	Androstane metabolites, estrogens, progesterone, and xenobiotics	NR1I3	(i) Induction of Insig-1, a protein with antilipogenic properties (ii) Interacts with PPARα during fasting (iii) Suppresses lipid metabolism and lowers serum triglyceride level by reducing SREBP-1 level

have proposed that not all individuals with NAFLD develop insulin resistance before the presence of a fatty liver [3, 8].

NAFLD is a cluster of metabolic, histological, and molecular disorders characterized by liver injury [9]. The purpose of this paper is to describe the complex working of NRs and their role in the hepatic accumulation of fat independent of excessive alcohol consumption.

NRs are ligand-activated transcription factors that have a broad range of metabolic, detoxifying, and regulatory functions. NRs are sensitive to the levels of many natural and synthetic ligands including hormones, biomolecules (lipids), vitamins, bile acids, metabolites, drugs, and xenobiotic toxins. In addition to their functions at the hepatic level, NRs also control hepatic inflammation, regeneration, fibrosis, and tumor formation [10]. These functions can be understood through a complex transcriptional network that allows them to maintain cellular nutrient homeostasis, to protect against toxins by limiting their uptake and facilitating their metabolism and excretion, and to play a role in several key steps in inflammation and fibrosis [11].

New knowledge about the functions of NRs helps clarify the pathogenesis and pathophysiology of a wide spectrum of hepatic disorders (see Table 1).

3. Nuclear Receptor Structure

The NRs are characterized by a central DNA-binding domain, which targets the receptor to specific DNA sequences known as hormone-response elements. The DNA-binding domain comprises two highly conserved zinc fingers that isolate the nuclear receptors from other DNA-binding proteins. The C-terminal half of the receptor encompasses the ligand-binding domain, which possesses the essential property of ligand recognition and ensures both specificity and selectivity of the physiological response [12, 13]. The predominant role of these receptors is the transcriptional regulation of enzymes and other proteins involved in energy homeostasis (Figure 1(a)).

4. Action Mode of Nuclear Receptors

NRs act in three steps [14]: repression, derepression, and transcription activation. Repression is characteristic of the apo-NR, which recruits a corepressor complex with histone deacetylase activity. Derepression occurs following ligand binding, which dissociates this complex and recruits the first coactivator complex, with histone acetyltransferase activity,

(a)

(b)

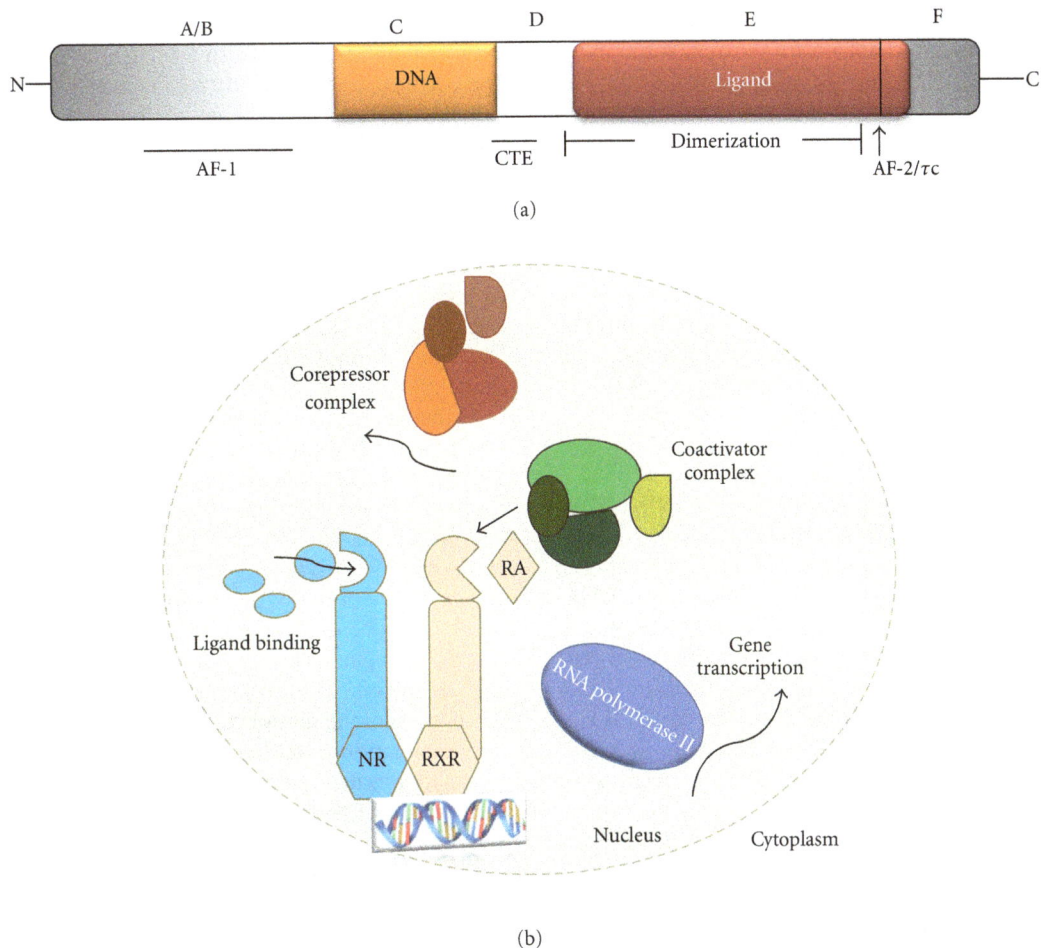

FIGURE 1: (a) Schematic representation of a typical nuclear receptor. Nuclear receptors may be divided into five regions based on structural and functional similarities (denoted A, B, C, D, E, and F). Regions C and E contain the conserved DNA-binding domains (DBDs) and ligand-binding domains (LBDs) that are the signature of this superfamily. In addition, the constitutive transport element (CTE) is a dimerization region within the LBD and two transactivation domains (denoted AF-1 and AF-2/τc). A second dimerization domain (not shown) exists in the DBD and is required for heterodimerization of receptors on response elements. (b) NR function. Ligand binding to NRs triggers changes in their conformation leading to the dissociation of corepressors and the recruitment of coactivators. After this exchange of coregulators, RNA polymerase II is recruited and mRNA transcription is initiated. Most NRs bind to their DNA response elements in a sequence-specific manner as dimers, functioning either as homodimers or as heterodimers with the RXR. RA: retinoic acid. Modified from [13, 94].

and causes chromatin decondensation, which is believed to be necessary, but not sufficient, for activation of the target gene. In the third step, transcription activation, the histone acetyltransferase complex dissociates to cause the assembly of a second coactivator, which can establish contact with the basal transcriptional machinery to activate the target gene [15] (Figure 1(b)).

Coactivators are molecules recruited by ligand-bound activated NRs (or other DNA-binding transcription factors) that increase gene expression. Coactivators contribute to the transcriptional process through a diverse array of enzymatic activities such as acetylation, methylation, ubiquitination, and phosphorylation, or as chromatin remodelers [16].

The result is the modulation of the expression of a wide array of physiologically important groups of genes involved in diverse pathological processes including cancer, inherited genetic diseases, metabolic disorders, and inflammation.

In contrast to the coactivator function, corepressors interact with NRs that are not bound to the ligand and repress transcription. Corepressor-associated proteins such as histone deacetylases enforce a local chromatin environment that opposes the transcription-promoting activities of coactivators [17].

5. Nuclear Receptors in the Liver

The hepatocyte is responsible for processes involved in providing for many of the body's metabolic needs, including the synthesis and control of the pathways involved in the metabolism of cholesterol, fatty acids, carbohydrates, amino acids, serum proteins, and bile acids, and the detoxification of drugs and xenobiotics.

The hepatocyte employs multiple levels of regulation to perform its functions and possesses self-protective processes to avoid self-destruction. Some members of the NR super-family provide hepatic mechanisms for self-regulation in hepatocytes [18].

Gene regulation by NRs is more complex than simply the presence of a potential DNA recognition sequence in a promoter. Rather, it is a complex and multilayered process that involves competition between agonists and antagonists, heterodimerization, coregulator recruitment, and NR protein modification.

The NR family comprises 48 family members and is the largest group of transcriptional regulators in the human. Because some NRs participate in the control of hepatic homeostasis, they may provide a new therapeutic target for the treatment of liver diseases such as NAFLD [19].

5.1. Liver X Receptor. The transcriptional factor liver X receptor (LXR) is involved in cholesterol metabolism. The LXR gene encodes two distinct products, LXRα and LXRβ, each with diverse patterns of expression but similar target DNA-binding elements and ligands. The human LXRα gene is located on chromosome 11p11.2, and the human LXRβ gene is located on chromosome 19q13.3. We will focus on LXRα because of its high expression in the liver, although it is also expressed at lower levels in the kidney, intestine, lung, fat, adrenal, spleen, and macrophages [20, 21]. The ligands for LXR are oxysterols. Once activated, LXR induces the expression of a cluster of genes that function in lipid metabolism; these functions are cholesterol absorption, efflux, transport, and excretion [22–24]. Besides its metabolic role, LXRs also modulate immune and inflammatory responses in macrophages [25].

Like most other nuclear receptors, LXR forms heterodimers with the retinoid X receptor (RXR) within the nucleus. Binding of the RXR to LXR leads to the formation of a complex with corepressors such as silencing mediator of retinoic acid, thyroid hormone receptor, and nuclear corepressor [26].

In the absence of a ligand, these corepressor interactions are maintained and the transcriptional activity of target genes is suppressed. Binding of a ligand to LXR causes a conformational change that facilitates inactivation of the corepressor complex and the transcription of target genes [27].

LXR is a key regulator of whole-body lipid and bile acid metabolism [20, 28] (Figure 2). LXR regulates a cluster of genes that participate in the transport of excess cholesterol in the form of high-density lipoprotein (HDL) from peripheral tissue to the liver—a process called reverse cholesterol transport. *In vivo* activation of LXR with a synthetic, high-affinity ligand increases the HDL level and net cholesterol secretion [29]. LXR positively regulates several enzymes involved in lipoprotein metabolism including lipoprotein lipase (LPL), human cholesteryl ester transport protein, and the phospholipid transfer protein [30]. LXR also regulates the crucial bile acid enzyme CYP7A1. In rodents, this enzyme contains an LXR response element that is upregulated in response to excess cholesterol in the diet. The enzymatic activation and conversion of cholesterol to bile acids is one mechanism for handling excess dietary cholesterol [31–33].

In addition to its ability to modulate cholesterol and bile acid metabolism, LXR is also a key regulator of hepatic lipogenesis. Its lipogenic activity results from the upregulation of the master regulator of hepatic lipogenesis sterol regulatory element-binding protein-c (SREBP-c) and from the induction of fatty acid synthase, acyl coenzyme A carboxylase, and stearoyl CoA desaturase 1, all leading to increased hepatic lipid levels [34, 35], one of the etiological agents in the pathogenesis of NAFLD. Moreover, LXR induces the carbohydrate-response element-binding protein, ChREBP [36]. ChREBP is a target gene of LXR and is a glucose-sensitive transcription factor that promotes the hepatic conversion of carbohydrates into lipids. Several important proteins might mediate the LXR-mediated hypertriglyceridemic effect. These include angiopoietin-like protein 3 (Angptl3) [37], a liver-secreted protein that increases the concentrations of both plasma triglycerides by inhibiting LPL activity in different tissues and free fatty acids by activating lipolysis in adipocytes and/or apoA-V. LXR activation increases Angptl3 expression and downregulates apoA-V expression [38]. The second "hit" in NAFLD is related to the proinflammatory molecules, whose expression is repressed by LXR. These include inducible nitric oxide synthase, cyclooxygenase 2, interleukin-6 (IL-6), IL-1β, chemokine monocyte chemoattractant protein-1, and chemokine monocyte chemoattractant protein-3 [39].

LXR-activated pathways play central roles in whole-body lipid metabolism by regulating multiple pathways in liver cells. Further investigation into the effects of synthetic LXR-specific agonists and/or antagonists may provide new therapeutic tools for the treatment of NAFLD.

5.2. Peroxisome Proliferator-Activated Receptors. NAFLD appears to be a link between insulin resistance and obesity. Several recent studies have shown that a family of transcription factors, named the peroxisome-proliferator-activated receptors (PPARs), improve several of the metabolic abnormalities associated with insulin resistance and impaired fat metabolism [40].

The PPARs are nuclear hormone receptors. Three isotypes have been identified in humans: PPARα, PPARβ/δ, and PPARγ [41]. These receptors exhibit different tissue distribution and functions and, to some extent, different ligand specificities. PPARα is highly expressed in the liver, brown adipose tissue, heart, skeletal muscle, kidney, and at lower levels in other organs. PPARγ is highly expressed in adipose tissues and is present in the colon and lymphoid organs. PPARβ/δ is expressed ubiquitously, but its levels may vary considerably [42, 43].

Mechanistically, the PPARs also form heterodimers with the RXR and activate transcription by binding to a specific DNA element, termed the peroxisome proliferator response element (PPRE), in the regulatory region of several genes encoding proteins that are involved in lipid metabolism and energy balance. Binding of agonists causes a conformational change that promotes the binding to transcriptional coactivators. Conversely, binding of antagonists induces a conformation that favors the binding of corepressors. Physiologically, PPAR-RXR heterodimers may bind to PPREs in the absence of a ligand, although the transcriptional activation depends on the ligand-bound PPAR-RXR [44, 45]. The predominant role of these receptors is the transcriptional regulation of enzymes and other proteins involved in energy homeostasis, some of which are in the liver. To explain their possible action in the development and treatment of NAFLD, a brief description of each PPAR follows [46, 47].

In the liver, PPARα promotes fatty acid oxidation. It is the target for the hypolipidemic fibrates, such as fenofibrate, clofibrate, and gemfibrozil, which are used in the treatment of hypertriglyceridemia [48].

The role of PPARα in hepatic fatty acid metabolism is especially prominent during fasting. In fasted PPARα-null mice, its absence is associated with pronounced hepatic steatosis, decreased levels of plasma glucose and ketone bodies, and elevated plasma free fatty acids levels, and hypothermia. These severe metabolic disturbances are the result of the decreased expression of many genes involved in hepatic lipid metabolism. The PPARα target genes are those for acyl CoA oxidase (ACO-OX), acyl CoA synthase (ACS), enoyl-CoA hydratase, malic enzyme, HMG CoA synthase, mitochondrial enzymes, liver-fatty-acid-binding protein, and fatty acid transport protein. PPARα can also regulate other genes such as LPL, which is involved in the degradation of triglycerides, and APOA1 and APOCIII, which are both downregulated by PPARα [49–55] (Figure 2).

Whereas PPARα controls lipid catabolism and homeostasis in the liver, PPARγ promotes the storage of lipids in adipose tissues and plays a pivotal role in adipocyte differentiation. It is a target of the insulin-sensitizing thiazolidinediones. Despite its relatively low expression levels in healthy liver, PPARγ is critical for the development of NAFLD [56].

In the liver, PPARβ/δ is protective against liver toxicity induced by environmental chemicals, possibly by downregulating the expression of proinflammatory genes. PPARβ/δ regulates glucose utilization and lipoprotein metabolism by promoting reverse cholesterol transport [57–60]. PPARs appear to be targets for the treatment of metabolic disorders. PPARα and PPARγ are already therapeutic targets for the treatment of hypertriglyceridemia and insulin resistance, respectively, disorders that relate directly to the progress of NAFLD. The discovery of more pathways may provide new treatments for hepatopathies.

5.3. Farnesoid X Receptor. The farnesoid X receptor (FXR), a member of the NR superfamily, has a typical NR structure and contains a hydrophobic pocket that accommodates lipophilic molecules such as bile acids [61]. Its gene is located on chromosome 12, and it is expressed predominantly in the liver, gut, kidneys, and adrenals and at lower levels in white adipose tissue [62, 63]. The FXR binds to specific response elements as a heterodimer with the RXR, although it has also been reported to bind DNA as a monomer [28, 64]. The main physiological role of the FXR is to act as a bile acid sensor in the enterohepatic tissues. FXR activation regulates the expression of various transport proteins and biosynthetic enzymes crucial to the physiological maintenance of bile acids and lipid and carbohydrate metabolism.

Bile acids bind to and activate this NR. The order of potency of FXR binding to bile acids is chenodeoxycholic acid > lithocholic acid = deoxycholic acid > cholic acid [65, 66].

In addition to their well-established roles in bile acid metabolism, recent data have demonstrated that activation of the FXR is also implicated in lipid metabolism. Activation of the FXR reduces both hepatic lipogenesis and plasma triglyceride and cholesterol levels, induces the genes implicated in lipoprotein metabolism/clearance, and represses hepatic genes involved in the synthesis of triglycerides [67]. The FXR promotes reverse transport of cholesterol by increasing hepatic uptake of HDL cholesterol via two independent mechanisms. The first is FXR-mediated suppression of hepatic lipase expression [68]. Hepatic lipase reduces HDL particle size by hydrolyzing its triglycerides and phospholipids in hepatic sinusoids, which facilitates hepatic uptake of HDL cholesterol. The second mechanism is the induction by the FXR of the expression of the gene for scavenger receptor B1, the HDL uptake transporter in the liver [69].

Activation of the FXR also increases the hepatic expression of receptors such as VLDL receptor and syndecan-1, which are involved in lipoprotein clearance, and increases the expression of ApoC-II, which coactivates lipoprotein lipase (LPL). FXR activation also decreases the expression of proteins such as ApoC-III and Angptl3 [70] that normally function as inhibitors of LPL. Finally, the FXR induces human PPARα [71], an NR that functions to promote fatty acid β-oxidation. Taken together, these data suggest that FXR activation lowers plasma triglyceride levels via both repressing SREBP1-c and triglyceride secretion and increasing the clearance of triglyceride-rich lipoproteins from the blood (Figure 2).

In carbohydrate metabolism, activation of the hepatic FXR regulates gluconeogenesis, glycogen synthesis, and insulin sensitivity [72]. The bile acid sensor FXR also has anti-inflammatory properties in the liver and intestine, mainly by interacting with NF-κB signaling. FXR agonists might therefore represent useful agents to reduce inflammation in cells with high FXR expression levels, such as hepatocytes, and to prevent or delay cirrhosis and cancer development in inflammation-driven liver diseases.

These data suggest that FXR activation by its ligands would reduce hepatic steatosis and that such activation may have a beneficial role in NAFLD by decreasing hepatic de novo lipogenesis, which constitutes the first "hit" of the disease. Inflammatory processes lead to the development of hepatitis and subsequent liver fibrosis. The hepatic FXR appears to be downregulated during the acute-phase response in rodents in a manner similar to that seen for other NRs such as PPARα and the LXR [73].

5.4. The Pregnant X Receptor and Constitutive Androstane Receptor. The pregnane X receptor (PXR) and constitutive androstane receptor (CAR) share some common ligands and have an overlapping target gene pattern. The CAR gene is the product of the NR1I3 gene located on chromosome 1, locus 1q23, whereas hPXR is the product of the NR1I2 gene,

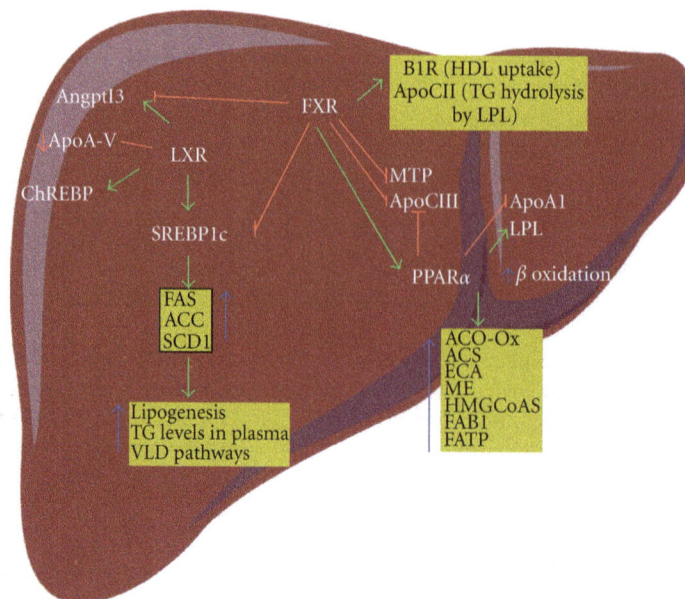

FIGURE 2: NRs as central regulators of hepatic lipid metabolism. Oxysterols activate the LXR, whereas bile acids (BA) stimulate SHP expression through the FXR (not shown). The LXR activates SREBP-1c and induces *de novo* fatty acid (FA) synthesis and hypertriglyceridemia by activating FAS, ACC, SCD1, and ChREBP (a glucose-sensitive transcription factor that promotes the hepatic conversion of carbohydrates into lipids). Several important proteins that could mediate the LXR-mediated hypertriglyceridemic effect are regulated. One protein is angiopoietin-like protein 3 (Angptl3), a liver-secreted protein that increases both plasma triglyceride level by inhibiting LPL activity in different tissues and free fatty acid level by activating lipolysis in adipocytes. LXR activation increases the expression of Angptl3 and LPL and downregulates apoA-V expression. Activation of the FXR leads to the repression of hepatic lipogenesis by reducing the expression of SREBP-1c. By increasing the expression of PPARα, the FXR also promotes FFA catabolism via β-oxidation, which induces ACO-OX, ACS, ECA, HMG-CoAS, FAB1, and FATP. By repressing the expression of MTP, an enzyme that controls VLDL assembly, the FXR reduces VLDL production. Activation of the FXR increases TG clearance by promoting LPL activity, via induction of ApoC-II and B1R. Activation of the FXR also reduces TG clearance by decreasing the expression of ApoC-III and Angptl3, two LPL inhibitors. PPARα can be activated by FXR and fibrates (not shown). PPAR activation leads to β-oxidation, which induces ACO-Ox, ACS, ECA, HMG-CoAS, FAB1, and FATP. Others genes are regulated. For example, LPL, which is involved in the degradation of TG, is activated, and APOA1 and APOCIII are both downregulated. The activation pathways are shown by green arrows, inhibitory pathways by red lines, and inhibited activation pathways by broken green arrows. Angptl3: angiopoietin-like protein 3; ACC: acetyl-CoA carboxylase; Apo: apolipoprotein; ChREBP: carbohydrate response element-binding protein; FAS: fatty acid synthase; FATP: fatty acid transport protein; FXR: farnesoid X receptor; LPL: lipoprotein lipase; LXR: liver X receptor; MTP: microsomal triglyceride transfer protein; PPAR: peroxisome proliferator-activated receptor; SCD1: stearoyl-coenzyme A desaturase 1; SREBP-1c: sterol regulatory element-binding protein-1c; TG: triglyceride. Arrows and stop bars indicate positive regulation or activation and negative regulation or repression, respectively.

which is located on chromosome 3, locus 3q12–q13.3 [74–76]. Like most other NRs, the PXR and CAR have an N-terminal DNA-binding domain and a C-terminal ligand-binding domain. PXR and CAR regulate gene expression by forming heterodimers with the RXR.

The PXR is located in the nucleus and has a low basal activity and is highly activated upon ligand binding [77, 78]. By contrast, in the noninduced state, the CAR resides in the cytoplasm. Compounds that activate the CAR and PXR are structurally very diverse; most are small and are highly lipophilic [79]. The PXR is activated by pregnanes, progesterone, and glucocorticoids [80, 81], whereas the CAR is affected both positively and negatively by androstane metabolites, estrogens, and progesterone [82, 83]. For this reason, in addition to functioning as xenobiotic receptors, the PXR and CAR are thought to be endobiotic receptors that influence physiology and diseases [84, 85].

For example, several studies have shown that the PXR induces lipogenesis in a SREBP-independent manner. Lipid accumulation and marked hepatic steatosis in PXR-transgenic mice are associated with increased expression of the fatty acid translocase CD36 (also called FAT) and several accessory lipogenic enzymes, such as SCD-1 and long-chain free fatty acid elongase. CD36, a multiligand scavenger receptor present on the surface of a number of cell types, may contribute to hepatic steatosis by facilitating the high-affinity uptake of fatty acids from the circulation [86]. The CD36 level in the liver correlates with hepatic triglyceride storage and secretion, suggesting that CD36 plays a causative role in the pathogenesis of hepatic steatosis [87]. PXR may also promote hepatic steatosis by increasing the expression of CD36 directly or indirectly through the PXR-mediated activation of PPARγ [86].

Interestingly, an independent study showed that hepatic triglyceride level decreases temporarily after short-term (10-hour) activation of the PXR [88]. PXR activation is also associated with upregulation of PPARγ, a positive regulator of CD36 and a master regulator of adipogenesis [89]. PXR

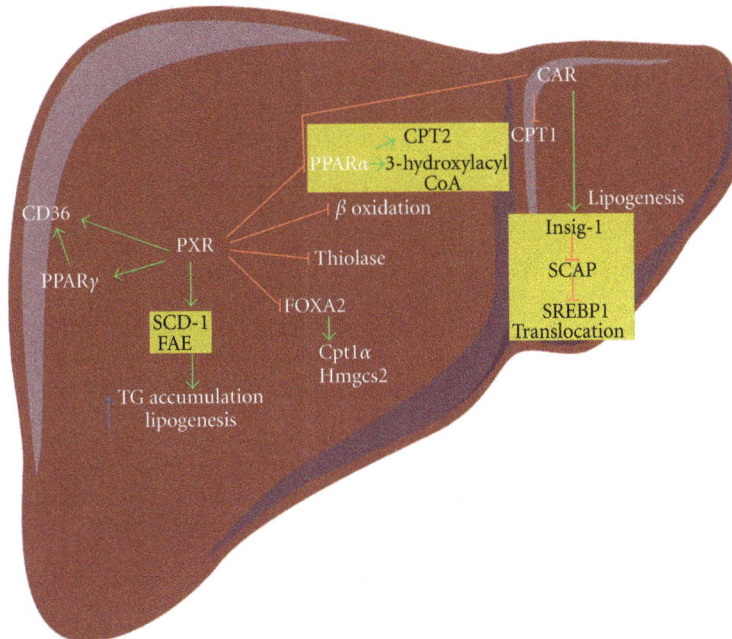

FIGURE 3: Activation of the PXR induces lipogenesis and inhibits fatty acid β-oxidation. The PXR induces lipogenesis through activation of CD36, PPARγ, SCD1, and FAE gene expression. The PXR inhibits fatty acid β-oxidation through its suppression of PPARα and thiolase gene expression. In addition, PXR binds to FoxA2, a key regulator of β-oxidation, and inhibits FoxA1-mediated activation of Cpt1a and Hmgcs2 gene expression. CAR activation inhibits lipogenesis by inducing Insig-1, a protein that plays a role in SREBP-mediated regulation of lipogenic genes. Insig proteins bind and trap SCAP, retaining it in the ER and preventing it from escorting SREBPs to the site of proteolytic activation in the Golgi complex (not shown). SREBPs are cleaved by two proteases in the Golgi complex, and the bHLH-Zip domain of SREBPs transfers from the membrane to the nucleus to bind the sterol response elements in the promoter region of the target genes (not shown). CAR inhibits fatty acid β-oxidation. CAR competes with PPARα for its binding site in the 3-hydroxyacyl CoA dehydrogenase gene promoter. Activation of CAR also decreases the expression of Cpt1, a rate-limiting enzyme of β-oxidation. Arrows and stop bars indicate positive regulation or activation and negative regulation or repression, respectively. Cpt1a: carnitine palmitoyltransferase 1a; FAE: long-chain free fatty acid elongase; FoxA2: forkhead box factor A2; Hmgcs2: mitochondrial 3-hydroxy-3-methylglutaryl CoA synthase 2; PPAR: peroxisome proliferator-activated receptor; SCAP: SREBP cleavage-activating protein; SCD1: stearoyl CoA desaturase 1; SREBP: sterol regulatory-element binding protein.

activation is also associated with suppression of several genes involved in fatty acid β-oxidation, such as PPARα and thiolase [90]. A study by Nakamura and colleagues showed that PXR represses β-oxidation-related genes such as carnitine palmitoyltransferase 1a (Cpt1a) and mitochondrial 3-hydroxy-3-methylglutaryl CoA synthase 2 (Hmgcs2) through crosstalk with the insulin-responsive forkhead box factor A2 (FoxA2) (Figure 3).

Activation of the CAR might suppress lipid metabolism and lower serum triglyceride levels by reducing the level of SREBP-1, a master regulator of lipid metabolism. The inhibitory effects of the CAR on lipid metabolism might also be attributed to induction of Insig-1, a protein with antilipogenic properties [88].

The CAR interacts with PPARα during fasting and has been reported to interfere with fatty acid metabolism by binding to DNA elements overlapping with the PPARα-binding site in the promoter region of 3-hydroxyacyl CoA dehydrogenase, an important enzyme in peroxisomal fatty acid β-oxidation [91] (Figure 3).

Finally, other studies indicate that the CAR might be involved in the pathogenesis of NASH [92] by regulating the response of serum triglyceride level to metabolic stress

[93]. The overlap of the activation of endogenous lipids by the CAR and PXR suggests a functional connection between these receptors in liver physiology. This knowledge might be useful in the development of new treatments to limit or prevent the pathogenesis of NAFLD by developing agonists or antagonists to prevent or lessen lipid accumulation within the liver parenchyma.

6. Conclusion

NAFLD encompasses a spectrum of conditions characterized histologically by hepatic steatosis ranging from simple fatty liver to NASH cirrhosis and HCC [4].

NRs control fatty acid transport from peripheral adipose tissue to the liver and regulate several critical metabolic steps involved in the pathogenesis of NAFLD, including fat storage, export, uptake, oxidation, and lipolysis [94]. The discovery that many ligands activate the whole family of NRs (FXR, LXR, PPARs, PXR, and CAR) and their possible interconnected mechanisms that control lipid metabolism suggests the possibility of developing novel therapies for the treatment of NAFLD. The LXR and PXR regulate several metabolically relevant pathways and clusters of genes that lead to hepatic lipogenesis and might be directly related to the pathogenesis of liver diseases. The FXR, PPARα, and CAR are activated by ligands to orchestrate a broad range of lipolytic activities. These might become future candidates for drugs designed to target metabolic liver disorders.

Acknowledgments

This study was supported by grant from the Mexican National Research Council (CONACYT 62460) and from Medica Sur Clinic & Foundation.

References

[1] N. Méndez-Sánchez, E. García-Villegas, B. Merino-Zeferino et al., "Liver diseases in Mexico and their associated mortality trends from 2000 to 2007: a retrospective study of the nation and the federal states," *Annals of Hepatology*, vol. 9, no. 4, pp. 428–438, 2010.

[2] N. Méndez-Sánchez, A. R. Villa, N. C. Chávez-Tapia et al., "Trends in liver disease prevalence in Mexico from 2005 to 2050 through mortality data," *Annals of Hepatology*, vol. 4, no. 1, pp. 52–55, 2005.

[3] N. Méndez-Sánchez, N. C. Chávez-Tapia, and M. Uribe, "Obesity and non-alcoholic steatohepatitis," *Gaceta Medica de Mexico*, vol. 140, supplement 2, pp. S67–S72, 2004.

[4] G. C. Farrell and C. Z. Larter, "Nonalcoholic fatty liver disease: from steatosis to cirrhosis," *Hepatology*, vol. 43, no. 2, pp. S99–S112, 2006.

[5] N. C. Chavez-Tapia, N. Méndez-Sánchez, and M. Uribe, "Role of nonalcoholic fatty liver disease in hepatocellular carcinoma," *Annals of Hepatology*, vol. 8, supplement 1, pp. S34–S39, 2009.

[6] N. Méndez-Sánchez, M. Arrese, D. Zamora-Valdés, and M. Uribe, "Current concepts in the pathogenesis of nonalcoholic fatty liver disease," *Liver International*, vol. 27, no. 4, pp. 423–433, 2007.

[7] M. Adiels, M. R. Taskinen, C. Packard et al., "Overproduction of large VLDL particles is driven by increased liver fat content in man," *Diabetologia*, vol. 49, no. 4, pp. 755–765, 2006.

[8] N. Méndez-Sánchez, N. C. Chavez-Tapia, D. Zamora-Valdés, and M. Uribe, "Adiponectin, structure, function and pathophysiological implications in non-alcoholic fatty liver disease," *Mini-Reviews in Medicinal Chemistry*, vol. 6, no. 6, pp. 651–656, 2006.

[9] C. A. Matteoni, Z. M. Younossi, T. Gramlich, N. Boparai, Y. C. Liu, and A. J. McCullough, "Nonalcoholic fatty liver disease: a spectrum of clinical and pathological severity," *Gastroenterology*, vol. 116, no. 6, pp. 1413–1419, 1999.

[10] M. Robinson-Rechavi, H. E. Garcia, and V. Laudet, "The nuclear receptor superfamily," *Journal of Cell Science*, vol. 116, no. 4, pp. 585–586, 2003.

[11] C. H. Lee and L. N. Wei, "Characterization of an inverted repeat with a zero spacer (IR0)-type retinoic acid response element from the mouse nuclear orphan receptor TR2-11 gene," *Biochemistry*, vol. 38, no. 27, pp. 8820–8825, 1999.

[12] D. J. Mangelsdorf, C. Thummel, M. Beato et al., "The nuclear receptor super-family: the second decade," *Cell*, vol. 83, no. 6, pp. 835–839, 1995.

[13] D. J. Mangelsdorf and R. M. Evans, "The RXR heterodimers and orphan receptors," *Cell*, vol. 83, no. 6, pp. 841–850, 1995.

[14] D. M. Lonard and B. W. O'Malley, "The expanding cosmos of nuclear receptor coactivators," *Cell*, vol. 125, no. 3, pp. 411–414, 2006.

[15] M. Beato, "Transcriptional control by nuclear receptors," *FASEB Journal*, vol. 5, no. 7, pp. 2044–2051, 1991.

[16] H. Gronemeyer, J. A. Gustafsson, and V. Laudet, "Principles for modulation of the nuclear receptor superfamily," *Nature Reviews Drug Discovery*, vol. 3, no. 11, pp. 950–964, 2004.

[17] D. M. Lonard, R. B. Lanz, and B. W. O'Malley, "Nuclear receptor coregulators and human disease," *Endocrine Reviews*, vol. 28, no. 5, pp. 575–587, 2007.

[18] S. J. Karpen, "Nuclear receptor regulation of hepatic function," *Journal of Hepatology*, vol. 36, no. 6, pp. 832–850, 2002.

[19] M. Arrese and S. J. Karpen, "Nuclear receptors, inflammation, and liver disease: insights for cholestatic and fatty liver diseases," *Clinical Pharmacology and Therapeutics*, vol. 87, no. 4, pp. 473–478, 2010.

[20] T. T. Lu, J. J. Repa, and D. J. Mangelsdorf, "Orphan nuclear receptors as eLiXiRs and FiXeRs of sterol metabolism," *Journal of Biological Chemistry*, vol. 276, no. 41, pp. 37735–37738, 2001.

[21] P. J. Willy, K. Umesono, E. S. Ong, R. M. Evans, R. A. Heyman, and D. J. Mangelsdorf, "LXR, a nuclear receptor that defines a distinct retinoid response pathway," *Genes and Development*, vol. 9, no. 9, pp. 1033–1045, 1995.

[22] S. B. Joseph, B. A. Laffitte, P. H. Patel et al., "Direct and indirect mechanisms for regulation of fatty acid synthase gene expression by liver X receptors," *Journal of Biological Chemistry*, vol. 277, no. 13, pp. 11019–11025, 2002.

[23] J. D. Horton, J. L. Goldstein, and M. S. Brown, "SREBPs: activators of the complete program of cholesterol and fatty acid synthesis in the liver," *Journal of Clinical Investigation*, vol. 109, no. 9, pp. 1125–1131, 2002.

[24] J. J. Repa, G. Liang, J. Ou et al., "Regulation of mouse sterol regulatory element-binding protein-1c gene (SREBP-1c) by oxysterol receptors, LXRα and LXRβ," *Genes and Development*, vol. 14, no. 22, pp. 2819–2830, 2000.

[25] A. Castrillo and P. Tontonoz, "Nuclear receptors in macrophage biology: at the crossroads of lipid metabolism and inflammation," *Annual Review of Cell and Developmental Biology*, vol. 20, pp. 455–480, 2004.

[26] J. D. Chen and R. M. Evans, "A transcriptional co-repressor that interacts with nuclear hormone receptors," *Nature*, vol. 377, no. 6548, pp. 454–457, 1995.

[27] C. K. Glass and M. G. Rosenfeld, "The coregulator exchange in transcriptional functions of nuclear receptors," *Genes and Development*, vol. 14, no. 2, pp. 121–141, 2000.

[28] P. A. Edwards, H. R. Kast, and A. M. Anisfeld, "BAREing it all: the adoption of LXR and FXR and their roles in lipid homeostasis," *Journal of Lipid Research*, vol. 43, no. 1, pp. 2–12, 2002.

[29] G. F. Lewis and D. J. Rader, "New insights into the regulation of HDL metabolism and reverse cholesterol transport," *Circulation Research*, vol. 96, no. 12, pp. 1221–1232, 2005.

[30] Y. Zhang, J. J. Repa, K. Gauthier, and D. J. Mangelsdorf, "Regulation of lipoprotein lipase by the oxysterol receptors, LXRα and LXRβ," *Journal of Biological Chemistry*, vol. 276, no. 46, pp. 43018–43024, 2001.

[31] J. J. Repa, S. D. Turley, J. M. A. Lobaccaro et al., "Regulation of absorption and ABC1-mediated efflux of cholesterol by RXR heterodimers," *Science*, vol. 289, no. 5484, pp. 1524–1529, 2000.

[32] J. Y. L. Chiang, R. Kimmel, and D. Stroup, "Regulation of cholesterol 7α-hydroxylase gene (CYP7A1) transcription by the liver orphan receptor (LXRα)," *Gene*, vol. 262, no. 1-2, pp. 257–265, 2001.

[33] T. T. Lu, M. Makishima, J. J. Repa et al., "Molecular basis for feedback regulation of bile acid synthesis by nuclear receptos," *Molecular Cell*, vol. 6, no. 3, pp. 507–515, 2000.

[34] B. A. Laffitte, J. J. Repa, S. B. Joseph et al., "LXRs control lipid-inducible expression of the apolipoprotein E gene in macrophages and adipocytes," *Proceedings of the National Academy of Sciences of the United States of America*, vol. 98, no. 2, pp. 507–512, 2001.

[35] Y. Luo and A. R. Tall, "Sterol upregulation of human CETP expression in vitro and in transgenic mice by an LXR element," *Journal of Clinical Investigation*, vol. 105, no. 4, pp. 513–520, 2000.

[36] J. Y. Cha and J. J. Repa, "The liver X receptor (LXR) and hepatic lipogenesis: the carbohydrate-response element-binding protein is a target gene of LXR," *Journal of Biological Chemistry*, vol. 282, no. 1, pp. 743–751, 2007.

[37] T. Inaba, M. Matsuda, M. Shimamura et al., "Angiopoietin-like protein 3 mediates hypertriglyceridemia induced by the liver X receptor," *Journal of Biological Chemistry*, vol. 278, no. 24, pp. 21344–21351, 2003.

[38] H. Jakel, M. Nowak, E. Moitrot et al., "The liver X receptor ligand T0901317 down-regulates APOA5 gene expression through activation of SREBP-1c," *Journal of Biological Chemistry*, vol. 279, no. 44, pp. 45462–45469, 2004.

[39] S. B. Joseph, A. Castrillo, B. A. Laffitte, D. J. Mangelsdorf, and P. Tontonoz, "Reciprocal regulation of inflammation and lipid metabolism by liver X receptors," *Nature Medicine*, vol. 9, no. 2, pp. 213–219, 2003.

[40] P. Dandona, A. Aljada, and A. Bandyopadhyay, "Inflammation: the link between insulin resistance, obesity and diabetes," *Trends in Immunology*, vol. 25, no. 1, pp. 4–7, 2004.

[41] J. P. Berger, T. E. Akiyama, and P. T. Meinke, "PPARs: therapeutic targets for metabolic disease," *Trends in Pharmacological Sciences*, vol. 26, no. 5, pp. 244–251, 2005.

[42] J. Auwerx, E. Baulieu, M. Beato et al., "A unified nomenclature system for the nuclear receptor superfamily," *Cell*, vol. 97, no. 2, pp. 161–163, 1999.

[43] B. P. Kota, T. H. W. Huang, and B. D. Roufogalis, "An overview on biological mechanisms of PPARs," *Pharmacological Research*, vol. 51, no. 2, pp. 85–94, 2005.

[44] B. Zingarelli, M. Sheehan, P. W. Hake, M. O'Connor, A. Denenberg, and J. A. Cook, "Peroxisome proliferator activator receptor-γ ligands, 15-deoxy-Δ12,14-prostaglandin J2 and ciglitazone, reduce systemic inflammation in polymicrobial sepsis by modulation of signal transduction pathways," *Journal of Immunology*, vol. 171, no. 12, pp. 6827–6837, 2003.

[45] S. Yu and J. K. Reddy, "Transcription coactivators for peroxisome proliferator-activated receptors," *Biochimica et Biophysica Acta*, vol. 1771, no. 8, pp. 936–951, 2007.

[46] T. Hashimoto, W. S. Cook, C. Qi, A. V. Yeldandi, J. K. Reddy, and M. S. Rao, "Defect in peroxisome proliferator-activated receptor α-inducible fatty acid oxidation determines the severity of hepatic steatosis in response to fasting," *Journal of Biological Chemistry*, vol. 275, no. 37, pp. 28918–28928, 2000.

[47] S. Kersten, J. Seydoux, J. M. Peters, F. J. Gonzalez, B. Desvergne, and W. Wahli, "Peroxisome proliferator-activated receptor α mediates the adaptive response to fasting," *Journal of Clinical Investigation*, vol. 103, no. 11, pp. 1489–1498, 1999.

[48] R. Hess, W. Stäubli, and W. Riess, "Nature of the hepatomegalic effect produced by ethyl-chlorophenoxy- isobutyrate in the rat," *Nature*, vol. 208, no. 5013, pp. 856–858, 1965.

[49] C. Qi, Y. Zhu, and J. K. Reddy, "Peroxisome proliferator-activated receptors, coactivators, and downstream targets," *Cell Biochemistry and Biophysics*, vol. 32, pp. 187–204, 2000.

[50] M. Ricote, A. C. Li, T. M. Willson, C. J. Kelly, and C. K. Glass, "The peroxisome proliferator-activated receptor-γ is a negative regulator of macrophage activation," *Nature*, vol. 391, no. 6662, pp. 79–82, 1998.

[51] N. Latruffe and J. Vamecq, "Peroxisome proliferators and peroxisome proliferator activated receptors (PPARs) as regulators of lipid metabolism," *Biochimie*, vol. 79, no. 2-3, pp. 81–94, 1997.

[52] G. Martin, H. Poirier, N. Hennuyer et al., "Induction of the fatty acid transport protein 1 and acyl-CoA synthase genes by dimer-selective rexinoids suggests that the peroxisome proliferator- activated receptor-retinoid X receptor heterodimer is their molecular target," *Journal of Biological Chemistry*, vol. 275, no. 17, pp. 12612–12618, 2000.

[53] S. Fourcade, S. Savary, S. Albet et al., "Fibrate induction of the adrenoleukodystrophy-related gene (ABCD2): promoter analysis and role of the peroxisome proliferator-activated receptor PPARα," *European Journal of Biochemistry*, vol. 268, no. 12, pp. 3490–3500, 2001.

[54] S. É. Michaud and G. Renier, "Direct regulatory effect of fatty acids on macrophage lipoprotein lipase: potential role of PPARs," *Diabetes*, vol. 50, no. 3, pp. 660–666, 2001.

[55] M. S. Rao and J. K. Reddy, "Peroxisomal β-oxidation and steatohepatitis," *Seminars in Liver Disease*, vol. 21, no. 1, pp. 43–55, 2001.

[56] S. Yu, K. Matsusue, P. Kashireddy et al., "Adipocyte-specific gene expression and adipogenic steatosis in the mouse liver due to peroxisome proliferator-activated receptor γ1 (PPARγ1) overexpression," *Journal of Biological Chemistry*, vol. 278, no. 1, pp. 498–505, 2003.

[57] G. D. Barish, V. A. Narkar, and R. M. Evans, "PPARδ: a dagger in the heart of the metabolic syndrome," *Journal of Clinical Investigation*, vol. 116, no. 3, pp. 590–597, 2006.

[58] J. B. Hansen, H. Zhang, T. H. Rasmussen, R. K. Petersen, E. N. Flindt, and K. Kristiansen, "Peroxisome proliferator-activated receptor δ (PPARδ)-mediated regulation of preadipocyte proliferation and gene expression is dependent on cAMP signaling," *Journal of Biological Chemistry*, vol. 276, no. 5, pp. 3175–3182, 2001.

[59] M. Schuler, F. Ali, C. Chambon et al., "PGC1α expression is controlled in skeletal muscles by PPARβ, whose ablation results in fiber-type switching, obesity, and type 2 diabetes," *Cell Metabolism*, vol. 4, no. 5, pp. 407–414, 2006.

[60] W. R. Oliver Jr., J. L. Shenk, M. R. Snaith et al., "A selective peroxisome proliferator-activated receptor δ agonist promotes reverse cholesterol transport," *Proceedings of the National Academy of Sciences of the United States of America*, vol. 98, no. 9, pp. 5306–5311, 2001.

[61] Y. Zhang, H. R. Kast-Woelbern, and P. A. Edwards, "Natural structural variants of the nuclear receptor farnesoid X receptor affect transcriptional activation," *Journal of Biological Chemistry*, vol. 278, no. 1, pp. 104–110, 2003.

[62] B. M. Forman, E. Goode, J. Chen et al., "Identification of a nuclear receptor that is activated by farnesol metabolites," *Cell*, vol. 81, no. 5, pp. 687–693, 1995.

[63] W. Seol, H. S. Choi, and D. D. Moore, "Isolation of proteins that interact specifically with the retinoid X receptor: two novel orphan receptors," *Molecular Endocrinology*, vol. 9, no. 1, pp. 72–85, 1995.

[64] N. Y. Kalaany and D. J. Mangelsdorf, "LXRs and FXR: the Yin and Yang of cholesterol and fat metabolism," *Annual Review of Physiology*, vol. 68, pp. 159–191, 2006.

[65] M. Makishima, A. Y. Okamoto, J. J. Repa et al., "Identification of a nuclear receptor for bite acids," *Science*, vol. 284, no. 5418, pp. 1362–1365, 1999.

[66] D. J. Parks, S. G. Blanchard, R. K. Bledsoe et al., "Bile acids: natural ligands for an orphan nuclear receptor," *Science*, vol. 284, no. 5418, pp. 1365–1368, 1999.

[67] Y. Zhang, L. W. Castellani, C. J. Sinal, F. J. Gonzalez, and P. A. Edwards, "Peroxisome proliferator-activated receptor-γ coactivator 1α (PGC-1α) regulates triglyceride metabolism by activation of the nuclear receptor FXR," *Genes and Development*, vol. 18, no. 2, pp. 157–169, 2004.

[68] A. Sirvent, T. Claudel, G. Martin et al., "The farnesoid X receptor induces very low density lipoprotein receptor gene expression," *FEBS Letters*, vol. 566, no. 1–3, pp. 173–177, 2004.

[69] G. Lambert, M. J. A. Amar, G. Guo, H. B. Brewer Jr., F. J. Gonzalez, and C. J. Sinal, "The farnesoid X-receptor is an essential regulator of cholesterol homeostasis," *Journal of Biological Chemistry*, vol. 278, no. 4, pp. 2563–2570, 2003.

[70] F. Y. Lee, H. Lee, M. L. Hubbert, P. A. Edwards, and Y. Zhang, "FXR, a multipurpose nuclear receptor," *Trends in Biochemical Sciences*, vol. 31, no. 10, pp. 572–580, 2006.

[71] I. P. Torra, T. Claudel, C. Duval, V. Kosykh, J. C. Fruchart, and B. Staels, "Bile acids induce the expression of the human peroxisome proliferator-activated receptor α gene via activation of the farnesoid X receptor," *Molecular Endocrinology*, vol. 17, no. 2, pp. 259–272, 2003.

[72] B. Cariou, K. van Harmelen, D. Duran-Sandoval et al., "The farnesoid X receptor modulates adiposity and peripheral insulin sensitivity in mice," *Journal of Biological Chemistry*, vol. 281, no. 16, pp. 11039–11049, 2006.

[73] M. S. Kim, J. Shigenaga, A. Moser, K. Feingold, and C. Grunfeld, "Repression of farnesoid X receptor during the acute phase response," *Journal of Biological Chemistry*, vol. 278, no. 11, pp. 8988–8995, 2003.

[74] L. B. Moore, D. J. Parks, S. A. Jones et al., "Orphan nuclear receptors constitutive androstane receptor and pregnane X receptor share xenobiotic and steroid ligands," *Journal of Biological Chemistry*, vol. 275, no. 20, pp. 15122–15127, 2000.

[75] W. Xie, J. L. Barwick, C. M. Simon et al., "Reciprocal activation of xenobiotic response genes by nuclear receptors SXR/PXR and CAR," *Genes and Development*, vol. 14, no. 23, pp. 3014–3023, 2000.

[76] P. Wei, J. Zhang, D. H. Dowhan, Y. Han, and D. D. Moore, "Specific and overlapping functions of the nuclear hormone receptors CAR and PXR in xenobiotic response," *Pharmacogenomics Journal*, vol. 2, no. 2, pp. 117–126, 2002.

[77] J. T. Moore, L. B. Moore, J. M. Maglich, and S. A. Kliewer, "Functional and structural comparison of PXR and CAR," *Biochimica et Biophysica Acta*, vol. 1619, no. 3, pp. 235–238, 2003.

[78] S. A. Kliewer, B. Goodwin, and T. M. Willson, "The nuclear pregnane X receptor: a key regulator of xenobiotic metabolism," *Endocrine Reviews*, vol. 23, no. 5, pp. 687–702, 2002.

[79] J. Chianale, L. Mulholland, P. G. Traber, and J. J. Gumucio, "Phenobarbital induction of cytochrome P-450 b,e genes is dependent on protein synthesis," *Hepatology*, vol. 8, no. 2, pp. 327–331, 1988.

[80] P. G. Quinn and D. Yeagley, "Insulin regulation of PEPCK gene expression: a model for rapid and reversible modulation," *Current Drug Targets: Immune, Endocrine and Metabolic Disorders*, vol. 5, no. 4, pp. 423–437, 2005.

[81] J. Nakae, T. Kitamura, D. L. Silver, and D. Accili, "The forkhead transcription factor Foxo1 (Fkhr) confers insulin sensitivity onto glucose-6-phosphatase expression," *Journal of Clinical Investigation*, vol. 108, no. 9, pp. 1359–1367, 2001.

[82] S. Kakizaki, Y. Yamazaki, D. Takizawa, and M. Negishi, "New insights on the xenobiotic-sensing nuclear receptors in liver diseases- CAR and PXR-," *Current Drug Metabolism*, vol. 9, no. 7, pp. 614–621, 2008.

[83] M. Qatanani, J. Zhang, and D. D. Moore, "Role of the constitutive androstane receptor in xenobiotic-induced thyroid hormone metabolism," *Endocrinology*, vol. 146, no. 3, pp. 995–1002, 2005.

[84] X. Ma, J. R. Idle, and F. J. Gonzalez, "The pregnane X receptor: from bench to bedside," *Expert Opinion on Drug Metabolism and Toxicology*, vol. 4, no. 7, pp. 895–908, 2008.

[85] A. Moreau, M. J. Vilarem, P. Maurel, and J. M. Pascussi, "Xenoreceptors CAR and PXR activation and consequences on lipid metabolism, glucose homeostasis, and inflammatory response," *Molecular Pharmaceutics*, vol. 5, no. 1, pp. 35–41, 2008.

[86] H. L. Jung, J. Zhou, and W. Xie, "PXR and LXR in hepatic steatosis: a new dog and an old dog with new tricks," *Molecular Pharmaceutics*, vol. 5, no. 1, pp. 60–66, 2008.

[87] Y. Konno, M. Negishi, and S. Kodama, "The roles of nuclear receptors CAR and PXR in hepatic energy metabolism," *Drug Metabolism and Pharmacokinetics*, vol. 23, no. 1, pp. 8–13, 2008.

[88] A. Roth, R. Looser, M. Kaufmann et al., "Regulatory cross-talk between drug metabolism and lipid homeostasis: constitutive androstane receptor and pregnane X receptor increase Insig-1 expression," *Molecular Pharmacology*, vol. 73, no. 4, pp. 1282–1289, 2008.

[89] R. K. Semple, V. K. K. Chatterjee, and S. O'Rahilly, "PPARγ and human metabolic disease," *Journal of Clinical Investigation*, vol. 116, no. 3, pp. 581–589, 2006.

[90] J. Zhou, Y. Zhai, Y. Mu et al., "A novel pregnane X receptor-mediated and sterol regulatory element-binding protein-independent lipogenic pathway," *Journal of Biological Chemistry*, vol. 281, no. 21, pp. 15013–15020, 2006.

[91] A. Kassam, C. J. Winrow, F. Fernandez-Rachubinski, J. P. Capone, and R. A. Rachubinski, "The peroxisome proliferator response element of the gene encoding the peroxisomal β-oxidation enzyme enoyl-CoA hydratase/3-hydroxyacyl-CoA dehydrogenase is a target for constitutive androstane receptor β/9-cis- retinoic acid receptor-mediated transactivation," *Journal of Biological Chemistry*, vol. 275, no. 6, pp. 4345–4350, 2000.

[92] Y. Yamazaki, S. Kakizaki, N. Horiguchi et al., "The role of the nuclear receptor constitutive androstane receptor in the pathogenesis of non-alcoholic steatohepatitis," *Gut*, vol. 56, no. 4, pp. 565–574, 2007.

[93] J. M. Maglich, D. C. Lobe, and J. T. Moore, "The nuclear receptor CAR (NR1I3) regulates serum triglyceride levels under conditions of metabolic stress," *Journal of Lipid Research*, vol. 50, no. 3, pp. 439–445, 2009.

[94] M. Wagner, G. Zollner, and M. Trauner, "Nuclear receptors in liver disease," *Hepatology*, vol. 53, no. 3, pp. 1023–1034, 2011.

2

Sphingolipid Metabolic Pathway: An Overview of Major Roles Played in Human Diseases

Raghavendra Pralhada Rao, Nanditha Vaidyanathan, Mathiyazhagan Rengasamy, Anup Mammen Oommen, Neeti Somaiya, and M. R. Jagannath

Biology Group, Connexios Life Sciences Private Limited, No.49 Shilpa Vidya, First Main Road, 3rd Phase JP Nagara, Bangalore 560078, India

Correspondence should be addressed to M. R. Jagannath; m.r.jagannath@connexios.com

Academic Editor: Philip W. Wertz

Sphingolipids, a family of membrane lipids, are bioactive molecules that participate in diverse functions controlling fundamental cellular processes such as cell division, differentiation, and cell death. Given that most of these cellular processes form the basis for several pathologies, it is not surprising that sphingolipids are key players in several pathological processes. This review discusses the role of the sphingolipid metabolic pathway in diabetes, Alzheimer's disease, and hepatocellular carcinoma, with a special emphasis on the changes in gene expression pattern in these disease conditions. For convenience, the sphingolipid metabolic pathway is divided into hypothetical compartments (modules) with each compartment representing a physiological process and changes in gene expression pattern are mapped to each of these modules. It appears that alterations in the gene expression pattern in these disease conditions are biased to manipulate the system in order to result in a particular disease.

1. Introduction

Sphingolipids are a class of natural lipids comprised of a sphingoid base backbone, sphingosine. Sphingosine N-acylated with fatty acids forms ceramide [1], a central molecule in the sphingolipid biology. A variety of charged, neutral, phosphorylated, or glycosylated moieties are attached to ceramide further forming complex sphingolipids [2] (see Figure 1 for details). For example, phosphoryl choline attached to ceramide makes the most abundant mammalian sphingolipid, sphingomyelin. These moieties result in both polar and nonpolar regions giving the molecules an amphipathic character which accounts for their tendency to aggregate into membranous structures. Furthermore, such variations found in their chemical structures allow them to play diverse roles in cellular metabolism. Research in the past decade has clearly indicated that sphingolipids are not just the structural components of cell membrane but also act as signaling molecules controlling a majority of cellular events including signal transduction, cell growth, differentiation, and apoptosis [3–5]. Ceramide, sphingosine, sphingosine-1-phosphate

(S1P), and ceramide-1-phosphate (C1P) have emerged as chief bioactive mediators in the context of sphingolipid biology.

Although sphingolipids contribute to only a small proportion of the total cellular lipid pool, their accumulation in certain cells may be a trigger for pathology of many diseases. Because of the presence of a highly integrated metabolic network among various bioactive sphingolipids, it can be implicated that manipulation of one enzyme or metabolite may result in unexpected changes in metabolite levels, enzyme activities, and cellular programs [6, 7].

Whilst the scientific literature has been enriched by articles focusing on structural diversity and cellular metabolism, this review focuses on how alterations in expression of genes involved in sphingolipid metabolism could result in the progression of severe diseases. We chose three most prevalent diseases: type 2 diabetes, Alzheimer's disease, and hepatocellular carcinoma and analyzed the nature of gene expression changes in these disease conditions. The gene expression changes were further translated into possible physiological effects and these effects were analyzed to check for any correlation to the key pathology of the disease in

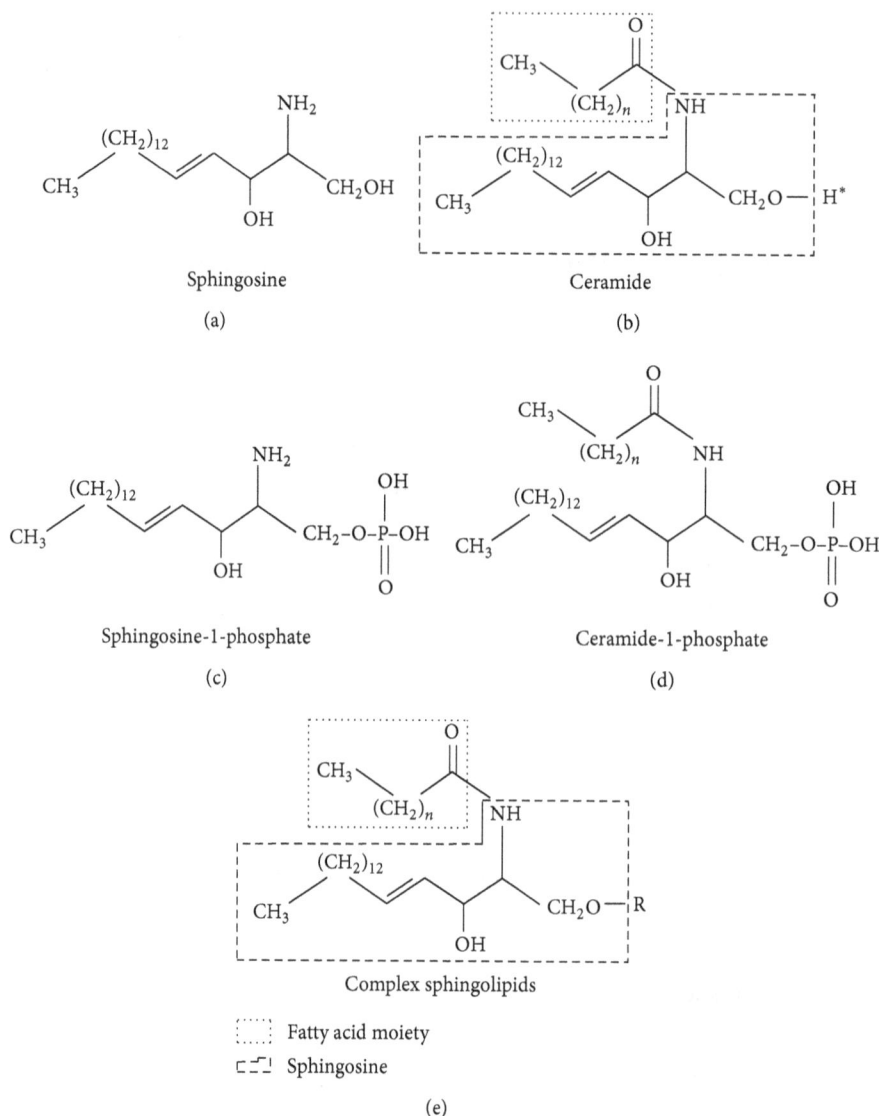

FIGURE 1: Structure of key sphingolipid molecules. All sphingolipids are comprised of a sphingoid base, and in mammals sphingosine (mainly C-18) is the major sphingoid base (a). A long chain fatty acid attached to sphingosine through amide linkage forms ceramide (b). Complex sphingolipids are obtained by replacement of hydrogen group of ceramide (H*) with various functional head groups (represented as R group in (e)). Complex sphingolipids vary in the nature of the polar head groups. For example, in sphingomyelin the head group is phosphocholine whereas in glycosphingolipids the head group could be one or more sugar residues. The phosphorylated derivatives, namely, sphingosine-1-phosphate (c) and ceramide 1 phosphate (d), are obtained by action of respective kinases on sphingosine and sphingosine-1-phosphate.

question. It turns out that, under these disease conditions at least, the sphingolipid metabolic pathway is modulated in such a way that the resultant changes in the physiology act as a significant contributor to the disease pathology.

2. Sphingolipid Metabolism: An Overview

The process of metabolism of sphingolipids has been studied extensively and most of the biochemical pathways of synthesis and degradation, including all the enzymes involved, have been determined successfully [8]. Sphingolipid metabolic pathway is an important cellular pathway that represents a

highly coordinated system linking together various pathways, where ceramide occupies a central position in both biosynthesis and catabolism, thereby crafting a metabolic hub [9]. The reaction sequences involved in the formation of ceramide and other sphingolipids are represented in Figure 2.

2.1. De Novo Synthesis. The first step in the *de novo* biosynthesis of sphingolipids is the condensation of serine and palmitoyl CoA, a reaction catalyzed by the rate-limiting enzyme, serine palmitoyltransferase (SPT, EC 2.3.1.50), to produce 3-ketodihydrosphingosine [10, 11]. Among various organisms, several metabolic divergences appear after the

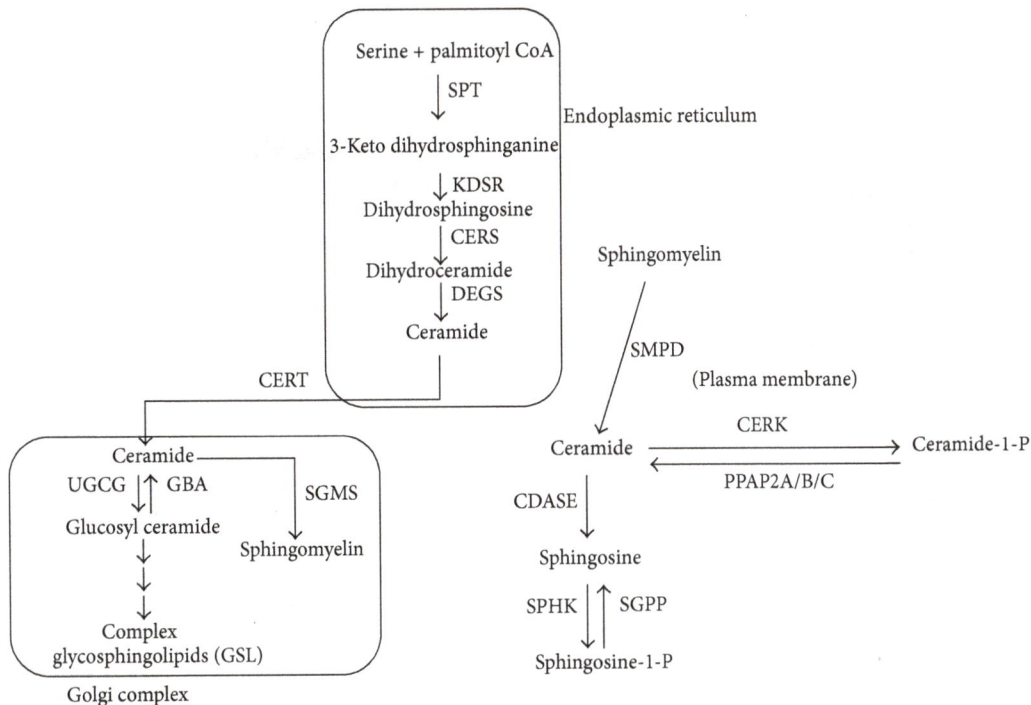

FIGURE 2: Key reactions involved in the sphingolipid metabolic pathway. Ceramide is produced in the ER and later transported to the Golgi complex for further conversion to complex sphingolipids. In addition to *de novo* synthesis, ceramide is also generated by hydrolysis of sphingomyelin. Ceramide is subject to conversion to various other sphingolipid intermediates like ceramide-1-phosphate, sphingosine, and sphingosine-1-phosphate. Cellular compartments are represented by boxes and enzymes are italicized. CDASE: ceramidase; CERK: ceramide kinase; CERS: ceramide synthase; CERT: ceramide transfer protein; DEGS: dihydroceramide desaturase; ER: endoplasmic reticulum; GBA: glucosyl ceramidase; GC: Golgi complex; KDSR: 3-keto dihydrosphinganine reductase; PM: plasma membrane; PPAP2A/B/C: phosphatidic acid phosphatase 2A/B/C; SGMS: sphingomyelin synthase; SGPP: sphingosine-1-phosphate phosphatase; SMPD: sphingomyelin phosphodiesterase; SPHK: sphingosine kinase; SPT: serine palmitoyll transferase; UGCG: UDP-glucose ceramide glucosyltransferase.

formation of sphinganine (dihydrosphingosine). In fungi and higher plants, sphinganine thus formed is first hydroxylated to phytosphingosine and then acylated to produce phytoceramide, whereas in animal cells sphinganine it is acylated to dihydroceramide which is later desaturated to form ceramide [12]. These reactions leading to generation of ceramide starting from serine take place in the endoplasmic reticulum. Ceramide thus generated needs to be transported to the Golgi complex, where it serves as a substrate for production of complex sphingolipids like sphingomyelin and glycosphingolipids (Figure 2). Both vesicular and non-vesicular transport mechanisms can mediate this process. The non-vesicular transport is mediated by the ceramide transfer protein (CERT) in mammals, in an ATP-dependent manner [13]. Once transported to the Golgi complex, several different head groups can be added to ceramide to form different classes of complex sphingolipids [14]. These complex sphingolipids will traverse different cellular locations mainly through vesicular transport.

2.2. Ceramide Homeostasis. Ceramide is considered as a molecule central to sphingolipid metabolic pathway and it serves as a branch point in the pathway. It acts as substrate not only for complex sphingolipids but also for the generation of ceramide-1-phosphate (C1P) and sphingosine, and

sphingosine can be further converted into sphingosine-1-phosphate (S1P) (reactions are outlined in Figure 2). Various secondary signaling intermediates produced by further conversion of ceramide can participate in diametrically opposite cellular processes; for example, ceramide and sphingosine are proapoptotic while their phosphorylated derivatives, C1P and S1P, are involved in progrowth activities [15]. Several studies have established that ceramide and sphingosine function as tumor-suppressor lipids mediating apoptosis, growth arrest, senescence, and differentiation. On the other hand, S1P and C1P are regarded as a tumor-promoting lipids involved in cell proliferation, migration, transformation, inflammation, and angiogenesis [16–19].

In addition to *de novo* biosynthesis, ceramide can be generated in the cell through hydrolysis of complex sphingolipids. The hydrolytic pathway controls the regeneration of ceramide from the complex sphingolipid pool, for example, from glycosphingolipids (GSLs) and sphingomyelin (SM), through the action of specific hydrolases and phosphodiesterases. Regeneration of ceramide from sphingomyelin is carried out by the plasma membrane bound enzyme SMPD (sphingomyelin phosphodiesterase) as presented in Figure 2. Ceramide regeneration from complex glycosphingolipids can be regulated by either lysosomal or nonlysosomal degradation. In lysosomal degradation, catabolism of GSLs occurs by cleavage of sugar residues which leads to the formation of glucosyl ceramide and galactosylceramide. Thereafter, specific β-glucosidases and galactosidases hydrolyze these lipids to produce ceramide that can later be deacylated by an acid ceramidase to form sphingosine [9, 16]. The sphingosine thus produced can further be salvaged to form ceramide (the intricate details of reactions involved in the degradative pathway of complex sphingolipids are well covered in the literature and are not discussed in this review; interested readers are recommended to refer to [16] from this article and relevant citations from this reference). Defects in the function of these enzymes lead to a variety of lysosomal storage disorders such as Gaucher, Sandhoff, and Tay-Sachs diseases [20]. Degradation of sphingolipids is also an indispensable component of lipid homeostasis; consequently SM levels are maintained by the catabolic action of sphingomyelinases (also called sphingomyelin phosphodiesterases, SMPD), releasing ceramide and the corresponding head group, phosphorylcholine.

2.2.1. Compartmental View of Sphingolipid Metabolic Pathway. It is evident that sphingolipid metabolic pathway is complex and involves several reactions and cellular organelles. Outcomes of these reactions in different organelles could lead to serous physiological consequences and this forms the basis for the role of sphingolipids in disease pathogenesis. In this review we have presented sphingolipid metabolic pathway as a combination of four sets of modules as depicted in Figure 3.

(i) The *de novo* biosynthesis of ceramide which occurs in ER is represented by the compartment C1.

(ii) The conversion of ceramide into complex sphingolipid like SM and glycosphingolipids is represented by compartment C2.

(iii) Hydrolysis of SM which produces ceramide is presented as compartment C3.

(iv) Conversion of ceramide into bioactive molecules such as C1P and S1P is represented by compartment C4.

Activities in each of these compartments can be mapped to physiological processes; for example, overall increase in the enzymes of *de novo* ceramide biosynthesis and hence increased levels of ceramide in compartment C1 can be designated as a proapoptotic scenario. Increased ceramide production in the ER followed by decreased CERT expression would result in an accumulation of ceramide in this organelle and hence can result in ER stress [21]. Decreased expression of enzymes involved in hydrolysis of complex sphingolipids as indicated in C3 can be characterized as sphingolipid storage. Increase in the components of sphingolipid rheostat (sphingosine kinase (SPHK1) and ceramide kinase (CERK)) thereby increasing the levels of C1P and S1P in compartment C4 can be taken to represent progrowth situations. Compartment C2 provides a channel wherein proapoptotic ceramide is converted into inert complex sphingolipids and this compartment can be considered as an adaptive channel since this channel is employed by drug resistant cancer cells as an adaptive mechanism. This approach provides

a more explanatory vision for finding links between severe pathological diseases and sphingolipid metabolism.

3. Sphingolipid Metabolism in Pathogenesis of Human Diseases

Human diseases resulting due to impaired sphingolipid metabolism are generally the outcome of defect in enzymes that degrade the sphingolipids [22]. In general, they are a group of relatively rare inborn errors of metabolism resulting in accumulation of sphingolipids (sphingolipidosis) caused by defects in the genes coding for proteins taking part in the lysosomal degradation of sphingolipids [23]. Historically, the pathological significance of sphingolipid related diseases is discussed with reference to sphingolipidosis. There has been a tremendous progress in the sphingolipid research in the recent past and it is now clear that sphingolipids can play a major role in pathogenesis of several diseases apart from traditionally studied sphingolipidosis. Deregulation of sphingolipid homeostasis is established as a key factor in the pathogenesis of several disorders like metabolic, neuronal, and proliferative disorders.

Recent studies have established the role of altered sphingolipid metabolism in brain cells in Alzheimer's disease [24]. The key role of sphingomyelinases in this disease is to promote apoptosis in neuronal cells through generation of proapoptotic molecule, ceramide [25]. In addition, serine palmitoyl transferase (SPT) has been shown to be downregulated by the amyloid precursor protein [26]. This exclusive physiological function of the amyloid precursor protein suggests the involvement of SPT and sphingolipid metabolism in Alzheimer's disease pathology.

It has been established for decades that patients with cirrhosis had increased levels of plasma long chain fatty acids (LCFA) as compared to controls, and the level of these substances is augmented with greater severity of the disease [27]. Furthermore, ceramide has been known as the prototype of sphingolipids that provokes cell death and its levels increase in response to apoptotic stimuli such as ionizing radiation or chemotherapy. In the liver, accumulation of ceramide may contribute to a variety of complications, leading to the substitution of steatosis to steatohepatitis [28], which can further develop into cirrhosis and hepatocellular carcinoma (HCC). Different studies have shown that either pharmacologic ceramide accumulation or systemic intravenous administration of liposomal ceramide is an effective approach against HCC [29].

Though a lot of research has already been done to emphasize the association of sphingolipids with several disorders, an understanding of the respective pathology in more detail with reference to sphingolipid metabolic pathway genes would be desirable. An analysis of the expression levels of genes involved in sphingolipid metabolic pathway presented below throws more light on the predisposition of sphingolipid pathway genes to execute the disease process.

Here we discuss the role of sphingolipid metabolism in three different disease conditions—type 2 diabetes mellitus

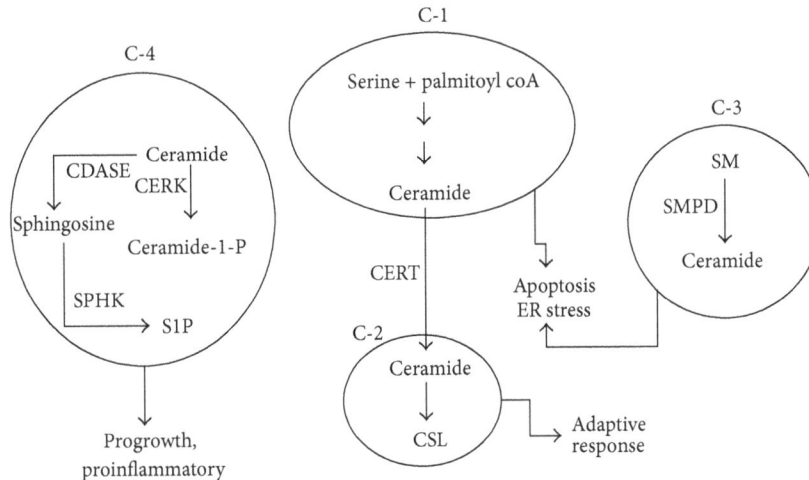

FIGURE 3: Hypothetical compartmentalization of sphingolipid metabolism. Reactions in the sphingolipid metabolic pathway can be isolated into different hypothetical compartments with each compartment representing ceramide either as a substrate or product of the reactions. Compartment C-1 is a *de novo* ceramide synthesis channel, compartment C-2 represents reactions leading to synthesis of complex sphingolipids, compartment C-3 is comprised of reaction leading to hydrolytic generation of ceramide from sphingomyelin, and compartment C-4 represents reactions in which ceramide is converted into other intermediates like sphingosine-1-phosphate and ceramide-1-phosphate. Activities/status of each compartment can be translated into different physiological events; accumulation of ceramide due to increased production of ceramide in C-1 would result in ER stress, increased generation of ceramide-1 phosphate, and sphingosine-1-phosphate in compartment C-4 results in progrowth and proinflammatory scenario in the cell. In different disease conditions, the status of each of these compartments contributes to the overall pathology of the disease.

(T2DM), Alzheimer's disease (AD), and hepatocellular carcinoma (HCC). For this purpose, we obtained the patterns in expression profiles of genes involved in sphingolipid metabolism under each of the diseased state from publicly available databases and mapped them to different physiological processes visualized through hypothetical compartments (Figure 3). Once the sphingolipid gene expression pattern from each study was mapped to a physiological process, possible effects of changes in the enzymes were predicted under diseased conditions. This is discussed in detail in the subsequent sections.

3.1. Role of Sphingolipid Metabolism in Pathogenesis of Type 2 Diabetes Mellitus (T2DM). T2DM is a metabolic disease characterized by insulin resistance primarily in adipose, liver, and muscle tissues. Earlier reports suggesting that sphingolipids might play a role in insulin signaling came from experiments indicating accumulation of ceramides in insulin-resistant tissues. Zucker obese rats, a common model to study insulin resistance, were found to have elevated ceramide levels within the liver and skeletal muscles [30]. Ceramide can influence insulin signaling pathway by two autonomous mechanisms: (1) through activation of protein phosphatase 2A (PP2A), which in turn can dephosphorylate and hence inhibit Akt/PKB [31] and (2) by inhibiting the translocation and activation of Akt/PKB through protein kinase C [32]. Ceramide analogs have been shown to inhibit insulin stimulated glucose uptake, GLUT4 translocation, and glycogen synthesis in cultured cells. They inhibit Akt/PKB in cultured muscle cells [33–35], adipocytes [36], and hepatocytes [37]. Decreasing the levels of ceramide by using

inhibitors can restore the insulin sensitivity in lipid induced insulin resistant cell culture [33, 34]. Ceramide levels have also been reported to increase in the muscles or liver of insulin-resistant rodents [30] or human [38, 39]. S1P having progrowth properties often opposes ceramide action, allowing researchers to purport the existence of a ceramide: S1P rheostat that controls cellular responses [40]. Sphingosine kinase (SPHK) hence is an important lipid kinase that maintains the balance between progrowth and proapoptotic precursors. Recently, it has been reported that such a rheostat is very important for both islet function and beta cell survival and may act as a possible therapeutic target to protect the beta cell from diabetes related complications and to improve pancreatic islet function [41].

We analyzed gene expression data obtained from separate studies (GSE15653, GSE22435, and GSE22309) and Figure 4 presents the status of different hypothetical compartments C1, C2, C3, and C4 in T2DM. Genes involved in *de novo* biosynthesis of ceramide (compartment C1) show an increased level of expression in T2DM which apparently indicate that ceramide levels are increased. In general, ceramide thus formed must be transported to the Golgi apparatus with the help of CERT protein in mammals [13, 14]. However, in T2DM, genes coding for CERT protein were found to be downregulated. Overall, this might present a scenario wherein there is an increased *de novo* ceramide biosynthesis in the ER and a decreased transport of ceramide due to decreased levels of CERT. Ceramide thus accumulated in the ER is a contributing factor for ER stress. ER stress in turn is known to inhibit insulin receptor signaling through the activation of c-Jun N-terminal kinase (JNK) and subsequent

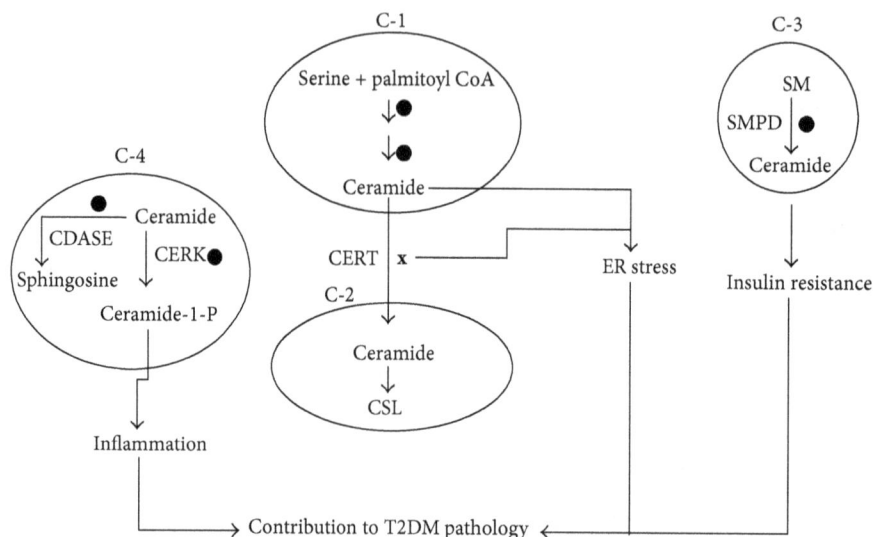

FIGURE 4: Status of different compartments in T2DM. In case of T2DM, the enzymes of the *de novo* ceramide synthesis, sphingomyelinase (SMPD), and ceramide kinase (CERK) genes are upregulated (●). CERT gene which is responsible for the transport of ceramide from ER to Golgi complex is downregulated (x). Consequently, this scenario might result in increased ER stress, insulin resistance, and inflammation, all of which are key pathologies associated with T2DM.

serine phosphorylation of insulin receptor substrate-1 (IRS-1) [42]. Studies indicate that mice deficient in X-box-binding protein-1 (XBP-1), a transcription factor that controls the ER stress response, develop insulin resistance [43]. Apparently elevated ceramide in the ER and subsequent ER stress contributes to insulin resistance—a major pathology of T2DM.

In addition to its *de novo* route, ceramide can also be produced by hydrolysis of sphingomyelin in the membrane. This reaction is carried out by sphingomyelinase. Several cellular insults have been shown to trigger the sphingomyelin hydrolysis leading to generation of ceramide, which in turn would bring about the subsequent effects. Two proinflammatory cytokines, tumor necrosis factor alpha (TNF-α) and interleukin 1 beta (IL-1β), play an important role in hydrolytic generation and accumulation of ceramide [44, 45]. TNF-α has been shown to be associated with the stimulation of insulin resistance [46–48]. Neutral and acid SMPD is reported to be elevated in adipose tissue of obese rodents, possibly through a TNF-α-regulated mechanism [49]. In T2DM, the genes responsible for hydrolytic generation of ceramide (SMPD) are upregulated in the compartment C3. This may again result in increased ceramide levels and hence concomitantly lead to insulin resistance.

Inflammation is one of the major components of the pathogenesis of T2DM and immunomodulatory strategies targeting inflammation are proposed as a therapeutic approach [50, 51]. Sphingolipids act as potential players in the process of inflammation and clinical data suggest a correlation between ceramide, inflammation, and insulin resistance [52, 53]. Studies have demonstrated that many of the proinflammatory effects of ceramide can be attributed to its phosphorylated derivative ceramide-1-phosphate [54]. Ceramide-1-phosphate can bind directly to phospholipase-A2 (PLA2) [55] and allosterically activate the enzyme leading to release of

arachidonic acid and subsequent prostaglandin formation [56]. Ceramide kinase (CERK), a key gene involved in the generation of ceramide-1-phosphate, is upregulated in T2DM condition (Figure 4). With obesity/T2DM being a case of chronic low grade inflammation, it is possible that increased CERK activity (and consequent increase in C1P levels) would serve to execute a proinflammatory scenario contributing towards pathology. Recently, the CERK null mice have been shown to be resistant to diet-induced obesity and glycemic dysregulation [57].

Overall in the context of T2DM, sphingolipid metabolic pathway contributes to ER stress and inflammation, two major contributors for insulin resistance.

3.2. Role of Sphingolipid Metabolism in Pathogenesis of Alzheimer's Disease. Sphingolipids form the integral component of the brain and a proper sphingolipid homeostasis is essential for the normal functioning of neurons. Several neurological disorders like Niemann-Pick disease (type I), Gaucher's disease, and Tay-Sacks disease result due to impaired activities of the enzymes that handle complex sphingolipids [14]. Studies suggest that even minor changes in sphingolipid balance may play significant roles in the development of neurodegenerative diseases including Alzheimer's disease [58], amyotrophic lateral sclerosis [59], Parkinson's disease [60], and dementia [61].

Alzheimer's disease (AD) is a neurodegenerative disorder characterized by a progressive decline in cognitive processes gradually leading to dementia. Extracellular deposition of Aβ peptide and neurofibrillary tangles are the well-known histopathological markers of AD [62]. Accumulation of abnormally folded Aβ in association with inability to catabolize Aβ peptide triggers the neuronal degeneration and this event is critical to the development of AD [63]. Sphingolipids

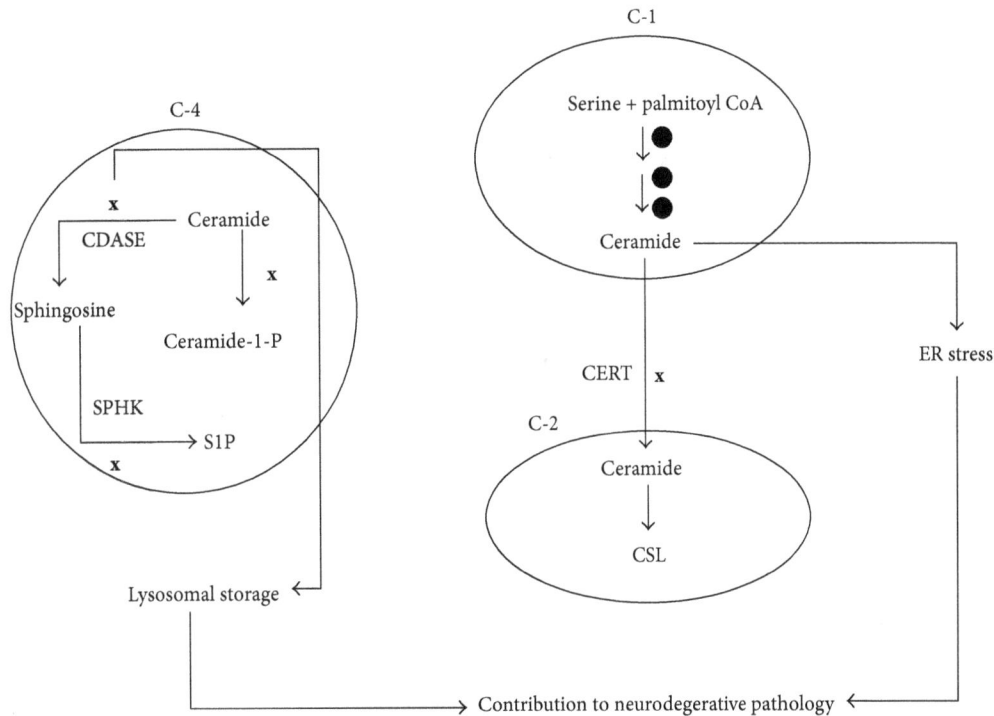

FIGURE 5: Status of different compartments in Alzheimer's disease. In Alzheimer's disease condition, there is an upregulation of genes involved in the *de novo* ceramide synthesis (●) and downregulation of CERT (**x**). This contributes towards ER stress. CERK gene which generates C1P a progrowth molecule is downregulated (**x**). Down regulation of ceramidase might lead to accumulation of ceramide which is a proapoptotic molecule. Overall this scenario would result in ER stress and proapoptotic phenotype.

are known to contribute to the development of AD at various steps during the progression of the disease.

We analyzed data obtained from gene expression studies (GSE29652, GSE5281, GSE16759, GDS1979, GSE34879, GSE30945, GSE15222, GSE4757, and GSE28146) and as depicted in Figure 5, *de novo* synthesis of ceramide (indicated by compartment C1) is increased significantly and CERT gene which is involved in transport of ceramide from ER to Golgi complex is downregulated. The overall effect may lead to ER stress due to ceramide accumulation in the ER. ER is an organelle involved in proper folding and sorting of proteins. Impaired protein folding and subsequent accumulation of neurotoxic peptides are major pathological factors in AD. Neuronal cells are vulnerable to perturbations which affect the homeostasis of ER and disturbances in redox and Ca2+ balances [64]. A number of studies have demonstrated ER stress as a major pathological driver in several neurodegenerative diseases [65–68]. Immunohistochemical studies have indicated that neurons of AD patients show prominent expression of ER stress markers [66]. The presence of oxidative stress, accumulation of neurofibrillary tangles, and intraneuronal amyloid-β aggregates [69] in AD point out the role of ER stress in this disease [70, 71]. It has been established that higher concentration of ceramide promotes the Aβ biogenesis by stabilizing the APP cleaving enzyme [72, 73], but diminished concentration of ceramide leads to a reduction in the secretion of APP and Aβ in human neuroblastoma cells [74]. Thus, it is imperative that ceramide

and Aβ may work together to encourage neuronal death in AD.

CERK and ceramidase (CDASE) are also downregulated in AD (compartment C4), which indicate decreased levels of progrowth molecule ceramide 1 phosphate and accumulation of ceramide. Diminished levels of C1P hence might reinforce a proapoptotic/degenerative scenario—a hall mark of AD. The decreased activity of ceramidase can reflect a scenario wherein ceramide accumulation in the lysosomes could lead to lysosomal dysfunction. In fact, lysosomal storage defect (LSD) is considered as one of the early histological changes associated with AD [75]. Accumulation of sphingolipids not only underlies the pathogenesis of LSDs but also elicits increased generation of Aβ and contributes to neurodegeneration in AD. Experimental evidences indicate that accumulation of sphingolipids decreases the lysosome dependent degradation of APP-CTFs (amyloid precursor protein c-terminal fragments) and increases γ-secretase activity. Both these activities result in increased generation of intracellular and secreted Aβ [74].

Thus, from sphingolipid perceptive, AD may be envisaged as an ER stress and lysosomal storage disorder arising through increased ceramide synthesis and decreased catabolism. Increased ceramide production due to increased *de novo* biosynthesis of ceramide can be a contributing factor for neurodegeneration. An earlier study reports that genes controlling *de novo* synthesis of ceramide are upregulated at early stages in AD disease progression [76]. Sphingosine,

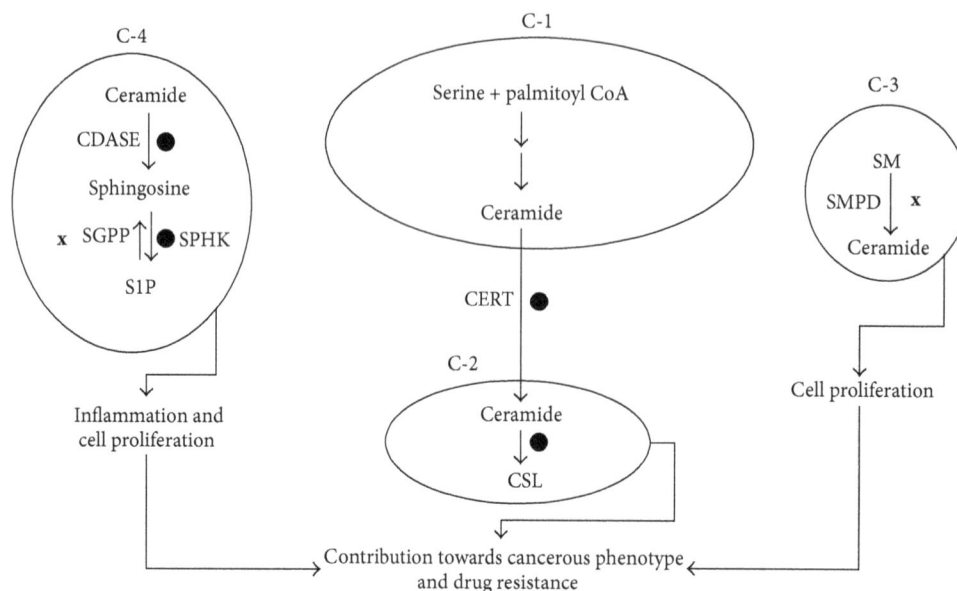

FIGURE 6: Status of different compartments in hepatocellular carcinoma. Hepatocellular carcinoma is a case of proliferative scenario mediated through downregulation of ceramide production (in compartment C-3(x)) and upregulation (●) of complex sphingolipid biosynthesis in order to quench the proapoptotic ceramide.

a proapoptotic sphingolipid is known to be accumulated in AD brain [77, 78]. One study has specified decreased S1P levels in cytosolic fractions of grey matter from the frontotemporal areas of AD patients [78].

4. Role of Sphingolipid Metabolism in Pathogenesis of Hepatocellular Carcinoma

Impaired sphingolipid homeostasis is a common theme for most of the cancers [79] and overexpression of SPHK1 has been identified in multiple cancer cells derived from breast, colon, lung, ovary, stomach, uterus, kidney, and rectum [80]. Central to the role of sphingolipids in cancer is the fact that ceramide generated through either the de novo/hydrolytic pathway is a proapoptotic molecule and sphingosine-1-phosphate and ceramide 1 phosphate resist the action of ceramide and promote cell proliferation [81]. Owing to the significant role of sphingolipids in cell death/survival regulation, the alteration in sphingolipid metabolism has a profound impact on cancer biology and therapy. While increased levels of progrowth sphingolipids would enhance the cell proliferation, channelizing the proapoptotic ceramide into other sphingolipid molecules would augment resistance to drug induced cell death. In fact many cancer cells suppress their ceramide biosynthetic machinery while upregulating the biosynthesis of progrowth sphingolipids like S1P. This modulation would serve to increase proliferation of these cells. Besides maintaining a proliferative phenotype, many cancer cell types resist therapy in several ways including escape from therapy-induced apoptosis [82]. Decreasing the levels of proapoptotic ceramide in the cells is one of the adaptive means to resist therapy-induced apoptosis.

Hepatocellular carcinoma is a common type of liver cancer resulting due to either viral causes or cirrhosis. We analyzed gene expression data obtained from several studies (GSE21362, GSE6764, GSE14323, GSE5975, GSE25097, GSE10459, and GSE14323) and Figure 6 demonstrates the status of four compartments in hepatocellular carcinoma (HCC). The expression of genes involved in the de novo ceramide synthesis is unaltered as depicted by compartment C1 and levels of CERT are upregulated. In compartment C2, markedly, there is an upregulation of the enzymes involved in the synthesis of complex sphingolipids. In compartment C3, hydrolytic generation of ceramide from sphingomyelin is downregulated once again ensuring lesser levels of ceramide.

The digression of proapoptotic ceramide to progrowth molecule S1P may modulate the future of a cell in response to cancer therapy [83, 84]. Ceramidases promote carcinogenesis by disturbing the ceramide/S1P ratio, permitting phosphorylation of sphingosine by SPHKs, and determine the efficacy of cancer therapy. For instance, inhibition of an acid CDASE by a newly developed ceramide analogue, B13, induces apoptosis in cultured human colon cancer cells and prevents liver metastases in vivo [85]. Concurrent to these findings, our analysis also revealed that, in compartment C4, the expression of CDASEs and SPHK, enzymes involved in the conversion of ceramide to sphingosine and later to sphingosine-1-phosphate are upregulated. Sphingosine-1-phosphate phosphatase (SGPP), an enzyme involved in dephosphorylation of progrowth molecule sphingosine-1-phosphate to proapoptotic sphingosine, was found to be downregulated. Apparently in HCC, sphingolipid metabolic pathway is driven towards decreasing the levels of ceramide and increasing progrowth sphingolipids. Increased expression of enzymes involved in the complex sphingolipid biosynthesis in compartment C2 serves as an adaptive channel to quench the proapoptotic ceramide. Elevated expression of CERT apparently will augment the transfer of ceramide from ER to Golgi complex, providing it as a substrate for complex sphingolipid biosynthesis. Earlier studies employing different cancer cell types have established the relationship between glucosyl ceramide synthase and chemoresistance. In studies comparing the nature of lipid intermediates between drug-sensitive and resistant cancer cell lines, glucosyl ceramide levels were found to be higher in drug-resistant cells [86, 87]. The role of GCS (glucosyl ceramide synthase) is further confirmed in cell model where overexpression of GCS resulted in resistance of HL-60 to doxorubicin-induced apoptosis [88].

In summary, in HCC sphingolipid, metabolism is manipulated at 3 levels.

(1) Channels generating the proapoptotic ceramide through hydrolytic machinery are decreased.

(2) The channel converting the proapoptotic ceramide into progrowth molecule is upregulated.

(3) Adaptive channel which converts proapoptotic ceramide into less effective complex sphingolipids is activated.

While the first two changes ensure that cell assumes a proliferative mode favoring the cancer phenotype, the adaptive channel in which proapoptotic ceramide is quenched into less effective complex sphingolipids contributes towards drug resistance.

5. Conclusions

Sphingolipids affect various aspects of cell physiology like cell proliferation, cell death, differentiation, and cell signaling and are known to contribute to key cellular pathologies like ER stress, insulin resistance, inflammation, and drug resistance. Analyzing the gene expression pattern in three different disease conditions—T2DM, Alzheimer's disease, and hepatocellular carcinoma, indicates that under different disease conditions sphingolipid metabolic pathway may employ slightly different routes to contribute towards the pathology. For example, in diabetes, sphingolipid metabolic pathway genes are manipulated in such a way that it leads to ER stress and inflammation, while Alzheimer's disease condition is a case of increased ceramide mediated apoptosis coupled with inability to degrade ceramide. In hepatocellular carcinoma, sphingolipid pathway is manipulated in such a way that reactions leading to quenching of proapoptotic molecule ceramide into inert complex sphingolipids and those leading to generation of progrowth sphingolipids are upregulated. Although a bias for the sphingolipid metabolic pathway is apparent under these disease conditions, further analysis is required to verify if such a phenomenon is universally applicable. Understanding the nature of changes in the sphingolipid metabolic pathway and their overall pathological effect under different disease conditions would be very useful in order to design therapeutic strategies.

6. Highlights

(i) The sphingolipid metabolic pathway is a major player in the pathology of human diseases.

(ii) This review focuses on changes in the sphingolipid pathway in 3 different diseases.

(iii) In metabolic disorder, this pathway contributes to endoplasmic reticulum stress (ER stress) and inflammation.

(iv) In Alzheimer's disease, the pathway contributes to ER stress and apoptosis.

(v) In hepatocellular carcinoma, the pathway contributes to proproliferative scenario.

Abbreviation

AD: Alzheimer's disease
APP: Amyloid precursor protein
$A\beta$:   Amyloid beta
C1P: Ceramide-1-phosphate
CDASE: Ceramidase
CERK: Ceramide kinase
CERT: Ceramide transfer protein
COX-2:   Cyclooxygenase-2
ER: Endoplasmic reticulum
GLUT-4:   Glucose transporter-4
GSL: Glycosphingolipid
HCC: Hepatocellular carcinoma
IL-1β: Interleukin-1β
IRS-1:   Insulin receptor substrate-1
JNK: c-Jun N-terminal kinase
LCFA: Long chain fatty acids
LSD: Lysosomal storage disorder
PGE-2:   Prostaglandin E-2
PKB: Protein kinase B
PLA-2:   Phospholipase A-2
PP2A: Protein phosphatase 2A
S1P: Sphingosine-1-phosphate
SL: Sphingolipid
SM: Sphingomyelin
SMPD: Sphingomyelin phosphodiesterase
SPHK: Sphingosine kinase
SPT: Serine palmitoyltransferase
T2DM: Type 2 diabetes mellitus
XBP-1:   X-box binding protein-1.

Conflict of Interests

The authors declare that there is no conflict of interests relevant to this article.

Authors' Contribution

Raghavendra, Nandita, and Mathiyazhagan were involved in collecting the gene expression data and analysis. Neeti was involved in collection of gene expression data. Raghavendra, Anup, and Jagannath were involved in the analysis of the data and writing.

Acknowledgments

The authors sincerely thank Dr. Suri Venkatachalam for the valuable comments and helpful discussions and Dr. Nishta Jain for scientific writing services offered during the preparation of this review article.

References

[1] Y. Chen, Y. Liu, M. C. Sullards, and A. H. Merrill Jr., "An introduction to sphingolipid metabolism and analysis by new technologies," *NeuroMolecular Medicine*, vol. 12, no. 4, pp. 306–319, 2010.

[2] A. H. Merrill Jr., M. D. Wang, M. Park, and M. C. Sullards, "(Glyco)sphingolipidology: an amazing challenge and opportunity for systems biology," *Trends in Biochemical Sciences*, vol. 32, no. 10, pp. 457–468, 2007.

[3] Y. A. Hannun and L. M. Obeid, "The ceramide-centric universe of lipid-mediated cell regulation: stress encounters of the lipid kind," *Journal of Biological Chemistry*, vol. 277, no. 29, pp. 25847–25850, 2002.

[4] J. Ohanian and V. Ohanian, "Sphingolipids in mammalian cell signalling," *Cellular and Molecular Life Sciences*, vol. 58, no. 14, pp. 2053–2068, 2001.

[5] J. C. M. Holthuis, T. Pomorski, R. J. Raggers, H. Sprong, and G. van Meer, "The organizing potential of sphingolipids in intracellular membrane transport," *Physiological Reviews*, vol. 81, no. 4, pp. 1689–1723, 2001.

[6] S. E. Brice and L. A. Cowart, "Sphingolipid metabolism and analysis in metabolic disease," *Advances in Experimental Medicine and Biology*, vol. 721, pp. 1–17, 2011.

[7] F. Alvarez-Vasquez, K. J. Sims, L. A. Cowart, Y. Okamoto, E. O. Voit, and Y. A. Hannun, "Simulation and validation of modelled sphingolipid metabolism in *Saccharomyces cerevisiae*," *Nature*, vol. 433, no. 7024, pp. 425–430, 2005.

[8] Y. A. Hannun and C. Luberto, "Lipid metabolism: ceramide transfer protein adds a new dimension," *Current Biology*, vol. 14, no. 4, pp. R163–R165, 2004.

[9] Y. A. Hannun and L. M. Obeid, "Principles of bioactive lipid signalling: lessons from sphingolipids," *Nature Reviews Molecular Cell Biology*, vol. 9, no. 2, pp. 139–150, 2008.

[10] K. Hanada, "Serine palmitoyltransferase, a key enzyme of sphingolipid metabolism," *Biochimica et Biophysica Acta*, vol. 1632, no. 1–3, pp. 16–30, 2003.

[11] U. Acharya and J. K. Acharya, "Enzymes of sphingolipid metabolism in *Drosophila melanogaster*," *Cellular and Molecular Life Sciences*, vol. 62, no. 2, pp. 128–142, 2005.

[12] Y. Sugimoto, H. Sakoh, and K. Yamada, "IPC synthase as a useful target for antifungal drugs," *Current Drug Targets*, vol. 4, no. 4, pp. 311–322, 2004.

[13] K. Hanada, K. Kumagai, N. Tomishige, and T. Yamaji, "CERT-mediated trafficking of ceramide," *Biochimica et Biophysica Acta*, vol. 1791, no. 7, pp. 684–691, 2009.

[14] R. P. Rao and J. K. Acharya, "Sphingolipids and membrane biology as determined from genetic models," *Prostaglandins and Other Lipid Mediators*, vol. 85, no. 1-2, pp. 1–16, 2008.

[15] Y. H. Zeidan and Y. A. Hannun, "Translational aspects of sphingolipid metabolism," *Trends in Molecular Medicine*, vol. 13, no. 8, pp. 327–336, 2007.

[16] T. Kolter and K. Sandhoff, "Principles of lysosomal membrane digestion: stimulation of sphingolipid degradation by sphingolipid activator proteins and anionic lysosomal lipids," *Annual Review of Cell and Developmental Biology*, vol. 21, pp. 81–103, 2005.

[17] C. F. Snook, J. A. Jones, and Y. A. Hannun, "Sphingolipid-binding proteins," *Biochimica et Biophysica Acta*, vol. 1761, no. 8, pp. 927–946, 2006.

[18] S. A. Summers, "Ceramides in insulin resistance and lipotoxicity," *Progress in Lipid Research*, vol. 45, no. 1, pp. 42–72, 2006.

[19] D. S. Menaldino, A. Bushnev, A. Sun et al., "Sphingoid bases and de novo ceramide synthesis: enzymes involved, pharmacology and mechanisms of action," *Pharmacological Research*, vol. 47, no. 5, pp. 373–381, 2003.

[20] H. Schulze and K. Sandhoff, "Lysosomal lipid storage diseases," *Cold Spring Harbor Perspectives in Biology*, vol. 3, no. 6, 2011.

[21] X. Wang, R. P. Rao, T. Kosakowska-Cholody et al., "Mitochondrial degeneration and not apoptosis is the primary cause of embryonic lethality in ceramide transfer protein mutant mice," *Journal of Cell Biology*, vol. 184, no. 1, pp. 143–158, 2009.

[22] T. Kolter and K. Sandhoff, "Sphingolipid metabolism diseases," *Biochimica et Biophysica Acta*, vol. 1758, no. 12, pp. 2057–2079, 2006.

[23] A. Raas-Rothschild, I. Pankova-Kholmyansky, Y. Kacher, and A. H. Futerman, "Glycosphingolipidoses: beyond the enzymatic defect," *Glycoconjugate Journal*, vol. 21, no. 6, pp. 295–304, 2004.

[24] N. J. Haughey, V. V. R. Bandaru, M. Bae, and M. P. Mattson, "Roles for dysfunctional sphingolipid metabolism in Alzheimer's disease neuropathogenesis," *Biochimica et Biophysica Acta*, vol. 1801, no. 8, pp. 878–886, 2010.

[25] A. Jana and K. Pahan, "Fibrillar amyloid-β-activated human astroglia kill primary human neurons via neutral sphingomyelinase: implications for Alzheimer's disease," *Journal of Neuroscience*, vol. 30, no. 38, pp. 12676–12689, 2010.

[26] M. O. W. Grimm, S. Grösgen, T. L. Rothhaar et al., "Intracellular APP domain regulates serine-palmitoyl-CoA transferase expression and is affected in Alzheimer's disease," *International Journal of Alzheimer's Disease*, vol. 2011, Article ID 695413, 8 pages, 2011.

[27] H. G. Wilcox, G. D. Dunn, and S. Schenker, "Plasma long chain fatty acids and esterified lipids in cirrhosis and hepatic encephalopathy," *American Journal of the Medical Sciences*, vol. 276, no. 3, pp. 293–303, 1978.

[28] L. Longato, K. Ripp, M. Setshedi et al., "Insulin resistance, ceramide accumulation, and endoplasmic reticulum stress in human chronic alcohol-related liver disease," *Oxidative Medicine and Cellular Longevity*, vol. 2012, Article ID 479348, 17 pages, 2012.

[29] A. Morales, M. Marí, C. García-Ruiz, A. Colell, and J. C. Fernández-Checa, "Hepatocarcinogenesis and ceramide/cholesterol metabolism," *Anti-Cancer Agents in Medicinal Chemistry*, vol. 12, no. 4, pp. 364–375, 2012.

[30] J. Turinsky, D. M. O'Sullivan, and B. P. Bayly, "1,2-diacylglycerol and ceramide levels in insulin-resistant tissues of the rat in vivo," *Journal of Biological Chemistry*, vol. 265, no. 28, pp. 16880–16885, 1990.

[31] S. Resjö, O. Göransson, L. Härndahl, S. Zolnierowicz, V. Manganiello, and E. Degerman, "Protein phosphatase 2A is the main phosphatase involved in the regulation of protein kinase B in rat adipocytes," *Cellular Signalling*, vol. 14, no. 3, pp. 231–238, 2002.

[32] D. J. Powell, E. Hajduch, G. Kular, and H. S. Hundal, "Ceramide disables 3-phosphoinositide binding to the pleckstrin homology domain of protein kinase B (PKB)/Akt by a PKCζ-dependent mechanism," *Molecular and Cellular Biology*, vol. 23, no. 21, pp. 7794–7808, 2003.

[33] J. A. Chavez, T. A. Knotts, L.-P. Wang et al., "A role for ceramide, but not diacylglycerol, in the antagonism of insulin signal transduction by saturated fatty acids," *Journal of Biological Chemistry*, vol. 278, no. 12, pp. 10297–10303, 2003.

[34] D. J. Powell, S. Turban, A. Gray, E. Hajduch, and H. S. Hundal, "Intracellular ceramide synthesis and protein kinase Cζ activation play an essential role in palmitate-induced insulin resistance in rat L6 skeletal muscle cells," *Biochemical Journal*, vol. 382, no. 2, pp. 619–629, 2004.

[35] C. Schmitz-Peiffer, D. L. Craig, and T. J. Biden, "Ceramide generation is sufficient to account for the inhibition of the insulin-stimulated PKB pathway in C2C12 skeletal muscle cells pretreated with palmitate," *Journal of Biological Chemistry*, vol. 274, no. 34, pp. 24202–24210, 1999.

[36] S. A. Summers, L. A. Garza, H. Zhou, and M. J. Birnbaum, "Regulation of insulin-stimulated glucose transporter GLUT4 translocation and Akt kinase activity by ceramide," *Molecular and Cellular Biology*, vol. 18, no. 9, pp. 5457–5464, 1998.

[37] W. L. Holland, T. A. Knotts, J. A. Chavez, L.-P. Wang, K. L. Hoehn, and S. A. Summers, "Lipid mediators of insulin resistance," *Nutrition Reviews*, vol. 65, no. 6, pp. S39–S46, 2007.

[38] J. M. Adams II, T. Pratipanawatr, R. Berria et al., "Ceramide content is increased in skeletal muscle from obese insulin-resistant humans," *Diabetes*, vol. 53, no. 1, pp. 25–31, 2004.

[39] M. Straczkowski, I. Kowalska, A. Nikolajuk et al., "Relationship between insulin sensitivity and sphingomyelin signaling pathway in human skeletal muscle," *Diabetes*, vol. 53, no. 5, pp. 1215–1221, 2004.

[40] B. T. Bikman and S. A. Summers, "Ceramides as modulators of cellular and whole-body metabolism," *Journal of Clinical Investigation*, vol. 121, no. 11, pp. 4222–4230, 2011.

[41] C. F. Jessup, C. S. Bonder, S. M. Pitson, and P. T. H. Coates, "The sphingolipid rheostat: a potential target for improving pancreatic islet survival and function," *Endocrine, Metabolic and Immune Disorders*, vol. 11, no. 4, pp. 262–272, 2011.

[42] Y. H. Lee, J. Giraud, R. J. Davis, and M. F. White, "c-Jun N-terminal kinase (JNK) mediates feedback inhibition of the insulin signaling cascade," *Journal of Biological Chemistry*, vol. 278, no. 5, pp. 2896–2902, 2003.

[43] U. Özcan, Q. Cao, E. Yilmaz et al., "Endoplasmic reticulum stress links obesity, insulin action, and type 2 diabetes," *Science*, vol. 306, no. 5695, pp. 457–461, 2004.

[44] K. A. Dressler, S. Mathias, and R. N. Kolesnick, "Tumor necrosis factor-α activates the sphingomyelin signal transduction pathway in a cell-free system," *Science*, vol. 255, no. 5052, pp. 1715–1618, 1992.

[45] K. Wiegmann, S. Schütze, T. Machleidt, D. Witte, and M. Krönke, "Functional dichotomy of neutral and acidic sphingomyelinases in tumor necrosis factor signaling," *Cell*, vol. 78, no. 6, pp. 1005–1015, 1994.

[46] G. S. Hotamisligil, N. S. Shargill, and B. M. Spiegelman, "Adipose expression of tumor necrosis factor-α: direct role in obesity-linked insulin resistance," *Science*, vol. 259, no. 5091, pp. 87–91, 1993.

[47] C. Hofmann, K. Lorenz, S. S. Braithwaite et al., "Altered gene expression for tumor necrosis factor-α and its receptors during drug and dietary modulation of insulin resistance," *Endocrinology*, vol. 134, no. 1, pp. 264–270, 1994.

[48] J. M. Stephens, J. Lee, and P. F. Pilch, "Tumor necrosis factor-α-induced insulin resistance in 3T3-L1 adipocytes is accompanied by a loss of insulin receptor substrate-1 and GLUT4 expression without a loss of insulin receptor-mediated signal transduction," *Journal of Biological Chemistry*, vol. 272, no. 2, pp. 971–976, 1997.

[49] F. Samad, K. D. Hester, G. Yang, Y. A. Hannun, and J. Bielawski, "Altered adipose and plasma sphingolipid metabolism in obesity: a potential mechanism for cardiovascular and metabolic risk," *Diabetes*, vol. 55, no. 9, pp. 2579–2587, 2006.

[50] M. Y. Donath and S. E. Shoelson, "Type 2 diabetes as an inflammatory disease," *Nature Reviews Immunology*, vol. 11, no. 2, pp. 98–107, 2011.

[51] S. E. Shoelson, J. Lee, and A. B. Goldfine, "Inflammation and insulin resistance," *Journal of Clinical Investigation*, vol. 116, no. 7, pp. 1793–1801, 2006.

[52] V. D. F. de Mello, M. Lankinen, U. Schwab et al., "Link between plasma ceramides, inflammation and insulin resistance: association with serum IL-6 concentration in patients with coronary heart disease," *Diabetologia*, vol. 52, no. 12, pp. 2612–2615, 2009.

[53] J. M. R. Gill and N. Sattar, "Ceramides: a new player in the inflammation-insulin resistance paradigm?" *Diabetologia*, vol. 52, no. 12, pp. 2475–2477, 2009.

[54] A. Gómez-Munoz, P. Gangoiti, M. H. Granado, L. Arana, and A. Ouro, "Ceramide-1-phosphate in cell survival and inflammatory signaling," *Advances in Experimental Medicine and Biology*, vol. 688, pp. 118–130, 2010.

[55] P. Subramanian, M. Vora, L. B. Gentile, R. V. Stahelin, and C. E. Chalfant, "Anionic lipids activate group IVA cytosolic phospholipase A2 via distinct and separate mechanisms," *Journal of Lipid Research*, vol. 48, no. 12, pp. 2701–2708, 2007.

[56] B. J. Pettus, A. Bielawska, S. Spiegel, P. Roddy, Y. A. Hannun, and C. E. Chalfant, "Ceramide kinase mediates cytokine- and calcium ionophore-induced arachidonic acid release," *Journal of Biological Chemistry*, vol. 278, no. 40, pp. 38206–38213, 2003.

[57] S. Mitsutake, T. Date, H. Yokota, M. Sugiura, T. Kohama, and Y. Igarashi, "Ceramide kinase deficiency improves diet-induced obesity and insulin resistance," *FEBS Letters*, vol. 586, no. 9, pp. 1300–1305, 2012.

[58] M. M. Mielke and C. G. Lyketsos, "Alterations of the sphingolipid pathway in Alzheimer's disease: new biomarkers and treatment targets?" *NeuroMolecular Medicine*, vol. 12, no. 4, pp. 331–340, 2010.

[59] R. G. Cutler, W. A. Pedersen, S. Camandola, J. D. Rothstein, and M. P. Mattson, "Evidence that accumulation of ceramides and cholesterol esters mediates oxidative stress-induced death of motor neurons in amyotrophic lateral sclerosis," *Annals of Neurology*, vol. 52, no. 4, pp. 448–457, 2002.

[60] V. France-Lanord, B. Brugg, P. P. Michel, Y. Agid, and M. Ruberg, "Mitochondrial free radical signal in ceramide-dependent apoptosis: a putative mechanism for neuronal death in Parkinson's disease," *Journal of Neurochemistry*, vol. 69, no. 4, pp. 1612–1621, 1997.

[61] N. J. Haughey, R. G. Cutler, A. Tamara et al., "Perturbation of sphingolipid metabolism and ceramide production in HIV-dementia," *Annals of Neurology*, vol. 55, no. 2, pp. 257–267, 2004.

[62] I. Y. Tamboli, H. Hampel, N. T. Tien et al., "Sphingolipid storage affects autophagic metabolism of the amyloid precursor protein and promotes Aβ generation," *Journal of Neuroscience*, vol. 31, no. 5, pp. 1837–1849, 2011.

[63] M. P. Murphy and H. Levine III, "Alzheimer's disease and the amyloid-β peptide," *Journal of Alzheimer's Disease*, vol. 19, no. 1, pp. 311–323, 2010.

[64] A. Salminen, A. Kauppinen, T. Suuronen, K. Kaarniranta, and J. Ojala, "ER stress in Alzheimer's disease: a novel neuronal trigger for inflammation and Alzheimer's pathology," *Journal of Neuroinflammation*, vol. 6, article 41, 2009.

[65] D. Lindholm, H. Wootz, and L. Korhonen, "ER stress and neurodegenerative diseases," *Cell Death and Differentiation*, vol. 13, no. 3, pp. 385–392, 2006.

[66] J. J. M. Hoozemans, E. S. van Haastert, D. A. T. Nijholt, A. J. M. Rozemuller, P. Eikelenboom, and W. Scheper, "The unfolded protein response is activated in pretangle neurons in Alzheimer's disease hippocampus," *American Journal of Pathology*, vol. 174, no. 4, pp. 1241–1251, 2009.

[67] T. Katayama, K. Imaizumi, T. Manabe, J. Hitomi, T. Kudo, and M. Tohyama, "Induction of neuronal death by ER stress in Alzheimer's disease," *Journal of Chemical Neuroanatomy*, vol. 28, no. 1-2, pp. 67–78, 2004.

[68] R. Resende, E. Ferreiro, C. Pereira, and C. R. Oliveira, "ER stress is involved in $A\beta$-induced GSK-3β activation and tau phosphorylation," *Journal of Neuroscience Research*, vol. 86, no. 9, pp. 2091–2099, 2008.

[69] D. J. Selkoe, "Alzheimer's disease: genes, proteins, and therapy," *Physiological Reviews*, vol. 81, no. 2, pp. 741–766, 2001.

[70] R. E. Tanzi and L. Bertram, "Twenty years of the Alzheimer's disease amyloid hypothesis: a genetic perspective," *Cell*, vol. 120, no. 4, pp. 545–555, 2005.

[71] F. M. LaFerla, K. N. Green, and S. Oddo, "Intracellular amyloid-β in Alzheimer's disease," *Nature Reviews Neuroscience*, vol. 8, no. 7, pp. 499–509, 2007.

[72] S. Patil, J. Melrose, and C. Chan, "Involvement of astroglial ceramide in palmitic acid-induced Alzheimer-like changes in primary neurons," *European Journal of Neuroscience*, vol. 26, no. 8, pp. 2131–2141, 2007.

[73] L. Puglielli, B. C. Ellis, A. J. Saunders, and D. M. Kovacs, "Ceramide stabilizes β-site amyloid precursor protein-cleaving enzyme 1 and promotes amyloid β-peptide biogenesis," *Journal of Biological Chemistry*, vol. 278, no. 22, pp. 19777–19783, 2003.

[74] I. Y. Tamboli, K. Prager, E. Barth, M. Heneka, K. Sandhoff, and J. Walter, "Inhibition of glycosphingolipid biosynthesis reduces secretion of the β-amyloid precursor protein and amyloid β-peptide," *Journal of Biological Chemistry*, vol. 280, no. 30, pp. 28110–28117, 2005.

[75] R. A. Nixon, A. M. Cataldo, and P. M. Mathews, "The endosomal-lysosomal system of neurons in Alzheimer's disease pathogenesis: a review," *Neurochemical Research*, vol. 25, no. 9-10, pp. 1161–1172, 2000.

[76] P. Katsel, C. Li, and V. Haroutunian, "Gene expression alterations in the sphingolipid metabolism pathways during progression of dementia and Alzheimer's disease: a shift toward ceramide accumulation at the earliest recognizable stages of Alzheimer's disease?" *Neurochemical Research*, vol. 32, no. 4-5, pp. 845–856, 2007.

[77] Y. Huang, H. Tanimukai, F. Liu, K. Iqbal, I. Grundke-Iqbal, and C.-X. Gong, "Elevation of the level and activity of acid ceramidase in Alzheimer's disease brain," *European Journal of Neuroscience*, vol. 20, no. 12, pp. 3489–3497, 2004.

[78] X. He, Y. Huang, B. Li, C.-X. Gong, and E. H. Schuchman, "Deregulation of sphingolipid metabolism in Alzheimer's disease," *Neurobiology of Aging*, vol. 31, no. 3, pp. 398–408, 2010.

[79] L. K. Ryland, T. E. Fox, X. Liu, T. P. Loughran, and M. Kester, "Dysregulation of sphingolipid metabolism in cancer," *Cancer Biology and Therapy*, vol. 11, no. 2, pp. 138–149, 2011.

[80] W. C. Huang, C. L. Chen, Y. S. Lin, and C. F. Lin, "Apoptotic sphingolipid ceramide in cancer therapy," *Journal of Lipids*, vol. 2011, Article ID 565316, 15 pages, 2011.

[81] B. Oskouian and J. D. Saba, "Cancer treatment strategies targeting sphingolipid metabolism," *Advances in Experimental Medicine and Biology*, vol. 688, pp. 185–205, 2010.

[82] M. M. Gottesman, "Mechanisms of cancer drug resistance," *Annual Review of Medicine*, vol. 53, pp. 615–627, 2002.

[83] R. Kolesnick, "The therapeutic potential of modulating the ceramide/sphingomyelin pathway," *Journal of Clinical Investigation*, vol. 110, no. 1, pp. 3–8, 2002.

[84] B. Ogretmen and Y. A. Hannun, "Biologically active sphingolipids in cancer pathogenesis and treatment," *Nature Reviews Cancer*, vol. 4, no. 8, pp. 604–616, 2004.

[85] M. Selzner, A. Bielawska, M. A. Morse et al., "Induction of apoptotic cell death and prevention of tumor growth by ceramide analogues in metastatic human colon cancer," *Cancer Research*, vol. 61, no. 3, pp. 1233–1240, 2001.

[86] Y. Lavie, H.-T. Cao, S. L. Bursten, A. E. Giuliano, and M. C. Cabot, "Accumulation of glucosylceramides in multidrug-resistant cancer cells," *Journal of Biological Chemistry*, vol. 271, no. 32, pp. 19530–19536, 1996.

[87] H. Morjani, N. Aouali, R. Belhoussine, R. J. Veldman, T. Levade, and M. Manfait, "Elevation of glucosylceramide in multidrug-resistant cancer cells and accumulation in cytoplasmic droplets," *International Journal of Cancer*, vol. 94, no. 2, pp. 157–165, 2001.

[88] M. Itoh, T. Kitano, M. Watanabe et al., "Possible role of ceramide as an indicator of chemoresistance: decrease of the ceramide content via activation of glucosylceramide synthase and sphingomyelin synthase in chemoresistant leukemia," *Clinical Cancer Research*, vol. 9, no. 1, pp. 415–423, 2003.

3

Sphingosine 1-Phosphate Distribution in Human Plasma: Associations with Lipid Profiles

Samar M. Hammad,[1] Mohammed M. Al Gadban,[1] Andrea J. Semler,[2] and Richard L. Klein[2,3]

[1] Department of Regenerative Medicine and Cell Biology, Medical University of South Carolina, Charleston, SC 29425, USA
[2] Division of Endocrinology, Metabolism, and Medical Genetics, Department of Medicine,
 Medical University of South Carolina, Charleston, SC 29425, USA
[3] Research Service, Ralph H. Johnson VA Medical Center, United States Department of Veterans Affairs, Charleston, SC 29401, USA

Correspondence should be addressed to Samar M. Hammad, hammadsm@musc.edu and Richard L. Klein, kleinrl@musc.edu

Academic Editor: Philip W. Wertz

The physiological significance of sphingosine 1-phosphate (S1P) transport in blood has been debated. We have recently reported a comprehensive sphingolipid profile in human plasma and lipoprotein particles (VLDL, LDL, and HDL) using HPLC-MS/MS (Hammad et al., 2010). We now determined the relative concentrations of sphingolipids including S1P in the plasma subfraction containing lipoproteins compared to those in the remaining plasma proteins. Sphingomyelin and ceramide were predominantly recovered in the lipoprotein-containing fraction. Total plasma S1P concentration was positively correlated with S1P concentration in the protein-containing fraction, but not with S1P concentration in the lipoprotein-containing fraction. The percentage of S1P transported in plasma lipoproteins was positively correlated with HDL cholesterol (HDL-C) concentration; however, S1P transport in lipoproteins was not limited by the concentration of HDL-C in the individual subject. Thus, different plasma pools of S1P may have different contributions to S1P signaling in health and disease.

1. Introduction

Sphingolipids have been implicated in diseases such as cancer, obesity, and atherosclerosis; however, efforts addressing blood sphingolipids as biomarkers of disease or targets for therapeutics are still in their infancy. Sphingosine-1-phosphate (S1P) is a bioactive lipid that has been shown to play major roles in immunity, inflammation, and cardiovascular physiology [1–6]. S1P is found in plasma at relatively high concentrations (>200 nM) and is transported in blood bound both to lipoproteins (~65%) and albumin [7–9] with the bulk of the lipoprotein-associated S1P found in HDL (~54%), especially the smaller diameter HDL3 subfraction [10, 11]. Plasma HDL cholesterol levels correlate positively with those of plasma S1P [12], but HDLs do not appear to be merely inert carriers of S1P as the S1P constituent of HDL is biologically active and contributes to numerous metabolic effects of HDL [10, 11, 13–16].

We have recently reported a comprehensive sphingolipid profile in "normal" human plasma and lipoprotein particles (VLDL, LDL, and HDL) using HPLC-MS/MS [9]. As an extension of this study we now identify the nonlipoprotein (albumin)-associated versus the lipoprotein-associated sphingolipids. It is established that the concentration of S1P in plasma/serum is much higher than the half-maximal concentration of S1P needed to stimulate its receptors. Nevertheless, it was shown for instance that the inositol phosphate response mediated by S1P receptors was much smaller than the response expected from the total amount of S1P introduced to cells [8]. This response to exogenous S1P was markedly attenuated in the presence of lipoprotein-deficient serum, and was associated with the trapping of exogenous S1P [8]. Importantly, HDL and LDL showed a stronger activity for trapping S1P than lipoprotein-deficient serum [8]. As such, these findings suggest that the "tight" binding of S1P to lipoproteins may interfere with the S1P binding to its receptors and thereby attenuate the S1P-receptor-mediated cell responses.

The physiological significance of S1P transport in HDL versus bound to albumin has been debated and this has led to the suggestion that S1P may be atheroprotective when bound to HDL but proatherogenic when localized to the plasma protein fraction of blood bound to albumin [1, 7, 16–18]. Most importantly, the distribution of S1P between the plasma lipoprotein and protein fractions of blood is reportedly altered in patients with coronary artery disease [19]. We determined the concentrations of sphingolipids in total plasma, in the subfraction of plasma containing all the lipoprotein fractions, and in the plasma subfraction containing the remaining plasma proteins which were separated using ultracentrifugation of plasma. In addition, we investigated the distribution of S1P between the lipoprotein and plasma protein S1P pools and the influence of plasma lipoprotein and lipid concentrations on this distribution.

2. Materials and Methods

2.1. Study Subjects. Subjects aged 30–58 years were screened before participating in the study and those with known heart disease, kidney disease, diabetes, cancer, or serious current illness were excluded. Smokers, pregnant females, and subjects who take daily multivitamins or antioxidants were also excluded. A conventional lipid panel (total cholesterol (C), HDL-C, LDL-C, VLDL-C, and triglycerides) was determined in each subject (Cholestech LDX, Cholestech Corporation, Hayward, CA, USA). The plasma lipid and lipoprotein profiles for each subject are summarized in Table 1. Participating study subjects ($n = 6$) were asked to fast overnight for at least 10 h before collecting the blood sample for analysis. The study was approved by the Institutional Review Board (IRB) at the Medical University of South Carolina (MUSC) and proper consent was obtained from each subject.

2.2. Blood Sample Collection. Blood was collected in the presence of a lipoprotein preservative cocktail EDTA (0.1% w/v), chloramphenicol (20 μg/ml), gentamycin sulfate (50 μg/ml), epsilon aminocaproic acid (0.13% w/v), and dithiobisnitrobenzoic acid (0.04% w/v) to inhibit LCAT activity (final concentrations). Blood was then centrifuged (2,400 ×g, 20 min, 4°C) to obtain plasma. Plasma samples were stored at 4°C until analyzed further.

2.3. Conventional Plasma Lipoprotein Profiles. Plasma conventional lipid profiles were determined by Cholestech LDX (Cholestech Corporation, Hayward, CA, USA).

2.4. Separation of Lipoproteins from Plasma Proteins. The fraction of plasma containing all the lipoproteins (VLDL, LDL, and HDL) was separated from the plasma subfraction containing the nonlipoprotein, plasma proteins using isopycnic density ultracentrifugation. Plasma solvent density was adjusted to $d = 1.225$ g/ml with solid potassium bromide (KBr) and all the lipoproteins in plasma were isolated by preparative ultracentrifugation of the plasma samples (10 ml) using a type SW41Ti rotor (Beckman Coulter, Fullerton, CA, USA) spun at 41,000 rpm (288,000 ×g) for 40 h at 4°C in an Optima TM XL-100K ultracentrifuge.

The floating lipoprotein subfraction in the ultracentrifugal supernatant was separated from the sedimented plasma lipoproteins in the subnatant after slicing each tube exactly 1.25 inch from the meniscus. The lipoprotein supernatant and plasma protein subnatant fractions were quantitatively harvested and the volume of each subfraction was adjusted to 10 ml using saline/EDTA (150 mM NaCl, 300 μM EDTA, pH 7.4) and the fractions were stored at 4°C until used. Cholesterol levels in the total lipoprotein preparation in the ultracentrifugal supernatant and in the subnatant containing the nonlipoprotein plasma proteins were measured using gas chromatography as described previously [20]. Whole plasma and the ultracentrifugally prepared plasma subfractions were analyzed using electrophoresis in agarose gels to assess the purity of the subfractions (Beckman Paragon LIPOEPG, Beckman-Coulter, Fullerton, CA, USA).

2.5. Sphingolipid Extraction and Analysis. Analysis of endogenous sphingoid bases, sphingoid base 1-phosphates, ceramide, and sphingomyelin (SM) species were conducted in the Lipidomics Core Facility at MUSC as previously described [9]. Briefly, 100 μl of each plasma sample, and the ultracentrifuge supernatant and subnatant subfractions of plasma were spiked with internal standards and the sphingolipid complement in each sample was quantitatively extracted as described previously [9]. Analyses of sphingolipids were performed by high-performance liquid chromatography-tandem mass spectrometry (LC-ESI-MS/MS) as described previously [9]. The equipment consisted of a Thermo Scientific Accela Autosampler and Quaternary Pump (Waltham, MA, USA). A Thermo Scientific Quantum Access triple quadrupole mass spectrometer equipped with an Electrospray Ion Source (ESI) operating in multiple reaction monitoring (MRM) positive ion mode was used. Chromatographic separations were obtained under a gradient elution of a Peeke Scientific (Redwood City, CA, USA), Spectra C8SR 150 × 3.0 mm, 3-μm particle size column. Quantitative analyses were based on calibration curves generated by injecting known amounts of the target analytes and an equal amount of the internal standards. Final concentrations of analytes in samples were determined using the appropriate corrections for sample loss based on internal standard recovery calculations. Sphingolipids with no available standards were quantified using the calibration curve of its closest counterpart. The resulting data were then normalized to the volume of sample analyzed.

3. Statistics

Data were analyzed using a commercially available statistical analysis package (SigmaStat v3.0, SPSS, Chicago, IL, USA).

4. Results

We selectively recruited six subjects whose plasma lipid and lipoprotein levels (Table 1) were representative of normolipidemia (subjects 1–4), combined hyperlipoproteinemia

TABLE 1: Plasma lipid and lipoprotein profiles in study subjects.

Subject	Gender	Age[a]	Total cholesterol (TC)	Total triglycerides	LDL-C	HDL-C	VLDL-C	TC/HDL-C
1	Female	56	4.1 (156)[b]	0.6 (50)	1.7 (66)	2.1 (80)	0.3 (10)	2.0
2	Male	30	4.2 (160)	1.4 (129)	2.1 (80)	1.4 (54)	0.7 (26)	3.0
3	Female	42	4.3 (165)	1.3 (117)	1.6 (60)	2.1 (81)	0.6 (23)	2.0
4	Female	48	4.5 (174)	1.8 (165)	2.6 (99)	1.1 (42)	0.9 (33)	4.1
5	Female	54	6.0 (231)	2.4 (220)	3.6 (138)	1.2 (48)	1.1 (44)	4.8
6	Male	58	2.5 (97)	4.5 (408)	0.5 (19)[c]	0.7 (28)	1.3 (50)	3.5

[a] Years.
[b] mM (mg/dL).
[c] Lipoprotein cholesterol determined directly.

(subject 5), and hypertriglyceridemia (subject 6). Most importantly, these subjects provided the potential to investigate the concentrations and distribution of sphingolipids over a broad range of plasma HDL-C (0.7–2.1 mM) concentrations.

We determined the concentrations of sphingolipids in the fraction of plasma containing all the plasma lipoproteins distinct from plasma fraction containing the remaining plasma proteins. These two plasma fractions were prepared by ultracentrifugation of whole plasma to float all the lipoprotein classes to separate them from the remaining plasma proteins which sedimented to the tube bottom. There were no detectable lipid staining bands with mobility comparable to plasma lipoproteins in the ultracentrifuge subnatant fractions after electrophoresis of the ultracentrifuge super- and subnatants in agarose gels (Figure 1(a)) confirming that plasma lipoproteins had been quantitatively removed from the fraction containing the plasma proteins. We also determined the concentrations of total cholesterol in each plasma fraction using gas chromatography (Figure 1(b)). Cholesterol levels in the plasma protein-containing fraction averaged only 1.8 ± 0.6 percent of the total cholesterol in plasma. Collectively, these data suggest that using the methods described, lipoproteins (lipoprotein cholesterol) were quantitatively removed from the plasma fraction containing the remaining plasma proteins.

The concentrations of the molecular species of ceramides measured in whole plasma, in the fraction of plasma containing all the plasma lipoproteins, and the plasma fraction containing the plasma proteins are shown in Table 2. Ceramide concentrations in the lipoprotein-containing fraction averaged $96 \pm 3\%$ of total plasma ceramide content in agreement with their role as an integral lipoprotein structural constituent and further support our conclusion that we quantitatively separated lipoproteins from the remaining plasma proteins. The concentrations of the ceramide species in the lipoprotein- and plasma protein-containing fractions averaged $97 \pm 5\%$ of the total ceramide concentration in plasma in support of our conclusion that recovery of sphingolipid in the plasma fractions after ultracentrifugation was quantitative. The ceramides present in highest concentration were the C_{24}, $C_{24:1}$, C_{20}, and C_{22} species in agreement with our previous reports [9, 11]. We also analyzed the plasma

concentrations of the SM species which averaged (mean $\pm$ SD, μM) 10.7 ± 5.4 (C_{14}-SM), 89.7 ± 26.9 (C_{16}-SM), 5.8 ± 1.2 (C_{18}-SM), 2.3 ± 0.9 ($C_{18:1}$-SM), 6.1 ± 1.6 (C_{20}-SM), 1.5 ± 0.6 ($C_{20:1}$-SM), 10.8 ± 2.0 (C_{22}-SM), 7.9 ± 2.6 ($C_{22:1}$-SM), 8.4 ± 2.3 (C_{24}-SM), and 17.1 ± 4.2 ($C_{24:1}$-SM). SM species distribution in the lipoprotein-containing fraction averaged $96.4 \pm 1.5\%$ of total plasma SM in agreement with our findings for ceramide species distribution. This data provides additional support of our conclusion that the separation of lipoproteins from the plasma protein fraction after ultracentrifugation was quantitative.

The concentrations of sphingoid bases, sphingosine and dihydrosphingosine, and their 1-phosphates (S1P and dhS1P) measured in whole plasma, in the fraction of plasma containing all the plasma lipoproteins, and in the plasma fraction containing the plasma proteins are summarized in Table 3. The percentage of S1P concentration in plasma localized to the lipoprotein-containing fraction was significantly, positively correlated ($R^2 = 0.7117$; $P < 0.05$) with HDL cholesterol concentration in the six subjects (Figure 1(c)). The total concentration of S1P in plasma varied significantly among the six subjects studied (Table 3). We wished to determine if increases in S1P mass in plasma were transported primarily in lipoproteins or were localized to the plasma protein-containing fraction. Increasing concentrations of S1P in plasma were significantly and positively correlated with the concentrations of S1P in the plasma protein-containing fraction (Figure 1(d)) ($R^2 = 0.65$; $P < 0.05$), but not with the S1P concentrations in the lipoprotein-containing fraction (Figure 1(e)). Because the bulk of the lipoprotein-associated S1P is found in HDL and HDL concentration in the six subjects investigated varied by almost 2.9-fold (Table 1), we investigated if the S1P concentration in the lipoprotein-containing fraction in each subject was limited by the HDL concentration in the subject. Therefore, we normalized the S1P concentration in the lipoprotein-containing fraction by the HDL cholesterol concentration in the subject (Table 3). There was a significant, positive association ($R^2 = 0.6551$; $P < 0.05$) between total S1P concentration in plasma with the "normalized" amount of S1P in the lipoprotein-containing fraction when the concentration of S1P was normalized by the HDL cholesterol concentration in the subject (Figure 1(f)).

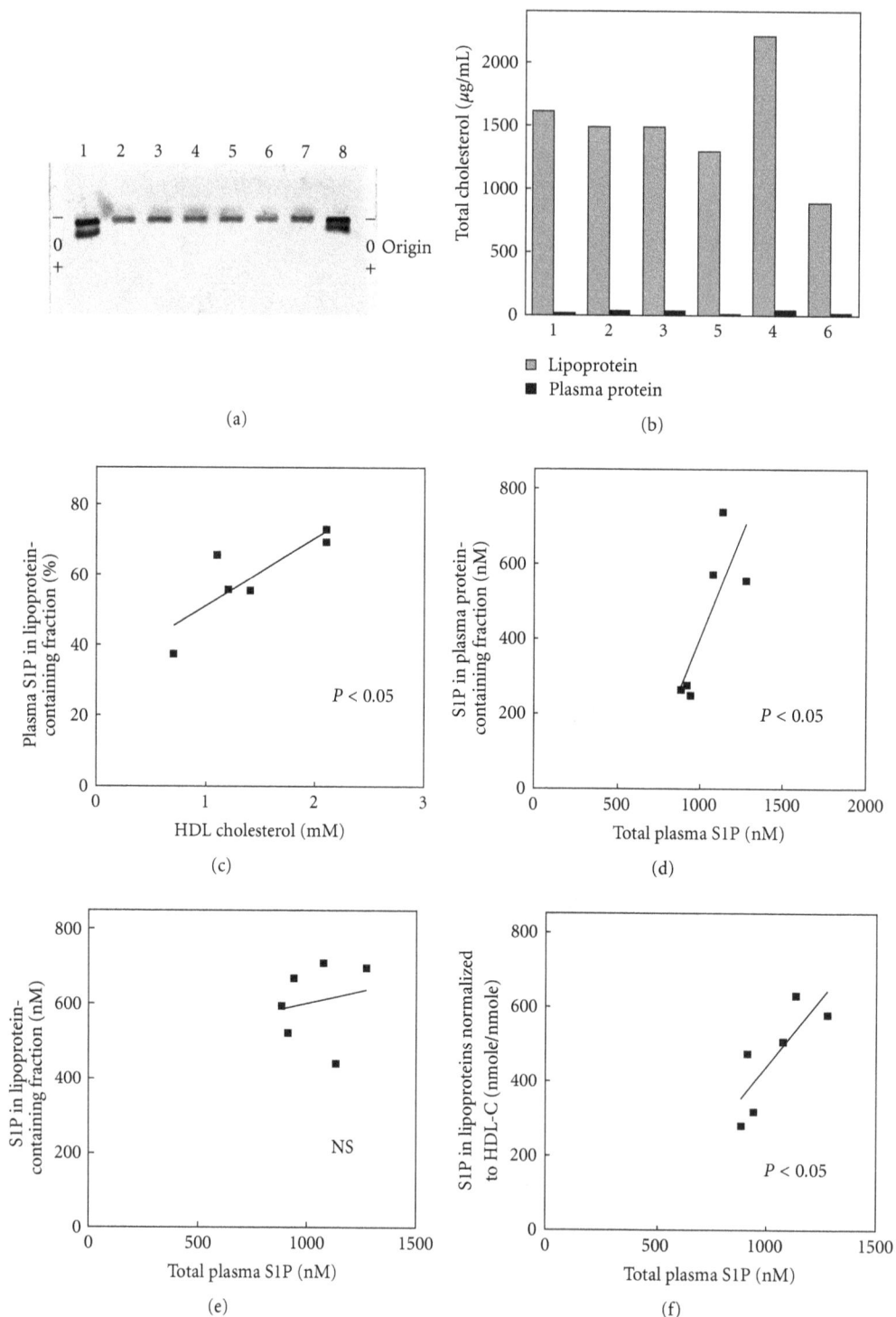

FIGURE 1: Analyses of human plasma sphingosine 1-phosphate (S1P) in lipoprotein- and protein-containing fractions. (a): agarose gel electrophoretogram of $d > 1.21$ g/ml plasma sub-fraction from each subject (lanes 2–7) with whole plasma (lane 1) and $d < 1.21$ g/ml plasma subfraction from a representative subject (lane 8) shown for reference. (b): total cholesterol concentration in the lipoprotein- and plasma protein-containing fractions from each subject determined by gas chromatography. (c): association between HDL cholesterol concentration and the percentage of S1P in plasma localized to the lipoprotein-containing fraction. (d): association between total plasma S1P concentration and the S1P concentration in the plasma protein-containing fraction. (e): association between total plasma S1P concentration and the S1P concentration in the ultracentrifugally isolated lipoprotein-containing fraction. (f): association between total plasma S1P concentration and the S1P concentration in the lipoprotein-containing fraction normalized by the HDL cholesterol concentration in the subject.

TABLE 2: (a) Concentrations of ceramide species in plasma. (b) Concentrations of ceramide species in lipoprotein-containing fraction ($d < 1.21$ g/mL ultracentrifuge fraction). (c) Concentrations of ceramide species in plasma protein-containing fraction ($d > 1.21$ g/mL ultracentrifuge fraction).

(a)

Subject	C14-Cer	C16-Cer	C18-Cer	C18:1-Cer	C20-Cer	C20:1-Cer	C22-Cer	C22:1-Cer	C24-Cer	C24:1-Cer	C26-Cer	C26:1-Cer	dhC16-Cer
							Concentration (nM)						
1	118	445	87	45	282	14	340	83	2498	1090	112	46	55
2	48	158	80	26	279	13	377	85	2773	1035	117	45	53
3	77	146	68	35	297	8	322	62	2548	728	70	21	53
4	50	250	84	33	156	10	383	78	2994	1325	91	39	62
5	170	548	138	80	420	20	486	111	3975	1476	181	66	103
6	77	182	114	22	493	11	465	83	3001	1084	88	46	36

(b)

Subject	C14-Cer	C16-Cer	C18-Cer	C18:1-Cer	C20-Cer	C20:1-Cer	C22-Cer	C22:1-Cer	C24-Cer	C24:1-Cer	C26-Cer	C26:1-Cer	dhC16-Cer
							Concentration (nM)						
1	121	459	89	66	455	20	405	99	3388	1354	147	57	42
2	57	135	76	24	274	15	394	85	2347	1046	86	41	49
3	71	161	99	38	352	13	311	63	2351	687	61	22	52
4	62	332	56	31	214	10	383	80	2570	1193	88	33	64
5	183	774	116	66	537	27	598	129	4819	1786	201	71	113
6	88	178	121	24	540	13	477	84	3130	1106	94	50	48

(c)

Subject	C14-Cer	C16-Cer	C18-Cer	C18:1-Cer	C20-Cer	C20:1-Cer	C22-Cer	C22:1-Cer	C24-Cer	C24:1-Cer	C26-Cer	C26:1-Cer	dhC16-Cer
							Concentration (nM)						
1	ND	5	8	3	6	ND	27	2	28	16	1.5	0.3	1.3
2	ND	8	9	3	10	0.6	28	6	39	32	2.0	0.8	1.3
3	1.6	11	12	1	4	0.1	24	2	19	13	1.4	ND	1.7
4	ND	2	4	1	3	0.4	25	2	17	22	1.2	ND	0.5
5	2.7	10	17	1	13	0.9	38	7	75	36	1.5	0.6	0.9
6	7.5	15	15	4	14	0.8	27	5	31	28	2.1	0.4	3.5

ND: not detected, Cer: ceramide.

TABLE 3: (a) Concentrations of sphingoid bases and their 1-phosphates in plasma. (b) Concentrations of sphingoid bases and their 1-phosphates in lipoprotein-containing fraction ($d <$ 1.21 g/mL ultracentrifuge fraction). (c) Concentrations of sphingoid bases and their 1-phosphates in plasma protein-containing fraction ($d >$ 1.21 g/mL ultracentrifuge fraction).

(a)

Subject	Concentration (nM)			
	dhSph	dhS1P	Sph	S1P
1	13	157	8	939
2	19	114	19	1072
3	10	132	18	884
4	14	138	20	913
5	7	171	23	1272
6	19	147	24	1132

(b)

Subject	Concentration (nM)				S1P normalized to HDL-C
	dhSph	dhS1P	Sph	S1P	(nmole/mmole)
1	6	33	11	669	32
2	19	21	15	710	51
3	4	31	9	595	28
4	3	34	11	523	48
5	17	26	12	696	58
6	19	12	18	442	63

(c)

Subject	Concentration (nM)				S1P in plasma protein fraction (%)
	dhSph	dhS1P	Sph	S1P	
1	7	105	7	248	27
2	8	133	8	571	45
3	8	111	10	263	31
4	7	79	10	215	34
5	11	150	7	555	44
6	8	117	19	919	63

Sph: sphingosine, dhSph: dihydrosphingosine, their 1-phosphates S1P and dhS1P.

5. Discussion

We investigated the associations of plasma lipid and lipoprotein concentrations with the total plasma concentrations of ceramides and sphingoid bases, sphingosine, and dihydrosphingosine, and their 1-phosphates. We also determined the distribution of these sphingolipids between the pool localized to plasma lipoproteins compared to that found in the remaining plasma proteins mainly bound to albumin. We determined that S1P concentration in plasma is highly variable and determined further that S1P in plasma was nonuniformly distributed between the plasma lipoprotein and plasma protein pools of S1P in the six subjects investigated. We determined further that increasing concentrations of S1P in plasma were localized primarily to the plasma protein-containing fraction (Figures 1(d) versus 1(e)) in subjects exhibiting a broad range of plasma lipid concentrations (Table 1).

The results of our studies in these select subjects confirmed and extended those reporting previously [12] that S1P concentrations in the lipoprotein-containing fraction of plasma were positively correlated with HDL cholesterol concentrations (Figure 1(c)). Because HDL is the primary lipoprotein which transports S1P, and because HDL-C concentration varied widely in the six subjects (Table 1), we also normalized the levels of S1P in the lipoprotein-containing fraction to the HDL-C concentration in each subject (Figure 1(f)). We determined that this "normalized" concentration of S1P transported in the lipoprotein-containing fraction was positively correlated to the total S1P concentration in plasma, which suggests that the S1P content of HDL is variable and that S1P transport in lipoproteins was not limited by the concentration of HDL-C in the individual subject. In fact, we have demonstrated previously that per particle, the larger VLDL particle contains the highest content of S1P, and the smallest lipoprotein particle, HDL3, contains higher S1P levels than LDL particles [9]. Our current data may allow subsequent research on mechanisms that mediate binding of the different S1P carriers to S1P receptors, knowing that only 5% of HDL particles carry S1P molecules [21].

Recent studies have reported that HDL-associated S1P is bound specifically to apolipoprotein M (apoM) and, furthermore, is transported selectively in apoM-containing HDL particles [22]. We did not measure apoM concentration or distribution in this select subject set and thus cannot infer if differences in apoM metabolism between the subjects were associated with the observed differences in S1P metabolism. Alternatively, we [9, 11] and others [8, 10] have determined that S1P levels in HDL are significantly higher in the smaller sized HDL3 subfraction compared to S1P levels in HDL2. The smaller HDL3 particles were shown also to be enriched in apoM [23]. Despite the very strong evidence from apoM knockout and apoM-transgenic mice that apoM determines S1P concentrations in plasma and HDL [22, 23], there was no statistically significant correlation between S1P and apoM concentrations in human patients with different monogenic disorders of HDL metabolism [23]. Thus, the relative concentrations of HDL subfractions may have influenced the distribution of S1P in the individual subjects.

More recently, the crystal structure of a main S1P G protein-coupled receptor, S1P1, was revealed, and it was found that extracellular access to the binding pocket of this receptor is occluded by the aminoterminus and extracellular loops of the receptor [24]. Interestingly, access is gained by ligands entering laterally between helices I and VII within the transmembrane region of the receptor [24]. In the context of previous findings, we postulate that effects of HDL-bound S1P could be related to (1) HDL particle size and the ability to modulate cell lipid rafts and comprised receptors, (2) S1P content of HDL particles being higher in the smaller HDL particles, and (3) HDL role as scavenger/reservoir for biologically active lipids including S1P.

A confounding factor in recent studies addressing the cardioprotective role of S1P levels in HDL is the assumption that 30% of S1P is recovered in the lipoprotein-deficient serum [25]. As shown in Table 3, S1P distribution in the plasma protein-containing fraction approximated 30% in only three of the six subjects studied. Thus, even in this limited number of subjects, S1P distribution in 50% of the subjects does not support this often cited assumption. Figure 1(e) clearly shows that there is no significant association between total plasma S1P and S1P concentration in plasma lipoprotein-containing fraction, which suggests a crucial pathophysiological role of the S1P bound to plasma proteins. Furthermore, the procedure in which apolipoprotein B-depleted plasma is prepared and S1P concentration is subsequently determined in the "HDL-containing" fraction as performed in two recent studies [21, 26] also does not discriminate between HDL- and albumin-associated S1P. Thus, when this methodology is employed, assumptions regarding S1P distribution in plasma may result in erroneous estimates.

We have shown previously that the secretion of plasminogen activator inhibitor-1 (PAI-1) from cultured adipocytes is significantly increased when the cells were incubated with increasing S1P concentration in the media regardless of whether the S1P is bound to HDL or transported by albumin [11]. Previous studies have determined that triglyceride level was independently related to plasma PAI-1 activity level in both subjects with hypertriglyceridemia and in age-matched normotriglyceridemic subjects [27]. The factor(s) contributing to this phenomenon are still unclear. In this limited study, plasma S1P concentrations were increased in the two subjects exhibiting elevated plasma triglyceride levels. Clearly, additional studies are warranted in this area.

In summary, we determined that S1P concentration in plasma is highly variable, S1P in plasma is nonuniformly distributed between the plasma lipoprotein and plasma protein pools of S1P and increasing concentrations of S1P in plasma are localized primarily to the plasma protein-containing fraction. We also determined that the S1P content of HDL, the major lipoprotein carrier of S1P, is variable and that S1P transport in lipoproteins is not limited by the concentration of HDL-C. The data further show that there is no significant association between total plasma S1P and S1P concentration in plasma lipoprotein-containing fraction, which suggests a crucial pathophysiological role of the S1P bound to plasma proteins.

Abbreviations

S1P: sphingosine 1-phosphate
VLDL: very low-density lipoproteins
LDL: low-density lipoproteins
HDL: high-density lipoproteins
SM: sphingomyelin
C: total cholestero
dhS1P: dihydrosphingosine 1-phosphate
PAI-1: plasminogen activator inhibitor-1.

Acknowledgments

This work was supported by the National Institutes of Health Grants HL-079274 and R0-1HL079274-04S1 (American Recovery and Reinvestment Act), the Southeastern Clinical and Translational Research Institute, the South Carolina Center of Biomedical Research Excellence (COBRE) in Lipidomics and Pathobiology Grant no. P20 RR17677 from National Center for Research Resources (NCCR), and the Pilot Research funding from the Hollings Cancer Center's Cancer Center Support Grant at MUSC Grant no. P30 CA13813 (to S. M. Hammad), and the Department of Veterans Affairs Merit Review Program (to R. L. Klein). The contents presented herein do not represent the views of the Department of Veterans Affairs or the United States Government. The authors gratefully acknowledge the Lipidomics Shared Resource Facility at MUSC for their continued expert analytical contributions, without which this work could not have been conducted. We are particularly grateful to Mr. Kent Smith for technical assistance.

References

[1] A. E. Alewijnse and S. L. M. Peters, "Sphingolipid signalling in the cardiovascular system: good, bad or both?" *European Journal of Pharmacology*, vol. 585, no. 2-3, pp. 292–302, 2008.

[2] S. Spiegel and S. Milstien, "Sphingosine-1-phosphate: anenigmatic signalling lipid," *Nature Reviews Molecular Cell Biology*, vol. 4, no. 5, pp. 397–407, 2003.

[3] T. Hla, "Physiological and pathological actions of sphingosine 1-phosphate," *Seminars in Cell and Developmental Biology*, vol. 15, no. 5, pp. 513–520, 2004.

[4] K. Watterson, H. Sankala, S. Milstien, and S. Spiegel, "Pleiotropic actions of sphingosine-1-phosphate," *Progress in Lipid Research*, vol. 42, no. 4, pp. 344–357, 2003.

[5] Y. Takuwa, Y. Okamoto, K. Yoshioka, and N. Takuwa, "Sphingosine-1-phosphate signaling and biological activities in the cardiovascular system," *Biochimica et Biophysica Acta*, vol. 1781, no. 9, pp. 483–488, 2008.

[6] H. Rosen, P. J. Gonzalez-Cabrera, M. G. Sanna, and S. Brown, "Sphingosine 1-phosphate receptor signaling," *Annual Review of Biochemistry*, vol. 78, pp. 743–768, 2009.

[7] F. Okajima, "Plasma lipoproteins behave as carriers of extracellular sphingosine 1-phosphate: is this an atherogenic mediator or an anti-atherogenic mediator?" *Biochimica et Biophysica Acta*, vol. 1582, no. 1–3, pp. 132–137, 2002.

[8] N. Murata, K. Sato, J. Kon et al., "Interaction of sphingosine 1-phosphate with plasma components, including lipoproteins, regulates the lipid receptor-mediated actions," *Biochemical Journal*, vol. 352, no. 3, pp. 809–815, 2000.

[9] S. M. Hammad, J. S. Pierce, F. Soodavar et al., "Blood sphingolipidomics in healthy humans: impact of sample collection methodology," *Journal of Lipid Research*, vol. 51, no. 10, pp. 3074–3087, 2010.

[10] A. Kontush, P. Therond, A. Zerrad et al., "Preferential sphingosine-1-phosphate enrichment and sphingomyelin depletion are key features of small dense HDL3 particles: relevance to antiapoptotic and antioxidative activities," *Arteriosclerosis, Thrombosis, and Vascular Biology*, vol. 27, no. 8, pp. 1843–1849, 2007.

[11] M. H. Lee, S. M. Hammad, A. J. Semler, L. M. Luttrell, M. F. Lopes-Virella, and R. L. Klein, "HDL3, but not HDL2,

stimulates plasminogen activator inhibitor-1 release from adipocytes: the role of sphingosine-1-phosphate," *Journal of Lipid Research*, vol. 51, no. 9, pp. 2619–2628, 2010.

[12] B. Zhang, H. Tomura, A. Kuwabara et al., "Correlation of high density lipoprotein (HDL)-associated sphingosine 1-phosphate with serum levels of HDL-cholesterol and apolipoproteins," *Atherosclerosis*, vol. 178, no. 1, pp. 199–205, 2005.

[13] J. R. Nofer, M. Van Der Giet, M. Tölle et al., "HDL induces NO-dependent vasorelaxation via the lysophospholipid receptor S1P3," *Journal of Clinical Investigation*, vol. 113, no. 4, pp. 569–581, 2004.

[14] C. Rodríguez, M. Gonzalez-Diez, L. Badimon, and J. Martínez-González, "Sphingosine-I-phosphate: a bioactive lipid that confers high-density lipoprotein with vasculoprotection mediated by nitric oxide and prostacyclin," *Thrombosis and Haemostasis*, vol. 101, no. 4, pp. 665–673, 2009.

[15] G. Theilmeier, C. Schmidt, J. Herrmann et al., "High-density lipoproteins and their constituent, sphingosine-1-phosphate, directly protect the heart against ischemia/reperfusion injury in vivo via the S1P3 lysophospholipid receptor," *Circulation*, vol. 114, no. 13, pp. 1403–1409, 2006.

[16] K. Sattler and B. Levkau, "Sphingosine-1-phosphate as a mediator of high-density lipoprotein effects in cardiovascular protection," *Cardiovascular Research*, vol. 82, no. 2, pp. 201–211, 2009.

[17] J. S. Karliner, "Sphingosine kinase and sphingosine 1-phosphate in cardioprotection," *Journal of Cardiovascular Pharmacology*, vol. 53, no. 3, pp. 189–197, 2009.

[18] S. M. Hammad, "Blood sphingolipids in homeostasis and pathobiology," *Sphingolipids and Metabolic Disease*, vol. 721, pp. 57–66, 2011.

[19] K. J. E. Sattler, Ş. Elbasan, P. Keul et al., "Sphingosine 1-phosphate levels in plasma and HDL are altered in coronary artery disease," *Basic Research in Cardiology*, vol. 105, no. 6, pp. 821–832, 2010.

[20] R. L. Klein, T. J. Lyons, and M. F. Lopes-Virella, "Metabolism of very low- and low-density lipoproteins isolated from normolipidaemic Type 2 (non-insulin-dependent) diabetic patients by human monocyte-derived macrophages," *Diabetologia*, vol. 33, no. 5, pp. 299–305, 1990.

[21] R. Karuna, R. Park, A. Othman et al., "Plasma levels of sphingosine-1-phosphate and apolipoprotein M in patients with monogenic disorders of HDL metabolism," *Atherosclerosis*, vol. 219, no. 2, pp. 855–863, 2011.

[22] C. Christoffersen, H. Obinata, S. B. Kumaraswamy et al., "Endothelium-protective sphingosine-1-phosphate provided by HDL-associated apolipoprotein M," *Proceedings of the National Academy of Sciences of the United States of America*, vol. 108, no. 23, pp. 9613–9618, 2011.

[23] W. S. Davidson, R. A. G. D. Silva, S. Chantepie, W. R. Lagor, M. J. Chapman, and A. Kontush, "Proteomic analysis of defined hdl subpopulations reveals particle-specific protein clusters: relevance to antioxidative function," *Arteriosclerosis, Thrombosis, and Vascular Biology*, vol. 29, no. 6, pp. 870–876, 2009.

[24] M. A. Hanson, C. B. Roth, E. Jo et al., "Crystal structure of a lipid G protein-coupled receptor," *Science*, vol. 335, no. 6070, pp. 851–855, 2012.

[25] N. Murata, K. Sato, J. Kon et al., "Interaction of sphingosine 1-phosphate with plasma components, including lipoproteins, regulates the lipid receptor-mediated actions," *Biochemical Journal*, vol. 352, no. 3, pp. 809–815, 2000.

[26] K. M. Argraves, A. A. Sethi, P. J. Gazzolo et al., "S1P, dihydro-S1P and C24:1-ceramide levels in the HDL-containing fraction of serum inversely correlate with occurrence of ischemic heart disease," *Lipids in Health and Disease*, vol. 10, p. 70, 2011.

[27] A. Asplund-Carlson, A. Hamsten, B. Wiman, and L. A. Carlson, "Relationship between plasma plasminogen activator inhibitor-1 activity and VLDL triglyceride concentration, insulin levels and insulin sensitivity: studies in randomly selected normo- and hypertriglyceridaemic men," *Diabetologia*, vol. 36, no. 9, pp. 817–825, 1993.

4

Novel Associations of Nonstructural Loci with Paraoxonase Activity

Ellen E. Quillen, David L. Rainwater, Thomas D. Dyer, Melanie A. Carless, Joanne E. Curran, Matthew P. Johnson, Harald H. H. Göring, Shelley A. Cole, Sue Rutherford, Jean W. MacCluer, Eric K. Moses, John Blangero, Laura Almasy, and Michael C. Mahaney

Department of Genetics, Texas Biomedical Research Institute, P.O. Box 760549, San Antonio, TX 78245-0549, USA

Correspondence should be addressed to Ellen E. Quillen, equillen@txbiomedgenetics.org

Academic Editor: Mira Rosenblat

The high-density-lipoprotein-(HDL-) associated esterase paraoxonase 1 (PON1) is a likely contributor to the antioxidant and antiatherosclerotic capabilities of HDL. Two nonsynonymous mutations in the structural gene, *PON1*, have been associated with variation in activity levels, but substantial interindividual differences remain unexplained and are greatest for substrates other than the eponymous paraoxon. PON1 activity levels were measured for three substrates—organophosphate paraoxon, arylester phenyl acetate, and lactone dihydrocoumarin—in 767 Mexican American individuals from San Antonio, Texas. Genetic influences on activity levels for each substrate were evaluated by association with approximately one million single nucleotide polymorphism (SNPs) while conditioning on *PON1* genotypes. Significant associations were detected at five loci including regions on chromosomes 4 and 17 known to be associated with atherosclerosis and lipoprotein regulation and loci on chromosome 3 that regulate ubiquitous transcription factors. These loci explain 7.8% of variation in PON1 activity with lactone as a substrate, 5.6% with the arylester, and 3.0% with paraoxon. In light of the potential importance of *PON1* in preventing cardiovascular disease/events, these novel loci merit further investigation.

1. Introduction

More than 2,200 Americans die from cardiovascular disease each day with 75% of those deaths attributable to atherosclerosis [1]. Atherosclerosis is characterized by the buildup of fatty lesions, inflammation, and scarring of arterial walls with oxidative stress as a primary contributing factor. Paraoxonase 1 (PON1) is a high-density-lipoprotein-(HDL-) associated esterase which appears to contribute to the antioxidant and antiatherosclerotic capabilities of HDL. PON1 is synthesized in the liver and secreted into the bloodstream where it is capable of breaking down both man-made and naturally occurring compounds. Named for its ability to hydrolyze organophosphates like paraoxon [2, 3] found in insecticides, PON1 is also able to hydrolyze *N*-acyl-homo-serine, a lactone used by pathogenic bacteria [4], and lipid peroxides, thereby inhibiting the formation of foam cells known to contribute to atherosclerosis [5, 6].

PON1 has been widely studied following evidence that high activity levels decrease systemic oxidative stress and are associated with a lower incidence of cardiovascular events [7]. PON1 levels have been tied to a number of other disorders including type 1 and 2 diabetes [8, 9], thyroid dysfunction [10], uremia [11], renal failure [12], and inflammatory response [13].

The structural gene *PON1* is by far the largest contributor to variation in serum PON1 activity levels with four known single nucleotide polymorphisms (SNPs) in the promoter region [14–16] and two nonsynonymous substitutions in the coding region of the gene [17, 18] shown to significantly influence activity levels. Amino acid substitution 192Q > R (rs662) specifies 2 allozymes [17] whose differences in activity are substrate dependent. The R allozyme shows greater activity for the organophosphates paraoxon and fenitroxon while the Q form more efficiently hydrolyzes other organophosphates including diazoxon, soman, and sarin.

Phenyl acetate is hydrolyzed at the same rate by both forms. [19–21] The 192Q > R substitution is associated with up to 13-fold interindividual differences in PON1 activity [22] and an adjusted hazard ratio for major cardiac events of 1.5 [23]. The 55L > M (rs854560) substitution has also been associated with variation in serum PON1 activity levels, but has a smaller effect size. These polymorphisms have also been linked to Parkinson's disease [24], inflammatory bowel diseases [25], and, controversially, to Alzheimer's disease and vascular dementia [26].

PON1 is part of a family of genes including PON2 and PON3 located within a 140 kb region at 7q21.3. Although PON2 and PON3 also synthesize paraoxonase proteins, PON2 is not excreted into the blood and any effect by either protein on atherosclerosis or cardiovascular disease is small [27]. Although the PON region explains a large degree of the variation in PON1 activity, PON1 activity levels are still better predictors of disease than PON1 genotypes alone [22, 28]. This supports the existence of additional, unidentified polymorphisms associated with PON1 activity as well as potential epigenetic contributors.

Despite the large body of literature on the PON loci, this study is the first to take a genome-wide association approach to identify additional genomic regions contributing to interindividual variation in PON1 activity. Previous studies of this sample identified QTLs for PON1 activity with paraoxon as a substrate on chromosomes 7 (PON1), 12, 17, and 19 using whole-genome multipoint linkage analysis [29]. Further investigation, which included alternate substrates and conditioned on the major QTL at chromosome 7, located additional QTLs on chromosomes 1, 3, and 14 [30].

2. Methods

2.1. Subjects.
Samples were drawn from the San Antonio Family Heart Study (SAFHS) which is composed of 1414 individuals (837 females, 577 males) belonging to 42 extended pedigrees originating with probands randomly ascertained with respect to disease status and phenotype. All probands were Mexican-American individuals between the ages of 40 and 60 at the time of ascertainment, living in San Antonio, TX, with a minimum of six offspring and/or siblings who were at least sixteen years of age and also living in the area. After giving their informed consent, participants underwent a physical examination, demographic and lifestyle interview, and provided blood samples for genotyping and blood chemistry analysis. The study protocol was approved by the Institutional Review Board at the University of Texas Health Science Center in San Antonio and is described in more detail in a previous publication [31].

2.2. Paraoxonase Activity.
Based on previous evidence of genetic variation giving rise to different activity levels in a substrate-dependent fashion, PON1 activity was assessed on an organophosphate, an arylester, and a lactone. Activity was calculated for 767 individuals based on standard spectrophotometric assays described previously [32]. Briefly, PON1-para activity was determined from the rate of con-

version of paraoxon to p-nitrophenol, PON1-aryl activity was calculated from the conversion of phenyl acetate to phenol, and PON1-lact activity was based on the conversion of dihydrocoumarin to 3-(2-hydroxyphenyl)propionate. The underlying shared genetic relationship between the activity levels was estimated by calculating the shared genetic covariance (ρ_G) for each pair of activity levels. PON-aryl showed significant evidence of a shared genetic contribution with both PON-lact ($\rho_G = 0.54$, $P = 1.9e - 19$) and PON-para ($\rho_G = 0.65$, $P = 5.5e - 23$); however, the genetic correlation between the activity levels for PON-lact and PON-para was essentially zero. This suggests that there are likely to be independent variants influencing activity levels for the different substrates.

2.3. Statistical Genetic Analyses.
DNA was extracted from buffy coats and used for genotyping on a series of Illumina microarrays (Illumia, Inc., San Diego, CA). 931,219 SNPs passed quality control and were included in the genome-wide association (GWA) analysis. Where it could be done with a high degree of certainty, known pedigree relationships were used to infer missing genotype data using the PEDSYS routine INFER [33]. For ambiguous genotypes, a weighted average of the possible genotypes was used.

Association was assessed for each measurement of paraoxonase activity using the measured genotype test implemented in the program SOLAR (Sequential Oligogenic Linkage Analysis Routines) [34] which takes into account relationships among family members. For all analyses, paraoxonase activity and age were normalized and sex, age, age², and the interaction of sex by age were used as covariates. Additional covariates considered but not found to be associated with PON1 activity in this sample ($P > 0.1$) include dietary measures (proportion of saturated fats, monounsaturated fats, polyunsaturated fats, and fat calories reported in diet), alcohol and cigarette consumption, body mass index (BMI), and total metabolic equivalents as a measure of activity level. To minimize the risk of false associations due to stratification in this admixed sample, principal component analysis was performed on the full set of approximately one million genotypes to capture the total genetic variation in the sample and the first four principal components (accounting for only 2.8% of the variation in the sample) were included as covariates. The efficacy of this correction for stratification was examined by calculating λ from the distribution of the lower 90% of P values for each GWA. There was no evidence of overdistribution due to stratification as all λ values were less than 1.02. Manhattan and Q-Q plots for each GWA can be found in Supplementary File 1 available online at doi:10.1155/2012/189681.

To identify genes contributing to the residual variation, genome-wide association was assessed while including genotypes at the two major PON1 substitutions 192QR and 55LM (rs662 and rs854560) as covariates. Despite the inclusion of these known variants, other SNPs in the PON region of chromosome 7 still showed association with PON1 activity for all three substrates. This suggests that additional variation in PON1 or nearby genes is contributing to the variation in PON1 activity levels. To remove all effects of the *PON* loci, four additional SNPs—rs854522, rs854534, rs757158, and rs7803148—each tagging a haploblock in the region surrounding the three *PON* genes—were included as covariates for the GWA analysis. With the inclusion of these covariates, there is 95% power to detect variants with an effect size (R_G^2) of at least 0.041 for PON1-aryl, 0.041 for PON1-lact, and 0.024 for PON1-para.

2.4. Candidate Gene Identification. The program SUSPECTS was used to identify candidate genes in the region surrounding each SNP showing significant association with PON1 activity [35]. SUSPECTS determined candidate genes on the basis of similarities in structure, gene ontology, InterPro domains, and/or gene expression with genes known to be related to atherosclerosis, cholesterol regulation, or heart disease. Additionally, all genes within 250 kb of the associated SNP were investigated for potential functional relationships to PON1 activity. A 250 kb region is sufficient to encompass more than 97% of haploblocks in Mexican populations [36]. To further contextualize the results of the GWA analysis, gene set and pathway enrichment analysis was performed using the Database for Annotation, Visualization, and Integrated Discovery (DAVID) [37, 38]. All markers were annotated and the genes found in the top 1% of associations were clustered based on similarity of gene ontology (GO) terms and compared to the genes represented by the remaining 99% of markers.

Additionally, the relationship between gene expression levels measured on an Illumina Sentrix Human Whole Genome (WG-6) Series I BeadChip array [39] and associated SNPs was calculated from the genetic covariance in a polygenic model including sex, age, age^2, sex*age, 192QR, 55LM, and the additional associated loci in the *PON* region.

3. Results

The heritability of paraoxonase activity varies with substrate, $h^2 = 0.65$ for phenyl acetate (PON1-aryl), $h^2 = 0.73$ for paraoxon (PON1-para), and $h^2 = 0.79$ for dihydrocoumarin (PON1-lact) after the inclusion of sex and age in the model. Similarly, the phenotypic variation explained by the covariates and the *PON1* alleles differs among the substrates (Table 1). Cumulatively, age, sex, and their interaction effect explain less than 4% of the variation in PON1 activity for all substrates. In this sample, known *PON1* variants (192QR and 55LM) explain 50.4% of the variation in PON1-para activity but only 9.6% of the variation in PON1-aryl and 12.4% of the variation in PON1-lact activities. Previous research in individuals of Korean ancestry reported that 192QR explained 65.8% of variation in PON1-para activity [40]. A similar study in individuals of European ancestry estimates that 192QR explained 46% and 55LM explained 16% of the PON1-para variation [41]. Although the haplotype blocks in the Mexican American sample are similar in location to those seen in the CEPH Europeans downloaded from HapMap, the linkage disequilibrium (LD) is more extreme, likely due to admixture. This increase in

TABLE 1: Proportion of variance explained by covariates. The proportion of variation explained by each significant variable was calculated for the three substrates by adding each to a polygenic model in SOLAR.

Covariate	Variation explained		
	PON1-aryl	PON1-lact	PON1-para
Sex	0.4%	0.0%	0.4%
Age	1.1%	1.4%	0.4%
Age^2	1.1%	1.9%	0.1%
Sex*age	0.7%	0.4%	0.7%
192QR	3.9%	8.3%	36.5%
55LM	5.7%	4.1%	13.9%
rs854522	2.7%	1.8%	1.1%
rs854534	3.4%	3.8%	2.1%
rs757158	11.5%	9.2%	4.1%
rs7803148	0.9%	1.1%	0.6%

LD would be expected to inflate the explanatory power of the known polymorphisms, so it is unclear why 192QR and 55LM explain less variation in this sample.

To assess the residual variation in the *PON* region of chromosome 7, haplotype blocks were identified using the solid spine of linkage disequilibrium method implemented in Haploview [42]. The SNP rs757158 tags a haploblock including the 5′ promoter region of *PON1* and explains the greatest proportion of remaining variation in PON1 activity for all substrates. This is in line with previous reports of promoter polymorphisms. As seen for 192QR and 55LM, the promoter polymorphisms have different effects depending on the substrate—this region explains a larger proportion of the variation in PON1-aryl and PON1-lact than in PON1-para. Polymorphisms in the remaining haploblocks are also associated with variation in PON1 activity. rs854534 tags a haplotype block including the majority of the genic region of *PON1*, rs854522 tags a region downstream of all the *PON* genes, and rs7803148 lies in the genic region of *PON2* and tags a haplotype block that includes the entirety of that gene and the majority of *PON3*. Variation in these three haploblocks cumulatively explains less variation in PON1 activity than the haploblock tagged by rs757158. Following the inclusion of these markers as covariates, there are no significant associations in the *PON* regions for PON1-para or PON1-lact; however, PON1-aryl activity is suggestively associated with rs2299262, an intronic SNP in *PON1* that explains 2.8% of the variation in PON1-aryl activity. It should be noted that while the proportion of variation described here and subsequently is useful for comparing the relative contributions of the loci to the variation in activity levels, there can be a substantial upward bias in these estimates unless replicated in an independent sample [43].

The variation in effect size of the polymorphism by substrate bolsters the previously identified substrate-specific effect and indicates additional genetic variants must be contributing to differences in activity levels for PON1-aryl and PON1-lact in particular. The relatively large amount of variation captured by including additional SNPs in the

TABLE 2: Summary of Significant and Suggestive Associations. SNP associations varied among substrates with overlap only at rs1078701 on chromosome 4. For each SNP, the chromosomal location, proportion of variation in PON1 activity explained, measured genotype test P-value, and minor allele frequency are listed. Shoulder SNPs are SNPs ranking in the top 5% of associations located within 500 kb of the significantly associated SNP. Genes identified by SUSPECTS are within 7.5 Mb of the candidate SNP and have a similarity score greater than 15. Bolded genes are discussed in the texts as the most likely contributors to PON1 activity based on known function, however, other genes may have unknown functions important in the regulation of PON1. The two associated SNPs on chromosome 3 are in perfect linkage disequilibrium and explain 2.7% of the variation in PON1-lact cumulatively.

Substrate	SNP	Chr	Position	Variation Explained	MG p	MAF	Shoulder SNPs	SUSPECTS Genes
para	rs12083993	1	64,691,396	1.2%	$1.8E-07$	0.3%	34	*ALG6*, ***ANGPTL3***, ***CYP2J2***, *IL12RB2*, *INSL5*, ***LEPROT***, *OMA1*, *PTGER3*
lact	rs13322362	3	76,613,497	2.7%	$4.8E-07$	5.4%	6	*PROK2*, *ROBO1*, *ROBO2*
lact	rs11915977	3	76,613,530		$4.8E-07$	5.5%		
para				1.8%	$4.7E-09$		32	
aryl	rs1078701	4	8,241,119	2.8%	$3.0E-08$	2.2%	19	***ADD1***, *CPZ*, *DGKQ*, *FGFBP1*, *FGFR3*, *FGFRL1*, *HGFAC*, *HS3ST1*, *LRPAP1*, *MXD4*
lact				2.3%	$3.8E-07$		15	
aryl	rs2299262	7	94,787,864	2.8%	$1.3E-07$	43.9%	24	*PON1*
lact	rs7225624	17	47,811,373	2.9%	$4.7E-09$	0.3%	4	*ABCD4*, *ADAM11*, *DGKE*, *GRN*, *HOXB2*, *NGFR*, *NMT1*, *OSBPL7*, *PLCD3*, ***PCTP***, *SCPEP1*, *SLC35B1*

region once the known *PON1* variants are included in the model suggests finer-scale analysis of this region may identify additional contributing polymorphisms.

Association was assessed for each SNP using a genome-wide significance threshold of $P < 5E-8$ with $P < 5E-7$ considered a suggestive association. As this study assesses associations in Mexican-American families, it should be noted that there is wider linkage disequilibrium than would be found in a population of randomly mating, unadmixed individuals, making this value of α conservative. Using these thresholds, two SNPs are associated with PON1-para activity, four with PON1-lact activity (two of which are in perfect LD), and two with PON1-aryl activity after conditioning on the associated SNPs in the *PON* region (Table 2, Supplemental File 1). Only a single SNP (rs1078701) on chromosome 4 is significantly associated with activity on all two substrates and suggestively associated for the third. The proportion of variation explained by rs1078701 varies among the different substrates, ranging from 1.8–2.8%.

Located less than 200 kb from rs1078701, *ACOX3* is a strong positional candidate gene as well as a potential contributor to cholesterol regulation. This gene encodes a peroxisomal pristanoyl-CoA oxidase essential for the catabolism of branched fatty acids into precursors for cholesterol biosynthesis [44]. However, the degree to which peroxisomal fatty acid metabolism contributes to circulating cholesterol is unclear [45]. Also found near rs1078701 is *LRPAP1* which produces a glycoprotein that has been linked with gallstone

disease caused by an excess of cholesterol [46] and with cholesterol-related brain disorders such as dementia [47] and Alzheimer's disease [48]. *LRPAP1* regulates the amount of LDL-receptor-related protein expressed in the liver and brain and may also act as a chaperone for lipoprotein lipase [49]. A final candidate gene from this region, *ADD1*, is localized to the erythrocyte membrane and is involved in renal sodium handling and hypertension [50]. It has been implicated in blood pressure, adipogenesis, and coronary heart disease [51, 52]. The other significantly associated SNP is rs7225624 on chromosome 17 which explains 2.9% of the observed variation in PON1-lact. *PCTP* is 6Mbp away from rs7225624, but has a strong SUSPECTS score due the involvement of this gene in cholesterol metabolism and transport as well as lipid binding. *PCTP* is a transfer protein found in macrophages, which are pervasive in atherosclerotic legions, and work in model organisms indicates *PCTP* regulates lipid efflux into the blood stream [53, 54].

Two additional loci show suggestive associations with PON1 activity. Located on chromosome 1, rs12083993 is associated with variation in PON1-para and replicates a previous linkage result for PON1-aryl [32]. This polymorphism explains 1.2% of the variation in PON1-para and SUSPECTS prioritizes three candidate genes involved in lipid metabolism. *ANGPTL3* is predominantly expressed in the liver, but suppresses lipoprotein lipase in the blood stream which in turn hydrolyzes HDL [55]. Polymorphisms in this gene are also associated with increased arterial wall thickness

[56]. *LEPROT* is involved in the cell-surface expression of the leptin receptor, regulation of growth hormones linked to obesity in mice, and cell signaling in response to circulating nutrient levels [57, 58]. Finally, *CYP2J2* is a member of the cytochrome P450 gene family which is widely involved in the oxidation of organic substances and metabolism. *CYP2J2* is primarily expressed in the aorta and coronary artery and has been linked to hypertension risk [59, 60]. The most likely mechanism for this relationship is the metabolism of arachidonic acids to epoxyeicosatrienoic acids (EETs) which are vasodilating agents capable of inhibiting inflammatory response and promoting fibrinolysis [61, 62]. Because PON1 can be inactivated under oxidative conditions [63] similar to those present in the absence of functional *CYP2J2* [64, 65], this association may be related to the realized activity of inactivated PON1 enzyme rather than the basal concentration of PON1 or the activity level under normal plasma conditions.

A suggestive association was also found for PON1-lact with two SNPs in perfect LD (rs13322362 and rs11915977) on chromosome 3. Jointly, these explain 2.7% of the overall variation in PON1-lact. Although a region of chromosome 3 was identified in previous linkage analyses [32], that QTL is more than 44 Mbp from rs13322362. This region of the genome contains few genes but may contain one or more transcription factors indicated by the significant association between rs13322362 and transcript levels of three genes including *SLC25A26*, a mitochondrial transport gene [66], *PROK2*, which regulates circadian rhythms [67], and *RYBP*, a broadly expressed binding protein essential for development [68].

The candidate genes identified in this analysis are consistent with the evidence of enrichment of the top 1% of associations for genes analyzed in DAVID. These results suggest the involvement of pathways related to vasculature development and angiogenesis, cell junctions and signaling, cell adhesion, transmembrane glycoproteins and ion transport, and immunoglobulin.

4. Discussion

The use of known polymorphisms in *PON1* as covariates in this genome-wide association analysis is an unusual but essential method for identifying regions of the genome with smaller effects. By accounting for more than 40% of the variation in PON1 activity in this way, four additional regions of the genome showing an association with residual PON1 activity were identified that could not otherwise be isolated.

The pathways through which candidate genes influence PON1 activity are, for many of the genes, unclear. Frequently, this is due to a lack of full, functional understanding of the genes themselves. Genes that are associated with cholesterol levels and associated syndromes, but have no clear mechanism, may be contributing to the PON1 pathway but further research would be required to demonstrate this. Because PON1 binds to, and is carried by, nonoxidized HDL molecules, these regions may play a role in plasma HDL

concentration which would indirectly increase PON1 concentration and activity. Additionally, it is important to recognize that PON activity is likely influenced by the tertiary structure of the PON1 protein itself or the circulating levels of PON in the blood, which may be influenced by the lipoprotein milieu.

It is necessary, therefore, to consider these results within the broader network of genes and proteins involved in the regulation of lipid metabolism in the blood stream. However, lipid metabolic pathways are not the only ones implicated in these association results. Ubiquitous transcription factors and genes related to oxidative stress were also identified and could play substantial roles in the regulation of PON1 concentration in the blood or the inactivation of PON1 which would decrease activity even in the presence of high levels of the enzyme. Considering the number of factors associated with the activity of single enzyme, future work on the genetic underpinnings and biochemical regulation of atherosclerosis, hypertension, cholesterolemia, and inflammatory diseases more broadly must be understood in the broadest biochemical context.

Acknowledgments

The authors would like to thank the volunteers who participated in this study and three anonymous reviewers for their comments. Perry H. Moore, Jr., performed the paraoxonase enzyme activity assays. This work was supported in part by Grants HL045522, MH059490, P01HL028972, and MH083824 from the National Institutes of Health. Parts of this investigation were conducted in facilities constructed with support from the Research Facilities Improvement Program (C06 RR013556 and C06 RR017515) from the National Center for Research Resources, National Institutes of Health. The AT&T Genomics Computing Center supercomputing facilities used for this work were supported in part by a gift from the AT&T Foundation.

References

[1] V. L. Roger, A. S. Go, D. M. Lloyd-Jones et al., "Heart disease and stroke statistics-2011 update: a report from the American Heart Association," *Circulation*, vol. 123, no. 4, pp. e18–e209, 2010.

[2] W. N. Aldridge, "Serum esterases. I. Two types of esterase (A and B) hydrolysing p-nitrophenyl acetate, propionate and butyrate, and a method for their determination," *The Biochemical journal*, vol. 53, no. 1, pp. 110–117, 1953.

[3] W. N. Aldridge, "Serum esterases. II. An enzyme hydrolysing diethyl p-nitrophenyl phosphate (E600) and its identity with the A-esterase of mammalian sera," *The Biochemical journal*, vol. 53, no. 1, pp. 117–124, 1953.

[4] D. I. Draganov, J. F. Teiber, A. Speelman, Y. Osawa, R. Sunahara, and B. N. La Du, "Human paraoxonases (PON1, PON2, and PON3) are lactonases with overlapping and distinct substrate specificities," *Journal of Lipid Research*, vol. 46, no. 6, pp. 1239–1247, 2005.

[5] M. I. Mackness, S. Arrol, and P. N. Durrington, "Paraoxonase prevents accumulation of lipoperoxides in low-density lipoprotein," *FEBS Letters*, vol. 286, no. 1-2, pp. 152–154, 1991.

[6] M. Aviram and M. Rosenblat, "Paraoxonases 1, 2, and 3, oxidative stress, and macrophage foam cell formation during atherosclerosis development," *Free Radical Biology and Medicine*, vol. 37, no. 9, pp. 1304–1316, 2004.

[7] B. Mackness, P. Durrington, P. McElduff et al., "Low paraoxonase activity predicts coronary events in the Caerphilly Prospective Study," *Circulation*, vol. 107, no. 22, pp. 2775–2779, 2003.

[8] M. Boemi, I. Leviev, C. Sirolla, C. Pieri, M. Marra, and R. W. James, "Serum paraoxonase is reduced in type 1 diabetic patients compared to non-diabetic, first degree relatives; influence on the ability of HDL to protect LDL from oxidation," *Atherosclerosis*, vol. 155, no. 1, pp. 229–235, 2001.

[9] M. Inoue, T. Suehiro, T. Nakamura, Y. Ikeda, Y. Kumon, and K. Hashimoto, "Serum arylesterase/diazoxonase activity and genetic polymorphisms in patients with type 2 diabetes," *Metabolism: Clinical and Experimental*, vol. 49, no. 11, pp. 1400–1405, 2000.

[10] F. Azizi, F. Raiszadeh, M. Solati, A. Etemadi, M. Rahmani, and M. Arabi, "Serum paraoxonase 1 activity is decreased in thyroid dysfunction," *Journal of Endocrinological Investigation*, vol. 26, no. 8, pp. 703–709, 2003.

[11] S. Biasioli, R. Schiavon, L. Petrosino et al., "Paraoxonase activity and paraoxonase 1 gene polymorphism in patients with uremia," *ASAIO Journal*, vol. 49, no. 3, pp. 295–299, 2003.

[12] T. F. Dantoine, J. Debord, J. P. Charmes et al., "Decrease of serum paraoxonase activity in chronic renal failure," *Journal of the American Society of Nephrology*, vol. 9, no. 11, pp. 2082–2088, 1998.

[13] B. J. Van Lenten, A. C. Wagner, M. Navab, and A. M. Fogelman, "Oxidized phospholipids induce changes in hepatic paraoxonase and apoJ but not monocyte chemoattractant protein-1 via interleukin-6," *Journal of Biological Chemistry*, vol. 276, no. 3, pp. 1923–1929, 2001.

[14] I. Leviev and R. W. James, "Promoter polymorphisms of human paraoxonase PON1 gene and serum paraoxonase activities and concentrations," *Arteriosclerosis, Thrombosis, and Vascular Biology*, vol. 20, no. 2, pp. 516–521, 2000.

[15] V. H. Brophy, R. L. Jampsa, J. B. Clendenning, L. A. McKinstry, G. P. Jarvik, and C. E. Furlong, "Effects of 5' regulatory-region polymorphisms on paraoxonase-gene (PON1) expression," *American Journal of Human Genetics*, vol. 68, no. 6, pp. 1428–1436, 2001.

[16] S. Deakin, I. Leviev, M. C. Brulhart-Meynet, and R. W. James, "Paraoxonase-1 promoter haplotypes and serum paraoxonase: a predominant role for polymorphic position—107, implicating the Sp1 transcription factor," *Biochemical Journal*, vol. 372, no. 2, pp. 643–649, 2003.

[17] S. Adkins, K. N. Gan, M. Mody, and B. N. La Du, "Molecular basis for the polymorphic forms of human serum paraoxonase/arylesterase: glutamine or arginine at position 191, for the respective A or B allozymes," *American Journal of Human Genetics*, vol. 52, no. 3, pp. 598–608, 1993.

[18] R. Humbert, D. A. Adler, C. M. Disteche, C. Hassett, C. J. Omiecinski, and C. E. Furlong, "The molecular basis of the human serum paraoxonase activity polymorphism," *Nature Genetics*, vol. 3, no. 1, pp. 73–76, 1993.

[19] H. G. Davies, R. J. Richter, M. Keifer, C. A. Broomfield, J. Sowalla, and C. E. Furlong, "The effect of the human serum paraoxonase polymorphism is reversed with diazoxon, soman and sarin," *Nature genetics*, vol. 14, no. 3, pp. 334–336, 1996.

[20] M. I. Mackness, S. Arrol, B. Mackness, and P. N. Durrington, "Alloenzymes of paraoxonase and effectiveness of high-density lipoproteins in protecting low-density lipoprotein against lipid peroxidation," *Lancet*, vol. 349, no. 9055, pp. 851–852, 1997.

[21] B. MacKness, M. I. MacKness, S. Arrol, W. Turkie, and P. N. Durrington, "Effect of the human serum paraoxonase 55 and 192 genetic polymorphisms on the protection by high density lipoprotein against low density lipoprotein oxidative modification," *FEBS Letters*, vol. 423, no. 1, pp. 57–60, 1998.

[22] R. J. Richter and C. E. Furlong, "Determination of paraoxonase (PON1) status requires more than genotyping," *Pharmacogenetics*, vol. 9, no. 6, pp. 745–753, 1999.

[23] T. Bhattacharyya, S. J. Nicholls, E. J. Topol et al., "Relationship of paraoxonase 1 (PON1) gene polymorphisms and functional activity with systemic oxidative stress and cardiovascular risk," *Journal of the American Medical Association*, vol. 299, no. 11, pp. 1265–1276, 2008.

[24] E. Zintzaras and G. M. Hadjigeorgiou, "Association of paraoxonase 1 gene polymorphisms with risk of Parkinson's disease: a meta-analysis," *Journal of Human Genetics*, vol. 49, no. 9, pp. 474–481, 2004.

[25] A. Karban, C. Hartman, R. Eliakim et al., "Paraoxonase (PON)1 192R allele carriage is associated with reduced risk of inflammatory bowel disease," *Digestive Diseases and Sciences*, vol. 52, no. 10, pp. 2707–2715, 2007.

[26] G. Paragh, P. Balla, E. Katona, I. Seres, A. Égerházi, and I. Degrell, "Serum paraoxonase activity changes in patients with Alzheimer's disease and vascular dementia," *European Archives of Psychiatry and Clinical Neuroscience*, vol. 252, no. 2, pp. 63–67, 2002.

[27] S. T. Reddy, A. Devarajan, N. Bourquard, D. Shih, and A. M. Fogelman, "Is it just paraoxonase 1 or are other members of the paraoxonase gene family implicated in atherosclerosis?" *Current Opinion in Lipidology*, vol. 19, no. 4, pp. 405–408, 2008.

[28] R. J. Richter, G. P. Jarvik, and C. E. Furlong, "Paraoxonase 1 status as a risk factor for disease or exposure," *Advances in Experimental Medicine and Biology*, vol. 660, pp. 29–35, 2010.

[29] D. A. Winnier, D. L. Rainwater, S. A. Cole et al., "Multiple QTLs influence variation in paraoxonase 1 activity in Mexican Americans," *Human Biology*, vol. 78, no. 3, pp. 341–352, 2006.

[30] D. L. Rainwater, M. C. Mahaney, X. L. Wang, J. Rogers, L. A. Cox, and J. L. Vandeberg, "Determinants of variation in serum paraoxonase enzyme activity in baboons," *Journal of Lipid Research*, vol. 46, no. 7, pp. 1450–1456, 2005.

[31] J. W. MacCluer, M. P. Stern, L. Almasy et al., "Genetics of atherosclerosis risk factors in Mexican Americans," *Nutrition Reviews*, vol. 57, no. 5, pp. S59–S65, 1999.

[32] D. L. Rainwater, S. Rutherford, T. D. Dyer et al., "Determinants of variation in human serum paraoxonase activity," *Heredity*, vol. 102, no. 2, pp. 147–154, 2009.

[33] B. Dyke, *PEDSYS: A Pedigree Data Management System User's Manual*, San Antonio, Tex, USA, 1995.

[34] L. Almasy and J. Blangero, "Multipoint quantitative-trait linkage analysis in general pedigrees," *American Journal of Human Genetics*, vol. 62, no. 5, pp. 1198–1211, 1998.

[35] E. A. Adie, R. R. Adams, K. L. Evans, D. J. Porteous, and B. S. Pickard, "Speeding disease gene discovery by sequence based candidate prioritization," *BMC Bioinformatics*, vol. 6, article no. 55, 2005.

[36] I. Silva-Zolezzi, A. Hidalgo-Miranda, J. Estrada-Gil et al., "Analysis of genomic diversity in Mexican Mestizo populations to develop genomic medicine in Mexico," *Proceedings of the National Academy of Sciences of the United States of America*, vol. 106, no. 21, pp. 8611–8616, 2009.

[37] D. W. Huang, B. T. Sherman, and R. A. Lempicki, "Systematic and integrative analysis of large gene lists using DAVID bioinformatics resources," *Nature Protocols*, vol. 4, no. 1, pp. 44–57, 2009.

[38] D. W. Huang, B. T. Sherman, and R. A. Lempicki, "Bioinformatics enrichment tools: paths toward the comprehensive functional analysis of large gene lists," *Nucleic Acids Research*, vol. 37, no. 1, pp. 1–13, 2009.

[39] H. H. H. Göring, J. E. Curran, M. P. Johnson et al., "Discovery of expression QTLs using large-scale transcriptional profiling in human lymphocytes," *Nature Genetics*, vol. 39, no. 10, pp. 1208–1216, 2007.

[40] S. Y. Eom, Y. S. Kim, C. J. Lee, C. H. Lee, Y. D. Kim, and H. Kim, "Effects of intronic and exonic polymorphisms of paraoxonase 1 (PON1) gene on serum PON1 activity in a Korean population," *Journal of Korean Medical Science*, vol. 26, no. 6, pp. 720–725, 2011.

[41] B. MacKness, M. I. MacKness, S. Arrol et al., "Serum paraoxonase (PON1) 55 and 192 polymorphism and paraoxonase activity and concentration in non-insulin dependent diabetes mellitus," *Atherosclerosis*, vol. 139, no. 2, pp. 341–349, 1998.

[42] J. C. Barrett, B. Fry, J. Maller, and M. J. Daly, "Haploview: analysis and visualization of LD and haplotype maps," *Bioinformatics*, vol. 21, no. 2, pp. 263–265, 2005.

[43] H. H. H. Göring, J. D. Terwilliger, and J. Blangero, "Large upward bias in estimation of locus-specific effects from genomewide scans," *American Journal of Human Genetics*, vol. 69, no. 6, pp. 1357–1369, 2001.

[44] M. A. K. Westin, M. C. Hunt, and S. E. H. Alexson, "Peroxisomes contain a specific phytanoyl-CoA/pristanoyl-CoA thioesterase acting as a novel auxiliary enzyme in α- and β-oxidation of methyl-branched fatty acids in mouse," *Journal of Biological Chemistry*, vol. 282, no. 37, pp. 26707–26716, 2007.

[45] R. J. A. Wanders and H. R. Waterham, "Biochemistry of mammalian peroxisomes revisited," *Annual Review of Biochemistry*, vol. 75, pp. 295–332, 2006.

[46] M. Dixit, G. Choudhuri, L. J. Keshri, and B. Mittal, "Association of low density lipoprotein receptor related protein-associated protein (LRPAP1) gene insertion/deletion polymorphism with gallstone disease," *Journal of Gastroenterology and Hepatology*, vol. 21, no. 5, pp. 847–849, 2006.

[47] P. Pandey, S. Pradhan, and B. Mittal, "LRP-associated protein gene (LRPAP1) and susceptibility to degenerative dementia," *Genes, Brain and Behavior*, vol. 7, no. 8, pp. 943–950, 2008.

[48] D. Rabinowitz, "Variation in the LRP-associated protein gene (LRPAP1) is associated with late-onset Alzheimer disease," *American Journal of Medical Genetics*, vol. 105, no. 1, pp. 76–78, 2001.

[49] S. Page, A. Judson, K. Melford, and A. Bensadoun, "Interaction of lipoprotein lipase and receptor-associated protein," *Journal of Biological Chemistry*, vol. 281, no. 20, pp. 13931–13938, 2006.

[50] C. Lanzani, L. Citterio, M. Jankaricova et al., "Role of the adducin family genes in human essential hypertension," *Journal of Hypertension*, vol. 23, no. 3, pp. 543–549, 2005.

[51] G. Bianchi, P. Ferrari, and J. A. Staessen, "Adducin polymorphism: detection and impact on hypertension and related disorders," *Hypertension*, vol. 45, no. 3, pp. 331–340, 2005.

[52] R. Sztrolovics, S. P. Wang, P. Lapierre, H. S. Chen, M. F. Robert, and G. A. Mitchell, "Hormone-sensitive lipase (Lipe): sequence analysis of the 129Sv mouse Lipe gene," *Mammalian Genome*, vol. 8, no. 2, pp. 86–89, 1997.

[53] J. M. Baez, S. E. Barbour, and D. E. Cohen, "Phosphatidylcholine transfer protein promotes apolipoprotein A-I-mediated lipid efflux in Chinese hamster ovary cells," *Journal of Biological Chemistry*, vol. 277, no. 8, pp. 6198–6206, 2002.

[54] J. M. Baez, I. Tabas, and D. E. Cohen, "Decreased lipid efflux and increased susceptibility to cholesterol-induced apoptosis in macrophages lacking phosphatidylcholine transfer protein," *Biochemical Journal*, vol. 388, no. 1, pp. 57–63, 2005.

[55] M. Shimamura, M. Matsuda, H. Yasumo et al., "Angiopoietin-like protein3 regulates plasma HDL cholesterol through suppression of endothelial lipase," *Arteriosclerosis, Thrombosis, and Vascular Biology*, vol. 27, no. 2, pp. 366–372, 2007.

[56] S. Hatsuda, T. Shoji, K. Shinohara et al., "Association between plasma angiopoietin-like protein 3 and arterial wall thickness in healthy subjects," *Journal of Vascular Research*, vol. 44, no. 1, pp. 61–66, 2007.

[57] C. Couturier, C. Sarkis, K. Séron et al., "Silencing of OB-RGRP in mouse hypothalamic arcuate nucleus increases leptin receptor signaling and prevents diet-induced obesity," *Proceedings of the National Academy of Sciences of the United States of America*, vol. 104, no. 49, pp. 19476–19481, 2007.

[58] T. Touvier, F. Conte-Auriol, O. Briand et al., "LEPROT and LEPROTL1 cooperatively decrease hepatic growth hormone action in mice," *Journal of Clinical Investigation*, vol. 119, no. 12, pp. 3830–3838, 2009.

[59] L. M. King, J. V. Gainer, G. L. David et al., "Single nucleotide polymorphisms in the CYP2J2 and CYP2C8 genes and the risk of hypertension," *Pharmacogenetics and Genomics*, vol. 15, no. 1, pp. 7–13, 2005.

[60] T. C. DeLozier, G. E. Kissling, S. J. Coulter et al., "Detection of human CYP2C8, CYP2C9, and CYP2J2 in cardiovascular tissues," *Drug Metabolism and Disposition*, vol. 35, no. 4, pp. 682–688, 2007.

[61] M. Spiecker, H. Darius, T. Hankeln et al., "Risk of coronary artery disease associated with polymorphism of the cytochrome P450 epoxygenase CYP2J2," *Circulation*, vol. 110, no. 15, pp. 2132–2136, 2004.

[62] I. Fleming, "Vascular cytochrome P450 enzymes: physiology and pathophysiology," *Trends in Cardiovascular Medicine*, vol. 18, no. 1, pp. 20–25, 2008.

[63] S. D. Nguyen and D. E. Sok, "Oxidative inactivation of paraoxonase1, an antioxidant protein and its effect on antioxidant action," *Free Radical Research*, vol. 37, no. 12, pp. 1319–1330, 2003.

[64] R. C. Zangar, D. R. Davydov, and S. Verma, "Mechanisms that regulate production of reactive oxygen species by cytochrome P450," *Toxicology and Applied Pharmacology*, vol. 199, no. 3, pp. 316–331, 2004.

[65] M. S. Wolin, M. Ahmad, and S. A. Gupte, "The sources of oxidative stress in the vessel wall," *Kidney International*, vol. 67, no. 5, pp. 1659–1661, 2005.

[66] G. Agrimi, M. A. Di Noia, C. M. T. Marobbio, G. Fiermonte, F. M. Lasorsa, and F. Palmieri, "Identification of the human mitochondrial S-adenosylmethionine transporter: bacterial expression, reconstitution, functional characterization and tissue distribution," *Biochemical Journal*, vol. 379, no. 1, pp. 183–190, 2004.

[67] M. Y. Cheng, C. M. Bullock, C. Li et al., "Prokineticin 2 transmits the behavioural circadian rhythm of the suprachiasmatic nucleus," *Nature*, vol. 417, no. 6887, pp. 405–410, 2002.

[68] R. Arrigoni, S. L. Alam, J. A. Wamstad, V. J. Bardwell, W. I. Sundquist, and N. Schreiber-Agus, "The Polycomb-associated protein Rybp is a ubiquitin binding protein," *FEBS Letters*, vol. 580, no. 26, pp. 6233–6241, 2006.

Use of the Signature Fatty Acid 16:1ω5 as a Tool to Determine the Distribution of Arbuscular Mycorrhizal Fungi in Soil

Christopher Ngosong,[1,2] Elke Gabriel,[3,4] and Liliane Ruess[1]

[1] *Institute of Biology, Ecology Group, Humboldt-Universität zu Berlin, Philippstraße 13, 10115 Berlin, Germany*
[2] *Department of Natural Resource Sciences, MacCampus, McGill University, 21,111 Lakeshore Road, Ste Anne de Bellevue, QC, Canada H9X 3V9*
[3] *Institute of Vegetable and Ornamental Crops Großbeeren, Theodor-Echtermeyer-Weg 1, 14979 Großbeeren, Germany*
[4] *Faculty of Food and Agriculture, UAE University, Jimi 1 Campus, Building 52, P.O. Box 17555, Al Ain, Abu Dhabi, UAE*

Correspondence should be addressed to Christopher Ngosong, ngosongk@yahoo.com

Academic Editor: Paul R. Herron

Biomass estimation of arbuscular mycorrhiza (AM) fungi, widespread plant root symbionts, commonly employs lipid biomarkers, predominantly the fatty acid 16:1ω5. We briefly reviewed the application of this signature fatty acid, followed by a case study comparing biochemical markers with microscopic techniques in an arable soil following a change to AM non-host plants after 27 years of continuous host crops, that is, two successive cropping seasons with wheat followed by amaranth. After switching to the non-host amaranth, spore biomass estimated by the neutral lipid fatty acid (NLFA) 16:1ω5 decreased to almost nil, whereas microscopic spore counts decreased by about 50% only. In contrast, AM hyphal biomass assessed by the phospholipid (PLFA) 16:1ω5 was greater under amaranth than wheat. The application of PLFA 16:1ω5 as biomarker was hampered by background level derived from bacteria, and further enhanced by its incorporation from degrading spores used as microbial resource. Meanwhile, biochemical and morphological assessments showed negative correlation for spores and none for hyphal biomass. In conclusion, the NLFA 16:1ω5 appears to be a feasible indicator for AM fungi of the Glomales group in the complex field soils, whereas the use of PLFA 16:1ω5 for hyphae is unsuitable and should be restricted to controlled laboratory studies.

1. Introduction

The chemotaxonomic use of lipids has a long tradition in microbiological research [1–3]. Due to the different enzymatic capabilities in lipid metabolism, fatty acids display a great structural diversity and biological specificity, providing an integrated and quantitative measure of microbial biomass and community structure in different environments [4]. In particularly, PLFAs have been employed in soil ecosystems as biomarkers for bacteria, saprotrophic fungi, and AM fungi; see Zelles [5] and Joergensen and Wichern [6] for detailed reviews. Moreover, as the lipid pattern of faunal consumers reflects the fatty acid composition of their diet, trophic biomarker fatty acids for major food resources in soil decomposers have been assigned [7].

Bacteria and fungi are important drivers of soil processes, predominantly nutrient mineralization and transfer to plants. Among the different mycorrhizal types, arbuscular fungi that form symbiosis with the roots of about 80% of all vascular plants are the dominant fungal symbionts that support plant growth [8, 9]. The AM fungal extraradical mycelium (ERM) spreads beyond the rhizosphere of host plants, providing additional surface area for the acquisition of phosphorus and nitrogen [10–12]. In recent years, global interest in sustainable agricultural practices has led to increase in the production and application of AM fungal inoculum in arable soils, which requires reliable methods for their quantification. AM fungi may occur naturally in arable soils, but their density and diversity may be increased by farm management practices such as fertilization or crop types [13–15].

Despite the importance of AM fungi for soil nutrient cycling, information on their distribution is inadequate due

TABLE 1: Application of the marker fatty acid 16:1ω5 to assess the distribution of arbuscular mycorrhiza fungi in artificial and natural soil systems. NLFA: neutral lipid fatty acid, PLFA: phospholipid fatty acid.

Authors	Lipid fraction and application (marker for)	Soil system
Olsson et al. [21]	NLFA-storage lipids	Plant mesocosms with γ radiated or autoclaved soil
Olsson et al. [22]	PLFA hyphae; conversion factor 38 for nmol PLFA to fungal hyphal length is given	
Larsen and Bødker [24] Van Aarleand Olsson [25]	NLFA-energy storage (vesicles)	Mycelia structures in plant roots
	PLFA-membrane constituents (hyphae, arbuscles)	
	NLFA/PLFA-storage status of fungi	
Olsson and Wilhelmsson [26]		
Hedlund [27]		
Balser et al. [28]		
Hebel et al. [29]	PLFA-hyphal biomass	Grassland, mixed-wood forest stands, arable land, sand dunes; burned forest soil
Huang et al. [30]		
Royer-Tardif et al. [31]		
Marshall et al. [32]		
Olsson and Wilhelmsson [26] Hedlund [27]	NLFA-storage lipids	Arable land, sand dunes
van Groenigen et al. [33] Yao and Wu [34]	NLFA-fungal biomass	Arable land, grassland
Bradley et al. [35]	Total lipids (NLFA + PLFA) fungal biomass	Grassland
Aliasgharzad et al. [36]	NLFA/PLFA separation between arbuscular mycorrhizal fungi (high ratio) and bacteria (low ratio)	Seminatural sandy grassland
Olsson et al. [37] Schnoor et al. [38]	NFLA-^{13}C allocation in fungal storage lipids	Pot soil with plants in greenhouse

to analytical difficulties that are limiting the ability to study processes at the microscale. Four main approaches for quantifying fungal contribution to soil microbial communities are commonly used in soil ecology: (1) microscopic methods, (2) selective inhibition, (3) specific cell membrane components, and (4) specific cell wall components, with microscopic methods generally recording the lowest values [6]. Classical morphological estimation of AM fungal communities in soil includes membrane filter [16], grid line intersection [17], spore counts [3], and aqueous filtration extraction method and quantification of extraradical mycelium based on morphological criteria [18]. Molecular techniques have also been developed to determine AM fungi [19, 20], but difficulties involved in the sequencing as well as the absence of a sequence for some species constrain the use of this approach. More advanced, and particularly for quantification of AM fungal biomass, is the lipid biomarker technique that is being applied to determine the distribution of AM fungi in soil over the last two decades.

Olsson et al. [21, 22] were the first to use signature fatty acid 16:1ω5 regularly to assess growth and interactions of AM fungi in experimental soil-plant systems (Table 1). They proposed the 16:1ω5 as marker fatty acid, with the PLFA fraction assigning viable fungal hyphal biomass, and the NLFA fraction determining storage lipids such as spores [23]. This

biomarker approach was also adopted for mycelial structures within plant roots, that is, the NLFA for energy storage in vesicles and the PLFA for membrane constituents such as intraradical mycelium or arbuscles [24, 25]. Based on this, both the 16:1ω5 PLFA and NLFA are widely used as indicators for AM fungi across soil ecosystems (Table 1). In a recent review, Joergensen and Wichern [6] proposed 345 as conversion factor of nmol PLFA to biomass C of AM fungi. This factor, a weighted mean based on literature data originating from four studies, can be criticized. However, it demonstrates the common use of 16:1ω5 PLFA in AM investigations, even though its biomarker value may be hampered.

Despite the proposed primary origin of 16:1ω5 PLFA, it may additionally be found, although in smaller amounts, in the lipids of other organisms, particularly in soil-inhabiting bacteria [39, 40]. On the other hand, the 16:1ω5 NLFA is not only present in spores but also forms the transport vehicle for carbon between intraradical and extraradical mycelium [41]. As assigned by ^{13}C labelling studies, the fungus converts sugars taken up in the root compartment into lipids [42, 43], which can be used to assess the shift of carbon from roots into associated microbial communities without extraction, purification and identification of fungal mycelium [44, 45]. These stable isotope studies indicated that the NLFA 16:1ω5 is a good tool to assign AM fungal biomass in soil but showed no evidence for the application of the PLFA fraction. Despite these discrepancies several recent studies employed the PLFA marker for AM extraradical mycelium in field soils (Table 1).

This general application of 16:1ω5 as biomarker for AM fungi across different soil ecosystems calls for deeper insight to ensure reliability in the quantification of the fungal hyphae or spore biomass. We therefore performed a case study on the dynamics of AM fungal spores and extraradical mycelium comparing lipid biomarker and morphological approaches to determine the distribution of AM fungi in an arable field soil. A long-term fertilizer experiment was used, with the shift to a nonmycorrhizal host plant after 27 years of host crop cultivation. This unique experimental design offers the possibility to assess whether biochemical and microscopic techniques yield similar results under distinct changes in the mycorrhizal symbiosis *in situ*, without artificial manipulations (e.g., selective inhibition technique). Investigating the same plots in two successive vegetative periods allowed (i) to assign the diminishing of AM fungi in the absence of a host plant, (ii) to screen the background signal from the remaining vegetation period using both methods, and (iii) to determine correlations between the changes within biochemical marker and morphological assessments.

2. Materials and Methods

2.1. Field Site.
The study was conducted at a long-term arable field site established in 1980 at the Institute of Biodynamic Research (IBDF) Darmstadt in Germany. The field site is located at 49°N, 8°E, and 100 m above sea level, with annual mean air temperature of 9.5°C and precipitation of 590 mm. The soil type is haplic cambisol comprising 87% sand, 8% silt, and 5% clay in the topsoil. The experimental setup was a two-factorial design amended with mineral (NPK) and organic (cattle manure + biodynamic preparations) fertilizers applied at low and high amounts. These were implemented in a split block design with four replicate plots of 5 m × 5 m each. Except fertilization, all other farming practices such as irrigation, tillage, and crop rotation were similar across the 27 years since the establishment of the long-term field experiment. Plots amended with mineral fertilizer received N, P, K applied in rates of kg ha^{-1}y^{-1} as 60, 50, 75 (low), and 140, 100, and 125 (high), respectively. Organic plots received composted cattle manure with the addition of biodynamic preparations spread as solid fertilizer before ploughing and milling of the soil. The application rate was calculated to achieve similar nitrogen input as at the mineral fertilizer plots, which resulted in the variation of phosphorus and potassium amounts depending on the manure properties of a given year. On average, organic plots received less P (−25% for low and −38% for high) and 26% more K than mineral plots. For more details on farm management, see Ngosong et al. [46].

2.2. Sampling.
Since the long-term field site was established, there has been 27 years continuous mycorrhizal host crop rotation including lupine angustifolius, winter rye, potato, spring wheat, and clover. The present investigations were conducted during two successive cropping seasons with

the cultivation of spring wheat (*Triticum aestivum* cv. Passat) in 2007 and a shift to amaranth (*Amaranthus hypochondriacus*) in 2008. The former is a well-known host plant for AM fungi while the latter is recognized as non-host [8, 47]. Soil at the wheat plots was sampled four months after sowing and three weeks before crop harvest, while amaranth plots were sampled two months after sowing and two months before crop harvest. For examination of AM fungal morphological structures (hyphal length, spores) and lipids (PLFAs, NLFAs), one soil sample was taken from each replicate plot (*n* = 4 per treatment) at 0–5 and 5–10 cm depth using 5 cm diameter soil core. Soil was stored at −20°C prior to analyses. Additionally, random samples of wheat and amaranth roots were collected from the respective plots and analysed for infection by AM fungi.

2.3. Morphological AM Fungal Investigation.
Morphological assessment of AM fungal structures comprised the colonization of crop roots, length of extraradical mycelium, and number of spores in the bulk soil. For the assessment of root infection rate, fine roots (0.7–1.0 g) were stained with trypan blue in lactic acid and the colonized roots assessed by modified intersection method with 250–300 intersections counted per sample [17]. Fungal spores were isolated from 80 g air-dried soil by sieving and decanting method, with subsequent sucrose gradient centrifugation [48], and counted using the agar film technique [49].

The hyphal length was estimated from soil using a modified membrane filtration technique. Soil sample (1.0 g) was homogenized with 100 mL deionized water in a laboratory mixer (Waring Commercial; Connecticut, USA) for 60 seconds. The suspension was poured through a 40-micrometer filter and washed carefully with water to eliminate fine soil particles. Remaining material was transferred into a petri dish and stained with a few drops of 0.05% trypan blue in lactic acid. The suspension was transferred into a glass beaker and diluted to 300 mL volume. A subsample of the suspension was filtered on a 0.45 μm mesh width membrane filter (MicronSep; GE Water & Process Technologies, USA) using a bottleneck filtration unit (NALGENE Reusable Bottle Top Filter Unit; Nalge Company, New York, USA). The membrane filter was mounted onto microscopic slides and observed under the microscope at 200x magnification, and the AM fungal hyphal length estimated by a modified gridline intersection method [50, 51].

2.4. Fatty Acid Analysis.
Lipids were extracted from 4 g soil (wet weight) using the modified Bligh and Dyer method according to Frostegård et al. [52]. Fractionation into NLFAs, glycolipids, and PLFAs was performed using silica acid columns (HF BOND ELUT–SI, Varian Inc.), and the different fractions were eluded with chloroform, acetone, and methanol, respectively. Lipid methanolysis of PLFA and NLFA fractions was conducted in 0.2 M methanolic KOH, and methyl nonadecanoate (19:0) was added as internal standard; for more details see Ngosong et al. [53].

Fatty acid methyl esters (FAMEs) were identified by chromatographic retention time comparison with a standard mixture composed of 37 different FAMEs ranging from

TABLE 2: Arbuscular mycorrhizal (AM) extraradical mycelium length and hyphal biomass in amaranth plots estimated morphologically and by phospholipid fatty acid biomarker 16:1ω5 (% DW soil ± SD), respectively, at 0–5 and 5–10 cm soil depths, amended with mineral (NPK) and organic (cattle manure + biodynamic preparations) fertilizers, applied at low and high amounts.

	Mineral		Organic	
	Low	High	Low	High
0–5 cm				
PLFA 16:1ω5 (nmol g^{-1} dry soil)	0.43 ± 0.22	0.55 ± 0.22	0.63 ± 0.07	1.13 ± 0.66
Hyphal length (m g^{-1} dry soil)	3.96 ± 2.39	1.53 ± 0.70	2.08 ± 1.12	2.35 ± 1.36
5–10 cm				
PLFA 16:1ω5 (nmol g^{-1} dry soil)	0.51 ± 0.18	0.53 ± 0.3	0.76 ± 0.17	0.71 ± 0.14
Hyphal length (m g^{-1} dry soil)	2.46 ± 1.37	1.78 ± 1.16	1.82 ± 1.01	1.99 ± 1.71

C11 to C24 (Sigma-Aldrich, St Louis, MO, USA). Analysis was performed by gas chromatography using a GC-FID Clarus 500 (PerkinElmer Corporation, Norwalk, USA) equipped with HP-5 capillary column (30 m × 0.32 mm i.d., film thickness 0.25 μm). To verify correct identification of FAMEs (chain length and saturation), a range of soil samples were analyzed by mass spectrometry using a 3400/Saturn4 Diontrap GC/MS system (Varian, Darmstadt, Germany), equipped with a HP-5 capillary column (50 m × 0.32 mm i.d., film thickness 0.17 μm). A mass range from 50 to 500 m/z was monitored twice a second in Scan mode; for more details see Ngosong et al. [46]. The signature fatty acid 16:1ω5 was used as biomarker for AM fungi, where the PLFA fraction represents fungal extraradical mycelium and the NLFA spore for biomass [23, 54, 55].

2.5. Statistical Analysis. The effects of crop plant shift on AM fungal fatty acid marker and morphological estimations were tested using STATISTICA 6.0 for Windows [56]. Data were subjected to nonparametric statistics using Kruskal-Wallis. Significant effects ($P < 0.05$) of the different factors are indicated in figures. Additionally, the Spearman Rank Order Correlations between biochemical and microscopically derived results were performed.

3. Results

The morphological examination of crop roots revealed 32–67% colonization of wheat by AM fungi, whereas the nonmycorrhizal host amaranth was not infected (data not presented). This clearly indicates the absence of an active amaranth-fungal symbiosis. In conformity, in the absence of a host plant, AM fungal spore biomass assessed by the NLFA 16:1ω5 almost disappeared in amaranth soils, with less than 0.1 nmol g^{-1} DW, compared to 0.9–7.9 nmol g^{-1} DW for wheat soils across depths (Figure 1(a)). Similarly, the microscopic counted spore numbers decreased, but only by 55%, and ranged between 121 and 205 and 87 and 125 spores g^{-1} DW soil for wheat and amaranth plots, respectively (Figure 1(b)). The relationship between NLFA and microscopic spore estimates was negatively correlated ($r = -70, P < 0.05$) across fertilizers, depths, and crop plant. Overall, the estimation of AM fungi in the arable soil using signature fatty acid and microscopic techniques mirrored the same trend, but to a different extent.

In contrast to AM fungal spores, hyphal biomass assessed by the marker PLFA 16:1ω5 increased significantly under amaranth with 0.4–1.1 nmol g^{-1} DW compared to 0.1–0.8 nmol g^{-1} DW for wheat soil; see Ngosong et al. [46]. This corresponds to an increase by 24–65% in the upper soil and 39–79% in the lower soil layer at amaranth plots. Meanwhile, AM hyphal length under amaranth determined morphologically ranged between 1.5 and 4.0 m g^{-1} DW and 1.8 and 2.5 m g^{-1} DW soil at 0–5 and 5–10 cm depth, respectively (Table 2). This contradicts the absence of amaranth root infection by the fungus, and the strong decrease in spore numbers at those plots. Nonetheless, there was no observed correlation ($r = -0.13, P > 0.05$) between morphological AM hyphal length and biochemical PLFA 16:1ω5 hyphal biomass under amaranth across fertilizer types and soil depth. In addition, there was no correlation ($r = -0.10, P > 0.05$) between wheat root infection rate and the PLFA 16:1ω5 at wheat plots.

4. Discussion

The present investigation focuses on the correlation of morphological and biochemical estimates of AM fungal dynamics in light of the shift from host to non-host crops. The response of microbial communities, including AM fungi, in relation to fertilizer type and amount as demonstrated by lipid data is discussed in detail elsewhere [46]. When comparing the NLFA signature fatty acid with microscopic estimations, both approaches mirrored the same trend but to a different extent. For the fatty acid, the decline in spore biomass without a host plant was severe with almost nil left, whereas the number of spores remaining was about 50%. Firstly, these differences may be due to low NLFA yield since Olsson [23] suggested that, for efficient extraction of lipids, the spore wall must be broken. On the other hand, Madan et al. [57] reported only small and nonsignificant impact when spores were crushed before analysis. Secondly, Olsson and Johansen [58] demonstrated that AM fungal hyphae contain a significant portion of the NLFA 16:1ω5 used for carbon transport in lipids. Since hyphae are decomposed much faster than spores, this may have contributed to the diminishing of the NLFA signal within one crop cycle. However, as spores form 90% of the external fungal tissue and 20% of spore mass is NLFA [58], the impact of lipids from extraradical hyphae appears rather low. Thirdly, the signature

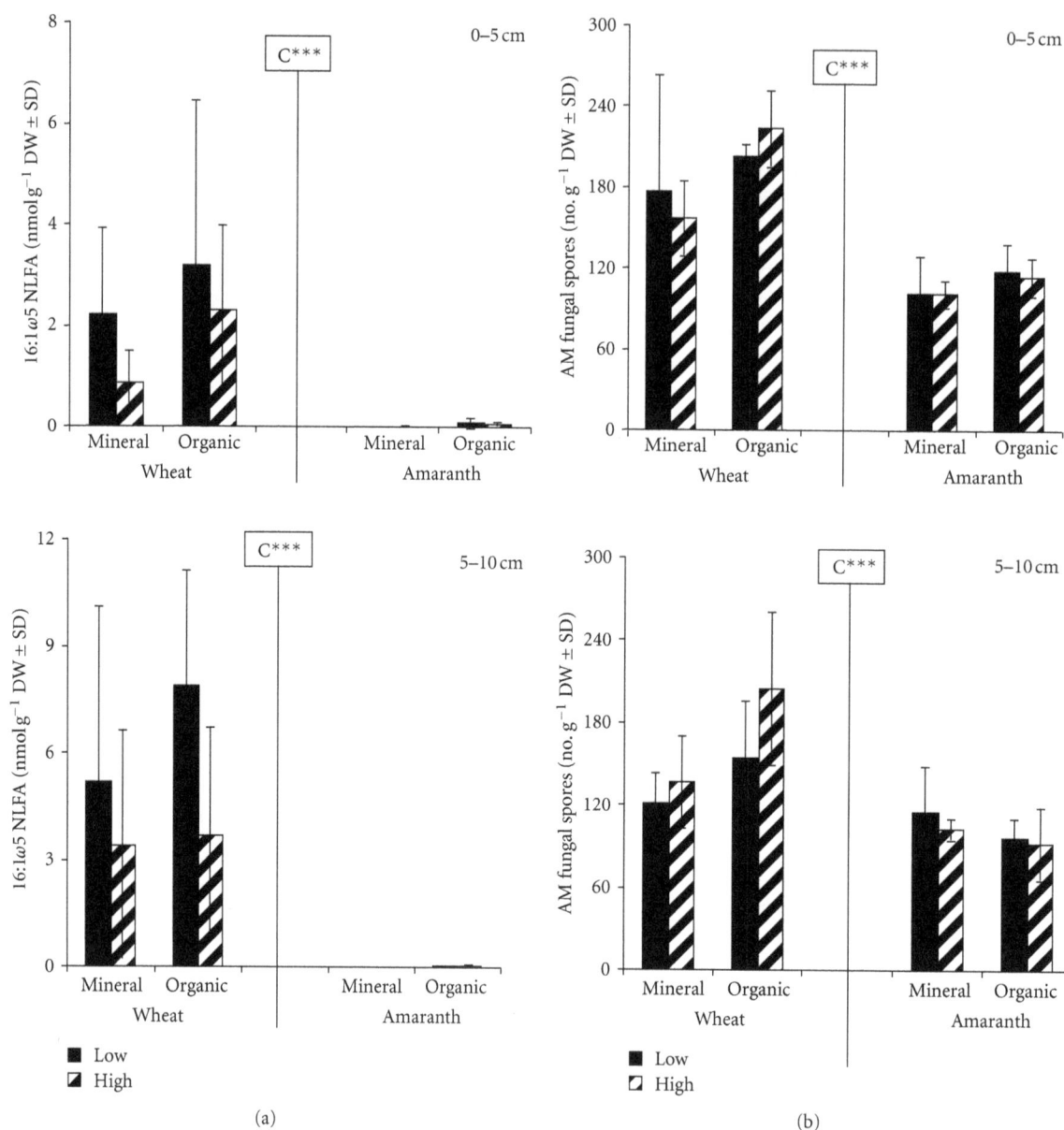

FIGURE 1: Comparison of microscopic and biochemical estimations of arbuscular mycorrhiza spores in an arable field soil cultivated with wheat or amaranth as crop; 0–5 and 5–10 cm soil depths; amendments with mineral (NPK) and organic (cattle manure + biodynamic preparations) fertilizers applied at low and high amounts; C = crop type; Kruskal-Wallis with *** for $P < 0.001$. (a) AM spores assessed by neutral lipid fatty acid $16:1\omega5$ (nmol g^{-1} DW soil ± SD), (b) Spore density counts (No. g^{-1} DW soil ± SD).

fatty acid $16:1\omega5$ is common in *Glomales*, whereas it is rare or lacking in other groups such as *Gigaspora* [54, 59]. For the latter, several long chain fatty acids such as $20:1\omega9$ $20:2\omega6$ and $22:1\omega9$ have been proposed as biomarkers [57, 60]. Hence, the signature fatty acid partially reflects the dynamics of AM in soil, but not of the entire fungal population. This is supported by the observation of larger spores (e.g., *Gigaspora*) during microscopic examination, which cannot be detected by $16:1\omega5$. Fourthly, microscopic counts are constrained by the fact that newly formed fungal spores are not distinguishable from those formed earlier in the season [61, 62], resulting in potential leftovers from the previous

crop. In sum, the NLFA $16:1\omega5$ reflected the decline of AM fungal spores after the change to a non-host crop, but it predominantly represented the *Glomalen* species within the population. Hence, it represents a reliable quantitative estimate of the fungal spore biomass when used in that regard, which is in line with recent studies that applied stable isotopes to assess carbon transfer from roots to AM fungi [44, 45].

In contrast to AM fungal spores, hyphal biomass assessed by the PLFA $16:1\omega5$ biomarker was higher in the non-host compared to the host crop soils. This is surprising as PLFAs are easily decomposed through enzymatic actions in soil and thus are assumed to reflect the occurrence of living organisms [2]. Meanwhile, the longevity of AM hyphae in soil has rarely been measured although it is assumed to be short. Staddon et al. [63] assigned a high turnover rate with an extraradical hyphal live from 5 to 6 days only. On the other hand, Steinberg and Rillig [64] reported that even under relatively favorable conditions for decomposition (18°C; 15% moisture) about 60% of hyphal length were still present 150 days after being separated from their host. However, there was no correlation of the PLFA estimates to hyphal length in amaranth soil, or to root infection rate at wheat plots, indicating a weak relationship between morphological and biochemical measurements. The significant increase of the PLFA 16:1ω5 at amaranth compared to wheat plots by up to 79% [46] suggests that bacteria used degrading spores as carbon source, thereby assimilating the marker fatty acid. Such trophic transfer of lipids between microorganisms and their substrates was frequently reported [7]. The extraradical mycelium has been assigned as large and rapid mycorrhizal pathway of carbon into other rhizosphere microorganisms [44, 45, 63]. Our results indicate that also fungal spores are attractive resources that form a considerable microbial carbon pool in the bulk soil.

In conclusion, investigating the development of AM fungi in an arable soil following the change of host crop revealed strong analytical discrepancies between biochemical and microscopic techniques. For the application of PLFA 16:1ω5, the background concentration derived from other soil organisms, and particularly bacteria can be too high to correctly quantify mycelium in microbial active soils. Meanwhile, the NLFA 16:1ω5 appears to be a reliable marker for AM fungal storage lipids such as spores, yet it cannot assign other than the *Glomales* group. Moreover, the occurrence of NLFA 16:1ω5 in extraradical mycelium in soil can superimpose on the overall signal. Clearly, the interpretation that NLFA biomarker arises solely from spores, and PLFA biomarker from mycelia is gross oversimplification. Meanwhile, the approach to combine both the phospholipid and neutral lipid fractions, as marker for AM fungi is no remedy, since it is hampered by the assimilation of signature fatty acid by other decomposers. Overall, these results strongly challenge the use of AM biomarkers, necessitating more comparative *in situ* based studies to identify their structural and functional origin, in order to effectively assign the dynamics of 16:1ω5 in complex field soils.

Acknowledgments

The authors extend gratitude to the counties Brandenburg, Thüringen, and Leibniz Society for financial support. They are grateful to J. Raupp for the possibility to use the long-term field experiment at the Institute of Biodynamic Research Darmstadt (IBDF) for our studies. They also thank M. Oltmanns (IBDF Darmstadt), M. Jarosch (IBDF Darmstadt), and S. Jeserigk (IGZ Großbeeren) for field and laboratory assistances, respectively.

References

[1] H. Lechevalier and M. P. Lechevalier, "Chemotaxonomic use of lipids-an overview," in *Microbial Lipids*, C. Ratledge and S. G. Wilkinson, Eds., pp. 869–902, Academic Press, London, UK, 1988.

[2] A. Tunlid and D. C. White, "Use of lipid biomarkers in environmental samples," in *Analytical Microbiology Methods, Chromatography and Mass Spectrometry*, A. Fox, S. L. Morgan, L. Larsson, and G. Odham, Eds., Plenum Press, New York, NY, USA, 1990.

[3] R. E. Koske and J. N. Gemma, "A modified procedure for staining roots to detect VA mycorrhizas," *Mycological Research*, vol. 92, pp. 486–505, 1989.

[4] D. C. White, J. O. Stair, and D. B. Ringelberg, "Quantitative comparisons of in situ microbial biodiversity by signature biomarker analysis," *Journal of Industrial Microbiology*, vol. 17, no. 3-4, pp. 185–196, 1996.

[5] L. Zelles, "Fatty acid patterns of phospholipids and lipopolysaccharides in the characterisation of microbial communities in soil: a review," *Biology and Fertility of Soils*, vol. 29, no. 2, pp. 111–129, 1999.

[6] R. G. Joergensen and F. Wichern, "Quantitative assessment of the fungal contribution to microbial tissue in soil," *Soil Biology and Biochemistry*, vol. 40, no. 12, pp. 2977–2991, 2008.

[7] L. Ruess and P. M. Chamberlain, "The fat that matters: soil food web analysis using fatty acids and their carbon stable isotope signature," *Soil Biology and Biochemistry*, vol. 42, no. 11, pp. 1898–1910, 2010.

[8] S. E. Smith and D. J. Read, *Mycorrhizal Symbioses*, Academic Press, London, UK, 3rd edition, 2008.

[9] S. D. Veresoglou, G. Menexes, and M. C. Rillig, "Do arbuscular mycorrhizal fungi affect the allometric partition of host plant biomass to shoots and roots? A meta-analysis of studies from 1990 to 2010," *Mycorrhiza*, vol. 22, pp. 227–235, 2011.

[10] E. Neumann and E. George, "Colonisation with the arbuscular mycorrhizal fungus *Glomus mosseae* (Nicol. & Gerd.) enhanced phosphorus uptake from dry soil in *Sorghum bicolor* (L.)," *Plant and Soil*, vol. 261, no. 1-2, pp. 245–255, 2004.

[11] M. Govindarajulu, P. E. Pfeffer, H. Jin et al., "Nitrogen transfer in the arbuscular mycorrhizal symbiosis," *Nature*, vol. 435, no. 7043, pp. 819–823, 2005.

[12] Y. Tanaka and K. Yano, "Nitrogen delivery to maize via mycorrhizal hyphae depends on the form of N supplied," *Plant, Cell and Environment*, vol. 28, no. 10, pp. 1247–1254, 2005.

[13] M. H. Ryan, G. A. Chilvers, and D. C. Dumaresq, "Colonisation of wheat by VA-mycorrhizal fungi was found to be higher on a farm managed in an organic manner than on a conventional neighbour," *Plant and Soil*, vol. 160, no. 1, pp. 33–40, 1994.

[14] M. E. Gavito and M. H. Miller, "Changes in mycorrhiza development in maize indeed by crop management practices," *Plant and Soil*, vol. 198, no. 2, pp. 185–192, 1998.

[15] F. Oehl, E. Sieverding, P. Mäder et al., "Impact of long-term conventional and organic farming on the diversity of arbuscular mycorrhizal fungi," *Oecologia*, vol. 138, no. 4, pp. 574–583, 2004.

[16] L. K. Abbott, A. D. Robson, and G. De Boer, "The effects of phosphorus on the formation of hyphae in soil by the vesicular arbuscular mycorrhizal fungus *Glomus fasiculatum*," *New Phytologist*, vol. 97, pp. 437–446, 1984.

[17] P. Kormanik and A. C. McGraw, "Quantification of vesicular-arbuscular mycorrhizae in plant roots," in *Methods and*

Principals of Mycorrhizal Research, N. C. Schenck, Ed., pp. 37–45, The American Phytopathological Society, Minn, USA, 1982.

[18] P. A. Olsson, I. M. Van Aarle, M. E. Gavito, P. Bengtson, and G. Bengtsson, "^{13}C incorporation into signature fatty acids as an assay for carbon allocation in arbuscular mycorrhiza," *Applied and Environmental Microbiology*, vol. 71, no. 5, pp. 2592–2599, 2005.

[19] D. Redecker, "Specific PCR primers to identify arbuscular mycorrhizal fungi within colonized roots," *Mycorrhiza*, vol. 10, no. 2, pp. 73–80, 2000.

[20] D. Schwarzott and A. Schüßler, "A simple and reliable method for SSU rRNA gene dna extraction, amplification, and cloning from single AM fungal spores," *Mycorrhiza*, vol. 10, no. 4, pp. 203–207, 2001.

[21] P. A. Olsson, E. Bååth, I. Jakobsen, and B. Soderstrom, "The use of phospholipid and neutral lipid fatty acids to estimate biomass of arbuscular mycorrhizal fungi in soil," *Mycological Research*, vol. 99, no. 5, pp. 623–629, 1995.

[22] P. A. Olsson, R. Francis, D. J. Read, and B. Söderström, "Growth of arbuscular mycorrhizal mycelium in calcareous dune sand and its interaction with other soil microorganisms as estimated by measurement of specific fatty acids," *Plant and Soil*, vol. 201, no. 1, pp. 9–16, 1998.

[23] P. A. Olsson, "Signature fatty acids provide tools for determination of the distribution and interactions of mycorrhizal fungi in soil," *FEMS Microbiology Ecology*, vol. 29, no. 4, pp. 303–310, 1999.

[24] J. Larsen and L. Bødker, "Interactions between pea root-inhabiting fungi examined using signature fatty acids," *New Phytologist*, vol. 149, no. 3, pp. 487–493, 2001.

[25] I. M. Van Aarle and P. A. Olsson, "Fungal lipid accumulation and development of mycelial structures by two arbuscular mycorrhizal fungi," *Applied and Environmental Microbiology*, vol. 69, no. 11, pp. 6762–6767, 2003.

[26] P. A. Olsson and P. Wilhelmsson, "The growth of external AM fungal mycelium in sand dunes and in experimental systems," *Plant and Soil*, vol. 226, no. 2, pp. 161–169, 2000.

[27] K. Hedlund, "Soil microbial community structure in relation to vegetation management on former agricultural land," *Soil Biology and Biochemistry*, vol. 34, no. 9, pp. 1299–1307, 2002.

[28] T. C. Balser, K. K. Treseder, and M. Ekenler, "Using lipid analysis and hyphal length to quantify AM and saprotrophic fungal abundance along a soil chronosequence," *Soil Biology and Biochemistry*, vol. 37, no. 3, pp. 601–604, 2005.

[29] C. L. Hebel, J. E. Smith, and K. Cromack, "Invasive plant species and soil microbial response to wildfire burn severity in the Cascade Range of Oregon," *Applied Soil Ecology*, vol. 42, no. 2, pp. 150–159, 2009.

[30] Y. Huang, K. Michel, S. An, and S. Zechmeister-Boltenstern, "Changes in microbial-community structure with depth and time in a chronosequence of restored grassland soils on the Loess Plateau in north-west China," *Journal of Plant Nutrition Soil Science*, vol. 174, pp. 765–774, 2011.

[31] S. Royer-Tardif, R. L. Bradley, and W. F. J. Parsons, "Evidence that plant diversity and site productivity confer stability to forest floor microbial biomass," *Soil Biology and Biochemistry*, vol. 42, no. 5, pp. 813–821, 2010.

[32] C. B. Marshall, J. R. McLaren, and R. Turkington, "Soil microbial communities resistant to changes in plant functional group composition," *Soil Biology and Biochemistry*, vol. 43, no. 1, pp. 78–85, 2011.

[33] K. J. van Groenigen, J. Bloem, E. Bååth et al., "Abundance, production and stabilization of microbial biomass under conventional and reduced tillage," *Soil Biology and Biochemistry*, vol. 42, no. 1, pp. 48–55, 2010.

[34] H. Yao and F. Wu, "Soil microbial community structure in cucumber rhizosphere of different resistance cultivars to fusarium wilt," *FEMS Microbiology Ecology*, vol. 72, no. 3, pp. 456–463, 2010.

[35] K. Bradley, R. A. Drijber, and J. Knops, "Increased N availability in grassland soils modifies their microbial communities and decreases the abundance of arbuscular mycorrhizal fungi," *Soil Biology and Biochemistry*, vol. 38, no. 7, pp. 1583–1595, 2006.

[36] N. Aliasgharzad, L. M. Mårtensson, and P. A. Olsson, "Acidification of a sandy grassland favours bacteria and disfavours fungal saprotrophs as estimated by fatty acid profiling," *Soil Biology and Biochemistry*, vol. 42, no. 7, pp. 1058–1064, 2010.

[37] P. A. Olsson, J. Rahm, and N. Aliasgharzad, "Carbon dynamics in mycorrhizal symbioses is linked to carbon costs and phosphorus benefits," *FEMS Microbiology Ecology*, vol. 72, no. 1, pp. 125–131, 2010.

[38] T. K. Schnoor, L. M. Mårtensson, and P. A. Olsson, "Soil disturbance alters plant community composition and decreases mycorrhizal carbon allocation in a sandy grassland," *Oecologia*, vol. 167, pp. 809–819, 2011.

[39] P. Nichols, B. K. Stulp, J. G. Jones, and D. C. White, "Comparison of fatty acid content and DNA homology of the filamentous gliding bacteria *Vitreoscilla, Flexibacter, Filibacter*," *Archives of Microbiology*, vol. 146, no. 1, pp. 1–6, 1986.

[40] L. Zelles, "Phospholipid fatty acid profiles in selected members of soil microbial communities," *Chemosphere*, vol. 35, no. 1-2, pp. 275–294, 1997.

[41] B. Bago, P. E. Pfeffer, W. Zipfel, P. Lammers, and Y. Shachar-Hill, "Tracking metabolism and imaging transport in arbuscular mycorrhizal fungi," *Plant and Soil*, vol. 244, no. 1-2, pp. 189–197, 2002.

[42] P. E. Pfeffer, D. D. Douds, G. Bécard, and Y. Shachar-Hill, "Carbon uptake and the metabolism and transport of lipids in an arbuscular mycorrhiza," *Plant Physiology*, vol. 120, no. 2, pp. 587–598, 1999.

[43] M. E. Gavito and P. A. Olsson, "Allocation of plant carbon to foraging and storage in arbuscular mycorrhizal fungi," *FEMS Microbiology Ecology*, vol. 45, no. 2, pp. 181–187, 2003.

[44] P. A. Olsson and N. C. Johnson, "Tracking carbon from the atmosphere to the rhizosphere," *Ecology Letters*, vol. 8, no. 12, pp. 1264–1270, 2005.

[45] B. Drigo, A. S. Pijl, H. Duyts et al., "Shifting carbon flow from roots into associated microbial communities in response to elevated atmospheric CO_2," *Proceedings of the National Academy of Sciences of the United States of America*, vol. 107, no. 24, pp. 10938–10942, 2010.

[46] C. Ngosong, M. Jarosch, J. Raupp, E. Neumann, and L. Ruess, "The impact of farming practice on soil microorganisms and arbuscular mycorrhizal fungi: crop type *versus* long-term mineral and organic fertilization," *Applied Soil Ecology*, vol. 46, no. 1, pp. 134–142, 2010.

[47] R. L. Peterson, A. E. Ashford, and W. G. Allaway, "Vesicular-arbuscular mycorrhizal associations of vascular plants on Heron Island, a Great Barrier Reef coral cay," *Australian Journal of Botany*, vol. 33, no. 6, pp. 669–676, 1985.

[48] D. C. Ianson and M. F. Allen, "The effect of soil texture on extraction of vesicular-arbuscular mycorrhizal spores from arid soils," *Mycologia*, vol. 78, pp. 168–164, 1986.

[49] A. Thomas, D. P. Nicholas, and D. Parkinson, "Modifications of the agar film technique for assaying lengths of mycelium in soil," *Nature*, vol. 205, no. 4966, p. 105, 1965.

[50] E. I. Newman, "A method of estimating the total length of root in a sample," *Journal of Applied Ecology*, vol. 3, pp. 139–145, 1966.

[51] D. Tennant, "A test of a modified line intersect method of estimating root length," *Journal of Ecology*, vol. 63, pp. 995–1001, 1975.

[52] Å. Frostegård, A. Tunlid, and E. Bååth, "Phospholipid fatty acid composition, biomass, and activity of microbial communities from two soil types experimentally exposed to different heavy metals," *Applied and Environmental Microbiology*, vol. 59, no. 11, pp. 3605–3617, 1993.

[53] C. Ngosong, J. Raupp, S. Scheu, and L. Ruess, "Low importance for a fungal based food web in arable soils under mineral and organic fertilization indicated by Collembola grazers," *Soil Biology and Biochemistry*, vol. 41, no. 11, pp. 2308–2317, 2009.

[54] J. H. Graham, N. C. Hodge, and J. B. Morton, "Fatty acid methyl ester profiles for characterization of glomalean fungi and their endomycorrhizae," *Applied and Environmental Microbiology*, vol. 61, no. 1, pp. 58–64, 1995.

[55] P. A. Olsson, L. Larsson, B. Bago, H. Wallander, and I. M. Van Aarle, "Ergosterol and fatty acids for biomass estimation of mycorrhizal fungi," *New Phytologist*, vol. 159, no. 1, pp. 7–10, 2003.

[56] StatSoft, Statistica. Version 6.0 for windows, StatSoft, Tusla, Okla, USA, 2001.

[57] R. Madan, C. Pankhurst, B. Hawke, and S. Smith, "Use of fatty acids for identification of AM fungi and estimation of the biomass of AM spores in soil," *Soil Biology and Biochemistry*, vol. 34, no. 1, pp. 125–128, 2002.

[58] P. A. Olsson and A. Johansen, "Lipid and fatty acid composition of hyphae and spores of arbuscular mycorrhizal fungi at different growth stages," *Mycological Research*, vol. 104, no. 4, pp. 429–434, 2000.

[59] S. P. Bentivenga and J. B. Morton, "Stability and heritability of fatty acid methyl ester profiles of glomalean endomycorrhizal fungi," *Mycological Research*, vol. 98, no. 12, pp. 1419–1426, 1994.

[60] K. Sakamoto, T. Iijima, and R. Higuchi, "Use of specific phospholipid fatty acids for identifying and quantifying the external hyphae of the arbuscular mycorrhizal fungus *Gigaspora rosea*," *Soil Biology and Biochemistry*, vol. 36, no. 11, pp. 1827–1834, 2004.

[61] Lee Pau Ju and R. E. Koske, "*Gigaspora gigantea*: seasonal abundance and ageing of spores in a sand dune," *Mycological Research*, vol. 98, no. 4, pp. 453–457, 1994.

[62] J. P. Clapp, J. P. Young, J. W. Merryweather, and A. H. Fitter, "Diversity of fungal symbionts in arbuscular mycorrhizas from a natural community," *New Phytologist*, vol. 130, no. 2, pp. 259–265, 1995.

[63] P. L. Staddon, C. B. Ramsey, N. Ostle, P. Ineson, and A. H. Fitter, "Rapid turnover of hyphae of mycorrhizal fungi determined by AMS microanalysis of ^{14}C," *Science*, vol. 300, no. 5622, pp. 1138–1140, 2003.

[64] P. D. Steinberg and M. C. Rillig, "Differential decomposition of arbuscular mycorrhizal fungal hyphae and glomalin," *Soil Biology and Biochemistry*, vol. 35, no. 1, pp. 191–194, 2003.

6

PPARγ Networks in Cell Signaling: Update and Impact of Cyclic Phosphatidic Acid

Tamotsu Tsukahara

Department of Integrative Physiology and Bio-System Control, Shinshu University School of Medicine, 3-1-1 Asahi, Matsumoto, Nagano 390-8621, Japan

Correspondence should be addressed to Tamotsu Tsukahara; ttamotsu@shinshu-u.ac.jp

Academic Editor: Robert Salomon

Lysophospholipid (LPL) has long been recognized as a membrane phospholipid metabolite. Recently, however, the LPL has emerged as a candidate for diagnostic and pharmacological interest. LPLs include lysophosphatidic acid (LPA), alkyl glycerol phosphate (AGP), cyclic phosphatidic acid (cPA), and sphingosine-1-phosphate (S1P). These biologically active lipid mediators serve to promote a variety of responses that include cell proliferation, migration, and survival. These LPL-related responses are mediated by cell surface G-protein-coupled receptors and also intracellular receptor peroxisome proliferator-activated receptor gamma (PPARγ). In this paper, we focus mainly on the most recent findings regarding the biological function of nuclear receptor-mediated lysophospholipid signaling in mammalian systems, specifically as they relate to health and diseases. Also, we will briefly review the biology of PPARγ and then provide an update of lysophospholipids PPARγ ligands that are under investigation as a therapeutic compound and which are targets of PPARγ relevant to diseases.

1. Introduction

Peroxisome proliferator-activated receptor gamma (PPARγ) is a member of the nuclear hormone receptor superfamily, many of which function as ligand-activated transcription factors [1]. Synthetic agonists of PPARγ include the thiazolidinedione (TZD) class of drugs, which are widely used to improve insulin sensitivity in type II diabetes. Despite the beneficial effects of PPARγ on glucose and lipid homeostasis, excess PPARγ activity can be deleterious. These classical PPARγ agonists elicit a variety of side effects, including weight gain, edema, increased fat mass, and tumor formation in rodents [2]. In contrast, there have been many reports in which the putative physiological agonists of PPARγ have been identified [3–5]. LPA is a naturally occurring phospholipid with growth-like effects in almost every mammalian cell type. LPAs elicit their biological responses through eight plasma membrane receptors [6] and intracellularly through the PPARγ [3, 4]. Although LPA derived from hydrolysis of plasma membrane phospholipids is established as a ligand for G-coupled cell surface LPA receptor, studies suggested that LPA might also enter cells to activate PPARγ. PPARγ plays a role in regulating lipid and glucose homeostasis, cell proliferation, apoptosis, and inflammation [7, 8]. These pathways have a direct impact on human diseases in obesity, diabetes, atherosclerosis, and cancer [9–11]. On the other hand, cyclic phosphatidic acid (cPA), similar in structure to LPA, can be generated by phospholipase D2 (PLD2) and negatively regulate PPARγ functions (Figure 1). cPA shows several unique actions compared to those of LPA. cPA inhibits cell proliferation, whereas LPA stimulates it [12–16]. It has been reported that cPA attenuates cancer cell invasion; moreover, metabolically stabilized derivative of cPA suppressed cancer cell metastasis [17, 18]. cPA is a second messenger and a physiological inhibitor of PPARγ, revealing that PPARγ is regulated by agonists as well as by antagonists.

2. Receptors and Signaling

2.1. Intracellular Receptor of PPARγ. PPARs are members of the nuclear hormone receptor superfamily, many of which function as lipid-activated transcription factors [1]. There are three PPAR isoforms that include PPARα, β/δ, and γ that differ in ligand specificity, tissue distribution, and

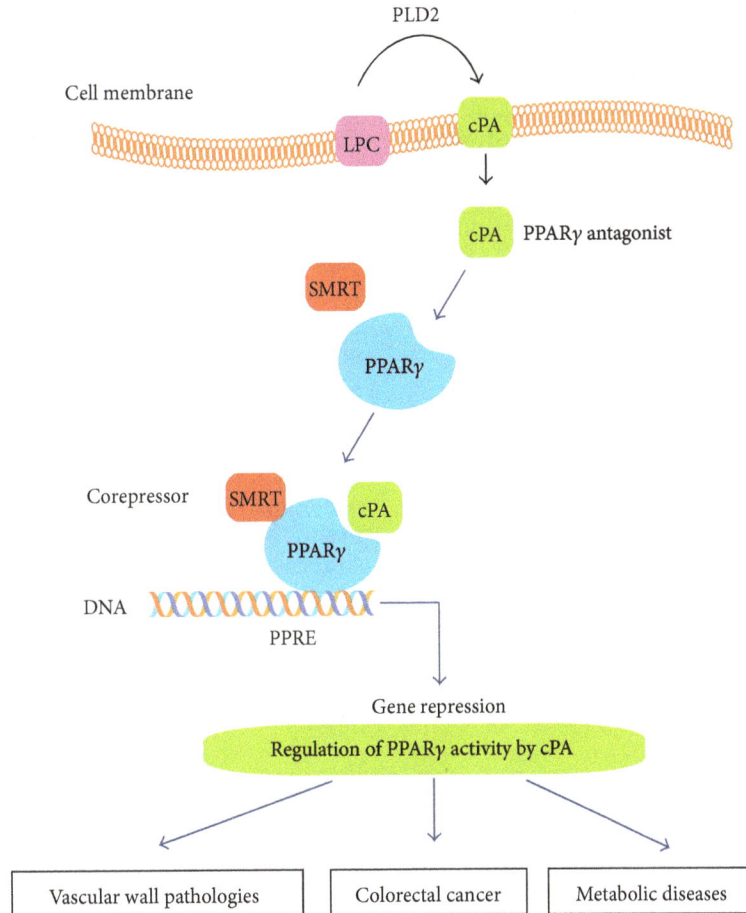

FIGURE 1: Regulation of PPARγ activity by cPA. cPA is generated intracellularly in a stimulus-coupled manner by the PLD2 enzyme. cPA stabilize interactions with corepressor, such as SMRT, that act to repress gene transcription. This endogenous cPA regulates PPARγ function required for vascular wall pathologies, colorectal cancer cell growth, and metabolic diseases.

developmental expression [19]. PPARγ, the most extensively studied among the three PPAR subtypes, plays an important role in regulating lipid metabolism, glucose homeostasis, cell differentiation, and motility [10, 20]. There are 2 PPARγ isoforms, PPARγ$_1$ and PPARγ$_2$. PPARγ2 has 30 additional amino acids at the N-terminus in human caused by differential promoter usage and alternative splicing [21]. Genetic deletion of PPARγ$_1$ causes embryonic mortality [9]. In contrast, deletion of PPARγ$_2$ causes minimal alterations in lipid metabolism [22]. PPARγ$_1$ is expressed in almost all tissues, whereas PPARγ$_2$ is highly expressed in only the adipose tissue [21]. PPARγ is comprised of four functional parts: the N-terminal A/B region bears a ligand-independent transcription-activating motif AF-1; C region binds response elements; D region binds to various transcription cofactors; and E/F region has an interface for dimerizing with retinoid X receptor α (RXRα), an AF-2 ligand-dependent transcription-activating motif, and a ligand binding domain (LBD) [23]. PPARγ heterodimerizes with the retinoid X receptor α (RXRα), and it is the ligand binding domain (LBD) of PPARγ that interacts with its agonists, including LPA [3]. The PPARγ-RXRα heterodimer binds to the peroxisome proliferator response element (PPRE) in the promoter region

of the target genes. In the absence of ligands, the corepressors, nuclear receptor corepressor (NCoR) and silencing mediator of retinoid (SMRT) and thyroid hormone, bind to the heterodimer to suppress the target gene activation [24]. Upon ligand binding, PPARγ undergoes a conformational change that facilitates the dissociation of the corepressors and recruits coactivators. According to their mechanism of action, coactivators can be divided into two large families: the former includes steroid receptor coactivator (SRC-1) and CBP/p300, that act in part as molecular scaffolds and in the other part by acetylating divers substrates. The latter, including peroxisome proliferator-activated receptor 1α (PGC-1α), does not act by remodeling chromatin [25]. It has been reported that DNA methylation and histone modification serve as epigenetic markers for active or inactive chromatin [26]. A variety of putative physiological PPARγ agonists have been identified [5, 27]. Since then, we and other authors have reported that selected forms of lysophospholipids, such as unsaturated LPA and alkyl glycerophosphate (AGP, 1-alkyl-2-hydroxy-sn-glycerol-3-phosphate), are physiological agonists of PPARγ [3, 4]. The different molecular species of LPA contain either saturated or unsaturated fatty acids. Saturated LPA species including palmitoyl (16:0) and stearoyl (18:0)

LPA are inactive. Among these ligands, AGP stands out with an equilibrium binding constant of 60 nM [4] that is similar to that of thiazolidinedione (TZD) class of synthetic agonists. Interestingly, some of the residues required for PPARγ activation by AGP are different from those required by TZD drug. H323 and 449 within the LBD of PPARγ are required for the binding and activation by rosiglitazone but are not required by AGP. R288 is an important residue for the binding of the AGP but not the rosiglitazone. Y273 is required for activation by both agonists [4]. AGP is unique in that its potency far exceeds that of LPA in activating PPARγ [4]. The reason why AGP and unsaturated acyl-LPA species are the best activators of PPARγ may reflect the differential delivery of these LPA analogs to PPARγ versus saturated LPA species, which are inactive. Together, these data help to explain why PPARγ binds the unsaturated LPA and AGP but not saturated LPA. On the other hand, we showed that cPA negatively regulates PPARγ functions by stabilizing the SMRT-PPARγ complex [15] and blocks TZD-stimulated adipogenesis and lipid accumulation. This ligand-dependent corepressor exchange results in transcriptional repression of genes involved in the control of insulin action as well as a diverse range of other functions.

3. Targets of PPARγ Relevant to Diseases

3.1. LPA-Mediated Activation of PPARγ and Vascular Wall Pathologies.

It has been reported that unsaturated LPA-elicited neointima was not mediated by the LPA GPCRs LPA$_1$ and LPA$_2$, which are the major LPA receptor subtypes expressed in the vessel wall [28]. LPA has been identified as a bioactive lipid and is produced in serum after the activation of multiple biochemical pathways [6, 29, 30]. Some clinical studies have shown the correlation between plasma LPA and vascular diseases [31]. Neointima formation is a characteristic feature of common vascular pathologies, such as atherosclerosis [32]. Atherosclerosis is a complex disease to which many factors contribute. Neointima lesions are characterized by accumulation of cells within the arterial wall and are an early step in the pathogenesis of atherosclerosis [33]. It is caused by a buildup of plaque in the inner lining of artery and made up of deposits of fatty substances and cholesterol [34]. Topical application of unsaturated LPA species into the noninjured carotid artery of rodents induces arterial wall remodeling [35, 36], and this response requires PPARγ. PPARγ plays an important role in the cardiovascular system. PPARγ is expressed in all cell types of vessel wall, as well as monocytes and macrophages [37]. Macrophages play essential roles in immunity and lipid homeostasis [37]. PPARγ is induced during the differentiation of monocytes into macrophages and is highly expressed in activated macrophages including the foam cells in atherosclerotic lesions [36]. CD36 is a one of PPARγ response genes. PPARγ activation upregulates CD36 expression which results in increased lipid uptake in macrophages [3]. In macrophages, oxidized low-density lipoprotein (oxLDL) uptake through CD36 results in the development of foam cells. Accumulation of foam cells in the arterial wall is a key event of the early atherosclerotic lesion [38]. The initial steps of foam cell formation have been extensively studied. A CD36-dependent signaling cascade is necessary for macrophage foam cell formation. Moore et al. reported that oxLDL uptake is decreased in PPARγ deficient macrophages due to the loss of CD36 [39]. CD36 is a member of the class B scavenger receptor family of cell surface protein [38, 40]. These receptors are a group of receptors that recognize modified LDL by oxidation or acethylation [41]. It has been reported that LPA and AGP are an agonist of the PPARγ and has been implicated in atherogenesis [36]. When AGP (18:1) was infused to an injured carotid artery, neointima thickening was augmented, although TZD drug, rosiglitazone- (Rosi-) attenuated neointima, induced by mechanical injury. However, noninjury model, Rosi, induces neointima when applied intraluminally into the carotid artery [15]. These results suggest that mechanisms underlying neointima formation in the chemically induced model are likely to be different from those in the injury-induced models. Coronary artery disease, the most common type of heart disease and leading cause of death among cardiovascular diseases, is almost always the result of atherosclerosis. Hence, the present results raise the possibility of utilizing this phospholipid scaffold as a lead for the development of new treatment acting on PPARγ.

3.2. PPARγ Ligand and Colorectal Cancer.

Colon cancer is a malignancy that develops in colon and rectal tissues. Colon cancer cells can also spread to other parts of body. The prognosis for metastatic colon cancer is associated with high mortality [42, 43]. It has been reported that prognosis for metastatic colon cancer remains poor; therefore, new therapeutic options are needed to reduce cancer mortality. It has been reported that PPARγ may provide a molecular link between a high-fat diet and increased risk of colon polyp formation during PPARγ activation [44]. Two studies have shown that administration of a synthetic PPARγ ligand to APC$^{Min/+}$ mice resulted in these mice developing more frequent colon cancers than those animals which did not receive PPARγ ligand [45]. APC$^{Min/+}$ mice have a mutation of APC, which is a major regulator of β-catenin activation and represent a model of adenomatous polyposis coli (APC) [46]. Mutations of PPARγ in colon cancer lead to the loss of ligand binding and suppression of cell growth. This may indicate that functional PPARγ is required for the normal growth properties of colon cells [47]. The *PPARγ* gene is expressed in many tissues, including high levels of expression in normal colonic mucosa, colorectal adenocarcinomas, and colon cancer cell lines [11, 48]. Recently, several studies reported that PPARγ agonists inhibit cancer cell proliferation, survival, and invasion [16, 49]. Although PPARγ is expressed at significant levels in human colon cancer cells and tissues, the role of PPARγ activation in colon cancer is still controversial. Furthermore, the role of PPARγ activation in cancer remains unclear. Some reports indicate that PPARγ is expressed at considerable levels in human colon cancer cells and tissues and that treatment with PPARγ agonists and antagonists reduces the cell growth rate [16, 50, 51]. Because PPARγ ligands have been shown to have a variety of PPARγ-dependent and -independent effects [52]. Our recent reports suggest that endogenous LPA agonist, cPA, which is a *bona fide* second messenger and a physiological inhibitor of PPARγ [15] has emerged as a potential therapeutic target in the treatment of colon cancer [16]. cPA is a structural analog of LPA, which is one of the simplest phospholipids in cells. cPA is a generated by phospholipase D-catalyzed-transphosphatidylation of lysophosphatidylcholine (LPC) [15]. LPA is a PPARγ agonist that induces cell proliferation and invasion, but cPA exerts the opposite effects in cancer cells [16]. cPA suppresses PPARγ activation both by preventing binding of exogenous agonist to PPARγ and by inducing a specific conformational change that suppresses PPARγ activation [15, 16, 53]. cPA binding to and inhibition of PPARγ might be involved in cPA-induced inhibition of colon cancer cell growth [16]. This study demonstrates the potential applications of these methods for colon cancer treatment.

3.3. cPA in the Treatment of Metabolic Diseases. Obesity and its associated conditions such as insulin resistance, type II diabetes, termed as the metabolic syndrome, is a worldwide health problem and occurs as a result of adipose tissue enlargement caused by store excess energy intake [54]. Obesity is a condition in which adipocytes accumulate a large amount of body fat and became enlarged [55]. Adipose differentiation is a complex process by which fibroblast-like undifferentiated cells are converted into cells that accumulate lipid droplets [56]. PPARγ agonists are known to induce the differentiation of preadipocytes into mature adipocyte. TZD drugs are widely used in type II diabetes mellitus to improve insulin sensitivity by inducing the expression of genes involved in adipocyte differentiation, lipid and glucose uptake, and fatty acid storage [19]. Our recent observation suggests that PPARγ activation in adipogenesis that can be blocked by treatment with cPA participates in adipocyte function through inhibition of PDE3B expression [57]. cPA reduced intracellular triglyceride levels and inhibited the phosphodiesterase 3B (PDE3B) expression in 3T3-L1 adipocytes [57]. Treatment of 3T3-L1 cells with cPA significantly increased the amount of free glycerol. This suggests that triglyceride was hydrolyzed in adipocytes to free fatty acid and glycerol through the lipolysis. Adipose tissue lipolysis is dependent on the intracellular concentration of cAMP [58, 59], which is determined at the levels of both synthesis and degradation. Hydrolysis of cAMP is accomplished by PDEs [60]. Investigation on PDE has been focused on hormone-sensitive PDE3B activity and its expression. PDE3B is expressed in insulin-sensitive cells and has been shown to be important in regulating antilipolysis [61]. These findings contribute to the participation of cPA on the lipolytic activity in adipocytes. cPA might be a therapeutic compound in the treatment of obesity and obesity-related diseases including type II diabetes and high blood pressure.

4. Conclusion

Clearly, the genomic response to activation and inhibition of PPARγ is complex and will be highly dependent on cellular context. PPARγ agonists and antagonists participate in the regulation of lipid metabolism, they play an important role during atherosclerosis, diabetes, and they also have a critical role in the regulation of growth of cancer cells. It has been suggested that PPARγ ligands with agonistic and antagonistic effects may have useful role in the treatment of PPARγ-mediated diseases. We can expect many promising results in this area in the near future.

Acknowledgments

This work was supported by research grants from the Astellas Foundation for Research on Metabolic Disorders and Takeda Science Foundation and Grants-in-Aid for Scientific Research (C) (22591482) from the Japan Society for the Promotion of Science (JSPS) to T. Tsukahara.

References

[1] R. M. Evans, "The nuclear receptor superfamily: a Rosetta stone for physiology," *Molecular Endocrinology*, vol. 19, no. 6, pp. 1429–1438, 2005.

[2] H. E. Lebovitz, "Differentiating members of the thiazolidinedione class: a focus on safety," *Diabetes/Metabolism Research and Reviews*, vol. 18, supplement 2, pp. S23–S29, 2002.

[3] T. M. McIntyre, A. V. Pontsler, A. R. Silva et al., "Identification of an intracellular receptor for lysophosphatidic acid (LPA): LPA is a transcellular PPARγ agonist," *Proceedings of the National Academy of Sciences of the United States of America*, vol. 100, no. 1, pp. 131–136, 2003.

[4] T. Tsukahara, R. Tsukahara, S. Yasuda et al., "Different residues mediate recognition of 1-O-oleyl-lysophosphatidic acid and rosiglitazone in the ligand binding domain of peroxisome proliferator-activated receptor," *Journal of Biological Chemistry*, vol. 281, no. 6, pp. 3398–3407, 2006.

[5] F. J. Schopfer, Y. Lin, P. R. S. Baker et al., "Nitrolinoleic acid: an endogenous peroxisome proliferator-activated receptor γ ligand," *Proceedings of the National Academy of Sciences of the United States of America*, vol. 102, no. 7, pp. 2340–2345, 2005.

[6] W. H. Moolenaar, "Lysophosphatidic acid, a multifunctional phospholipid messenger," *Journal of Biological Chemistry*, vol. 270, no. 22, pp. 12949–12952, 1995.

[7] M. Ricote and C. K. Glass, "PPARs and molecular mechanisms of transrepression," *Biochimica et Biophysica Acta*, vol. 1771, no. 8, pp. 926–935, 2007.

[8] P. Tontonoz and B. M. Spiegelman, "Fat and beyond: the diverse biology of PPARγ," *Annual Review of Biochemistry*, vol. 77, pp. 289–312, 2008.

[9] T. A. Cock, S. M. Houten, and J. Auwerx, "Peroxisome proliferator-activated receptor-γ: too much of a good thing causes harm," *EMBO Reports*, vol. 5, no. 2, pp. 142–147, 2004.

[10] C. Duval, G. Chinetti, F. Trottein, J. C. Fruchart, and B. Staels, "The role of PPARs in atherosclerosis," *Trends in Molecular Medicine*, vol. 8, no. 9, pp. 422–430, 2002.

[11] W. L. Yang and H. Frucht, "Activation of the PPAR pathway induces apoptosis and COX-2 inhibition in HT-29 human colon cancer cells," *Carcinogenesis*, vol. 22, no. 9, pp. 1379–1383, 2001.

[12] K. Murakami-Murofushi, A. Uchiyama, Y. Fujiwara et al., "Biological functions of a novel lipid mediator, cyclic phosphatidic acid," *Biochimica et Biophysica Acta*, vol. 1582, no. 1-3, pp. 1–7, 2002.

[13] K. Murakami-Murofushi, S. Kobayashi, K. Onimura et al., "Selective inhibition of DNA polymerase-α family with chemically synthesized derivatives of PHYLPA, a unique Physarum lysophosphatidic acid," *Biochimica et Biophysica Acta*, vol. 1258, no. 1, pp. 57–60, 1995.

[14] Y. Fujiwara, "Cyclic phosphatidic acid—a unique bioactive phospholipid," *Biochimica et Biophysica Acta*, vol. 1781, no. 9, pp. 519–524, 2008.

[15] T. Tsukahara, R. Tsukahara, Y. Fujiwara et al., "Phospholipase D2-dependent inhibition of the nuclear hormone receptor PPARγ by cyclic phosphatidic acid," *Molecular Cell*, vol. 39, no. 3, pp. 421–432, 2010.

[16] T. Tsukahara, S. Hanazawa, T. Kobayashi, Y. Iwamoto, and K. Murakami-Murofushi, "Cyclic phosphatidic acid decreases proliferation and survival of colon cancer cells by inhibiting peroxisome proliferator-activated receptor γ," *Prostaglandins and Other Lipid Mediators*, vol. 93, no. 3-4, pp. 126–133, 2010.

[17] M.] Mukai, F. Imamura, M. Ayaki et al., "Inhibition of tumor invasion and metastasis by a novel lysophosphatidic acid (cyclic LPA)," *International Journal of Cancer*, vol. 81, pp. 918–922, 1999.

[18] M. Mukai, T. Iwasaki, M. Tatsuta et al., "Cyclic phosphatidic acid inhibits RhoA-mediated autophosphorylation of FAK at Tyr-397 and subsequent tumor-cell invasion," *International Journal of Oncology*, vol. 22, no. 6, pp. 1247–1256, 2003.

[19] J. Berger and D. E. Moller, "The mechanisms of action of PPARs," *Annual Review of Medicine*, vol. 53, pp. 409–435, 2002.

[20] B. Kieć-Wilk, A. Dembińska-Kieć, A. Olszanecka, M. Bodzioch, and K. Kawecka-Jaszcz, "The selected pathophysiological aspects of PPARs activation," *Journal of Physiology and Pharmacology*, vol. 56, no. 2, pp. 149–162, 2005.

[21] P. Tontonoz, E. Hu, and B. M. Spiegelman, "Stimulation of adipogenesis in fibroblasts by PPARγ2, a lipid-activated transcription factor," *Cell*, vol. 79, no. 7, pp. 1147–1156, 1994.

[22] G. Medina-Gomez, S. Virtue, C. Lelliott et al., "The link between nutritional status and insulin sensitivity is dependent on the adipocyte-specific peroxisome proliferator-activated receptor-γ2 isoform," *Diabetes*, vol. 54, no. 6, pp. 1706–1716, 2005.

[23] D. S. Lala, R. Mukherjee, I. G. Schulman et al., "Activation of specific RXR heterodimers by an antagonist of RXR homodimers," *Nature*, vol. 383, no. 6599, pp. 450–453, 1996.

[24] R. N. Cohen, "Nuclear receptor corepressors and PPARgamma," *Nucl Recept Signal*, vol. 4, article e003, 2006.

[25] P. Puigserver and B. M. Spiegelman, "Peroxisome proliferator-activated receptor-γ coactivator 1α (PGC-1α): transcriptional coactivator and metabolic regulator," *Endocrine Reviews*, vol. 24, no. 1, pp. 78–90, 2003.

[26] T. S. Mikkelsen, M. Ku, D. B. Jaffe et al., "Genome-wide maps of chromatin state in pluripotent and lineage-committed cells," *Nature*, vol. 448, no. 7153, pp. 553–560, 2007.

[27] B. M. Forman, P. Tontonoz, J. Chen, R. P. Brun, B. M. Spiegelman, and R. M. Evans, "15-deoxy-Δ12, 14-prostaglandin J2 is a ligand for the adipocyte determination factor PPARγ," *Cell*, vol. 83, no. 5, pp. 803–812, 1995.

[28] G. Tigyi, "Aiming drug discovery at lysophosphatidic acid targets," *British Journal of Pharmacology*, vol. 161, no. 2, pp. 241–270, 2010.

[29] G. Tigyi and R. Miledi, "Lysophosphatidates bound to serum albumin activate membrane currents in Xenopus oocytes and neurite retraction in PC12 pheochromocytoma cells," *Journal of Biological Chemistry*, vol. 267, no. 30, pp. 21360–21367, 1992.

[30] G. Tigyi and A. L. Parrill, "Molecular mechanisms of lysophosphatidic acid action," *Progress in Lipid Research*, vol. 42, no. 6, pp. 498–526, 2003.

[31] A. Schober and W. Siess, "Lysophosphatidic acid in atherosclerotic diseases," *British Journal of Pharmacology*, vol. 167, pp. 465–482, 2012.

[32] G. A. Fishbein and M. C. Fishbein, "Arteriosclerosis: rethinking the current classification," *Archives of Pathology and Laboratory Medicine*, vol. 133, no. 8, pp. 1309–1316, 2009.

[33] N. Shibata and C. K. Glass, "Regulation of macrophage function in inflammation and atherosclerosis," *Journal of Lipid Research*, vol. 50, supplement, pp. S277–S281, 2009.

[34] R. G. Gerrity, "The role of the monocyte in atherogenesis. I. Transition of blood-borne monocytes into foam cells in fatty lesions," *American Journal of Pathology*, vol. 103, no. 2, pp. 181–190, 1981.

[35] K. Yoshida, W. Nishida, K. Hayashi et al., "Vascular remodeling induced by naturally occurring unsaturated lysophosphatidic acid in vivo," *Circulation*, vol. 108, no. 14, pp. 1746–1752, 2003.

[36] C. Zhang, D. L. Baker, S. Yasuda et al., "Lysophosphatidic acid induces neointima formation through PPARγ activation," *Journal of Experimental Medicine*, vol. 199, no. 6, pp. 763–774, 2004.

[37] C. H. Lee and R. M. Evans, "Peroxisome proliferator-activated receptor-γ in macrophage lipid homeostasis," *Trends in Endocrinology and Metabolism*, vol. 13, no. 8, pp. 331–335, 2002.

[38] M. Febbraio, D. P. Hajjar, and R. L. Silverstein, "CD36: a class B scavenger receptor involved in angiogenesis, atherosclerosis, inflammation, and lipid metabolism," *Journal of Clinical Investigation*, vol. 108, no. 6, pp. 785–791, 2001.

[39] P. C. Moore, M. A. Ugas, D. K. Hagman, S. D. Parazzoli, and V. Poitout, "Evidence against the involvement of oxidative stress in fatty acid inhibition of insulin secretion," *Diabetes*, vol. 53, no. 10, pp. 2610–2616, 2004.

[40] N. Ohgami, R. Nagai, M. Ikemoto et al., "CD36, a member of the class B scavenger receptor family, as a receptor for advanced glycation end products," *Journal of Biological Chemistry*, vol. 276, no. 5, pp. 3195–3202, 2001.

[41] M. S. Brown, S. K. Basu, and J. R. Falck, "The scavenger cell pathway for lipoprotein degradation: specificity of the binding site that mediates the uptake of negatively-charged LDL by macrophages," *Journal of Supramolecular and Cellular Biochemistry*, vol. 13, no. 1, pp. 67–81, 1980.

[42] D. J. Gallagher and N. Kemeny, "Metastatic colorectal cancer: from improved survival to potential cure," *Oncology*, vol. 78, no. 3-4, pp. 237–248, 2010.

[43] A. Manzano and P. Perez-Segura, "Colorectal cancer chemoprevention: is this the future of colorectal cancer prevention?" *Scientific World Journal*, vol. 2012, Article ID 327341, 8 pages, 2012.

[44] E. Saez, P. Tontonoz, M. C. Nelson et al., "Activators of the nuclear receptor PPARγ enhance colon polyp formation," *Nature Medicine*, vol. 4, no. 9, pp. 1058–1061, 1998.

[45] H. P. Koeffler, "Peroxisome proliferator-activated receptor γ and cancers," *Clinical Cancer Research*, vol. 9, no. 1, pp. 1–9, 2003.

[46] G. D. Girnun, W. M. Smith, S. Drori et al., "APC-dependent suppression of colon carcinogenesis by PPARγ," *Proceedings of the National Academy of Sciences of the United States of America*, vol. 99, no. 21, pp. 13771–13776, 2002.

[47] P. Sarraf, E. Mueller, W. M. Smith et al., "Loss-of-function mutations in PPARγ associated with human colon cancer," *Molecular Cell*, vol. 3, no. 6, pp. 799–804, 1999.

[48] H. Sato, S. Ishihara, K. Kawashima et al., "Expression of peroxisome proliferator-activated receptor (PPAR)γ in gastric cancer and inhibitory effects of PPARγ agonists," *British Journal of Cancer*, vol. 83, no. 10, pp. 1394–1400, 2000.

[49] D. Shen, C. Deng, and M. Zhang, "Peroxisome proliferator-activated receptor γ agonists inhibit the proliferation and invasion of human colon cancer cells," *Postgraduate Medical Journal*, vol. 83, no. 980, pp. 414–419, 2007.

[50] T. Tsukahara and H. Haniu, "Peroxisome proliferator-activated receptor gamma overexpression suppresses proliferation of human colon cancer cells," *Biochemical and Biophysical Research Communications*, vol. 424, pp. 524–529, 2012.

[51] G. Lee, F. Elwood, J. McNally et al., "T0070907, a selective ligand for peroxisome proliferator-activated receptor γ, functions as an antagonist of biochemical and cellular activities," *Journal of Biological Chemistry*, vol. 277, no. 22, pp. 19649–19657, 2002.

[52] T. Tsukahara, "The role of PPARgamma in the transcriptional control by agonists and antagonists," *PPAR Research*, vol. 2012, Article ID 362361, 2012.

[53] T. Tsukahara and K. Murakami-Murofushi, "Release of cyclic phosphatidic acid from gelatin-based hydrogels inhibit colon cancer cell growth and migration," *Scientific Reports*, vol. 2, p. 687, 2012.

[54] M. Krotkiewski, P. Bjorntorp, L. Sjostrom, and U. Smith, "Impact of obesity on metabolism in men and women. Importance of regional adipose tissue distribution," *Journal of Clinical Investigation*, vol. 72, no. 3, pp. 1150–1162, 1983.

[55] P. Bjorntorp, "Abdominal obesity and the metabolic syndrome," *Annals of Medicine*, vol. 24, no. 6, pp. 465–468, 1992.

[56] H. Green and M. Meuth, "An established pre adipose cell line and its differentiation in culture," *Cell*, vol. 3, no. 2, pp. 127–133, 1974.

[57] T. Tsukahara, S. Hanazawa, and K. Murakami-Murofushi, "Cyclic phosphatidic acid influences the expression and regulation of cyclic nucleotide phosphodiesterase 3B and lipolysis in 3T3-L1 cells," *Biochemical and Biophysical Research Communications*, vol. 404, no. 1, pp. 109–114, 2011.

[58] J. Ostman, P. Arner, P. Engfeldt, and L. Kager, "Regional differences in the control of lipolysis in human adipose tissue," *Metabolism: Clinical and Experimental*, vol. 28, no. 12, pp. 1198–1205, 1979.

[59] G. Y. Carmen and S. M. Víctor, "Signalling mechanisms regulating lipolysis," *Cellular Signalling*, vol. 18, no. 4, pp. 401–408, 2006.

[60] Y. H. Jeon, Y. S. Heo, C. M. Kim et al., "Phosphodiesterase: overview of protein structures, potential therapeutic applications and recent progress in drug development," *Cellular and Molecular Life Sciences*, vol. 62, no. 11, pp. 1198–1220, 2005.

[61] E. Degerman, F. Ahmad, Y. W. Chung et al., "From PDE3B to the regulation of energy homeostasis," *Current Opinion in Pharmacology*, vol. 11, pp. 676–682, 2011.

7

The Multifaceted Effects of Omega-3 Polyunsaturated Fatty Acids on the Hallmarks of Cancer

J. A. Stephenson,[1,2] O. Al-Taan,[1,3] A. Arshad,[1,3] B. Morgan,[1,2] M. S. Metcalfe,[3] and A. R. Dennison[3]

[1] *Department of Cancer Studies and Molecular Medicine, University of Leicester, Leicester Royal Infirmary, Leicester LE1 5WW, UK*
[2] *Department of Imaging, Leicester Royal Infirmary, Leicester LE1 5WW, UK*
[3] *Department of Surgery, University Hospitals of Leicester, Leicester General Hospital, Leicester LE5 4PW, UK*

Correspondence should be addressed to J. A. Stephenson; jastephenson@doctors.org.uk

Academic Editor: Angel Catala

Omega-3 polyunsaturated fatty acids, in particular eicosapentaenoic acid, and docosahexaenoic acid have been shown to have multiple beneficial antitumour actions that affect the essential alterations that dictate malignant growth. In this review we explore the putative mechanisms of action of omega-3 polyunsaturated fatty acid in cancer protection in relation to self-sufficiency in growth signals, insensitivity to growth-inhibitory signals, apoptosis, limitless replicative potential, sustained angiogenesis, and tissue invasion, and how these will hopefully translate from bench to bedside.

1. Introduction

Fatty acids (FAs) are a diverse group of molecules. The fatty acyl structure represents the major building block of complex lipids and FAs should be regarded as one of the most fundamental categories of biological lipids [1]. Fatty acids are key nutrients that affect early growth and development, as well as chronic disease in later life. The benefits and potential risks of FAs go well beyond their defined role as fuel [2].

An FA containing more than one carbon double bond is termed polyunsaturated fatty acid (PUFA). The most important families in human metabolism are omega-6 (n-6) and omega-3 (n-3) PUFAs. Specific n-6 and n-3 PUFAs are essential nutrients, while the eicosanoids and docosanoids they derive have distinct biological activities affecting the prevalence and severity of cardiovascular disease, diabetes, inflammation, cancer, and age-related functional decline [1, 2].

Important n-3 PUFAs involved in human nutrition are α-linolenic acid (ALA or 18 : 3n-3), eicosapentaenoic acid (EPA or 20 : 5n-3), docosapentaenoic acid (n-3 DPA or 20 : 5n-3), and docosahexaenoic acid (DHA or 22 : 6n-3).

ALA is the parent FA of the n-3 PUFA family. ALA is mainly found in the plant kingdom with high concentrations in flaxseed oil and perilla oil. It is also found in canola oil, soybean oil, and vegetable oils from where humans derive it in their diet. The human body is unable to readily synthesize ALA, which makes ALA, like linoleic acid (LA or 18 : 2n-6), the parent of the n-6 PUFA family, an essential fatty acid [1].

LA and ALA are converted to their respective n-6 and n-3 PUFA families by a series of independent reactions. However both pathways require the same enzymes for desaturation and elongation. This leads to competition between n-6 and n-3 PUFA for their metabolic conversion. The first step in the pathway requires $\Delta 6$ Desaturase [3, 4] which has a higher affinity for ALA than LA but due to the typically higher intake and concentration of LA there is greater conversion of n-6 PUFA producing the predominant product of the n-6 pathway, arachidonic acid (AA or 20 : 4n-6) [1, 5–7]. Thus the capacity of human metabolism to derive EPA and DHA by the desaturation of ALA is negligible in normal circumstances [1]. The efficiency of conversion is particularly poor in relation to DHA [6, 8]. The concentration of EPA and DHA in tissues can however be enhanced by direct ingestion of either oily fish or as a fish oil (FO) supplement or when competing amounts of n-6 PUFAs are relatively small [8–10].

Fish are able to build up large concentrations of n-3 PUFAs in their tissues by consuming algae and plankton and are therefore the main dietary source of essential n-3 PUFAs in humans. In particular cold-water oily fish such as mackerel, salmon, herring, anchovies, sardines, and smelt provide relatively large amounts of EPA and DHA [7].

2. Physiological Effects of Omega-6 and Omega-3 Polyunsaturated Fatty Acids

n-6 and n-3 PUFAs have a number of vital functions in the human body [11, 12]. As components of structural phospholipids in the cell membrane, they modulate cellular signaling, cellular interaction, and membrane fluidity [13].

They regulate the immune system by acting as precursors for eicosanoids-potent immunoregulatory metabolites. Eicosanoids are synthesised from the n-6 PUFA arachidonic acid (AA, 20 : 4n-6) and the n-3 PUFA, EPA. AA and EPA are metabolised by cyclooxygenase (COX) or lipoxygenase (LOX) enzymes into immunoregulatory metabolites prostaglandins (PGs), thromboxanes (TXs), and leukotrienes (LTs) [13]. As cell membrane phospholipids generally contain significantly higher levels of AA than EPA [14], AA is the most common eicosanoid precursor and gives rise to 2-series PGs and TXs and 4-series LTs. EPA gives rise to 3-series PGs and TXs, 5-series LTs, and E-series resolvins [13, 15].

DHA is a poor substrate for COX and LOX and it was thought that DHA did not produce bioactive COX and LOX mediators. However, Serhan and others identified bioactive docosanoids, named D-series resolvins and protectins [15–17].

AA and EPA also compete for the COX and LOX enzymes. Again, n-3 PUFAs are preferentially used, so supplementation with n-3 PUFAs will have a considerable impact on the production of eicosanoids and docosanoids. Thus, increased intake of n-3 PUFAs results in decreased generation of AA-derived eicosanoids and increased EPA derived eicosanoids and DHA docosanoids [18–21].

It is considered that the eicosanoids and docosanoids produced from EPA and DHA have less biological activity. Therefore have the advantage of being less pro-inflammatory in their action than the potent pro-inflammatory AA-derived mediators [13, 16, 22]. It is also suggested that they also have properties which are anti-inflammatory [15–17].

This theoretical benefit is the rationale for the use of FO supplements in chronic inflammatory disease such as asthma [23] and rheumatoid arthritis [22]. It is also why there is significant interest in the use of n-3 PUFA supplementation in critically ill patients and in patients undergoing major surgery [24–38].

3. The Role of Polyunsaturated Fatty Acids in Tumourigenesis

Hanahan and Weinberg in their landmark review "The hallmarks of cancer" and the subsequent "Hallmarks of the Cancer: the next generation" suggested that the vast catalog of cancer cell genotypes is a manifestation of essential alterations in cell physiology that collectively dictate malignant growth [39, 40].

The original six essential alterations described are self-sufficiency in growth signals, insensitivity to growth-inhibitory (antigrowth) signals, evasion of programmed cell death (apoptosis), limitless replicative potential, sustained angiogenesis, and tissue invasion and metastasis. This results in the cancerous cell having the predatory properties that allow it to survive, invade, and multiply where it should not. Recently the addition of reprogramming of energy metabolism and evading immune destruction has been suggested. Each of these physiologic changes (novel capabilities acquired during tumor development) represents the successful breaching of anticancer defense mechanisms. They proposed that these capabilities are shared in common by most and perhaps all types of human tumors and must be satisfied for tumour growth to occur within the tumour microenvironment [39, 40].

EPA and DHA have been shown to have multiple antitumour actions that affect all of the original six essential alterations that dictate malignant growth. This is a result of various pathways including inhibition of AA metabolism and independent effects on various cytokines involved in tumourigenesis. n-6 PUFA derived eicosanoids have promoting effects in cancer cell growth [41, 42], angiogenesis [42], and invasion [43]. As previously discussed n-3 PUFAs can also be metabolized to resolvins and protectins [15, 44]. These compounds possess immunoregulatory actions [45] and it is well documented that inflammation plays an important role in the development of numerous human malignancies [46–48]. Thus one of the possible mechanisms for inhibition of tumor growth by n-3 PUFAs is via immunoregulation through production of 5 series leukotrienes (LT), 3 series prostaglandins (PG) and thromboxanes (TX), and resolvins—Figure 1 [49].

4. Effects of Omega-3 Polyunsaturated Fatty Acids on Growth Signals

Normal cells are unable to proliferate in the absence of stimulatory signals from transmembrane receptors, which are activated by growth factors, extracellular matrix components, and cell-cell interaction molecules [39]. Conversely, tumour cells however have a reduced dependence on such exogenous growth signals. Cancerous cells often bypass this step by synthesizing their own growth factors [50], overexpressing cell surface receptors which transmit growth-stimulatory signals [50, 51] or switching integrins to ones which favour growth signal transition [39–52]. Also many oncogenes mimic normal growth signals, promoting proliferation [39, 53].

The overall result is that the cancer cell is self sufficient in stimulating its own multiplication. The reduced dependence on exogenous growth signals and stimulation from normal tissue microenvironment leads to unregulated and exponential growth.

4.1. In Vitro Evidence. The cell plasma membrane affects growth factor: receptor interaction and subsequent signal transduction. EPA and DHA have been shown to have beneficial effects on the plasma membrane in MDA-MB-231

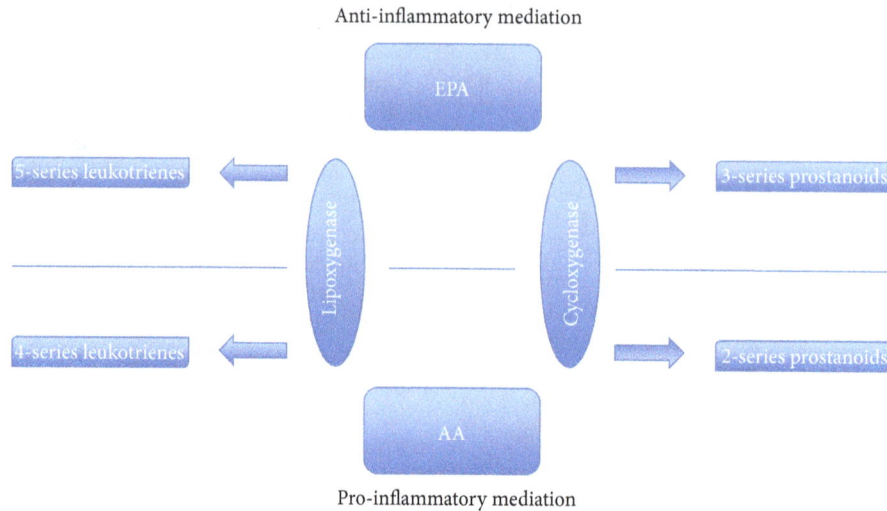

FIGURE 1: Inflammatory mediators derived from eicosapentaenoic acid and arachidonic acid. Adapted From Furst 2000.

breast cancer cells with a marked decrease of epidermal growth factor receptor (EGFR) in lipid rafts, leading to alteration in EGFR signaling in a way that decreases the growth of breast tumors [54].

n-3 PUFAs appear to downregulate protein kinase C $\beta2$ [55, 56], RAS [57], and nuclear factor $\kappa\beta$ (NF-kB) [58] which are important cell signaling mediators often found to be elevated in carcinogenesis.

DHA has also been shown to modulate heat shock proteins that act as "chaperones" in protein: protein interactions and in cell membrane transport. [59]. DHA is also known to modulate steroid receptors in human cancer cell lines [60].

Tumour derived nitric oxide (NO) has the ability to promote tumour growth by enhancing invasiveness of tumour cells [61, 62]. NO also increases PGE2 production, which is implicated in tumour growth and progression [62]. EPA and DHA suppress NO production in macrophage cell lines in a dose dependant fashion [63, 64].

EPA and DHA inhibited human colon adenocarcinoma Caco-2 cell proliferation. Cells cultured with EPA or DHA reached much lower final densities compared to cells cultured with LA. The authors proposed that low insulin growth factor II (IGF-II)/IGF binding protein-6 ratios may have resulted in less free IGF-II a potent cell proliferation promoter and, consequently, the slower proliferation of Caco-2 cells treated with EPA or DHA [65].

4.2. In Vivo Evidence. COX-2 over-expression has been reported in 90% of colon tumours and colonic adenomas [66]. COX-2 has direct and indirect effects on growth via upregulation of growth signals and prostaglandins, angiogenesis, apoptosis, and cell-cell interaction [67]. The specific effects will be discussed in each subsequent section. In relation to self-sufficiency, numerous studies have found that COX-2 and its active metabolite PGE2-levels are reduced by supplementation of n-3 PUFAs [66, 68, 69]. A prostate cancer cell xenograft in mice found that the reduced levels of COX-2 and PGE2 were related to a reduction in tumour growth rate, tumour volume, and serum PSA [70].

Protein kinase C (PKC) Δ has a tumour suppressor function. The carcinogen azoxymethane decreases levels of PKC Δ. This decrease has been shown to be ameliorated in rats feed FO [71]. PKC $\beta2$, which is induced early in colon carcinogenesis, leading to self-sufficiency, cancer promotion, and carcinogen induced epithelial hyper-proliferation [72–76], is significantly decreased in rats fed FO. This blocked PKC $\beta2$ hyperproliferation [74, 77].

The effect of n-3 PUFAs on growth signal transduction appears to be multi-faceted, with numerous putative pathways identified in the *in vitro* and *in vivo* setting (Figure 2). This suggests that any relationship of n-3 PUFA on tumourigenesis is complex.

5. Effects of Omega-3 Polyunsaturated Fatty Acids on Tumour Insensitivity to Growth-Inhibitory Signals

Tissue homeostasis and cellular quiescence is maintained in normal cells by anti-proliferative signals from growth inhibitory factors (tumour suppressor genes) and from cell-cell or cell-extracellular matrix interaction. These antigrowth signals are transmitted by cell surface receptors and may have 2 potential effects: (1) cells are forced into the quiescent (G0) state; (2) cells are induced into a post-mitotic state of permanent dormancy [39].

5.1. In Vitro Evidence. Investigation of the effect of EPA and DHA on colon cancer cell lines has shown decreases in cellular proliferation. Mengeaud demonstrated that cellular proliferation in HRT-18, HT-29, and Caco-2 cell lines is decreased by EPA [78]. This was replicated in a SIC oncogene transformant cell line by Tsai who also showed that DHA reduced cellular proliferation [79]. In two studies using HT-29 cell lines Clarke reported that EPA reduced cell proliferation and Chen demonstrated that DHA had a similar effect [80, 81]. Other studies have also shown decreases in cell proliferation in response to EPA and DHA [65, 82].

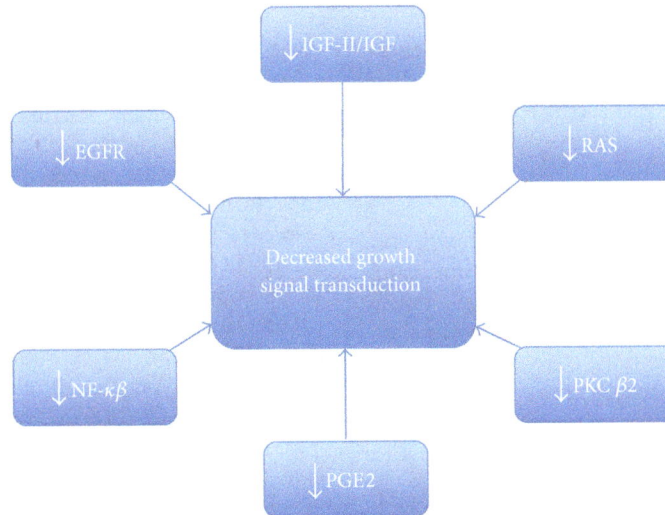

FIGURE 2: Multi-modal putative mechanisms of action of DHA and EPA on growth signal transduction.

5.2. In Vivo Evidence. In a rodent model of breast cancer, DHA induced a reduction in mammary tumours accompanied by a 60% upregulation of BRCA1 tumour suppressor protein [83].

Numerous studies have shown a decrease in tumour cellular proliferation in response to n-3 PUFAs; however the putative mechanisms are not well documented and further investigation is required.

6. Effects of Omega-3 Polyunsaturated Fatty Acids on Tumour Evasion of Programmed Cell Death

Apoptosis governs the rate of cell attrition. The ability of tumour cells to expand in number is governed by the balance of proliferation and apoptosis. Evasion of apoptosis allows tumour cell mass to increase dramatically and it is a hallmark of tumourigenesis.

6.1. In Vitro Evidence. The studies previously reported on HT-29 colon cell lines by Clarke and Chen also showed increased induction of apoptosis with n-3 PUFAs [80, 81]. Other studies have shown that DHA induces a dose-dependant effect upon cancer cell apoptosis [84–86].

DHA has been shown to induce cytochrome c release, which binds to apoptotic protease activating factor initiating cancer cell apoptosis [85, 87].

n-3 PUFAs alter peroxisome proliferator-activated receptors (PPARs) cell signaling by acting as direct ligands for the receptors. DHA has been shown to modulate PPAR receptor expression [88, 89] and induce cellular apoptosis [67, 90, 91]. This was mediated through the effect of PPAR on Syndecan-1 a protein product, which induces apoptosis [67, 91].

EPA and DHA have also been shown to modulate expression of the Bcl-2 family. They downregulate the expression of anti-apoptotic proteins Bcl-2 and Bcl-xL and increase levels of Bak and Bcl-xS pro-apoptotic proteins [92–97].

NFκβ, which has the ability to block programmed cell death potentiating tumour survival, is downregulated by n-3 PUFAs in murine macrophages, which decreases COX-2 expression restoring functional apoptosis [23, 98].

6.2. In Vivo Evidence. Hong showed that in a mouse model of colon carcinogenesis that initiation of tumour growth was restricted by increased apoptosis related to n-3 PUFA supplementation [99]. One way in which apoptosis may be regulated by n-3 PUFAs is via COX-2. COX-2 has been shown to decrease apoptosis via expression of the Bcl-2 gene. A reduction of COX-2 and COX-2 inhibition have been shown to repress the expression of Bcl-2 gene and its anti-apoptotic properties [67, 67, 69, 91].

The Bcl-2 family also has a pro-apoptotic member Bad. In its normal state Bad promotes cell death by displacing Bax from Bcl-2 [100, 101]. Phosphorylation of Bad prevents it from displacing Bax from Bcl-2 subsequently promoting cell survival [100, 102, 103]. A study by Berquin on Pten knockout mice showed that prostate tumours from mice with an enriched n-3 PUFA diet had lower levels of phosphorylated Bad and higher apoptotic indexes compared to mice on an n-6 PUFA diet. This led to reduced tumour growth, slowed histopathological progression, and increased survival rates [49].

Tumour evasion of programmed cell death is a complex and controlled by an intricate milieu of intra-cellular signal transduction pathways and external cytokines, survival factors, chemokines, growth factors, and death factors. Evidence suggests that DHA and EPA have effects on many of these pathways, which seem to be beneficial.

7. Effects of Omega-3 Polyunsaturated Fatty Acids on Limitless Replicative Potential of Tumours

Growth signal autonomy, insensitivity to antigrowth signals, and apoptotic evasion alone do not lead to expansive tumour growth because cells have the capacity for senescence, an intrinsic property that limits multiplication [39, 104]. Senescence can be circumvented by DNA damage and disabling tumour suppressor genes such as p53 and pRb, which eventually leads to immortalisation or the ability to multiply without limit [105].

7.1. In Vitro Evidence. AA may promote tumour growth and replication via activation of protein kinase C stimulating mitosis [106]. Studies in colonocytes and JB6 cells—mouse epidermal cells—have shown that growth promotion via the transcription factors RAS and AP1 is reduced by n-3 PUFAs [107, 108]. The second messengers of AA metabolism with COX and LOX also stimulate mitosis. Conversely EPA derived metabolites of COX and LOX have been shown to decrease growth of human breast cancer cell lines [42].

7.2. In Vivo Evidence. Numerous animal study models in colon cancer have demonstrated that n-3 PUFA supplementation leads to tumour growth suppression [68, 109–115].

It had been demonstrated in the rat colon that n-3 PUFAs reduced k-RAS mutations and decreased membrane RAS expression [116] and it has been suggested that this indicates that n-3 PUFAs may protect against colon carcinogenesis by decreasing DNA adduct formation and/or enhancing DNA repair [117]. In the study already discussed by Hong they also showed that there was a reduction in DNA adduct formation [99]. Reddy showed that initiation of experimentally induced colon cancer was reduced by the protective effect of n-3 PUFAs [118].

In xenografted rats carrying neuroblastoma tumours, Gleissman demonstrated that DHA-enriched diet prior to tumour cell injection delayed tumour formation and prevented tumour establishment [119]. In the same study Gleissman investigated the effect of DHA as a therapeutic agent in rats who had established tumours. Tumours in animals receiving high dose DHA showed partial response compared to animals receiving low dose DHA or control that showed stable disease and progressive disease, respectively [119].

Another therapeutic study performed in nude mice xenografted with BxPC-3 pancreatic cancer cells showed tumour inhibition by DHA. Interestingly the inhibition was increased in another group where DHA was combined with curcumin [120].

8. Effects of Omega-3 Polyunsaturated Fatty Acids on Sustained Angiogenesis

For a tumour to grow beyond 2 mm, angiogenesis and neovascular formation are required. The ability to induce and sustain angiogenesis from vascular quiescence is controlled by the "angiogenic switch." Tumors appear to activate the angiogenic switch by changing the balance of angiogenesis inducers and countervailing inhibitors [121]. This is seen with increased production, expression, and signal transduction of pro-angiogenic factors such as vascular endothelial growth factor (VEGF). n-3 PUFAs have been shown to have a profound effect on angiogenesis [122].

8.1. In Vitro Evidence. n-3 PUFAs have been shown to decrease sprouting angiogenesis by suppressing VEGF-stimulated endothelial cell proliferation, migration, and tube formation [123–125]. Tsuzuki and colleagues treated human umbilical vein endothelial cells with conjugated EPA and demonstrated that a reduction in sprouting angiogenesis tube formation and endothelial cell migration [123] was also seen in bovine aortic endothelial cells pre-treated with DPA. VEGF-Receptor (VEGF-R) 2 expression was also found to be suppressed [124]. The reduction in endothelial cell proliferation in response to EPA was shown to be dose dependant in bovine carotid artery endothelial cells [125]. The study by Yang et al. also elicited a dose dependant decrease in VEGF-1 (FlK-1) expression [125]. A reduction in VEGF/VEGF-R binding has also been demonstrated by Yuan et al. using an n-3 PUFA rich shark oil [126].

n-3 PUFAs also have stark effects on numerous other mediators involved in angiogenesis. Platelet derived growth factors (PDGF) play an important role in angiogenesis by stimulating fibroblast and vascular smooth muscle cell motility and acting as a chemo-attractant [127, 128]. As early as 1988 Fox and DiCorleto showed that *in vitro* production of PDGF was inhibited by n-3 PUFAs [129]. Investigating the effects of EPA and DHA on PDGF signal transduction Terano and colleagues demonstrated that EPA inhibited PDGF binding to its receptor and suppressed c-fos mRNA expression, a gene involved in receptor signal transduction. These effects led to inhibition of smooth muscle proliferation a prerequisite for angiogenesis [130].

As previously discussed PGE2 is formed from AA, catalysed by COX-2. There is well-defined link between E series prostaglandins and carcinogenesis [131]. Decreased levels of VEGF, COX-2, and PGE2 have been demonstrated in HT-29 colon cancer cell lines when cultured *in vitro* with EPA and DHA [112] and a synergistic inhibitory effect of n-3 PUFAs and COX-2 inhibitors on growth of human colon cancer cell lines has been shown [60, 132].

Nitric oxide (NO) promotes endothelial cell survival and proliferation and inhibits apoptosis [133, 134]. NO and COX-2 also regulate VEGF-mediated angiogenesis [135–137]. Inducible nitric oxide synthase (iNOS) stimulates NO production [136].

DHA has been demonstrated to inhibit NO production and iNOS expression in murine macrophages [63, 64, 138–140] and downregulate NO and nuclear factor kappa beta (NFKB) in human colon cancer cell lines [141].

In the study previously discussed by Tsuzuki et al., they also demonstrated that production of matrix metalloproteinases (MMP) 2 and 9—proteases which play a role in basement membrane proteolysis in the 3rd stage of sprouting angiogenesis—in human endothelial cells was inhibited by EPA [123].

It has also been demonstrated that DHA inhibits Beta-catenin—a transcriptional regulator of angiogenesis—production in colon cancer cells [142].

8.2. In Vivo Evidence. In a study where Fischer 344 rats were implanted with fibrosarcomas, the group with diets supplemented with EPA had tumours with significantly lower

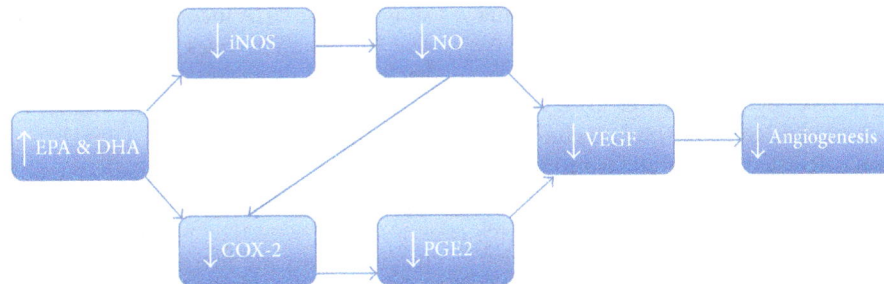

FIGURE 3: Pathways leading to the anti-angiogeneis effect of DHA and EPA.

tumour volume and decreased VEGF-alpha mRNA levels [143].

A study in nude mice supplemented with n-3 PUFA undergoing implantation of human colorectal carcinomas showed that tumour expression of VEGF, COX-2, and PGE2 was decreased compared to control [112]. Benefits were also seen in nude mice transplanted with breast carcinoma. Breast tumours in mice feed diets high in EPA and DHA had lower tumour microvessel density and VEGF levels compared to controls [144, 145].

Induction of vascular smooth muscle cell migration by PDGF, required for angiogenesis, is inhibited by EPA and DHA *in vivo* [130]. Several other small animal models have demonstrated that n-3 PUFA enriched diets inhibit COX-2 and PGE2 production [146] and reduced HT-29 colon cancer cell tumour growth and microvessel density after implantation into nude mice [112].

Factors such as PGE2, NO, COX-2, and NFKB have well-documented roles in both the inflammatory and angiogenic cascades with significant cross-relation in both pathways. This demonstrates the potential for n-3 FAs as anti-angiogenic agents via inhibition of these factors and others including VEGF and PDGF (Figure 3).

9. Effects of Omega-3 Polyunsaturated Fatty Acids on Tumour Tissue Invasion and Metastasis

Metastases are the cause of 90% of human cancer related deaths [147]. Like the formation of the initial tumour the above 5 characteristics are required for metastasis formation. Metastasis and tissue invasion also require loss of cell-cell adhesion-regulated by cell adhesion molecules (CAMs)—and cell-ECM interactions-regulated by integrins [39, 148]. E-cadherin, a CAM, is lost in the majority of epithelial cancers, which enables invasion and metastases [149]. Once at a new site tumour cells then shift the expression of integrins to facilitate preferential cell binding to allow the tumour to "seed" leading to subsequent distant growth.

9.1. In Vitro Evidence. DHA has been shown to reduce the induction of monocyte rolling, adhesion, and transmigration controlled by TNFα [150].

As previously discussed NO increases tumour growth and angiogenesis [61, 62, 151–153]. NO also plays an important role in tumour cell migration, which may be decreased by EPA and DHA supplementation due to suppression of tumour derived NO production [63, 64].

Cell-cell adhesion is modulated by DHA via down regulation of Rho GTPase, which inhibits cytoskeleton reorganisation [154], and reduction in ICAM-1 and VCAM-1 expression [155].

9.2. In Vivo Evidence. Again the COX-2 pathway plays an important role in each of the tumour development pathways. COX-2 reduces cell-cell and cell-matrix interactions leading to increased progression and metastases of gastric carcinoma [156, 157]. COX-2 inhibition has been shown to reduce invasiveness and depress metastases of gastric cancer in various animal models [67]. A xenograft animal model showed inhibition of tumour cell growth and invasion by n-3 PUFAs associated with decreased COX-2 and PGE2 levels [70]. n-3 PUFAs may act as a natural COX inhibitor [158]. Despite the wealth of evidence of the beneficial anti-tumour effects of DHA and EPA via downregulation of the COX-2 pathway a study by Boudreau has suggested that there are also COX-2 independent methods of protective action. In a colon cancer xenograft model, tumour formation was inhibited by n-3 PUFA supplementation in both COX-2 deficient and COX-2 overexpressing tumours [111].

Inhibition of metastases by n-3 PUFA enriched diets was demonstrated in both mouse and rat [159, 160] models of colorectal cancer. EPA and DHA have also been shown to suppress development of lung metastases due to reduced 72-kDa type IV collagenase gelatinolytic activity [161].

There is a large and growing body of evidence from laboratory-based studies that n-3 PUFAs have a marked beneficial effect on the hallmarks of cancer. However do these mechanistic studies translate into a clinical benefit?

10. Effects of Omega-3 Polyunsaturated Fatty Acids on Tumourigenesis in Humans

As well as the cellular mechanisms described in the *in-vitro* and *in-vivo* studies above, epidaemiological observations also appeared to suggest a benefit of n-3 PUFA in cancer prevention in humans. An example is an observational study in Inuits—Inuits have DHA levels several times higher than Caucasians [162]—which has demonstrated significantly lower levels of childhood cancer occurrence, particularly neuroblastoma—tenfold decrease—and Hodgkin lymphoma [92, 163].

However, a systematic review of the "Effects of omega-3 fatty acids on cancer risk" by MacLean et al. reviewed 38 articles published from 1966 to 2005 which included 65 estimates of association calculated over 20 differing cohorts for 11 different cancer types concluded that only 10 were statistically significant and that the body of literature does not provide evidence to suggest a significant association between n-3 PUFA and cancer incidence. They also stated that dietary supplementation with n-3 PUFAs is unlikely to prevent cancer [164]. Chen et al. raised concerns with the systematic review [165]. The studies included in this systematic review did not formally measure FA consumption-food frequency questionnaires and dietary records were used which correlate poorly with direct PUFA measurement [166]—The studies included in this systematic review did not formally measure FA consumption, they used food frequency questionnaires and dietary records, which correlate poorly with direct PUFA measurement [166] and do not differentiate between the source of FO consumption [165].

Animal data—some of which has been discussed above—is an invaluable tool for mechanistic studies and the models can closely mimic the clinical course of cancer progressions [165, 167, 168]. However the translation of animal data into the clinical arena is difficult due to the higher amounts of n-3 PUFAs used in relation to fat intake and percentage weight [164]. Inherent to the majority of animal studies is the use of high levels of dietary constituents [117] with n-3 PUFA intake between 18 and 48% of daily energy compared to 4–10% in human population based studies [169]. This is likely to be one of the reasons that only weak associations of PUFA intake and cancer are found in population-based studies [117]. Extrapolation of findings is also confounded by poor descriptions of experimental conditions and dose and purity of n-3 PUFA supplementation [164, 170].

However, a role is potentially being developed for n-3 PUFA in combination with current chemotherapeutic agents to augment their action [171]. Animal models have shown that the efficiency of doxorubicin [172] and mitomycin C [173] in inhibiting tumor growth and the inhibitory effect of tamoxifen in estrogen-dependent xenografts [174] are enhanced when combined with n-3 PUFA-enriched diets.

DHA supplementation with concurrent cytotoxic drug treatment is potentially a way in which to clinically utilise DHA in cancer treatment. DHA in combination with doxorubicin, irinotecan, cisplatin, melphalan, and vincristine on neuroblastoma cell survival shows additive or synergistic interactions [85, 92].

A therapeutic study in breast cancer patients where DHA was combined with the chemotherapeutic drugs epirubicin, cyclophosphamide, and 5-fluorouracil showed delayed time to tumour progression and longer overall survival. However these findings were only observed when patients were stratified into 2 groups of either high or low incorporation of DHA into plasma and erythrocytes. Patients who had high incorporation of DHA into plasma and erythrocytes benefitted compared to those who had low level DHA incorporation [175]. This observation correlates well with other studies showing that DHA incorporation differs between individuals due to dissimilar rates of metabolism, enzymatic activity, background diet, age, and sex [92, 176–178]. This is likely to be a recurrent problem in studies using oral n-3 PUFA supplementation.

11. Discussion

In the last decade there has been a growing interest in the role of FAs, especially PUFAs, in cancer development and progression. As discussed the link between FAs and cancer may relate to the synthesis of eicosanoids, which have wide-ranging diverse effects at a cellular level. There are currently several ongoing clinical trials to assess this, where n-3 PUFAs are being tested for cancer prevention, support, or therapy [158], but initial evidence suggests that researchers do not seem to be translating the profoundly beneficial results seen in the laboratory to the bedside. This is potentially due to the way in which n-3 PUFAs are being supplemented and we need to think about novel ways of overcoming the difficulties faced with FO supplementation to assess the true benefit of n-3 PUFAs in the fight against cancer. Maybe we also need to explore the broader therapeutic benefits of FO supplementation on areas such as cancer cachexia and aiding treatment tolerance as recently suggested by Murphy et al. [179]. However, ongoing and future clinical trials using intravenous n-3 PUFA infusions in cancer therapy are eagerly awaited.

Acknowledgment

A. R. Dennison, M. S. Metcalfe, J. A. Stephenson, O. Al-Taan, and A. Arshad have received educational travelling grants from B. Braun Melsungen.

References

[1] W. M. N. Ratnayake and C. Galli, "Fat and fatty acid terminology, methods of analysis and fat digestion and metabolism: a background review paper," *Annals of Nutrition and Metabolism*, vol. 55, no. 1–3, pp. 8–43, 2009.

[2] B. Burlingame, C. Nishida, R. Uauy, and R. Weisell, "Fats and fatty acids in human nutrition: introduction," *Annals of Nutrition and Metabolism*, vol. 55, no. 1–3, pp. 5–7, 2009.

[3] S. A. Moore, E. Hurt, E. Yoder, H. Sprecher, and A. A. Spector, "Docosahexaenoic acid synthesis in human skin fibroblasts involves peroxisomal retroconversion of tetracosahexaenoic acid," *Journal of Lipid Research*, vol. 36, no. 11, pp. 2433–2443, 1995.

[4] H. Sprecher, "The roles of anabolic and catabolic reactions in the synthesis and recycling of polyunsaturated fatty acids," *Prostaglandins Leukotrienes and Essential Fatty Acids*, vol. 67, no. 2-3, pp. 79–83, 2002.

[5] B. Lands, "A critique of paradoxes in current advice on dietary lipids," *Progress in Lipid Research*, vol. 47, no. 2, pp. 77–106, 2008.

[6] G. C. Burdge and P. C. Calder, "Conversion of α-linolenic acid to longer-chain polyunsaturated fatty acids in human adults," *Reproduction Nutrition Development*, vol. 45, no. 5, pp. 581–597, 2005.

[7] R. G. Ackman, "Fatty acids in fish and shellfish," in *Fatty Acids in Foods and Their Health Implications*, C. K. Chow, Ed., pp. 155–185, CRC Press, London, UK, 2008.

[8] J. T. Brenna, "Efficiency of conversion alpha-linolenic acid to long chain n-3 fatty acids in man," *Current Opinion in Clinical Nutrition & Metabolic Care*, vol. 5, pp. 127–132, 2002.

[9] P. L. L. Goyens, M. E. Spilker, P. L. Zock, M. B. Katan, and R. P. Mensink, "Conversion of α-linolenic acid in humans is influenced by the absolute amounts of α-linolenic acid and linoleic acid in the diet and not by their ratio," *American Journal of Clinical Nutrition*, vol. 84, no. 1, pp. 44–53, 2006.

[10] A. J. Sinclair, N. M. Attar-Bashi, and D. Lib, "What is the role of αt-linolenic acid for mammals?" *Lipids*, vol. 37, no. 12, pp. 1113–1123, 2003.

[11] B. Corridan and A. Wilson, "Health effects of n-3 polyunsaturated fatty acids," in *Encyclopedia of Human Nutrition*, J. Strain, M. Sadler, and B. Caballero, Eds., vol. 1, pp. 757–769, Academic Press, New York, NY, USA, 1998.

[12] C. E. Eberhart and R. N. Dubois, "Eicosanoids and the gastrointestinal tract," *Gastroenterology*, vol. 109, no. 1, pp. 285–301, 1995.

[13] C. E. Roynette, P. C. Calder, Y. M. Dupertuis, and C. Pichard, "n-3 polyunsaturated fatty acids and colon cancer prevention," *Clinical Nutrition*, vol. 23, no. 2, pp. 139–151, 2004.

[14] P. Yaqoob, H. S. Pala, M. Cortina-Borja, E. A. Newsholme, and P. C. Calder, "Encapsulated fish oil enriched in α-tocopherol alters plasma phospholipid and mononuclear cell fatty acid compositions but not mononuclear cell functions," *European Journal of Clinical Investigation*, vol. 30, no. 3, pp. 260–274, 2000.

[15] C. N. Serhan, S. Hong, K. Gronert et al., "Resolvins: a family of bioactive products of omega-3 fatty acid transformation circuits initiated by aspirin treatment that counter proinflammation signals," *Journal of Experimental Medicine*, vol. 196, no. 8, pp. 1025–1037, 2002.

[16] J. Lee and D. H. Hwang, "Dietary fatty acids and eicosanoids," in *Fatty Acids in Foods and Their Health Implications*, C. K. Chow, Ed., pp. 713–739, CRC Press, London, UK, 2008.

[17] N. G. Bazan, "Omega-3 fatty acids, pro-inflammatory signaling and neuroprotection," *Current Opinion in Clinical Nutrition & Metabolic Care*, vol. 10, pp. 136–141, 2007.

[18] P. C. Calder, "Polyunsaturated fatty acids and inflammation," *Prostaglandins Leukotrienes and Essential Fatty Acids*, vol. 75, pp. 197–202, 2006.

[19] P. C. Calder, "n-3 polyuusaturated fatty acids, inflammation, and inflammatory diseases," *The American Journal of Clinical Nutrition*, vol. 83, supplement 6, pp. 1505s–1519s, 2006.

[20] G. E. Caughey, E. Mantzioris, R. A. Gibson, L. G. Cleland, and M. J. James, "The effect on human tumor necrosis factor α and interleukin 1β production of diets enriched in n-3 fatty acids from vegetable oil or fish oil," *American Journal of Clinical Nutrition*, vol. 63, no. 1, pp. 116–122, 1996.

[21] R. I. Sperling, A. I. Benincaso, C. T. Knoell, J. K. Larkin, K. F. Austen, and D. R. Robinson, "Dietary ω-3 polyunsaturated fatty acids inhibit phosphoinositide formation and chemotaxis in neutrophils," *Journal of Clinical Investigation*, vol. 91, no. 2, pp. 651–660, 1993.

[22] P. C. Calder, "Dietary modification of inflammation with lipids," *Proceedings of the Nutrition Society*, vol. 61, pp. 345–358, 2002.

[23] S. A. Schwartz, A. Hernandez, and B. M. Evers, "The role of NF-kappaB/IkappaB proteins in cancer: implications for novel treatment strategies," *Surgical Oncology*, vol. 8, no. 3, pp. 143–153, 1999.

[24] V. M. Barbosa, E. A. Miles, C. Calhau, E. Lafuente, and P. C. Calder, "Effects of a fish oil containing lipid emulsion on plasma phospholipid fatty acids, inflammatory markers, and clinical outcomes in septic patients: a randomized, controlled clinical trial," *Critical Care*, vol. 14, no. 1, article R5, 2010.

[25] P. C. Calder, "Symposium 4: hot topics in parenteral nutrition Rationale for using new lipid emulsions in parenteral nutrition and a review of the trials performed in adults," *Proceedings of the Nutrition Society*, vol. 68, no. 3, pp. 252–260, 2009.

[26] P. C. Calder, G. L. Jensen, B. V. Koletzko, P. Singer, and G. J. A. Wanten, "Lipid emulsions in parenteral nutrition of intensive care patients: current thinking and future directions," *Intensive Care Medicine*, vol. 36, no. 5, pp. 735–749, 2010.

[27] S. Friesecke, C. Lotze, J. Köhler, A. Heinrich, S. B. Felix, and P. Abel, "Fish oil supplementation in the parenteral nutrition of critically ill medical patients: a randomised controlled trial," *Intensive Care Medicine*, vol. 34, no. 8, pp. 1411–1420, 2008.

[28] H. Grimm, N. Mertes, C. Goeters et al., "Improved fatty acid and leukotriene pattern with a novel lipid emulsion in surgical patients," *European Journal of Nutrition*, vol. 45, no. 1, pp. 55–60, 2006.

[29] A. R. Heller, T. Rössel, B. Gottschlich et al., "Omega-3 fatty acids improve liver and pancreas function in postoperative cancer patients," *International Journal of Cancer*, vol. 111, no. 4, pp. 611–616, 2004.

[30] A. R. Heller, S. Rössler, R. J. Litz et al., "Omega-3 fatty acids improve the diagnosis-related clinical outcome," *Critical Care Medicine*, vol. 34, pp. 972–979, 2006.

[31] M. M. Berger, L. Tappy, J. P. Revelly et al., "Fish oil after abdominal aorta aneurysm surgery," *European Journal of Clinical Nutrition*, vol. 62, no. 9, pp. 1116–1122, 2008.

[32] H. Grimm, "A balanced lipid emulsion—a new concept in parenteral nutrition," *Clinical Nutrition*, vol. 1, no. 3, pp. 25–30, 2005.

[33] M. Köller, M. Senkal, M. Kemen, W. König, V. Zumtobel, and G. Muhr, "Impact of omega-3 fatty acid enriched TPN on leukotriene synthesis by leukocytes after major surgery," *Clinical Nutrition*, vol. 22, no. 1, pp. 59–64, 2003.

[34] B. J. Morlion, E. Torwesten, H. Lessire et al., "The effect of parenteral fish oil on leukocyte membrane fatty acid composition and leukotriene-synthesizing capacity in patients with postoperative trauma," *Metabolism: Clinical and Experimental*, vol. 45, no. 10, pp. 1208–1213, 1996.

[35] K. Mayer, S. Gokorsch, C. Fegbeutel et al., "Parenteral nutrition with fish oil modulates cytokine response in patients with sepsis," *American Journal of Respiratory and Critical Care Medicine*, vol. 167, no. 10, pp. 1321–1328, 2003.

[36] E. Tsekos, C. Reuter, P. Stehle, and G. Boeden, "Perioperative administration of parenteral fish oil supplements in a routine clinical setting improves patient outcome after major abdominal surgery," *Clinical Nutrition*, vol. 23, no. 3, pp. 325–330, 2004.

[37] G. Weiss, F. Meyer, B. Matthies, M. Pross, W. Koenig, and H. Lippert, "Immunomodulation by perioperative administration of n-3 fatty acids," *British Journal of Nutrition*, vol. 87, no. 1, pp. S89–S94, 2002.

[38] M. W. Wichmann, P. Thul, H. D. Czarnetzki, B. J. Morlion, M. Kemen, and K. W. Jauch, "Evaluation of clinical safety and beneficial effects of a fish oil containing lipid emulsion (Lipoplus, MLF541): data from a prospective, randomized, multicenter trial," *Critical Care Medicine*, vol. 35, no. 3, pp. 700–706, 2007.

[39] D. Hanahan and R. A. Weinberg, "The hallmarks of cancer," *Cell*, vol. 100, no. 1, pp. 57–70, 2000.

[40] D. Hanahan and R. A. Weinberg, "Hallmarks of cancer: the next generation," *Cell*, vol. 144, no. 5, pp. 646–674, 2011.

[41] S. H. Abou-El-Ala, K. W. Prasse, R. L. Farrell, R. W. Carroll, A. E. Wade, and O. R. Bunce, "Effects of D,L-2-difluoromethylornithine and indomethacin on mammary tumor promotion in rats fed high n-3 and/or n-6 fat diets," *Cancer Research*, vol. 49, no. 6, pp. 1434–1440, 1989.

[42] D. P. Rose and J. M. Connolly, "Effects of fatty acids and inhibitors of eicosanoid synthesis on the growth of a human breast cancer cell line in culture," *Cancer Research*, vol. 50, no. 22, pp. 7139–7144, 1990.

[43] M. D. Brown, C. A. Hart, E. Gazi, S. Bagley, and N. W. Clarke, "Promotion of prostatic metastatic migration towards human bone marrow stoma by Omega 6 and its inhibition by Omega 3 PUFAs," *British Journal of Cancer*, vol. 94, no. 6, pp. 842–853, 2006.

[44] C. N. Serhan, M. Arita, S. Hong, and K. Gotlinger, "Resolvins, docosatrienes, and neuroprotectins, novel omega-3-derived mediators, and their endogenous aspirin-triggered epimers," *Lipids*, vol. 39, no. 11, pp. 1125–1132, 2004.

[45] C. N. Serhan and J. Savill, "Resolution of inflammation: the beginning programs the end," *Nature Immunology*, vol. 6, no. 12, pp. 1191–1197, 2005.

[46] M. Karin, "Nuclear factor-kappaB in cancer development and progression," *Nature*, vol. 441, pp. 431–436, 2006.

[47] L. M. Coussens and Z. Werb, "Inflammation and cancer," *Nature*, vol. 420, no. 6917, pp. 860–867, 2002.

[48] H. Clevers, "At the crossroads of inflammation and cancer," *Cell*, vol. 118, no. 6, pp. 671–674, 2004.

[49] I. M. Berquin, Y. Min, R. Wu et al., "Modulation of prostate cancer genetic risk by omega-3 and omega-6 fatty acids," *Journal of Clinical Investigation*, vol. 117, pp. 1866–1875, 2007.

[50] P. Fedi, S. R. Tronick, and S. A. Aaronson, "Growth factors," in *Cancer Medicine*, J. F. Holland, R. C. Bast, D. L. Morton, E. Frei, D. W. Kufe, and R. R. Weichselbaum, Eds., pp. 41–64, Williams and Wilkins, Baltimore, Md, USA, 1999.

[51] D. J. Slamon, G. M. Clark, S. G. Wong, W. J. Levin, A. Ullrich, and W. L. McGuire, "Human breast cancer: correlation of relapse and survival with amplification of the HER-2/neu oncogene," *Science*, vol. 235, no. 4785, pp. 177–182, 1987.

[52] M. E. Lukashev and Z. Werb, "ECM signalling: orchestrating cell behaviour and misbehaviour," *Trends in Cell Biology*, vol. 8, no. 11, pp. 437–441, 1998.

[53] K. W. Kinzler and B. Vogelstein, "Lessons from hereditary colorectal cancer," *Cell*, vol. 87, no. 2, pp. 159–170, 1996.

[54] P. D. Schley, D. N. Brindley, and C. J. Field, "(n-3) PUFA alter raft lipid composition and decrease epidermal growth factor receptor levels in lipid rafts of human breast cancer cells," *Journal of Nutrition*, vol. 137, no. 3, pp. 548–553, 2007.

[55] B. S. Reddy, B. Simi, N. Patel, C. Aliaga, and C. V. Rao, "Effect of amount and types of dietary fat on intestinal bacterial 7α-dehydroxylase and phosphatidylinositol-specific phospholipase C and colonic mucosal diacylglycerol kinase and PKC activities during different stages of colon tumor promotion," *Cancer Research*, vol. 56, no. 10, pp. 2314–2320, 1996.

[56] M. F. McCarty, "Fish oil may impede tumour angiogenesis and invasiveness by down-regulating protein kinase C and modulating eicosanoid production," *Medical Hypotheses*, vol. 46, no. 2, pp. 107–115, 1996.

[57] J. Singh, R. Hamid, and B. S. Reddy, "Dietary fat and colon cancer: modulating effect of types and amount of dietary fat on ras-p21 function during promotion and progression stages of colon cancer," *Cancer Research*, vol. 57, no. 2, pp. 253–258, 1997.

[58] T. E. Novak, T. A. Babcock, D. H. Jho, W. S. Helton, and N. J. Espat, "NF-κB inhibition by ω-3 fatty acids modulates LPS-stimulated macrophage TNF-α-transcription," *American Journal of Physiology*, vol. 284, no. 1, pp. L84–L89, 2003.

[59] N. K. Narayanan, B. A. Narayanan, M. Bosland, M. S. Condon, and D. Nargi, "Docosahexaenoic acid in combination with celecoxib modulates HSP70 and p53 proteins in prostate cancer cells," *International Journal of Cancer*, vol. 119, no. 7, pp. 1586–1598, 2006.

[60] N. K. Narayanan, B. A. Narayanan, and B. S. Reddy, "A combination of docosahexaenoic acid and celecoxib prevents prostate cancer cell growth in vitro and is associated with modulation of nuclear factor-kappaB, and steroid hormone receptors," *International Journal of Oncology*, vol. 26, no. 3, pp. 785–792, 2005.

[61] P. K. Lala and C. Chakraborty, "Role of nitric oxide in carcinogenesis and tumour progression," *Lancet Oncology*, vol. 2, no. 3, pp. 149–156, 2001.

[62] D. A. Wink, Y. Vodovotz, J. Laval, F. Laval, M. W. Dewhirst, and J. B. Mitchell, "The multifaceted roles of nitric oxide in cancer," *Carcinogenesis*, vol. 19, no. 5, pp. 711–721, 1998.

[63] T. Ohata, K. Fukuda, M. Takahashi, T. Sugimura, and K. Wakabayashi, "Suppression of nitric oxideproduction in lipopolysaccharide-stimulated macrophage cells by omega 3 polyunsaturated fatty acids," *Japanese Journal of Cancer Research*, vol. 88, no. 3, pp. 234–237, 1997.

[64] S. C. Larsson, M. Kumlin, M. Ingelman-Sundberg, and A. Wolk, "Dietary long-chain n-3 fatty acids for the prevention of cancer: a review of potential mechanisms," *The American Journal of Clinical Nutrition*, vol. 79, no. 6, pp. 935–945, 2004.

[65] E. J. Kim, W. Y. Kim, Y. H. Kang, Y. L. Ha, L. A. Bach, and J. H. Y. Park, "Inhibition of Caco-2 cell proliferation by (n-3) fatty acids: possible mediation by increased secretion of insulin-like growth factor binding protein-6," *Nutrition Research*, vol. 20, no. 10, pp. 1409–1421, 2000.

[66] J. Singh, R. Hamid, and B. S. Reddy, "Dietary fat and colon cancer: modulation of cyclooxygenase-2 by types and amount of dietary fat during the postinitiation stage of colon carcinogenesis," *Cancer Research*, vol. 57, no. 16, pp. 3465–3470, 1997.

[67] S. L. Fu, Y. L. Wu, Y. P. Zhang, M. M. Qiao, and Y. Chen, "Anticancer effects of COX-2 inhibitors and their correlation with angiogenesis and invasion in gastric cancer," *World Journal of Gastroenterology*, vol. 10, no. 13, pp. 1971–1974, 2004.

[68] M. Takahashi, M. Fukutake, T. Isoi et al., "Suppression of azoxymethane-induced rat colon carcinoma development by a fish oil component, docosahexaenoic acid (DHA)," *Carcinogenesis*, vol. 18, no. 7, pp. 1337–1342, 1997.

[69] C. V. Rao, Y. Hirose, C. Indranie, and B. S. Reddy, "Modulation of experimental colon tumorigenesis by types and amounts of dietary fatty acids," *Cancer Research*, vol. 61, no. 5, pp. 1927–1933, 2001.

[70] N. Kobayashi, R. J. Barnard, S. M. Henning et al., "Effect of altering dietary omega-6/ omega-3 fatty acid ratios on prostate cancer membrane composition, cyclooxygenase-2, and prostaglandin E2," *Clinical Cancer Research*, vol. 12, pp. 4662–4670, 2006.

[71] Y. H. Jiang, J. R. Lupton, and R. S. Chapkin, "Dietary fish oil blocks carcinogen-induced down-regulation of colonic protein

kinase C isozymes," *Carcinogenesis*, vol. 18, no. 2, pp. 351–357, 1997.

[72] Y. Gökmen-Polar, N. R. Murray, M. A. Velasco, Z. Gatalica, and A. P. Fields, "Elevated protein kinase C βII is an early promotive event in colon carcinogenesis," *Cancer Research*, vol. 61, no. 4, pp. 1375–1381, 2001.

[73] N. R. Murray, L. A. Davidson, R. S. Chapkin, W. C. Gustafson, D. G. Schattenberg, and A. P. Fields, "Overexpression of protein kinase C β(II) induces colonic hyperproliferation and increased sensitivity to colon carcinogenesis," *Journal of Cell Biology*, vol. 145, no. 4, pp. 699–711, 1999.

[74] N. R. Murray, C. Weems, L. Chen et al., "Protein kinase C betaII and TGFbetaRII in omega-3 fatty acid mediated inhibition of colon carcinogenesis," *The Journal of Cell Biology*, vol. 157, pp. 915–920, 2002.

[75] S. Sauma, Y. Zhongfa, S. Ohno, and E. Friedman, "Protein kinase Cβ1 and protein kinase Cβ2 activate p57 mitogen-activated protein kinase and block differentiation in colon carcinoma cells," *Cell Growth and Differentiation*, vol. 7, no. 5, pp. 587–594, 1996.

[76] M. L. Saxon, X. Zhao, and J. D. Black, "Activation of protein kinase C isozymes is associated with post-mitotic events in intestinal epithelial cells in situ," *Journal of Cell Biology*, vol. 126, no. 3, pp. 747–763, 1994.

[77] L. A. Davidson, R. E. Brown, W. C. L. Chang et al., "Morphodensitometric analysis of protein kinase C β(II) expression in rat colon: modulation by diet and relation to in situ cell proliferation and apoptosis," *Carcinogenesis*, vol. 21, no. 8, pp. 1513–1519, 2000.

[78] V. Mengeaud, J. L. Nano, S. Fournel, and P. Rampal, "Effects of eicosapentaenoic acid, gamma-linolenic acid and prostaglandin E1 on three human colon carcinoma cell lines," *Prostaglandins Leukotrienes and Essential Fatty Acids*, vol. 47, no. 4, pp. 313–319, 1992.

[79] W. S. Tsai, H. Nagawa, S. Kaizaki, T. Tsuruo, and T. Muto, "Inhibitory effects of n-3 polyunsaturated fatty acids on sigmoid colon cancer transformants," *Journal of Gastroenterology*, vol. 33, no. 2, pp. 206–212, 1998.

[80] Z. Y. Chen and N. W. Istfan, "Docosahexaenoic acid is a potent inducer of apoptosis in HT-29 colon cancer cells," *Prostaglandins Leukotrienes and Essential Fatty Acids*, vol. 63, no. 5, pp. 301–308, 2000.

[81] R. G. Clarke, E. K. Lund, P. Latham, A. C. Pinder, and I. T. Johnson, "Effect of eicosapentaenoic acid on the proliferation and incidence of apoptosis in the colorectal cell line HT29," *Lipids*, vol. 34, pp. 1287–1295, 1999.

[82] P. Palozza, G. Calviello, N. Maggiano, P. Lanza, F. O. Ranelletti, and G. M. Bartoli, "Beta-carotene antagonizes the effects of eicosapentaenoic acid on cell growth and lipid peroxidation in WiDr adenocarcinoma cells," *Free Radical Biology and Medicine*, vol. 28, no. 2, pp. 228–234, 2000.

[83] M. L. Jourdan, K. Mahéo, A. Barascu et al., "Increased BRCA1 protein in mammary tumours of rats fed marine omega-3 fatty acids," *Oncology Reports*, vol. 17, no. 4, pp. 713–719, 2007.

[84] H. Gleissman, R. Yang, K. Martinod et al., "Docosahexaenoic acid metabolome in neural tumors: identification of cytotoxic intermediates," *FASEB Journal*, vol. 24, no. 3, pp. 906–915, 2010.

[85] M. Lindskog, H. Gleissman, F. Ponthan, J. Castro, P. Kogner, and J. I. Johnsen, "Neuroblastoma cell death in response to docosahexaenoic acid: sensitization to chemotherapy and arsenic-induced oxidative stress," *International Journal of Cancer*, vol. 118, no. 10, pp. 2584–2593, 2006.

[86] S. Serini, E. Piccioni, N. Merendino, and G. Calviello, "Dietary polyunsaturated fatty acids as inducers of apoptosis: implications for cancer," *Apoptosis*, vol. 14, no. 2, pp. 135–152, 2009.

[87] K. Arita, H. Kobuchi, T. Utsumi et al., "Mechanism of apoptosis in HL-60 cells induced by n-3 and n-6 polyunsaturated fatty acids," *Biochemical Pharmacology*, vol. 62, no. 7, pp. 821–828, 2001.

[88] Y. Y. Fan, T. E. Spencer, N. Wang, M. P. Moyer, and R. S. Chapkin, "Chemopreventive n-3 fatty acids activate RXRα in colonocytes," *Carcinogenesis*, vol. 24, no. 9, pp. 1541–1548, 2003.

[89] C. Chambrier, J. P. Bastard, J. Rieusset et al., "Eicosapentaenoic acid induces mRNA expression of peroxisome proliferator-activated receptor γ," *Obesity Research*, vol. 10, no. 6, pp. 518–525, 2002.

[90] H. Zand, A. Rhimipour, M. Bakhshayesh, M. Shafiee, I. Nour Mohammadi, and S. Salimi, "Involvement of PPAR-γ and p53 in DHA-induced apoptosis in Reh cells," *Molecular and Cellular Biochemistry*, vol. 304, no. 1-2, pp. 71–77, 2007.

[91] B. Sun, Y. L. Wu, S. N. Wang et al., "The effects of sulindac on induction of apoptosis and expression of cyclooxygenase-2 and Bcl-2 in human hepatocellular carcinoma cells," *Zhonghua Xiaohua Zazhi*, vol. 22, pp. 338–340, 2002.

[92] H. Gleissman, J. I. Johnsen, and P. Kogner, "Omega-3 fatty acids in cancer, the protectors of good and the killers of evil?" *Experimental Cell Research*, vol. 316, no. 8, pp. 1365–1373, 2010.

[93] G. Calviello, F. Di Nicuolo, S. Serini et al., "Docosahexaenoic acid enhances the susceptibility of human colorectal cancer cells to 5-fluorouracil," *Cancer Chemotherapy and Pharmacology*, vol. 55, no. 1, pp. 12–20, 2005.

[94] N. Danbara, T. Yuri, M. Tsujita-Kyutoku et al., "Conjugated docosahexaenoic acid is a potent inducer of cell cycle arrest and apoptosis and inhibits growth of colo 201 human colon cancer cells," *Nutrition and Cancer*, vol. 50, no. 1, pp. 71–79, 2004.

[95] L. C. M. Chiu and J. M. F. Wan, "Induction of apoptosis in HL-60 cells by eicosapentaenoic acid (EPA) is associated with downregulation of bcl-2 expression," *Cancer Letters*, vol. 145, no. 1-2, pp. 17–27, 1999.

[96] B. A. Narayanan, N. K. Narayanan, and B. S. Reddy, "Docosahexaenoic acid regulated genes and transcription factors inducing apoptosis in human colon cancer cells," *International Journal of Oncology*, vol. 19, pp. 1255–1262, 2001.

[97] T. Yamagami, C. D. Porada, R. S. Pardini, E. D. Zanjani, and G. Almeida-Porada, "Docosahexaenoic acid induces dose dependent cell death in an early undifferentiated subtype of acute myeloid leukemia cell line," *Cancer Biology and Therapy*, vol. 8, no. 4, pp. 331–337, 2009.

[98] M. Tsujii and R. N. DuBois, "Alterations in cellular adhesion and apoptosis in epithelial cells overexpressing prostaglandin endoperoxide synthase 2," *Cell*, vol. 83, no. 3, pp. 493–501, 1995.

[99] M. Y. Hong, J. R. Lupton, J. S. Morris et al., "Dietary fish oil reduces O6-methylguanine DNA adduct levels in rat colon in part by increasing apoptosis during tumor initiation," *Cancer Epidemiology Biomarkers and Prevention*, vol. 9, no. 8, pp. 819–826, 2000.

[100] J. Zha, H. Harada, E. Yang, J. Jockel, and S. J. Korsmeyer, "Serine phosphorylation of death agonist BAD in response to survival factor results in binding to 14-3-3 not BCL-X(L)," *Cell*, vol. 87, no. 4, pp. 619–628, 1996.

[101] E. Yang, J. Zha, J. Jockel, L. H. Boise, C. B. Thompson, and S. J. Korsmeyer, "Bad, a heterodimeric partner for Bcl-x(L), and Bcl-2, displaces Bax and promotes cell death," *Cell*, vol. 80, no. 2, pp. 285–291, 1995.

[102] L. Del Peso, M. González-García, C. Page, R. Herrera, and G. Nuñez, "Interleukin-3-induced phosphorylation of BAD through the protein kinase Akt," *Science*, vol. 278, no. 5338, pp. 687–689, 1997.

[103] S. R. Datta, H. Dudek, T. Xu et al., "Akt phosphorylation of BAD couples survival signals to the cell-intrinsic death machinery," *Cell*, vol. 91, no. 2, pp. 231–241, 1997.

[104] L. Hayflick, "Mortality and immortality at the cellular level. A review," *Biochemistry*, vol. 62, no. 11, pp. 1180–1190, 1997.

[105] W. E. Wright, O. M. Pereira-Smith, and J. W. Shay, "Reversible cellular senescence: implications for immortalization of normal human diploid fibroblasts," *Molecular and Cellular Biology*, vol. 9, no. 7, pp. 3088–3092, 1989.

[106] G. C. Blobe, L. M. Obeid, and Y. A. Hannun, "Regulation of protein kinase C and role in cancer biology," *Cancer and Metastasis Reviews*, vol. 13, no. 3-4, pp. 411–431, 1994.

[107] E. D. Collett, L. A. Davidson, Y. Y. Fan, J. R. Lupton, and R. S. Chapkin, "n-6 and n-3 polyunsaturated fatty acids differentially modulate oncogenic Ras activation in colonocytes," *American Journal of Physiology*, vol. 280, no. 5, pp. C1066–C1075, 2001.

[108] G. Liu, D. M. Bibus, A. M. Bode, W. Y. Ma, R. T. Holman, and Z. Dong, "Omega 3 but not omega 6 fatty acids inhibit AP-1 activity and cell transformation in JB6 cells," *Proceedings of the National Academy of Sciences of the United States of America*, vol. 98, no. 13, pp. 7510–7515, 2001.

[109] F. Cannizzo Jr. and S. A. Broitman, "Postpromotional effects of dietary marine or safflower oils on large bowel or pulmonary implants of CT-26 in mice," *Cancer Research*, vol. 49, no. 15, pp. 4289–4294, 1989.

[110] J. E. Paulsen, I. K. Elvsaas, I. L. Steffensen, and J. Alexander, "A fish oil derived concentrate enriched in eicosapentaenoic and docosahexaenoic acid as ethyl ester suppresses the formation and growth of intestinal polyps in the Min mouse," *Carcinogenesis*, vol. 18, no. 10, pp. 1905–1910, 1997.

[111] M. D. Boudreau, K. H. Sohn, S. H. Rhee, S. W. Lee, J. D. Hunt, and D. H. Hwang, "Suppression of tumor cell growth both in nude mice and in culture by n-3 polyunsaturated fatty acids: mediation through cyclooxygenase-independent pathways," *Cancer Research*, vol. 61, no. 4, pp. 1386–1391, 2001.

[112] G. Calviello, F. Di Nicuolo, S. Gragnoli et al., "n-3 PUFAs reduce VEGF expression in human colon cancer cells modulating the COX-2/PGE2 induced ERK-1 and -2 and HIF-1α induction pathway," *Carcinogenesis*, vol. 25, no. 12, pp. 2303–2310, 2004.

[113] B. S. Reddy and H. Maruyama, "Effect of dietary fish oil on azoxymethane-induced colon carcinogenesis in male F344 rats," *Cancer Research*, vol. 46, no. 7, pp. 3367–3370, 1986.

[114] T. Minoura, T. Takata, M. Sakaguchi et al., "Effect of dietary eicosapentaenoic acid on azoxymethane-induced colon carcinogenesis in rats," *Cancer Research*, vol. 48, no. 17, pp. 4790–4794, 1988.

[115] C. W. Hendrickse, M. R. B. Keighley, and J. P. Neoptolemos, "Dietary ω-3 fats reduce proliferation and tumor yields at colorectal anastomosis in rats," *Gastroenterology*, vol. 109, no. 2, pp. 431–439, 1995.

[116] L. A. Davidson, J. L. Lupton, Y. H. Jiang, and R. S. Chapkin, "Carcinogen and dietary lipid regulate ras expression and localization in rat colon without affecting farnesylation kinetics," *Carcinogenesis*, vol. 20, no. 5, pp. 785–791, 1999.

[117] Y. E. M. Dommels, G. M. Alink, P. J. Van Bladeren, and B. Van Ommen, "Dietary n-6 and n-3 polyunsaturated fatty acids and colorectal carcinogenesis: results from cultured colon cells, animal models and human studies," *Environmental Toxicology and Pharmacology*, vol. 11, pp. 297–308, 2002.

[118] B. S. Reddy, C. Burill, and J. Rigotty, "Effect of diets high in ω-3 and ω-6 fatty acids on initiation and postinitiation stages of colon carcinogenesis," *Cancer Research*, vol. 51, no. 2, pp. 487–491, 1991.

[119] H. Gleissman, L. Segerström, M. Hamberg et al., "Omega-3 fatty acid supplementation delays the progression of neuroblastoma in vivo," *International Journal of Cancer*, vol. 128, no. 7, pp. 1703–1711, 2011.

[120] M. V. Swamy, B. Citineni, J. M. R. Patlolla, A. Mohammed, Y. Zhang, and C. V. Rao, "Prevention and treatment of pancreatic cancer by curcumin in combination with omega-3 fatty acids," *Nutrition and Cancer*, vol. 60, no. 1, pp. 81–89, 2008.

[121] D. Hanahan and J. Folkman, "Patterns and emerging mechanisms of the angiogenic switch during tumorigenesis," *Cell*, vol. 86, no. 3, pp. 353–364, 1996.

[122] L. Spencer, C. Mann, M. Metcalfe et al., "The effect of omega-3 FAs on tumour angiogenesis and their therapeutic potential," *European Journal of Cancer*, vol. 45, no. 12, pp. 2077–2086, 2009.

[123] T. Tsuzuki, A. Shibata, Y. Kawakami, K. Nakagawa, and T. Miyazawa, "Conjugated eicosapentaenoic acid inhibits vascular endothelial growth factor-induced angiogenesis by suppressing the migration of human umbilical vein endothelial cells," *Journal of Nutrition*, vol. 137, no. 3, pp. 641–646, 2007.

[124] M. Tsuji, S. I. Murota, and I. Morita, "Docosapentaenoic acid (22:5, n-3) suppressed tube-forming activity in endothelial cells induced by vascular endothelial growth factor," *Prostaglandins Leukotrienes and Essential Fatty Acids*, vol. 68, no. 5, pp. 337–342, 2003.

[125] S. P. Yang, I. Morita, and S. I. Murota, "Eicosapentaenoic acid attenuates vascular endothelial growth factor-induced proliferation via inhibiting flk-1 receptor expression in bovine carotid artery endothelial cells," *Journal of Cellular Physiology*, vol. 176, pp. 342–349, 1998.

[126] L. Yuan, M. Yoshida, and P. F. Davis, "Inhibition of pro-angiogenic factors by a lipid-rich shark extract," *Journal of Medicinal Food*, vol. 9, no. 3, pp. 300–306, 2006.

[127] R. Ross, J. Glomset, B. Kariya, and L. Harker, "A platelet dependent serum factor that stimulates the proliferation of arterial smooth muscle cells in vitro," *Proceedings of the National Academy of Sciences of the United States of America*, vol. 71, no. 4, pp. 1207–1210, 1974.

[128] C. H. Heldin and B. Westermark, "Mechanism of action and in vivo role of platelet-derived growth factor," *Physiological Reviews*, vol. 79, no. 4, pp. 1283–1316, 1999.

[129] P. L. Fox and P. E. DiCorleto, "Fish oils inhibit endothelial cell production of platelet-derived growth factor-like protein," *Science*, vol. 241, no. 4864, pp. 453–456, 1988.

[130] T. Terano, T. Shiina, and Y. Tamura, "Eicosapentaenoic acid suppressed the proliferation of vascular smooth muscle cells through modulation of various steps of growth signals," *Lipids*, vol. 31, no. 3, pp. S301–S304, 1996.

[131] R. Stein Werblowsky, "Prostaglandin and cancer," *Oncology*, vol. 30, no. 2, pp. 169–176, 1974.

[132] B. S. Reddy, J. M. Patlolla, B. Simi, S. H. Wang, and C. V. Rao, "Prevention of colon cancer by low doses of celecoxib, a cyclooxygenase inhibitor, administered in diet rich in ω-3 polyunsaturated fatty acids," *Cancer Research*, vol. 65, no. 17, pp. 8022–8027, 2005.

[133] S. Dimmeler, C. Hermann, J. Galle, and A. M. Zeiher, "Upregulation of superoxide dismutase and nitric oxide synthase mediates the apoptosis-suppressive effects of shear stress on endothelial cells," *Arteriosclerosis, Thrombosis, and Vascular Biology*, vol. 19, no. 3, pp. 656–664, 1999.

[134] L. Rössig, B. Fichtlscherer, K. Breitschopf et al., "Nitric oxide inhibits caspase-3 by S-nitrosation in vivo," *Journal of Biological Chemistry*, vol. 274, no. 11, pp. 6823–6826, 1999.

[135] F. Cianchi, C. Cortesini, P. Bechi et al., "Up-regulation of cyclooxygenase 2 gene expression correlates with tumor angiogenesis in human colorectal cancer," *Gastroenterology*, vol. 121, no. 6, pp. 1339–1347, 2001.

[136] F. Cianchi, C. Cortesini, O. Fantappiè et al., "Cyclooxygenase-2 activation mediates the proangiogenic effect of nitric oxide in colorectal cancer," *Clinical Cancer Research*, vol. 10, no. 8, pp. 2694–2704, 2004.

[137] M. Ziche, L. Morbidelli, R. Choudhuri et al., "Nitric oxide synthase mediates vascular endothelial growth factor but not basic fibroblast growth factor induced angiogenesis," *FASEB Journal*, vol. 11, no. 3, p. A196, 1997.

[138] W. Komatsu, K. Ishihara, M. Murata, H. Saito, and K. Shinohara, "Docosahexaenoic acid suppresses nitric oxide production and inducible nitric oxide synthase expression in interferon-γ plus lipopolysaccharide-stimulated murine macrophages by inhibiting the oxidative stress," *Free Radical Biology and Medicine*, vol. 34, no. 8, pp. 1006–1016, 2003.

[139] D. R. Jeyarajah, M. Kielar, J. Penfield, and C. Y. Lu, "Docosahexaenoic acid, a component of fish oil, inhibits nitric oxide production in vitro," *Journal of Surgical Research*, vol. 83, no. 2, pp. 147–150, 1999.

[140] V. Boutard, B. Fouqueray, C. Philippe, J. Perez, and L. Baud, "Fish oil supplementation and essential fatty acid deficiency reduce nitric oxide synthesis by rat macrophages," *Kidney International*, vol. 46, no. 5, pp. 1280–1286, 1994.

[141] B. A. Narayanan, N. K. Narayanan, B. Simi, and B. S. Reddy, "Modulation of inducible nitric oxide synthase and related proinflammatory genes by the omega-3 fatty acid docosahexaenoic acid in human colon cancer cells," *Cancer Research*, vol. 63, no. 5, pp. 972–979, 2003.

[142] B. A. Narayanan, N. K. Narayanan, D. Desai, B. Pittman, and B. S. Reddy, "Effects of a combination of docosahexaenoic acid and 1,4-phenylene bis(methylene) selenocyanate on cyclooxygenase 2, inducible nitric oxide synthase and β-catenin pathways in colon cancer cells," *Carcinogenesis*, vol. 25, no. 12, pp. 2443–2449, 2004.

[143] R. Tevar, D. H. Jho, T. Babcock, W. S. Helton, and N. J. Espat, "ω-3 fatty acid supplementation reduces tumor growth and vascular endothelial growth factor expression in a model of progressive non-metastasizing malignancy," *Journal of Parenteral and Enteral Nutrition*, vol. 26, no. 5, pp. 285–289, 2002.

[144] D. P. Rose and J. M. Connolly, "Antiangiogenicity of docosahexaenoic acid and its role in the suppression of breast cancer cell growth in nude mice," *International Journal of Oncology*, vol. 15, no. 5, pp. 1011–1015, 1999.

[145] M. Mukutmoni-Norris, N. E. Hubbard, and K. L. Erickson, "Modulation of murine mammary tumor vasculature by dietary n-3 fatty acids in fish oil," *Cancer Letters*, vol. 150, no. 1, pp. 101–109, 2000.

[146] A. Bommareddy, B. L. Arasada, D. P. Mathees, and C. Dwivedi, "Chemopreventive effects of dietary flaxseed on colon tumor development," *Nutrition and Cancer*, vol. 54, no. 2, pp. 216–222, 2006.

[147] M. B. Sporn, "The war on cancer," *Lancet*, vol. 347, pp. 1377–1381, 1996.

[148] A. E. Aplin, A. Howe, S. K. Alahari, and R. L. Juliano, "Signal transduction and signal modulation by cell adhesion receptors: the role of integrins, cadherins, immunoglobulin-cell adhesion molecules, and selectins," *Pharmacological Reviews*, vol. 50, no. 2, pp. 197–263, 1998.

[149] G. Christofori and H. Semb, "The role of the cell-adhesion molecule E-cadherin as a tumour-suppressor gene," *Trends in Biochemical Sciences*, vol. 24, no. 2, pp. 73–76, 1999.

[150] M. B. Schaefer, A. Wenzel, T. Fischer et al., "Fatty acids differentially influence phosphatidylinositol 3-kinase signal transduction in endothelial cells: impact on adhesion and apoptosis," *Atherosclerosis*, vol. 197, no. 2, pp. 630–637, 2008.

[151] S. Ambs, S. P. Hussain, and C. C. Harris, "Interactive effects of nitric oxide and the p53 tumor suppressor gene in carcinogenesis and tumor progression," *FASEB Journal*, vol. 11, no. 6, pp. 443–448, 1997.

[152] L. C. Jadeski, C. Chakraborty, and P. K. Lala, "Role of nitric oxide in tumour progression with special reference to a murine breast cancer model," *Canadian Journal of Physiology and Pharmacology*, vol. 80, no. 2, pp. 125–135, 2002.

[153] J. P. Cooke and D. W. Losordo, "Nitric oxide and angiogenesis," *Circulation*, vol. 105, no. 18, pp. 2133–2135, 2002.

[154] L. Yi, Q. Y. Zhang, and M. T. Mi, "Role of Rho GTPase in inhibiting metastatic ability of human prostate cancer cell line PC-3 by omega-3 polyunsaturated fatty acid," *Chinese Journal of Cancer*, vol. 26, no. 12, pp. 1281–1286, 2007.

[155] M. Goua, S. Mulgrew, J. Frank, D. Rees, A. A. Sneddon, and K. W. J. Wahle, "Regulation of adhesion molecule expression in human endothelial and smooth muscle cells by omega-3 fatty acids and conjugated linoleic acids: involvement of the transcription factor NF-κB?" *Prostaglandins Leukotrienes and Essential Fatty Acids*, vol. 78, no. 1, pp. 33–43, 2008.

[156] R. Ohno, K. Yoshinaga, T. Fujita et al., "Depth of invasion parallels increased cyclooxygenase-2 levels in patients with gastric carcinoma," *Cancer*, vol. 91, pp. 1876–1881, 2001.

[157] M. Joo, H. K. Lee, and Y. K. Kang, "Expression of E-cadherin, β-catenin, CD44s and CD44v6 in gastric adenocarcinoma: relationship with lymph node metastasis," *Anticancer Research B*, vol. 23, no. 2, pp. 1581–1588, 2003.

[158] I. M. Berquin, I. J. Edwards, and Y. Q. Chen, "Multi-targeted therapy of cancer by omega-3 fatty acids," *Cancer Letters*, vol. 269, no. 2, pp. 363–377, 2008.

[159] S. Iwamoto, H. Senzaki, Y. Kiyozuka et al., "Effects of fatty acids on liver metastasis of ACL-15 rat colon cancer cells," *Nutrition and Cancer*, vol. 31, no. 2, pp. 143–150, 1998.

[160] M. Kontogiannea, A. Gupta, F. Ntanios, T. Graham, P. Jones, and S. Meterissian, "ω-3 fatty acids decrease endothelial adhesion of human colorectal carcinoma cells," *Journal of Surgical Research*, vol. 92, no. 2, pp. 201–205, 2000.

[161] D. P. Rose, J. M. Connolly, and M. Coleman, "Effect of omega-3 fatty acids on the progression of metastases after the surgical excision of human breast cancer cell solid tumors growing in nude mice," *Clinical Cancer Research*, vol. 2, no. 10, pp. 1751–1756, 1996.

[162] M. Lucas, E. Dewailly, G. Muckle et al., "Gestational age and birth weight in relation to n-3 fatty acids among inuit (Canada)," *Lipids*, vol. 39, no. 7, pp. 617–626, 2004.

[163] A. P. Lanier, P. Holck, G. Ehrsam Day, and C. Key, "Childhood cancer among Alaska Natives," *Pediatrics*, vol. 112, no. 5, article e396, 2003.

[164] C. H. MacLean, S. J. Newberry, W. A. Mojica et al., "Effects of omega-3 fatty acids on cancer risk: a systematic review," *Journal of the American Medical Association*, vol. 295, no. 4, pp. 403–415, 2006.

[165] Y. Q. Chen, I. M. Berquin, L. W. Daniel et al., "Omega-3 fatty acids and cancer risk," *Journal of the American Medical Association*, vol. 296, no. 3, pp. 278–282, 2006.

[166] D. J. Hunter, E. B. Rimm, F. M. Sacks et al., "Comparison of measures of fatty acid intake by subcutaneous fat aspirate, food frequency questionnaire, and diet records in a free-living population of US men," *American Journal of Epidemiology*, vol. 135, no. 4, pp. 418–427, 1992.

[167] J. M. Ward and D. E. Devor-Henneman, "Mouse models of human familial cancer syndromes," *Toxicologic Pathology*, vol. 32, supplement 1, pp. 90–98, 2004.

[168] K. Maddison and A. R. Clarke, "New approaches for modelling cancer mechanisms in the mouse," *Journal of Pathology*, vol. 205, no. 2, pp. 181–193, 2005.

[169] P. L. Zock and M. B. Katan, "Linoleic acid intake and cancer risk: a review and meta-analysis," *American Journal of Clinical Nutrition*, vol. 68, no. 1, pp. 142–153, 1998.

[170] C. H. Maclean, S. J. Newberry, W. A. Mojica et al., "Effects of omega-3 fatty acids on cancer," *Evidence Report*, no. 113, pp. 1–4, 2005.

[171] R. S. Pardini, "Nutritional intervention with omega-3 fatty acids enhances tumor response to anti-neoplastic agents," *Chemico-Biological Interactions*, vol. 162, no. 2, pp. 89–105, 2006.

[172] W. E. Hardman, C. P. R. Avula, G. Fernandes, and I. L. Cameron, "Three percent dietary fish oil concentrate increased efficacy of doxorubicin against MDA-MB 231 breast cancer xenografts," *Clinical Cancer Research*, vol. 7, no. 7, pp. 2041–2049, 2001.

[173] Y. Shao, L. Pardini, and R. S. Pardini, "Dietary menhaden oil enhances mitomycin C antitumor activity toward human mammary carcinoma MX-1," *Lipids*, vol. 30, no. 11, pp. 1035–1045, 1995.

[174] J. Chen, E. Hui, T. Ip, and L. U. Thompson, "Dietary flaxseed enhances the inhibitory effect of tamoxifen on the growth of estrogen-dependent human breast cancer (MCF-7) in nude mice," *Clinical Cancer Research*, vol. 10, no. 22, pp. 7703–7711, 2004.

[175] P. Bougnoux, N. Hajjaji, M. N. Ferrasson, B. Giraudeau, C. Couet, and O. Le Floch, "Improving outcome of chemotherapy of metastatic breast cancer by docosahexaenoic acid: a phase II trial," *British Journal of Cancer*, vol. 101, no. 12, pp. 1978–1985, 2009.

[176] A. Rusca, A. F. D. Di Stefano, M. V. Doig, C. Scarsi, and E. Perucca, "Relative bioavailability and pharmacokinetics of two oral formulations of docosahexaenoic acid/eicosapentaenoic acid after multiple-dose administration in healthy volunteers," *European Journal of Clinical Pharmacology*, vol. 65, no. 5, pp. 503–510, 2009.

[177] C. E. Childs, M. Romeu-Nadal, G. C. Burdge, and P. C. Calder, "Gender differences in the n-3 fatty acid content of tissues," *Proceedings of the Nutrition Society*, vol. 67, no. 1, pp. 19–27, 2008.

[178] L. M. Arterburn, E. B. Hall, and H. Oken, "Distribution, inter-conversion, and dose response of n-3 fatty acids in humans," *American Journal of Clinical Nutrition*, vol. 83, supplement 6, pp. 1467S–1476S, 2006.

[179] R. A. Murphy, M. Mourtzakis, and V. C. Mazurak, "n-3 polyunsaturated fatty acids: the potential role for supplementation in cancer," *Current Opinion in Clinical Nutrition & Metabolic Care*, vol. 15, no. 3, pp. 246–251, 2012.

8

The Impairment of Macrophage-to-Feces Reverse Cholesterol Transport during Inflammation Does Not Depend on Serum Amyloid A

Maria C. de Beer,[1,2] Joanne M. Wroblewski,[1,3] Victoria P. Noffsinger,[1,3] Ailing Ji,[1,3] Jason M. Meyer,[1,4] Deneys R. van der Westhuyzen,[1,3,4,5] Frederick C. de Beer,[1,3] and Nancy R. Webb[1,3]

[1] Saha Cardiovascular Research Center, University of Kentucky Medical Center, Lexington, KY 40536, USA
[2] Department of Physiology, University of Kentucky Medical Center, Lexington, KY 40536, USA
[3] Department of Internal Medicine, University of Kentucky Medical Center, Lexington, KY 40536, USA
[4] Department of Molecular and Cellular Biochemistry, University of Kentucky Medical Center, Lexington, KY 40536, USA
[5] Department of Veterans Affairs Medical Center, Lexington, KY 40511, USA

Correspondence should be addressed to Maria C. de Beer; mdebeer@uky.edu

Academic Editor: Akihiro Inazu

Studies suggest that inflammation impairs reverse cholesterol transport (RCT). We investigated whether serum amyloid A (SAA) contributes to this impairment using an established macrophage-to-feces RCT model. Wild-type (WT) mice and mice deficient in SAA1.1 and SAA2.1 (SAAKO) were injected intraperitoneally with ^{3}H-cholesterol-labeled J774 macrophages 4 hr after administration of LPS or buffered saline. ^{3}H-cholesterol in plasma 4 hr after macrophage injection was significantly reduced in both WT and SAAKO mice injected with LPS, but this was not associated with a reduced capacity of serum from LPS-injected mice to promote macrophage cholesterol efflux in vitro. Hepatic accumulation of ^{3}H-cholesterol was unaltered in either WT or SAAKO mice by LPS treatment. Radioactivity present in bile and feces of LPS-injected WT mice 24 hr after macrophage injection was reduced by 36% ($P < 0.05$) and 80% ($P < 0.001$), respectively. In contrast, in SAAKO mice, LPS did not significantly reduce macrophage-derived ^{3}H-cholesterol in bile, and fecal excretion was reduced by only 45% ($P < 0.05$). Injection of cholesterol-loaded allogeneic J774 cells, but not syngeneic bone-marrow-derived macrophages, transiently induced SAA in C57BL/6 mice. Our study confirms reports that acute inflammation impairs steps in the RCT pathway and establishes that SAA plays only a minor role in this impairment.

1. Introduction

Epidemiological studies have identified a strong inverse relationship between the risk of cardiovascular disease and plasma HDL levels [1]. One of the primary antiatherogenic properties of HDL is thought to be its role in reverse cholesterol transport (RCT), a process whereby excess cholesterol is removed from peripheral tissues, including lipid-loaded macrophages in the vessel wall, and transported to the liver for excretion in the bile and feces [2]. Accelerated atherosclerosis has been associated with inflammatory diseases such as rheumatoid arthritis [3], and inflammatory biomarkers are increasingly used as predictors of cardiovascular disease progression [4, 5]. Inflammation gives rise to numerous metabolic and structural changes in lipoproteins, particularly HDL, which may impact the ability of HDL to mediate RCT [6, 7]. Inflammatory HDL undergoes structural and compositional changes, most notably an increase in its serum amyloid A (SAA) content, to the extent that SAA can become the major apolipoprotein of HDL [8]. SAA is a major hepatic acute phase reactant and can account for as much as 2.5% of the protein produced in the liver during severe inflammation [9]. During the acute phase, plasma SAA concentrations rise rapidly with peak concentrations exceeding 1 mg/mL.

Approximately 95% of plasma SAA is associated with HDL [8].

Two major acute phase SAA isoforms are expressed in mice, SAA1.1 and SAA2.1, which are synthesized in the liver upon inflammatory cytokine stimulation [10]. Other members of the murine SAA family include the constitutively expressed SAA4 [11], which is a minor HDL protein, and SAA3, which is primarily expressed extrahepatically, but does not comprise a major protein of acute phase HDL [12]. Studies on the impact of SAA on HDL metabolism during inflammation are confounded by the fact that the inflammatory mediators that induce SAA also impact numerous other metabolic systems, so that ascribing specific functions to SAA becomes challenging. To delineate the specific impact of acute phase SAAs on HDL metabolism and function, we recently generated gene-targeted mice lacking both SAA1.1 and SAA2.1 (SAAKO mice) [13].

According to previous reports, macrophage RCT is impaired in mice during an acute phase inflammatory response induced by LPS [14, 15] or zymosan [16]. In this study, we determined the extent to which acute phase SAA1.1 and SAA2.1 contribute to this impairment by quantifying *in vivo* macrophage RCT in wild-type (WT) and SAAKO mice in the presence or absence of LPS-elicited endotoxemia. Our results support the conclusion that inflammation impairs macrophage-to-feces RCT and that acute phase SAA has little impact on this impairment. We also report that administration of cholesterol-loaded J774 macrophages, a commonly used procedure in macrophage-to-feces RCT studies, independently induces a transient acute phase response in mice.

2. Materials and Methods

2.1. Animals. C57BL/6 mice were obtained from Jackson Laboratories. Mice lacking SAA1.1 and SAA2.1 were generated by targeted deletion of both mouse acute phase SAA genes *SAA1* and *SAA2* (InGenious Targeting Laboratory Inc., Stony Brook, NY) using embryonic stem cells derived from C57BL/6 ×129 SVEV mice as described previously [13]. Mice were maintained in a pathogen-free facility under equal light-dark cycles with free access to water and food. All procedures were carried out in accordance with PHS policy and approved by the Lexington Kentucky Veterans Affairs Medical Center Institutional Animal Care and Use Committee (Assurance number A3506-01).

2.2. Cell Culture

2.2.1. J774 Cells. J774 macrophages were kindly provided by Dr. G.H. Rothblat (University of Pennsylvania) and maintained in RPMI-1640 supplemented with 10% heat-inactivated FBS and 50 μg/mL gentamicin. For *in vivo* RCT experiments, cells were grown in suspension in HEPES-buffered RPMI-1640 containing 10% heat-inactivated FBS and 50 μg/mL gentamicin. The cells were cholesterol loaded for 24 hr in HEPES-buffered RPMI-1640 containing 1% FBS, 25 μg/mL acetylated LDL (acLDL), and 5 μCi/mL ^{3}H-cholesterol (35–50 Ci/mmol, Amersham Biosciences). On the day of injection, cells were washed, equilibrated for 2 hr

in HEPES-buffered RPMI-1640 containing 0.2% fatty acid free BSA, centrifuged, and resuspended in minimal essential media (MEM) prior to injection into mice (2.5×10^6 cells in 0.5 mL). For some studies, J774 cells were prepared for intraperitoneal injection into mice as described above, except that ^{3}H-cholesterol was omitted from the media during the cholesterol loading step.

2.2.2. Bone-Marrow-Derived Macrophages (BMMs). BMMs were harvested and cultured as described [17]. Briefly, BMMs were harvested from the tibias and femurs of C57BL/6 mice 5 days after intraperitoneal injection of 1 mL 2% biogel beads [18] by flushing the bone cavities with PBS. Bone-marrow was homogenized by drawing through an 18G needle and was maintained in sterile nontissue culture treated flasks using RPMI-1640 supplemented with 1% P/S, 10% FBS, and 15% L-cell-conditioned medium. The medium was changed on days 3 and 5, and on day 6 BMM were treated for 24 hr with RPMI 1640 containing 1% FBS, 1% P/S, and 25 μg/mL acLDL. On day 7 cells were equilibrated for 2 hr in RPMI-1640 containing 0.2% fatty acid-free BSA, dislodged from the tissue culture flask with nonenzymatic cell dissociation buffer (Sigma), centrifuged, and suspended in MEM prior to intraperitoneal injection of 2.5×10^6 cells/0.5 mL.

2.2.3. Rat Fu5AH Hepatoma Cells. Fu5AH cells, kindly provided by Dr. G.H. Rothblat (University of Pennsylvania), were maintained in MEM supplemented with 5% FBS and 50 μg/mL gentamicin.

2.3. Semiquantitative Real-Time PCR (RT-PCR). RT-PCR was performed as described previously [13]. Briefly, total RNA was isolated from mouse liver using the TRIzol Reagent (Molecular Research Center), and 10 μg was treated with TURBO DNA-Free DNAse (Ambion) according to the manufacturer's protocol. cDNA synthesis was performed using 500 ng RNA and the Applied Biosystems High Capacity cDNA kit. Amplification was carried out using the iCycler IQ5 system (Bio-Rad). Quantification was performed in duplicate using the standard curve method and normalized to GAPDH. The primers used are as follows: **mSRBI**, NM_016741: forward 5$'$-CTTCATGACACCCGAATCCT-3$'$, Reverse 5$'$-AATGCCTTCAAACACCCTTG-3$'$, 114bp; **mCyp7A1**, NM_007824 forward 5$'$-AGCAACTAAACA-ACCTGCCAGTACTA-3$'$, reverse 5$'$-GTCCGGATATTC-AAGGATGCA-3$'$, 84bp; **mABCA1**, NM_013454: forward 5$'$-AGCCAGAAGGGAGTGTCAGA-3$'$, reverse 5$'$-CAT-GCCATCTCGGTAAACCT-3$'$ 102bp; **mABCG1**, NM_009593: forward 5$'$-AGGCCTACTACCTGGCAAAGA-3$'$, reverse 5$'$-GCAGTAGGCCACAGGGAACA-3$'$, 68bp; **mABCG5**, NM_031884: forward 5$'$-TGGATCCAACAC-CTCTATGCTAAA-3$'$, reverse 5$'$-GGCAGGTTTTCTCGA-TGAACTG-3$'$ 77bp; **mABCG8**, NM_026180: forward 5$'$-TGCCCACCTTCCACATGTC-3$'$, reverse 5$'$-ATGAAG-CCGGCAGTAAGGTAGA-3$'$ 73bp; **mABCB11**, NM_021022: forward 5$'$-AAGCTACATCTGCCTTAG-

The Impairment of Macrophage-to-Feces Reverse Cholesterol Transport during Inflammation Does
Not Depend on Serum Amyloid A

67

ACACAGAAA-3', reverse 5'-CAATACAGGTCCGAC-CCTCTCT-3', 84bp.

2.4. In Vivo Macrophage RCT.

RCT studies were carried out in accordance with a well-established model [14, 19]. Three separate experiments were performed in ~16-week-old female SAAKO and WT mice (n = 4-5 per group per experiment) that were housed in individual cages and fed a normal rodent diet *ad libitum*. Anesthetized mice received a subcutaneous injection of 0.8 μg/g body weight lipopolysaccharide (LPS; from *E. coli* 0111:B4, Sigma L 2630) in ~100 μL PBS or an injection of PBS alone. After 4 hr, mice were administered intraperitoneally 2.5×10^6 radiolabeled J774 macrophages ($2.78–3.87 \times 10^6$ dpm/mouse) prepared as described above. Blood was collected via retroorbital bleed 4 hr after LPS administration and just prior to injection of macrophages and then again 4 hr after macrophage administration. Feces were collected throughout the course of the study. Mice were euthanized 24 hr after macrophage administration, and blood, bile, and liver were collected. To study whether intraperitoneal injection of macrophages induces an acute phase response in mice, BMM and J774 cells were prepared as described above, except that ³H-cholesterol was omitted from the media during the cholesterol loading step.

2.5. Ex Vivo Macrophage Cholesterol Efflux Experiments.

For efflux assays, plasma was obtained from WT and SAAKO mice 8 hr after subcutaneous injection of PBS or LPS (0.8 μg/g). Mouse serum was prepared as described [20]. Cellular cholesterol efflux experiments were carried out essentially as described [21, 22]. Briefly, J774 cells (~60% confluent) in 12-well plates were labeled for 48 hr with 0.2 μCi/mL [³H]cholesterol (35–50 Ci/mmol, Amersham Biosciences) in RPMI-1640 supplemented with 10% heat-inactivated FBS, 50 μg/mL gentamicin, and 50 μg/mL acLDL. Cells were then washed three times with PBS containing 1 mg/mL BSA (PBS-BSA) and equilibrated overnight in RPMI-1640 containing 0.2% fatty acid-free BSA (RPMI-BSA). Following two additional washes with PBS-BSA, cells were harvested to determine total dpm incorporated (time zero). Efflux to RPMI-BSA with or without 2.5% serum from control or LPS-treated mice was measured over 5 hr at 37°C. Following the incubation, cell media was collected and centrifuged to remove detached cells. Radioactivity in the medium was measured directly in a Packard β liquid scintillation counter. Adherent cells were washed at 4°C twice with PBS-BSA and twice with PBS and then solubilized in 0.1 N NaOH and counted for radioactivity. Efflux of cellular [³H]-cholesterol was expressed as the percentage of the total radioactivity incorporated into cells at time zero that was present in the media.

2.6. Cholesterol Influx into Hepatoma Cells.

Fu5AH hepatoma cells, maintained as described above, were seeded into 12-well plates. When cultures were more than 90% confluent, cells were pretreated for 1 hr with 10 μM block lipid transport-1 (BLT-1) in MEM containing 0.5% FAF-BSA, to inhibit SR-BI. Control cells were treated with the DMSO vehicle. Cells were subsequently incubated, in the presence or absence of BLT-1, in MEM alpha supplemented with 5% serum obtained from mice at the termination of RCT experiments (i.e., 24 hr after administration of [³H]-cholesterol-labeled macrophages) to monitor the uptake of [³H]-cholesterol which includes both free and esterified [³H]-cholesterol. After 6 hr, cells were washed at 4°C, 4 times with PBS containing 0.2% fatty acid-free BSA and twice with PBS, solubilized in 0.1 N NaOH, and counted for radioactivity.

2.7. Western Blot Analyses.

Total liver lysates and liver membrane protein extracts were prepared for western blot analyses essentially as described [23]. Briefly ~100 mg liver was homogenized with an Ultra Turrax T8 probe homogenizer in 1.2 mL 20 mM Tris pH 7.5, 2 mM MgCl₂, and 0.25 M sucrose containing protease inhibitors. After centrifugation at 12000 ×g for 10 min at 4°C, the supernatant (i.e., total liver lysate) was used for western blot analyses. Alternatively, membrane proteins were isolated by centrifuging the total liver lysate at 100,000 ×g for 30 min at 4°C; the membrane protein pellet was resuspended in the homogenization buffer described above. Protein concentrations of liver lysates and membrane protein suspensions were determined by BCA assay (Pierce). Liver proteins were size-fractionated by SDS-PAGE and immunoblotted with the following antibodies: anti-human/mouse SR-BI (1 : 1000) [24], anti-mouse/human ABCG1 (1 : 500; Novus NB400-132), anti-ABCG5 (1 : 10,000; a generous gift from Dr. Gregory Graf, University of Kentucky). For loading controls liver lysates/membranes were immunoblotted with anti-β actin (1 : 2000; Sigma, A5441) or anti-mouse calnexin (1 : 2000; Enzo Life Sciences). For western blot analyses of plasma proteins, aliquots of plasma (0.3–0.5 μL) were subjected to SDS-PAGE and immunoblotted using rabbit anti-mouse SAA (De Beer laboratory) or rabbit anti-mouse SAP (gift from Dr. Mark Pepys, University College of London, UK). Plasma SAA concentrations were determined by quantitative immunoblotting using acute phase mouse HDL with known SAA content as a standard.

2.8. Endotoxin Assay.

The Limulus Amebocyte Lysate kit (Genscript cat. no. L00351) was used to verify that plasma/serum samples used for *ex vivo* efflux or influx assays contained less than 0.005 EU/mL of endotoxin.

2.9. Statistical Analyses.

Data are expressed as the mean ± SEM. Results were analyzed by two-way ANOVA with Bonferroni posttest. Significance was defined as *: $P < 0.05$; **: $P < 0.01$; ***: $P < 0.001$.

3. Results

3.1. Mice Deficient in SAA1.1 and SAA2.1 Mount a Normal LPS-Induced Acute Phase Response.

According to previous reports, macrophage-to-feces RCT is reduced in mice during an LPS-elicited acute phase response [14, 15]. To investigate whether SAA contributes to this impairment, we carried out studies using our recently developed gene-targeted mice that lack both major acute phase SAA isoforms, SAA1.1 and SAA2.1 [13]. In 3 separate studies, the movement of

[3]H-cholesterol from macrophages to feces was monitored in WT and SAAKO mice under both normal and inflammatory conditions using an established RCT model [14, 19] (n = 4-5 per group per experiment). An acute phase response was evoked in mice by injecting a relatively modest dose of LPS subcutaneously (0.8 μg/g body weight). As expected, plasma SAA was readily detected in WT but not SAAKO mice 28 hr after LPS injection (Figure 1(a)), reaching values of ~130 μg/mL in the WT mice. The ability of SAAKO mice to mount an acute phase response was demonstrated by the marked induction of serum amyloid P component (SAP), which was present in the plasma at comparable levels in WT and SAAKO mice 28 hr after LPS injection (Figure 1(b)).

3.2. SAA Is Not Required for the Reduction in Macrophage to Plasma RCT during Endotoxemia. For the assessment of *in vivo* macrophage RCT, mice were injected intraperitoneally with ^{3}H-cholesterol-labeled macrophages 4 hr after administration of LPS or PBS control. The initial step of RCT (i.e., the movement of ^{3}H-cholesterol from macrophages to the plasma compartment) was monitored by quantifying plasma radioactivity 4 hr and 24 hr after macrophage injection (8 hr and 28 hr after LPS injection). In agreement with previous studies [14, 15], the amount of macrophage-derived ^{3}H-cholesterol present in plasma was lower in WT mice undergoing an acute phase response compared to control WT mice, an effect that was statistically significant at the earlier time point (Figure 2(a)). Similarly, there was significantly decreased radioactivity in plasma of LPS-injected SAAKO mice compared to control SAAKO mice when measured at 4 hr (Figure 2(a)). At 24 hr, the amount of radioactive tracer in plasma was similar for all four groups of mice (Figure 2(b)). To more directly assess whether SAA impacts macrophage cholesterol efflux, we carried out *in vitro* cholesterol efflux assays using cholesterol-loaded J774 cells and serum collected from WT and SAAKO mice 8 hr after LPS or saline injection (Figure 2(c)). Results from these assays indicated that the reduced amount of macrophage-derived ^{3}H-cholesterol in plasma of LPS-injected mice was not associated with a reduction in the capacity of acute phase serum from either WT or SAAKO mice to promote cholesterol efflux.

3.3. Hepatic Accumulation of Macrophage-Derived Cholesterol Is Not Altered during Endotoxemia, Regardless of the Presence of SAA. We next investigated whether hepatic accumulation of macrophage-derived ^{3}H-cholesterol is altered in WT or SAAKO mice during an acute inflammatory response. The amount of radioactivity in livers of LPS-injected WT and SAAKO mice 24 hr after macrophage injection was modestly lower compared to the corresponding control mice, but this difference did not reach statistical significance in either strain (Figure 3(a)). According to some [14] but not all [15] reports, endotoxemia results in transcriptional downregulation of SR-BI, an important HDL receptor in the liver that mediates selective lipid uptake from HDL. Our data indicate that SR-BI mRNA abundance was significantly reduced in mouse livers 28 hr after LPS injection, and this downregulation was not dependent on SAA (Table 1). However, SR-BI protein levels

were not significantly altered in either strain by LPS injection (Figure 3(b)). Since alterations in HDL-C uptake by the liver during endotoxemia may not necessarily be evident in static measures of hepatic ^{3}H-cholesterol content 24 hr following radiolabeled macrophage injection, we investigated whether there were any differences in uptake of the radioactive tracer from the sera of control or LPS-injected WT and SAAKO mice by Fu5AH hepatoma cells [22, 25]. Uptake studies were carried out in the presence and absence of 10 μM BLT-1, which has been shown to block SR-BI-mediated selective lipid uptake from HDL [26]. For all of the serum samples assayed, BLT-1 reduced the uptake of the radioactive tracer ~30% (Figure 3(c)), suggesting that only a portion of hepatic uptake of the macrophage-derived radioactive tracer was mediated by SR-BI. Furthermore, our data indicated no significant differences in the uptake of the radioactive tracer from serum collected from control or LPS-injected mice, regardless of the presence of SAA (Figure 3(c)).

3.4. SAA Has Limited Impact on Impaired Macrophage to Bile and Feces RCT during Endotoxemia. Endotoxemia significantly impaired biliary and fecal excretion of macrophage-derived ^{3}H-cholesterol in WT mice (Figures 4(a) and 4(b)), in line with previously published data [14, 15]. At 24 hr after macrophage injection, the amount of radioactivity present in the bile and feces of LPS-injected WT mice was reduced 36% and 80%, respectively, compared to control WT mice. On the other hand, endotoxemia did not significantly impact the amount of macrophage-derived ^{3}H-cholesterol present in bile of SAAKO mice, and fecal excretion was reduced by only 45% (Figures 4(a) and 4(b)). Thus, our results confirm previous findings that macrophage-to-feces RCT is significantly reduced in mice undergoing an acute phase response and establishes that SAA has limited impact on this impairment.

3.5. Deficiency of SAA Does Not Impact the Effect of Inflammation on Hepatic Expression of Enzymes and Transporters Involved in Cholesterol Flux. Previous studies have established that hepatic expression of genes involved in cholesterol transport and biliary excretion is altered in mice undergoing an acute phase response [14, 15]. In our studies, a moderate dose of LPS had no effect on hepatic expression of ABCA1 or Cyp7A1 mRNA (Table 1) in either WT or SAAKO mice when assessed 28 hr after LPS injection. On the other hand, ABCG1, ABCG5, ABCG8, and ABCB11 expression were all significantly suppressed after LPS treatment, and this suppression was not impacted by SAA deficiency (Table 1). There was no evidence that endotoxemia altered hepatic ABCG1 or ABCG5 protein in WT or SAAKO mice (Figures 4(c) and 4(d)).

3.6. Peritoneal Injection of Cholesterol-Loaded J774 Cells Induces a Transient Inflammatory Response in C57BL/6 Mice. In the course of our *in vivo* RCT experiments we routinely monitored the extent of induction of inflammation in LPS-injected mice. Since the SAAKO mice do not express the major mouse acute phase isoforms SAA1.1 and SAA2.1, we assessed the expression of serum amyloid P

The Impairment of Macrophage-to-Feces Reverse Cholesterol Transport during Inflammation Does
Not Depend on Serum Amyloid A

69

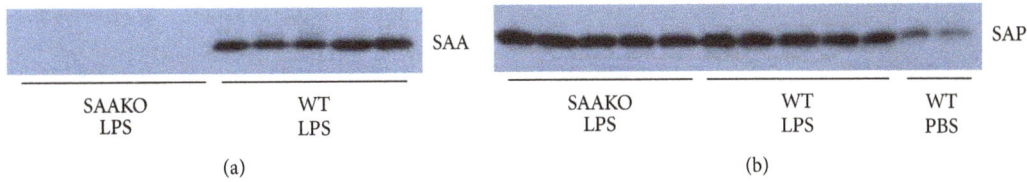

FIGURE 1: SAAKO mice are capable of mounting an acute phase response. WT and SAAKO mice were injected subcutaneously with 0.8 μg/g LPS followed 4 hr later by an intraperitoneal injection of ^{3}H-cholesterol-labeled J774 macrophages. Control WT mice received PBS only. Aliquots of plasma collected from individual mice 24 hr after macrophage administration (28 hr after LPS) were subjected to SDS-PAGE and immunoblotted using (a) rabbit anti-mouse SAA; and (b) rabbit anti-mouse SAP.

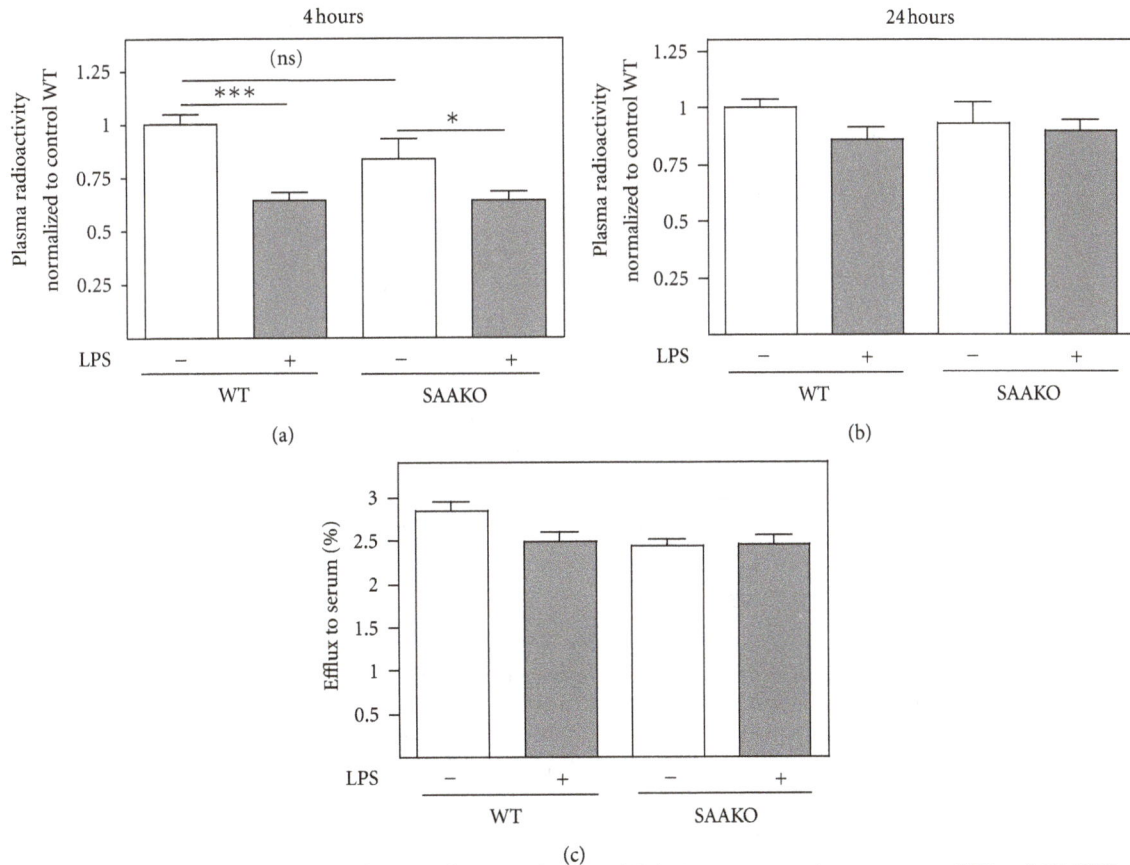

FIGURE 2: SAA does not contribute to reduced macrophage to plasma RCT during an acute phase response. WT and SAAKO mice were injected subcutaneously with PBS or 0.8 μg/g LPS as indicated, followed by an intraperitoneal injection of ^{3}H-cholesterol-labeled J774 macrophages 4 hr later. Radioactivity in plasma was determined 4 hr (a) and 24 hr (b) after macrophage injection (8 hr and 28 hr after LPS injection). The data shown in (a) and (b) were compiled from 3 separate experiments (n = 4-5 mice per group per experiment) after normalization to the PBS-injected WT mice. For the 3 experiments, radioactivity recovered in plasma of control WT mice at 4 and 24 hr ranged from 1.0% to 2.4% and 2.0% to 5.0% of the injected radioactivity, respectively. (c) Cholesterol efflux assays were carried out using J774 cells as described in Materials and Methods using serum (2.5% v : v) collected 8 hr after administration of LPS (0.8 μg/g body weight; n = 5). Values are the mean ± SEM; $^*P < 0.05$, $^{***}P < 0.001$.

component (SAP), another major acute phase reactant in mice. Unexpectedly, we readily detected SAP in the plasma of all mice injected with J774 macrophages, irrespective of LPS administration (data not shown). This finding prompted us to investigate the possibility that the experimental protocol itself produced an inflammatory response in mice. To this end, J774 macrophage foam cells were prepared according to the protocol for RCT and injected into WT mice 4 hr after administration of either PBS or LPS ("0 hr" on Figure 5(a)), and plasma samples were collected at selected intervals

during the course of the experiment and immunoblotted for SAA. As expected, LPS elicited a robust inflammatory response, as evidenced by a marked increase in plasma SAA that was detectable 4 hr after LPS injection and persisted for at least 48 hr (Figure 5(a)). Notably, intraperitoneal administration of J774 macrophages also evoked a robust inflammatory response that was detectable 4 hr after the injection of the cells, increased by 24 hr, and was still evident at 48 hr after administration (Figure 5(a)). This finding suggested that intraperitoneal administration of allogeneic macrophages

(a)

(b)

(c)

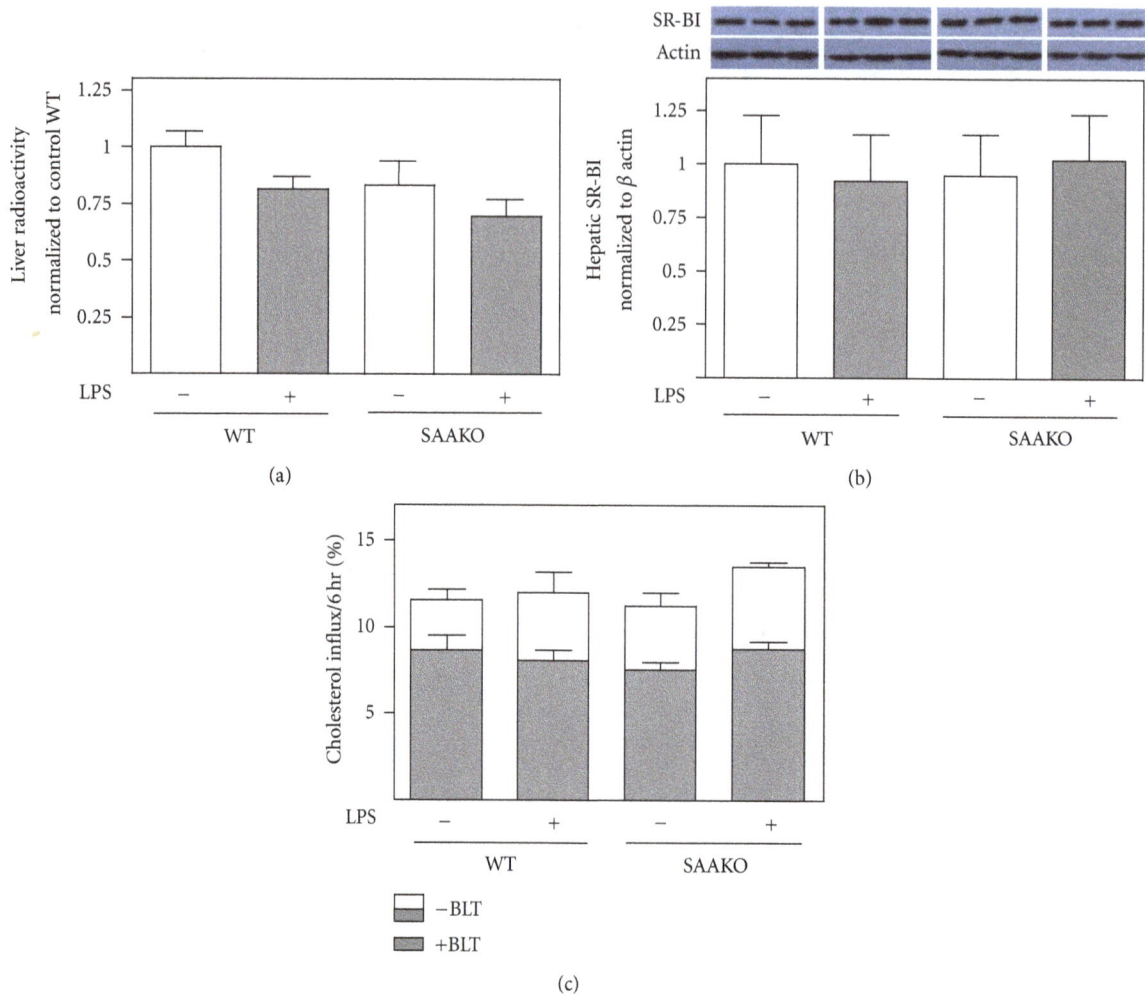

FIGURE 3: Endotoxemia does not alter hepatic accumulation of macrophage-derived ^{3}H-cholesterol, regardless of the presence of SAA. WT and SAAKO mice were injected subcutaneously with PBS or 0.8 μg/g LPS as indicated, followed by intraperitoneal injection of ^{3}H-cholesterol-labeled J774 macrophages 4 hr later. (a) Radioactivity in liver was determined 24 hr after macrophage injection (28 hr after LPS injection). The data shown were compiled from 3 separate experiments (n = 4-5 per group per experiment) after normalization to PBS-injected WT mice. Values are the mean ± SEM. Liver counts for PBS-injected WT and SAAKO mice were similar, ranging from 3.9% to 8.5% of the injected tracer for the 3 experiments. (b) SR-BI in total liver lysates (10 μg protein) was detected by immunoblotting and quantified by densitometry. Results from each of the 3 experiments (n = 4-5 per group per experiment) were expressed relative to PBS-injected WT mice after normalization to β-actin. A representative western blot is shown. (c) Fu5AH hepatoma cells were preincubated for 1 hr with or without 10 μM BLT-1 and then for 6 hr with media ± BLT-1 supplemented with 5% serum obtained from mice 24 hr after administration of ^{3}H-cholesterol-labeled macrophages (28 hr after LPS injection). The amount of radioactivity that was taken up by cells in the presence of BLT-1 is indicated by the shaded portion of the bars.

may promote an inflammatory response in mice (J774 cells were originally derived from the BALB/c strain). To investigate this possibility, we carried out parallel studies in which C57BL/6 mice were injected with either LPS, J774 cells, or syngeneic bone-marrow-derived macrophages. BMMs from C57BL/6 mice and J774 macrophages were converted to foam cells according to the protocol for RCT experiments, except that ^{3}H-cholesterol was omitted during cholesterol loading. The cells were then injected into C57BL/6 mice 4 hr after administration of either PBS or LPS, and plasma samples were immunoblotted for SAA 24 hr after injection of cells (Figure 5(b)), a time when SAA was markedly induced in the previous study (Figure 5(a)). Whereas injection of J774 cells elicited a marked increase in plasma SAA (corresponding to

~57% of the level evoked by a moderate dose (0.8 μg/g body weight) of LPS; Figure 5(c)), SAA was virtually undetectable in mouse plasma 24 hr after BMM injection, corresponding to ~8% of the amount elicited by LPS.

4. Discussion

Based on a number of large population studies, it is recognized that plasma levels of HDL and its major apolipoprotein apoA-I are inversely correlated with the risk of atherosclerosis. One of the most widely accepted mechanisms to explain HDL's cardioprotective effect is its role in RCT, whereby HDL promotes the removal of excess cholesterol from peripheral cells, including macrophage foam cells in the vessel wall,

The Impairment of Macrophage-to-Feces Reverse Cholesterol Transport during Inflammation Does
Not Depend on Serum Amyloid A

71

TABLE 1: Hepatic gene expression levels in control and LPS-treated mice.

Gene	WT		SAAKO	
	PBS	LPS	PBS	LPS
SR-BI	1.00 ± 0.06	0.42 ± 0.05[***]	0.77 ± 0.06	0.56 ± 0.07[*]
Cyp7A1	1.00 ± 0.20	0.28 ± 0.15	1.73 ± 0.35	0.94 ± 0.30
ABCA1	1.00 ± 0.06	1.22 ± 0.12	1.06 ± 0.11	1.01 ± 0.11
ABCG1	1.00 ± 0.05	0.43 ± 0.04[***]	0.99 ± 0.08	0.47 ± 0.06[***]
ABCG5	1.00 ± 0.07	0.38 ± 0.03[***]	1.24 ± 0.18	0.41 ± 0.04[***]
ABCG8	1.00 ± 0.10	0.23 ± 0.02[***]	1.24 ± 0.16	0.28 ± 0.04[***]
ABCB11	1.00 ± 0.05	0.22 ± 0.06[***]	0.97 ± 0.11	0.42 ± 0.09[***]

Values are expressed relative to control WT mice after normalization to GAPDH mRNA.
Data represent mean $\pm$ SEM, $n = 15$.
[*] $P < 0.05$, [***] $P < 0.001$, compared to corresponding PBS-treated mice.

and delivery of cholesterol to the liver for excretion. The RCT pathway involves several steps: (1) mobilization of cellular cholesterol and efflux to HDL or HDL-derived lipid-poor apolipoproteins; (2) esterification of free cholesterol in HDL to form cholesteryl ester (CE) by lecithin-cholesterol acyltransferase; (3) receptor-mediated uptake of HDL-CE by hepatocytes, either selectively by SR-BI or via whole particle uptake through a poorly defined pathway; and (4) deesterification and excretion of cholesterol into bile, either in the form of free cholesterol or as bile acids. In species that express cholesterol ester transfer protein (CETP), including humans but not mice, an alternate pathway for RCT is through the transfer of HDL-CE to apoB-containing lipoproteins and their subsequent uptake into the liver. The extent to which inflammation impacts the capacity of HDL to participate in RCT has significant clinical relevance, given the increasing incidence of chronic inflammatory conditions in humans, including rheumatoid arthritis, type 2 diabetes; and the metabolic syndrome; all of which are associated with an increased risk for atherosclerotic cardiovascular disease. Data from this study agrees with earlier reports [14–16] that acute inflammation impairs steps in the RCT pathway when assessed in an established macrophage-to-feces mouse model and determines that acute phase SAA has little impact on this impairment. We also determined that in the established RCT model, injection of J744 macrophages results in a transient inflammatory response.

In our studies, LPS-induced endotoxemia resulted in an ~35% decrease in the movement of ³H-cholesterol from macrophages to the plasma compartment when assessed 4 hr after macrophage injection. This result is consistent with previous reports that endotoxemia acutely impairs macrophage to plasma RCT and supports the conclusion that the extent of the impairment is related to the magnitude of the acute phase response [14]. McGillicuddy reported that while a relatively low dose of LPS (0.3 mg/kg s.c.) did not reduce ³H-cholesterol counts in plasma at any time point after macrophage injection, a high dose of LPS (3 mg/kg s.c.) resulted in a significant ~35% and ~20% decline in macrophage to plasma RCT when measured at 4 hr and 24 hr, respectively. In a study by another group, a similar dose of LPS administered i.p. resulted in a 33% and 27% reduction in the movement of macrophage-derived cholesterol to the

plasma at 6 and 24 hr, respectively [15]. On the other hand, the moderate dose of LPS used in our study (0.8 mg/kg s.c.) produced an effect that appeared to be more transient, since a significant difference in the amount of macrophage-derived cholesterol in plasma of control and LPS-injected mice was detected 4 hr, but not 24 hr, after macrophage injection. Notably, SAA deficiency did not alter the acute effect of endotoxemia to reduce macrophage to plasma RCT, suggesting that the duration of this impairment is not due to differences in the magnitude of the induction of SAA at different doses of LPS.

Previous studies have investigated whether SAA impacts the ability of HDL to carry out individual steps in the RCT pathway, including cellular cholesterol efflux. Studies investigating the impact of inflammation and SAA per se on cholesterol efflux have been conflicting, depending on the nature of the HDL and the cell system used in the assay. In several reports, SAA, either associated with HDL or in a lipid-free form, was shown to promote cholesterol efflux through both ABCA1-dependent and ABCA1-independent mechanisms [27–33]. With regards to ABCG1-dependent efflux, our group reported that inflammatory remodeling of mouse HDL leads to an increase in its capacity to promote efflux, but the presence of SAA did not play a role in this enhancing effect [33]. On the other hand, in two recent studies, SAA-enriched HDL from human subjects undergoing acute sepsis [15] or HDLs isolated from humans or mice subjected to experimental endotoxemia [14] showed a reduced capacity to promote macrophage cholesterol efflux. Using an overexpression approach, Annema et al. concluded that SAA does not alter the rate of movement of ³H-cholesterol from macrophages to mouse plasma in vivo in the absence of an acute phase response [15]. Results from the current study suggest that serum from mice collected 8 hr after LPS injection is not altered in its capacity to stimulate efflux from cholesterol-loaded J774 macrophages compared to control serum, regardless of the presence of SAA. Thus, at least in mice, the ability of SAA to modulate macrophage cholesterol efflux appears to be minimal.

Our data indicate that LPS-induced endotoxemia in either WT or SAAKO mice does not significantly alter macrophage to liver RCT, consistent with findings from

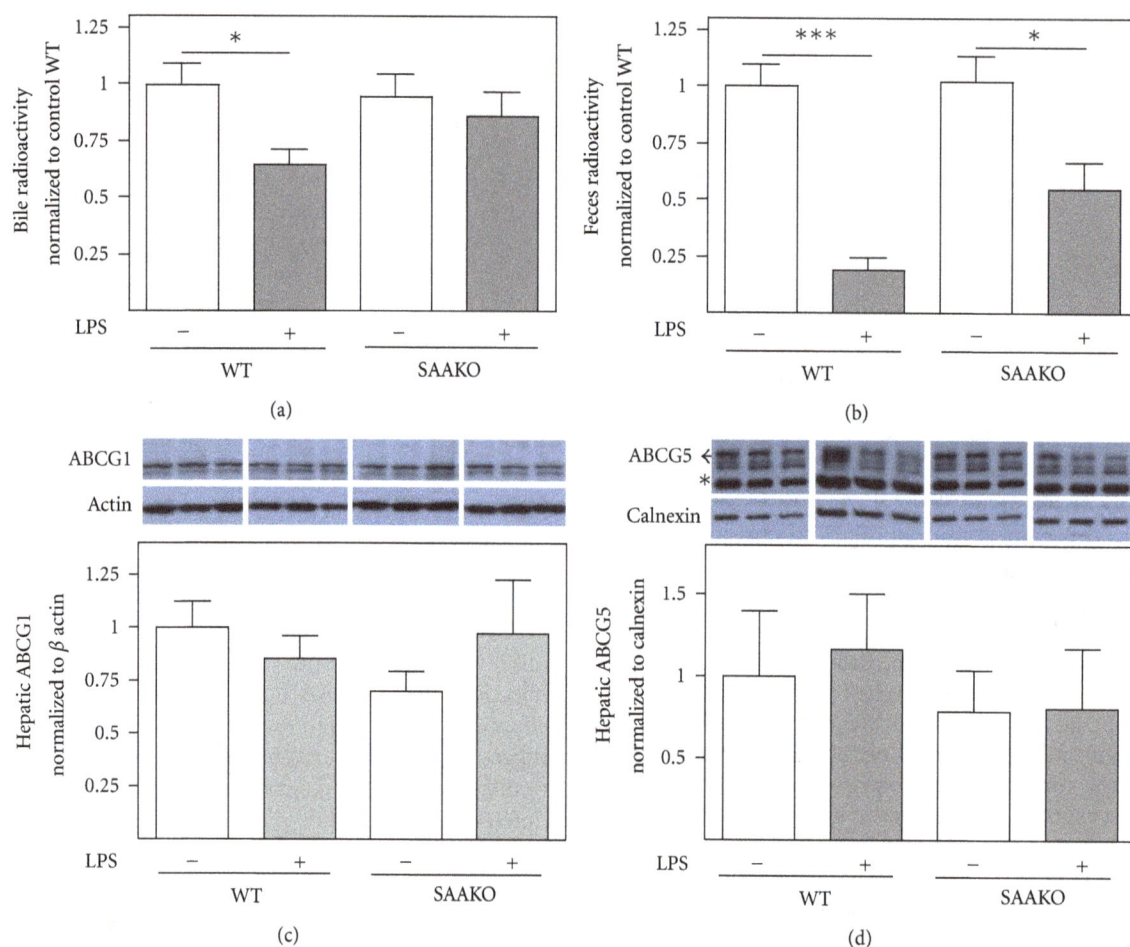

FIGURE 4: SAA has little impact on the impairment in biliary and fecal excretion of macrophage-derived ^{3}H-cholesterol during endotoxemia. WT and SAAKO mice were injected subcutaneously with PBS or 0.8 μg/g LPS as indicated, followed 4 hr later by intraperitoneal injection of ^{3}H-cholesterol-labeled J774 macrophages. (a) Radioactivity in bile was determined 24 hr after macrophage injection. The data shown are compiled from 3 separate experiments (n = 4-5 per group per experiment) and normalized to the PBS-injected WT mice. Values are the mean $\pm$ SEM; $^*P$ < 0.05%. For each of the 3 experiments, the amount of radioactivity recovered in the bile of PBS-injected WT and SAAKO mice was similar, ranging from 0.01% to 0.04% of injected tracer per μL of bile. (b) Feces were collected during the 24 hr RCT experiments, and the total amount of radioactivity was determined. Due to methodological issues in one of the experiments, data from only 2 experiments were compiled (n = 5 per group per experiment) and normalized to the PBS-injected WT mice. Values are the mean $\pm$ SEM; $^{***}P$ < 0.001; $^*P$ < 0.05. For each of the 2 experiments, the amount of radioactivity recovered in feces of PBS-injected WT and SAAKO mice was similar, ranging from 1.8%–2.6% of the injected tracer. (c) ABCG1 in liver lysates (40 μg protein) was identified by immunoblotting and quantified by densitometry. Results from each of the 3 experiments (n = 4-5 per group per experiment) were expressed relative to PBS-injected WT mice after normalization to β-actin. A representative western blot is shown. (d) ABCG5 in liver membranes (50 μg protein) was detected by immunoblotting and quantified by densitometry. Results from each of the 3 experiments (n = 4-5 per group per experiment) were expressed relative to PBS-injected WT mice after normalization to calnexin. A representative western blot is shown, with the immature (reticular) and mature (post-Golgi) forms of ABCG5 indicated with a *carat*. The band indicated by an *asterisk* is observed in G5G8-deficient mice suggesting that it represents a nonspecific signal [23].

earlier studies [14, 15]. While SR-BI mRNA expression appeared to be significantly reduced in livers of both WT and SAAKO mice 28 hr after LPS injection, SR-BI protein levels were not altered in either strain. Similarly, McGillicuddy et al. reported significantly reduced hepatic expression of SR-BI mRNA without a commensurate change in SR-BI protein 48 hr after high dose LPS injection [14]. Since hepatic accumulation of macrophage-derived ^{3}H-cholesterol at a single time point is influenced in opposite directions by the rates of flux into and out of the liver, we carried out *in vitro*

studies using Fu5AH hepatoma cells to determine whether hepatic uptake of ^{3}H-cholesterol from acute phase plasma was significantly different compared to control and the extent to which SAA might impact such uptake. Uptake by Fu5AH cells was measured in the presence or absence of BLT-1, a known inhibitor of SR-BI-mediated selective CE uptake [26]. Our data indicated no significant difference in either SR-BI-dependent or SR-BI-independent uptake of the radioactive tracer from control or LPS-injected mouse serum, regardless of the presence of SAA. In a previous study by our group we

The Impairment of Macrophage-to-Feces Reverse Cholesterol Transport during Inflammation Does
Not Depend on Serum Amyloid A

73

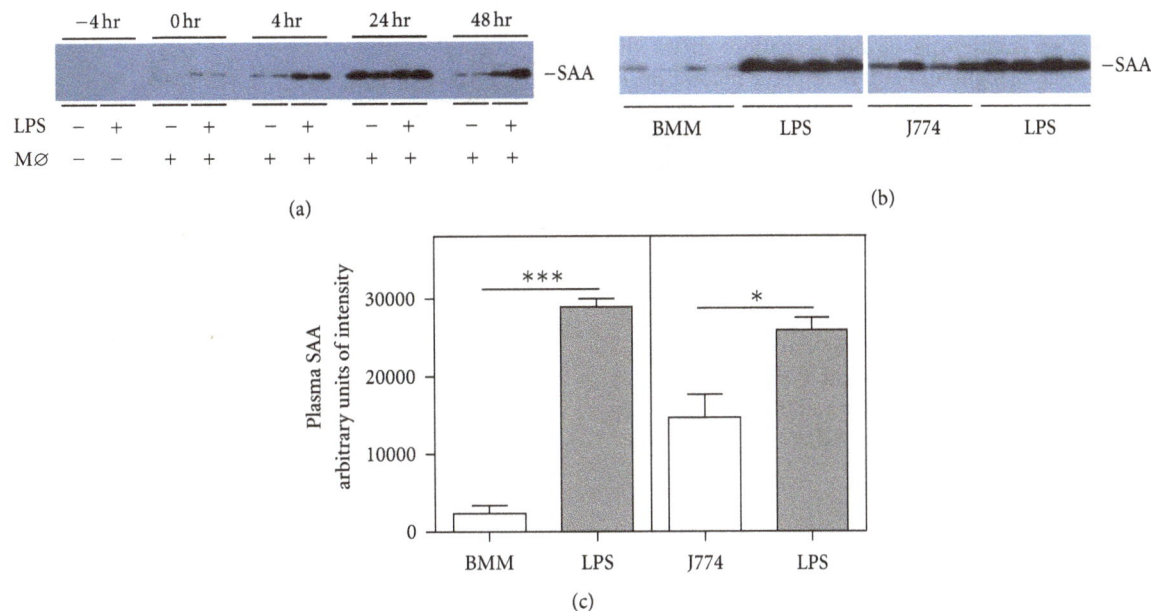

FIGURE 5: Intraperitoneal injection of J774 cells induces an acute phase response in C57BL/6 mice. (a) WT mice were injected subcutaneously with PBS or $0.8\,\mu g/g$ LPS ($-4\,hr$) followed 4 hr later by an intraperitoneal injection of ^{3}H-cholesterol-labeled J774 macrophages (0 hr); $n = 2$ per treatment. Aliquots of plasma ($0.25\,\mu L$) collected from individual mice at the indicated times were subjected to SDS-PAGE, followed by immunoblotting for SAA. (b) In two parallel studies, C57BL/6 mice were injected subcutaneously with PBS or $0.8\,\mu g/g$ LPS, and 4 hr later, the PBS-injected mice were injected i.p. with cholesterol-loaded J774 macrophages or C57BL/6 BMM, as indicated ($n = 4$ per treatment). Aliquots of plasma ($0.25\,\mu L$) collected from individual mice 28 hr after PBS/LPS injection (i.e., 24 hr after macrophage injection) were subjected to SDS-PAGE, followed by immunoblotting for SAA. (c) Relative plasma SAA concentrations 28 hr after PBS/LPS injection (i.e., 24 hr after macrophage injection) as determined by densitometry ($n = 4$). Values are the mean ± SEM; $^{***}P < 0.001$; $^{*}P < 0.05$.

concluded that SAA inhibits SR-BI-mediated selective lipid uptake from HDL [34]. The discrepant findings from the two studies may be due to differences in cell types used for uptake studies (Fu5AH cells versus stably transfected Chinese hamster ovary cells), the use of whole serum versus isolated HDL particles, the amount of SAA associated with the HDL fraction, and the fact that the earlier study investigated HDL enriched with SAA in the absence of inflammation.

McGillicuddy et al. [14] and Annema et al. [15] both reported that the most pronounced effect of LPS on RCT was on cholesterol flux through the liver to the bile and feces. The decrease in macrophage-to-feces RCT was associated with a significant reduction in the mRNA expression of genes involved in cholesterol transport and bile acid synthesis, including ABCG5, ABCG8, ABCB11, and CYP7A1, suggesting that one compensatory response to endotoxemia may be to inhibit the excretion of cholesterol out of the body. In our studies, the amount of macrophage-derived radiotracer present in the bile and feces of LPS-injected WT mice was reduced 36% and 80%, respectively, compared to control WT mice, confirming the previous studies. The impact of LPS on bile and feces RCT was only partially ameliorated in SAAKO mice. As reported previously [14, 15], we show that hepatic expression of ABCG5, ABCG8, and ABCB11 mRNAs is significantly reduced in WT mice 28 hr after LPS injection. However, SAA deficiency did not reverse this suppressive effect, nor did the reductions in ABCG5 mRNA coincide with reduced ABCG5 protein expression. Thus, our studies

would suggest that other mechanisms in addition to effects on hepatic expression of genes involved in biliary cholesterol excretion may also play a role in the impairment of RCT during acute inflammation. As pointed out by Malik et al. [16], perturbations in macrophage-to-feces RCT after LPS injection could in part be due to non-specific metabolic effects in the liver or intestine that could have a negative impact. Whether SAA modulates the potential metabolic effects of inflammatory stimuli is the subject of ongoing studies in our laboratory. A nonbiliary, transintestinal route for neutral sterol excretion has been described [35–38], but the role of this pathway in macrophage-to-feces RCT has not been clearly delineated [39, 40]. The possibility that inflammation or SAA modulates trans-intestinal cholesterol excretion merits further study.

An unanticipated outcome of our studies was the finding that C57BL/6 mice administered J774 macrophages undergo a transient acute phase response, as evidenced by a marked increase in plasma SAA. The inflammatory response was less severe and more short-lived than the response evoked by a modest dose of LPS. Interestingly, bone-marrow cells from syngeneic mice handled under identical conditions and with the same reagents did not produce an inflammatory response. Our conclusion that SAA has little impact on the impairment of RCT during inflammation is not negated by the finding that i.p. injection of J774 cells by itself leads to a modest acute phase response. Nevertheless, the use of syngeneic primary macrophages may be preferable over macrophage cell lines for *in vivo* RCT studies that are not intended to be carried out in the setting of inflammation, given the consistent finding that inflammation impairs macrophage-to-feces RCT in this commonly used surrogate model [14–16]. In summary, our findings clearly show that SAA does not substantially contribute to the impairment of RCT during inflammation.

Acknowledgments

The authors thank Michael Jansen and Darrell Robertson for their excellent technical assistance. This work was supported by National Institutes of Health Grant P01HL086670. The studies were supported with resources and facilities provided by the Lexington, Kentucky Veterans Affairs Medical Center.

References

[1] G. Assmann and A. M. Gotto Jr., "HDL cholesterol and protective factors in atherosclerosis," *Circulation*, vol. 109, no. 23, pp. III8–III14, 2004.

[2] M. Cuchel and D. J. Rader, "Macrophage reverse cholesterol transport: key to the regression of atherosclerosis?" *Circulation*, vol. 113, no. 21, pp. 2548–2555, 2006.

[3] C. P. Chung, A. Oeser, P. Raggi et al., "Increased coronary-artery atherosclerosis in rheumatoid arthritis: relationship to disease duration and cardiovascular risk factors," *Arthritis and Rheumatism*, vol. 52, no. 10, pp. 3045–3053, 2005.

[4] S. Tsimikas, J. T. Willerson, and P. M. Ridker, "C-reactive protein and other emerging blood biomarkers to optimize risk stratification of vulnerable patients," *Journal of the American College of Cardiology*, vol. 47, no. 8, pp. C19–C31, 2006.

[5] D. J. Rader, "Inflammatory markers of coronary risk," *The New England Journal of Medicine*, vol. 343, no. 16, pp. 1179–1182, 2000.

[6] W. Khovidhunkit, M. S. Kim, R. A. Memon et al., "Effects of infection and inflammation on lipid and lipoprotein metabolism: mechanisms and consequences to the host," *Journal of Lipid Research*, vol. 45, no. 7, pp. 1169–1196, 2004.

[7] D. R. Van Der Westhuyzen, F. C. De Beer, and N. R. Webb, "HDL cholesterol transport during inflammation," *Current Opinion in Lipidology*, vol. 18, no. 2, pp. 147–151, 2007.

[8] G. A. Coetzee, A. F. Strachan, and D. R. Van Der Westhuyzen, "Serum amyloid A-containing human high density lipoprotein 3. Density, size, and apolipoprotein composition," *Journal of Biological Chemistry*, vol. 261, no. 21, pp. 9644–9651, 1986.

[9] C. M. Uhlar and A. S. Whitehead, "Serum amyloid A, the major vertebrate acute-phase reactant," *European Journal of Biochemistry*, vol. 265, no. 2, pp. 501–523, 1999.

[10] J. Sipe, "Part 2: Revised nomenclature for serum amyloid a (SAA)," *Amyloid*, vol. 6, no. 1, pp. 67–70, 1999.

[11] A. S. Whitehead, M. C. De Beer, D. M. Steel et al., "Identification of novel members of the serum amyloid A protein superfamily as constitutive apolipoproteins of high density lipoprotein," *Journal of Biological Chemistry*, vol. 267, no. 6, pp. 3862–3867, 1992.

[12] T. Chiba, C. Y. Han, T. Valsar et al., "Serum amyloid A3 does not contribute to circulating SAA levels," *Journal of Lipid Research*, vol. 50, no. 7, pp. 1353–1362, 2009.

[13] M. C. De Beer, N. R. Webb, J. M. Wroblewski et al., "Impact of serum amyloid A on high density lipoprotein composition and levels," *Journal of Lipid Research*, vol. 51, no. 11, pp. 3117–3125, 2010.

[14] F. C. McGillicuddy, M. de la Llera Moya, C. C. Hinkle et al., "Inflammation impairs reverse cholesterol transport in vivo," *Circulation*, vol. 119, no. 8, pp. 1135–1145, 2009.

[15] W. Annema, N. Nijstad, M. Tölle et al., "Myeloperoxidase and serum amyloid A contribute to impaired in vivo reverse cholesterol transport during the acute phase response but not group IIA secretory phospholipase A2," *Journal of Lipid Research*, vol. 51, no. 4, pp. 743–754, 2010.

[16] P. Malik, S. Z. Berisha, J. Santore, C. Agatisa-Boyle, G. Brubaker, and J. D. Smith, "Zymosan-mediated inflammation impairs in vivo reverse cholesterol transport," *Journal of Lipid Research*, vol. 52, no. 5, pp. 951–957, 2011.

[17] S. Yona, S. E. M. Heinsbroek, L. Peiser, S. Gordon, M. Perretti, and R. J. Flower, "Impaired phagocytic mechanism in annexin 1 null macrophages," *British Journal of Pharmacology*, vol. 148, no. 4, pp. 469–477, 2006.

[18] Z. Zhao, M. C. De Beer, L. Cai et al., "Low-density lipoprotein from apolipoprotein E-deficient mice induces macrophage lipid accumulation in a CD36 and scavenger receptor class A-dependent manner," *Arteriosclerosis, Thrombosis, and Vascular Biology*, vol. 25, no. 1, pp. 168–173, 2005.

[19] Y. Zhang, I. Zanotti, M. P. Reilly, J. M. Glick, G. H. Rothblat, and D. J. Rader, "Overexpression of apolipoprotein A-I promotes reverse transport of cholesterol from macrophages to feces in vivo," *Circulation*, vol. 108, no. 6, pp. 661–663, 2003.

[20] B. F. Asztalos, M. De La Llera-Moya, G. E. Dallal, K. V. Horvath, E. J. Schaefer, and G. H. Rothblat, "Differential effects of HDL subpopulations on cellular ABCA1- and SR-BI-mediated cholesterol efflux," *Journal of Lipid Research*, vol. 46, no. 10, pp. 2246–2253, 2005.

[21] G. A. Francis, R. H. Knopp, and J. F. Oram, "Defective removal of cellular cholesterol and phospholipids by apolipoprotein A-I in Tangier disease," *Journal of Clinical Investigation*, vol. 96, no. 1, pp. 78–87, 1995.

[22] E. T. Alexander, C. Vedhachalam, S. Sankaranarayanan et al., "Influence of apolipoprotein A-I domain structure on macrophage reverse cholesterol transport in mice," *Arteriosclerosis, Thrombosis, and Vascular Biology*, vol. 31, no. 2, pp. 320–327, 2011.

[23] N. S. Sabeva, E. J. Rouse, and G. A. Graf, "Defects in the leptin axis reduce abundance of the ABCG5-ABCG8 sterol transporter in liver," *Journal of Biological Chemistry*, vol. 282, no. 31, pp. 22397–22405, 2007.

[24] N. R. Webb, P. M. Connell, G. A. Graf et al., "SR-BII, an isoform of the scavenger receptor BI containing an alternate cytoplasmic tail, mediates lipid transfer between high density lipoprotein and cells," *Journal of Biological Chemistry*, vol. 273, no. 24, pp. 15241–15248, 1998.

[25] E. T. Alexander, G. L. Weibel, M. R. Joshi et al., "Macrophage reverse cholesterol transport in mice expressing ApoA-I milano," *Arteriosclerosis, Thrombosis, and Vascular Biology*, vol. 29, no. 10, pp. 1496–1501, 2009.

[26] T. J. F. Nieland, M. Penman, L. Dori, M. Krieger, and T. Kirchhausen, "Discovery of chemical inhibitors of the selective transfer of lipids mediated by the HDL receptor SR-BI," *Proceedings of the National Academy of Sciences of the United States of America*, vol. 99, no. 24, pp. 15422–15427, 2002.

The Impairment of Macrophage-to-Feces Reverse Cholesterol Transport during Inflammation Does
Not Depend on Serum Amyloid A

75

[27] S. Hayat and J. G. Raynes, "Serum amyloid A has little effect on hight density lipoprotein (HDL) binding to U937 monocytes but may influence HDL mediated cholesterol transfer," *Biochemical Society Transactions*, vol. 25, no. 2, p. 348S, 1997.

[28] S. P. Tam, A. Flexman, J. Hulme, and R. Kisilevsky, "Promoting export of macrophage cholesterol: the physiological role of a major acute-phase protein, serum amyloid A 2.1," *Journal of Lipid Research*, vol. 43, no. 9, pp. 1410–1420, 2002.

[29] J. A. Stonik, A. T. Remaley, S. J. Demosky, E. B. Neufeld, A. Bocharov, and H. B. Brewer, "Serum Amyloid a promotes ABCA1-dependent and ABCA1-independent lipid efflux from cells," *Biochemical and Biophysical Research Communications*, vol. 321, no. 4, pp. 936–941, 2004.

[30] D. R. Van Der Westhuyzen, L. Cai, M. C. De Beer, and F. C. De Beer, "Serum amyloid A promotes cholesterol efflux mediated by scavenger receptor B-I," *Journal of Biological Chemistry*, vol. 280, no. 43, pp. 35890–35895, 2005.

[31] S. Abe-Dohmae, K. H. Kato, Y. Kumon et al., "Serum amyloid A generates high density lipoprotein with cellular lipid in an ABCA1- or ABCA7-dependent manner," *Journal of Lipid Research*, vol. 47, no. 7, pp. 1542–1550, 2006.

[32] G. Marsche, S. Frank, J. G. Raynes, K. F. Kozarsky, W. Sattler, and E. Malle, "The lipidation status of acute-phase protein serum amyloid A determines cholesterol mobilization via scavenger receptor class B, type I," *Biochemical Journal*, vol. 402, no. 1, pp. 117–124, 2007.

[33] M. C. De Beer, A. Ji, A. Jahangiri et al., "ATP binding cassette G1-dependent cholesterol efflux during inflammation," *Journal of Lipid Research*, vol. 52, no. 2, pp. 345–353, 2011.

[34] L. Cai, M. C. De Beer, F. C. De Beer, and D. R. Van Der Westhuyzen, "Serum amyloid a is a ligand for scavenger receptor class B type I and inhibits high density lipoprotein binding and selective lipid uptake," *Journal of Biological Chemistry*, vol. 280, no. 4, pp. 2954–2961, 2005.

[35] J. M. Brown, T. A. Bell III, H. M. Alger et al., "Targeted depletion of hepatic ACAT2-driven cholesterol esterification reveals a non-biliary route for fecal neutral sterol loss," *Journal of Biological Chemistry*, vol. 283, no. 16, pp. 10522–10534, 2008.

[36] J. K. Kruit, T. Plösch, R. Havinga et al., "Increased fecal neutral sterol loss upon liver X receptor activation is independent of biliary sterol secretion in mice," *Gastroenterology*, vol. 128, no. 1, pp. 147–156, 2005.

[37] J. N. van der Veen, T. H. van Dijk, C. L. J. Vrins et al., "Activation of the liver X receptor stimulates trans-intestinal excretion of plasma cholesterol," *Journal of Biological Chemistry*, vol. 284, no. 29, pp. 19211–19219, 2009.

[38] A. E. van der Velde, C. L. J. Vrins, K. van den Oever et al., "Direct intestinal cholesterol secretion contributes significantly to total fecal neutral sterol excretion in mice," *Gastroenterology*, vol. 133, no. 3, pp. 967–975, 2007.

[39] R. E. Temel, J. K. Sawyer, L. Yu et al., "Biliary sterol secretion is not required for macrophage reverse cholesterol transport," *Cell Metabolism*, vol. 12, no. 1, pp. 96–102, 2010.

[40] N. Nijstad, T. Gautier, F. Briand, D. J. Rader, and U. J. F. Tietge, "Biliary sterol secretion is required for functional in vivo reverse cholesterol transport in mice," *Gastroenterology*, vol. 140, no. 3, pp. 1043–1051, 2011.

9

Reduction of Cellular Lipid Content by a Knockdown of *Drosophila PDP1γ* and Mammalian Hepatic Leukemia Factor

Svetlana Dzitoyeva and Hari Manev

Department of Psychiatry, Psychiatric Institute, University of Illinois at Chicago, Chicago, IL 60612, USA

Correspondence should be addressed to Svetlana Dzitoyeva; sdzitoyeva@psych.uic.edu

Academic Editor: Philip W. Wertz

In exploring the utility of double-stranded RNA (dsRNA) injections for silencing the *PAR-domain protein 1 (Pdp1)* gene in adult *Drosophila*, we noticed a dramatic loss of fat tissue lipids. To verify that our RNAi approach produced the expected *Pdp1* knockdown, the abdominal fat tissues sections were stained with PDP1 antibodies. PDP1 protein immunostaining was absent in flies injected with dsRNA targeting a sequence common to all known *Pdp1* isoforms. Subsequent experiments revealed that lipid staining is reduced in flies injected with dsRNA against Pdp1γ (fat body specific) and not against *Pdp1ε* (predominantly involved in circadian mechanisms). *Drosophila* PDP1γ protein shows a high homology to mammalian thyrotroph embryonic factor (TEF), albumin D site-binding protein (DBP), and hepatic leukemia factor (HLF) transcription factors. In an in vitro model of drug- (olanzapine-) induced adiposity in mouse 3T3-L1 cells, the mRNA content of HLF but not TEF and DBP was increased by the drug treatment. A knockdown of the HLF mRNA by transfecting the cultures with HLF dsRNA significantly reduced their lipid content. Furthermore, the HLF RNAi prevented olanzapine from increasing the cell lipid content. These results suggest that the PDP1/HLF system may play a role in physiological and drug-influenced lipid regulation.

1. Introduction

Contrary to the previous belief that in adult *Drosophila* (fruit fly) RNA interference (RNAi) cannot occur by exogenous administration of double-stranded RNA (dsRNA), it was conclusively demonstrated that this type of systemic RNAi is operative in adult flies [1]. This mechanism explains the efficacy of dsRNA injections into adult *Drosophila*, applied as a tool for targeted gene knockdown in adult organisms [2–4]. The advantage of this RNAi method is that it avoids the unwanted developmental alterations and the possible side effects of genetic manipulations involved in alternative RNAi approaches. In the course of exploring the utility of dsRNA injections for silencing the *PAR-domain protein 1 (Pdp1)* gene in adult *Drosophila*, we noticed a peculiar RNAi-related phenotype, a dramatic loss of fat tissue lipids. Here, we report the follow-up study aimed at exploring this serendipitous discovery.

Drosophila Pdp1 encodes a transcription factor highly homologous to the proline- and acidic amino acid-rich (PAR) subfamily of mammalian bZIP transcription factors, albumin D site-binding protein (DBP), hepatic leukemia factor (HLF), and thyrotroph embryonic factor (TEF). PDP1 was originally identified as a regulator of the muscle activator region [5]. Subsequently, it was established that *Pdp1* is a component of the *Drosophila* circadian network and that its expression is directly activated by *dClock/Cycle* genes [6, 7]. The *Pdp1* gene encodes multiple transcripts, which are differentially expressed during embryogenesis [8]. Of the six *Pdp1* isoforms, *Pdp1ε* is the one that shows a circadian expression and is involved in the regulation of the *Drosophila* circadian behavior [9]. Whereas *Pdp1ε* is predominantly expressed in the nervous system, *Pdp1γ* is predominantly expressed in the fat body [8]. The mammalian homologous factors DBP, TEF, and HLF also show a circadian rhythm in their expression/accumulation; their absence results in epilepsy [10]. Furthermore, these mammalian proteins have recently been linked to a mechanism of fatty acid regulation [11]. To address possible similarities between *Drosophila Pdp1* and its mammalian homologues, in addition to the experiments in

TABLE 1: Primers used for the mRNA assay.

Target	Forward	Reverse
HLF	5′-gaaggagctgggcaaatgcaagaa-3′	5′-accagacaggaaacaagctgtcca-3′
Cyclophilin	5′-agcatacaggtcctggcatcttgt-3′	5′-aaacgctccatggcttccacaatg-3′
Pdp1 total	5′-acttccctcctcagttgccgg-3′	5′-tcgcagatggtctgtgtgta-3′
Pdp1ε	5′-ccacgctaacgtaccagaggattt-3′	5′-ctgaaatcgcgctttcaagctgtt-3′
Pdp1γ	5′-gagtttcagttacgcgttgttgct-3′	5′-gttgttctccgacaggaactcgtc-5′
RP49	5′-atgaccatccgcccagcataca-3′	5′-tgtgtattccgaccaggttac-3′

fruit flies, we employed a model of drug-induced adiposity in mammalian 3T3-L1 cells in vitro [12].

2. Material and Methods

2.1. Drosophila and Cell Culture. Male Canton-S flies were cultured for 5–7 days in a 12 h/12 h light/dark cycle [3]. They were CO_2 anesthetized, injected with 0.3-0.4 pmol dsRNA [3], and collected five days later. Mouse 3T3-L1 cells (American Type Culture Collection), grown to confluence in Dulbecco's modified Eagle's medium (GIBCO)/10% fetal bovine serum (Atlanta Biologicals), were differentiated into adipocytes by changing to a differentiation medium [12]. dsRNA transfection (10 pmol; TransPass R1 transfection reagent; New England Biolabs) started 24 h after addition of the differentiation medium; cells were collected 48 h later. Olanzapine (50 μM; Sequoia Research Products Ltd.) and vehicle (1 : 1000 dimethyl sulfoxide; Sigma) were added either together with the differentiation medium or 24 h after the initiation of dsRNA transfection.

2.2. Double-Stranded RNA (dsRNA). For *Drosophila* dsRNA RNAi, the corresponding RNAs were synthesized in vitro from DNA oligonucleotides with an attached T7 RNA polymerase promoter sequence (IDT DNA Technologies) targeting the aacttccctcctcagttgccgg cDNA region common to all *Pdp1* isoforms (AF172402, AF172403, AF172404, AF172405, AF172406, and AF209903), the agttacgcgttgttgct-gcgacc cDNA region unique to *Pdp1γ* (AF172404), and the aacgtaccagaggatttaccagg region unique to *Pdp1ε* (AF172406). Control dsRNA was based on the human 5-lipoxygenase cDNA, ttcatgcacatgttccagtctt (NM_000698; no matches to *Drosophila* genome sequences; *Drosophila* does not have 5-lipoxygenase homologues). The single-stranded sense and antisense DNA oligonucleotides were annealed in distilled water and 100 ng was used in 30 μL of in vitro transcription reaction. dsRNA molecules were formed by combining the corresponding transcribed sense and antisense RNAs (10 min at 65°C) [2, 3]. For mouse 3T3-L1 dsRNA RNAi, a 154 nucleotide fragment (1120–1274) of mouse HLF mRNA (NM_172563) was cloned into a pGEM-T vector (Promega). The insert was amplified with M13 primers and RNA was transcribed in vitro with T7 and SP6 RNA polymerases (New England Biolabs). An equal amount of the single-stranded sense and antisense RNA molecules was incubated for 10 min at 65°C and cooled on ice. Control dsRNA was prepared from a 239-nucleotide fragment (394–620) of the green fluorescent protein gene (X83959) amplified from the pGFP cloning plasmid DNA vector.

2.3. mRNA Assay. RNA was extracted from cells and *Drosophila* homogenates with TRIzol reagent (Invitrogen), reverse transcribed, and used for the quantitative real time PCR (Stratagene Mx3005P qPCR System; Agilent Technologies with Maxima SYBR Green qPCR Master Mix (MM); Fermentas Inc.), in a two-step PCR program [14]: the reaction mix: 10 μL 2x MM, 2 μL primer mix (0.2 μM final, each), and 8 μL reverse transcription mix. The corresponding primers are shown in Table 1. Data were normalized against the cyclophilin (3T3-L1 cells) or RP49 (*Drosophila*) internal control and presented as a coefficient of variation [14].

2.4. PDP1 Immunostaining. 30-micron *Drosophila* cryosections were fixed in 12% formaldehyde/phosphate buffered saline (PBS), rinsed in PBS, and incubated with PDP1 antibody (1 : 500, overnight, 4°C) [8] followed by incubation with secondary biotinylated anti-rabbit IgG (1 h; Vector Laboratories). Color was developed with DAB (Sigma).

2.5. Lipid Staining and Quantification. For lipid visualization in the *Drosophila* fat body, whole fly 30-micron cryosections were fixed, incubated with Oil Red O (ORO; Sigma; 0.5% in isopropanol diluted with water 3 : 2 and filtered through a 0.45 μm filter) for 30 min, rinsed, covered with 80% glycerol, and photographed. For quantitative measurements, flies were decapitated and their bodies cryosectioned and fixed. Free-floating sections (from 5 bodies/data point) were stained with ORO, washed, and dried, and the dye was recovered with isopropanol. 3T3-L1 cells were fixed and incubated (30 min) with ORO. After discarding this solution, the dye captured by intracellular lipids was recovered with 400 μL isopropanol, and the absorbance was measured (single wavelength, 520 nm filter; Bio-Rad, Model 550). The results are reported in units.

2.6. Statistics. Statistical analyses were performed using SPSS software (SPSS Inc.). Data (mean ± SEM) were analyzed by one-way analysis of variance followed by Student's *t*-test or Dunnett's test (significance at $P < 0.05$).

3. Results

3.1. Drosophila Pdp1 dsRNA RNAi. Our initial observation of the lipid-reducing effect of *Pdp1* RNAi was made by examining fly body sections without the help of any lipid

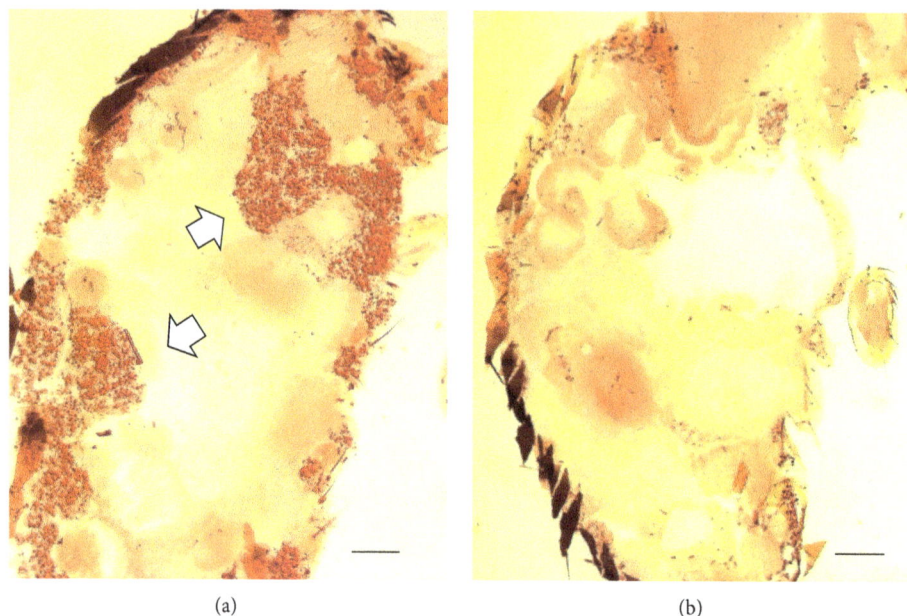

(a)   (b)

FIGURE 1: Effect of total *Pdp1* RNAi induced in adult *Drosophila* on lipid staining. Flies were injected with control dsRNA (panel (a)) and dsRNA targeted at a sequence common to all known *Pdp1* isoforms (total *Pdp1* RNAi; panel (b)). They were processed for lipid staining (red) five days later. Shown is a low magnification (objective 5x) sagittal abdominal section (size bar = 150 μm). Note the presence of lipid staining in the fat tissue cells of a control fly (indicated by white arrows) (a) and its absence in the *Pdp1* RNAi fly (b).

staining; that is, the effect was rather obvious. To verify this observation, we employed the ORO lipid staining. In these experiments, a total *Pdp1* RNAi was achieved by injecting flies with the dsRNA targeted at a sequence common to all known *Pdp1* isoforms. This resulted in the loss of ORO lipid staining as exemplified by the staining of *Drosophila* abdominal sections (Figure 1). To verify that our RNAi approach produced the expected *Pdp1* knockdown, we stained the whole fly sections including the abdominal fat tissue with PDP1 protein antibodies. PDP1 immunostaining revealed a robust nuclear PDP1 staining in the cells of flies injected with control dsRNA and the absence of PDP1 staining in *Pdp1* dsRNA injected flies (Figure 2). Also, in these sections, we confirmed the lipid-reducing effect of the total *Pdp1* dsRNA; the number and the size of lipid-containing droplets in fat tissue cells were reduced in the RNAi samples (Figure 2).

The previous results were obtained with the dsRNA targeted at a sequence common to all known *Pdp1* isoforms. In subsequent experiments, we investigated the *Pdp1* isoform specificity of the lipid-reducing phenotype. Hence, flies were injected with dsRNA targeted specifically against *Pdp1ε* (predominantly present in the nervous system and involved in circadian mechanisms) and *Pdp1γ* (expressed in the fat body), respectively, and their body lipid content was quantified. Only the *Pdp1γ* dsRNA significantly reduced the body lipid content (Figure 3).

3.2. Mammalian HLF dsRNA RNAi. Drosophila PDP1γ protein shows a high homology to TEF, DBP, and HLF members of the PAR subfamily of mammalian bZIP transcription factors (Figure 4) [13]. To explore possible similarities between

Pdp1γ and these factors, we selected the mouse 3T3-L1 preadipocytes in vitro. These cells have been used as a model for drug-induced adipogenic effects; that is, treatment of 3T3-L1 cells during differentiation into adipocytes with the antipsychotic drug olanzapine increases their lipid content [12]. In this model, we found that the adipogenic olanzapine treatment increases the mRNA content of HLF but not TEF and DBP (Figure 5). On the other hand, a knockdown of the endogenous HLF mRNA by transfecting the cultures with HLF dsRNA significantly reduced their lipid content (Figure 6). In an experiment in which olanzapine and vehicle treatments were initiated 24 h after the initiation of transfection and conducted for the next 24 h, we found that olanzapine treatment increased lipid content in naïve and sham-dsRNA transfected cells but not in HLF dsRNA-transfected cells (Figure 7).

4. Discussion

In this work, we confirmed and expanded our serendipitous observation that a systemic PDP1 knockdown in adult flies, induced by injections of *Pdp1* dsRNAs, leads to a significant lipid decrease, and we found that similar phenotype can be induced by HLF RNAi in mouse adipocytes in vitro. Collectively, our results demonstrated that a reduction of the PDP1γ/HLF transcription factor leads to a decreased lipid content.

Both mammalian HLF and *Drosophila* PDP1 are known components, that is, output regulators, of circadian cycles and as such have been linked to metabolic regulation [15]. The expression of *Drosophila* circadian genes (i.e., peripheral clocks) in the fat body has been shown to play a role in the

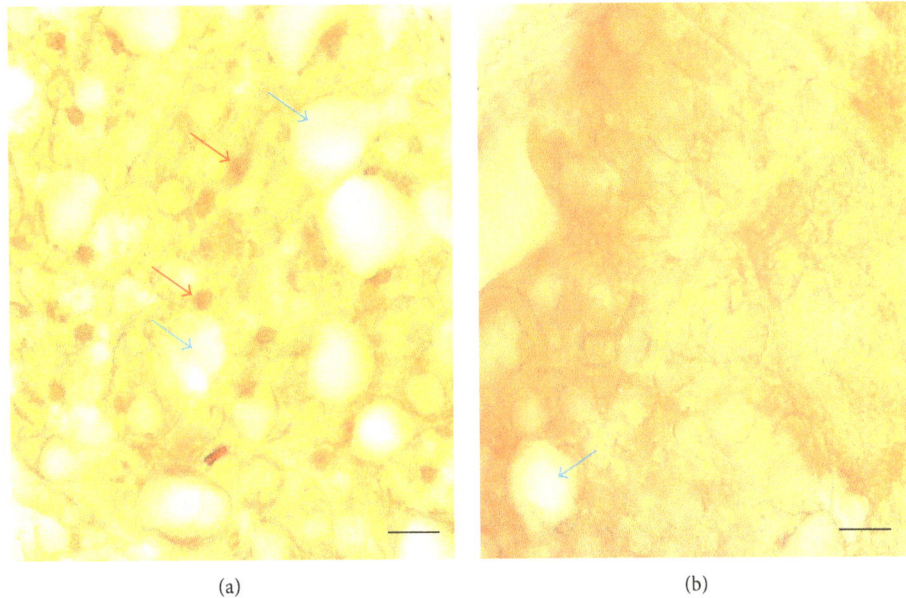

(a)   (b)

FIGURE 2: Effect of total *Pdp1* RNAi induced in adult *Drosophila* on PDP1 protein immunostaining in the fat tissue. Flies were injected with control dsRNA (panel (a)) and dsRNA targeted at a sequence common to all known *Pdp1* isoforms (total *Pdp1* RNAi; panel (b)). They were processed for PDP1 immunolabeling five days later. Shown is a high magnification (objective 20X) of abdominal fat tissue section (size bar = 20 μm). Note the presence of strong nuclear PDP1 protein immunolabeling (dark circles indicated by red arrows) in the control section (a) and its absence in the *Pdp1* RNAi section (b). Also, indicated (blue arrows) are multiple white circles of lipid droplet-containing cells in the control section and their reduced number after *Pdp1* RNAi.

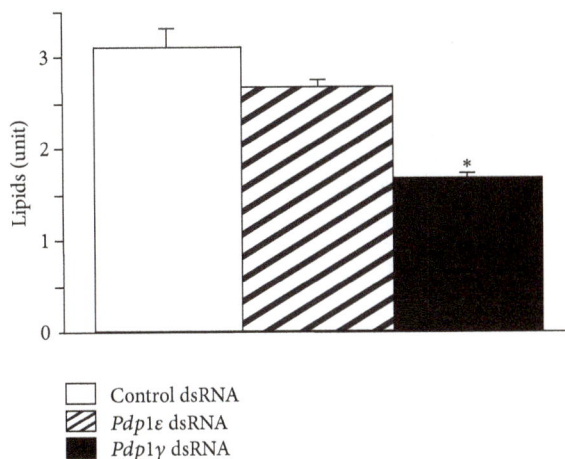

FIGURE 3: Quantitative assay of *Drosophila* lipid content following the isoform-specific *Pdp1* RNAi. Adult flies were injected with control, *Pdp1γ*, and *Pdp1ε* dsRNA and processed for quantitative lipid assay. *Pdp1γ* but not *Pdp1ε* dsRNA reduced body lipid content ($^{*}P < 0.001$ versus control; $n = 5$; mean $\pm$ standard error mean).

regulation of fly metabolism [16]. Of the six *Pdp1* isoforms, *Pdp1ε* is the one that is characterized by a prominent circadian expression and is involved in the regulation of the *Drosophila* circadian behavior [9]. One of the circadian functions of *Pdp1ε* is in regulating the circadian output gene *takeout* [17]. In male *Drosophila*, *takeout* is abundant in the fat body and plays a role in courtship behavior of these flies [18]. It was suggested that *Pdp1ε*-mediated regulation of the fat

body genes may influence this type of fly behavior [17]. In our *Drosophila* experiments, lipid staining was decreased by *Pdp1γ* and not *Pdp1ε* RNAi. Hence, it would be interesting to elucidate whether in addition to *Pdp1ε* also *Pdp1γ* regulates the expression of the output genes such as *takeout* and whether this mechanism mediates the observed lipid-decreasing effects of *Pdp1γ* knockdown.

Our experiments with adipocytes show that in addition to systemic PDP1/HLF alterations (e.g., systemic PDP1 knockdown in adult flies), also direct cellular HLF alterations (e.g., knockdown in 3T3-L1 cells) can reduce lipid content.

In our in vitro experiments, a drug-induced adiposity was accompanied by increased levels of HLF mRNA, whereas HLF RNAi was accompanied by decreased lipid content. Furthermore, it was previously reported that in a mouse model of severe reduction of lipid accumulation and severe loss of body weight, the liver HLF mRNA levels, along with the TEF and DBP mRNA levels, are significantly reduced [19]. Hence, HLF appears to be involved in the physiological regulation of cellular lipid levels.

In our experimental conditions, the RNAi-mediated HLF mRNA reduction significantly diminished a drug -induced adiposity (i.e., olanzapine). Olanzapine belongs to a class of drugs known as the second generation antipsychotic drugs (SGADs). All these compounds are capable of triggering significant weight gain associated with adverse metabolic alterations [20, 21]. It has been proposed that these side effects may occur by a direct stimulatory action of SGADs on adipocytes [12, 22, 23]. Our in vitro experiments confirmed the direct adipogenic action of olanzapine and found that this action can be diminished by HLF reduction. In

PDP1Y Drosophila   LGHAAG-LSLGLGH---------ITTKRERSPSPSDCISPDTLNPPSPAESTFSFASSGRD
HLF Mouse   PLHPGIPSPNCMQSPIRPGQLLPANRNTPSPIDPDTIQVPVGYEPDPADLALSSIPGQEM
HLF Rat   PLHPGIPSPNCMQNPIRPGQLLPANRNTPSPIDPDTIQVPVGYEPDPADLALSSIPGPEM
HLF Human   PLHPGIPSPNCMQSPIRPGQLLPANRNTPSPIDPDTIQVPVGYEPDPADLALSSIPGQEM
TEF Mouse   IFQPSETVSSTESS-------LEKERETPSPIDPSCVEVDVNFNPDPADLVLSSVPGGEL
TEF Rat   IFQPSETVSSTESS-------LEKERETPSPIDPNCVEVDVNFNPDPADLVLSSVPGGEL
TEF Human   IFQPSETVSSTESS-------LEKERETPSPIDPNCVEVDVNFNPDPADLVLSSVPGGEL
DBP Mouse   PGHAPARATLGAAGGHRAG---LTSRDTPSPVDPDTVEVLMTFEPDPADLALSSIPGHET
DBP Rat   PGHAPARATLGAAGGHRAG---LTSRDTPSPVDPDTVEVLMTFEPDPADLALSSIPGHET
DBP Human   PGHAPARAALGTASGHRAG---LTSRDTPSPVDPDTVEVLMTFEPDPADLALSSIPGHET

PDP1Y Drosophila   FDPRTRAFSDEELKPQPMIKKSRKQFVPDELK-DDKYWARRRKNNIAAKRSRDARRQKEN
HLF Mouse   FDPRKRKFSEEELKPQPMIKKARKVFIPDDLKQDDKYWARRRKNNMAAKRSRDARRLKEN
HLF Rat   FDPRKRKFSEEELKPQPMIKKARKVFIPDDLK-DDKYWARRRKNNMAAKRSRDARRLKEN
HLF Human   FDPRKRKFSEEELKPQPMIKKARKVFIPDDLK-DDKYWARRRKNNMAAKRSRDARRLKEN
TEF Mouse   FNPRKHRFAEEDLKPQPMIKKAKKVFVPDEQK-DEKYWTRRKKNNVAAKRSRDARRLKEN
TEF Rat   FNPRKHKFAEEDLKPQPMIKKAKKVFVPDEQK-DEKYWTRRKKNNVAAKRSRDARRLKEN
TEF Human   FNPRKHKFAEEDLKPQPMIKKAKKVFVPDEQK-DEKYWTRRKKNNVAAKRSRDARRLKEN
DBP Mouse   FDPRRHRFSEEELKPQPIMKKARKVQVPEEQK-DEKYWSRRYKNNEAAKRSRDARRLKEN
DBP Rat   FDPRRHRFSEEELKPQPIMKKARKVQVPEEQK-DEKYWSRRYKNNEAAKRSRDARRLKEN
DBP Human   FDPRRHRFSEEELKPQPIMKKARKIQVPEEQK-DEKYWSRRYKNNEAAKRSRDARRLKEN

PDP1Y Drosophila   QIAMRARYLEKENATLHQEVEQLKQENMDLRARLSKFQDV----
HLF Mouse   QIAIRASFLEKENSALRQEVADLRKELGKCKNILAKYEARHGPL
HLF Rat   QIAIRASFLEKENSALRQEVADLRKELGKCKNILAKYEARHGPL
HLF Human   QIAIRASFLEKENSALRQEVADLRKELGKCKNILAKYEARHGPL
TEF Mouse   QITIRAAFLEKENTALRTEVAELRKEVGKCKTIVSKYETKYGPL
TEF Rat   QITIRAAFLEKENTALRTEVAELRKEVGKCKTIVSKYETKYGPL
TEF Human   QITIRAAFLEKENTALRTEVAELRKEVGKCKTIVSKYETKYGPL
DBP Mouse   QISVRAAFLEKENALLRQEVVAVRQELSHYRAVLSRYQAQHGTL
DBP Rat   QISVRAAFLEKENALLRQEVVAVRQELSHYRAVLSRYQAQHGTL
DBP Human   QISVRAAFLEKENALLRQEVVAVRQELSHYRAVLSRYQAQHGAL

FIGURE 4: Homology between *Drosophila* PDP1γ and mammalian bZIP transcription factors HLF, TEF, and DBP. Shown is the protein sequence alignment in the b-ZIP structural regions with the highest homology, analyzed using the online CLUSTALW multiple sequences alignment tool [13]. The NCBI database accession information is as follows. *Drosophila* PDP1γ Q9TVQ4; HLF: mouse NP_766151, rat Q64709, and human NP_002117; TEF: mouse NP_059072, rat NP_062067, and human NP_003207; DBP: mouse NP_058670, rat NP_036675, and human NP_001343. Highlighted in yellow are the sequences of *Drosophila* PDP1γ and mouse HLF.

the therapy of psychiatric patients with SGADs, a better understanding of the mechanisms that lead to metabolic side effects is needed to identify the risk factors that facilitate and exacerbate this SGADs-associated clinical problem. Our results suggest for the first time that the HLF pathway could be such a mechanism. Furthermore, the observed direct susceptibility of the adipocyte HLF and lipids to regulation by drugs (e.g., olanzapine-increased HLF mRNA and lipid contents) suggests that future pharmacological tools could be tailored specifically to the adipocyte HLF pathway to interfere therapeutically with the mechanisms of adiposity.

Abbreviations

DBP: Albumin D site-binding protein
dsRNA: Double-stranded RNA
HLF: Hepatic leukemia factor
PAR: Proline- and acidic amino acid rich
PDP1:   PAR-domain protein 1
RNAi: RNA interference
SGADs: Second generation antipsychotic drugs
TEF: Thyrotroph embryonic factor.

Conflict of Interests

The authors declare that they have no conflict of interests.

Acknowledgments

This work was supported by the Psychiatric Institute, University of Illinois at Chicago. The authors thank Nikola

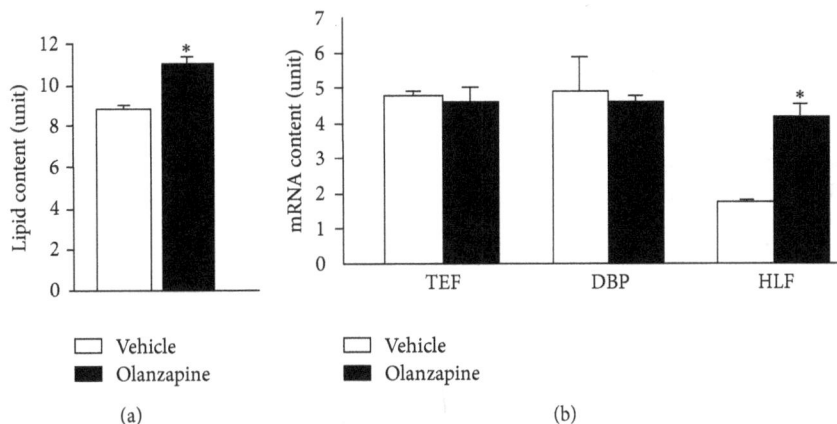

(a)

(b)

FIGURE 5: Effect of olanzapine on lipid content (a) and mRNA content of mammalian *Pdp1* homologues (b) in mouse 3T3-L1 cells. Vehicle (open bars) and olanzapine (50 μM; closed bars) were present in the culture medium for 24 h. Olanzapine increased lipid content (a) and HLF mRNA content (normalized to internal control mRNA) (b) ($^*P < 0.001$ versus corresponding control; $n = 6$; mean ± standard error mean).

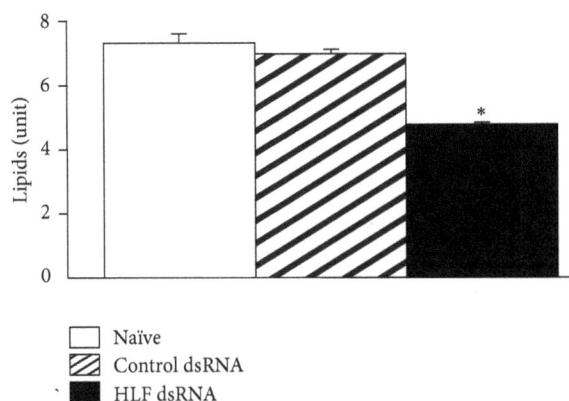

FIGURE 6: Effect of mammalian HLF RNAi on lipid content in mouse 3T3-L1 cells. Cells in culture were transfected with control and HLF dsRNA and processed for lipid assay 48 h later (see text for details). Lipid content in 3T3-L1 cells was reduced by HLF dsRNA treatment ($^*P < 0.001$ versus naïve; $n = 6$; mean ± standard error mean), which also reduced HLF mRNA content (not shown).

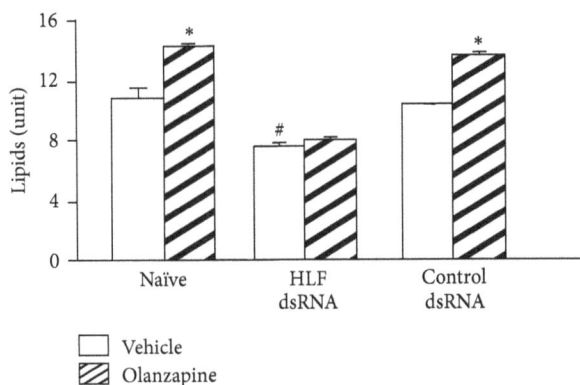

FIGURE 7: HLF RNAi prevents the adipogenic effect of olanzapine in mouse 3T3-L1 cells. Olanzapine and vehicle treatments were initiated 24 h after the initiation of transfection and conducted for the next 24 h. At the end of the treatment, cells were assayed for lipid content ($^*P < 0.001$ versus corresponding vehicle; $^\#P < 0.001$ versus naïve and control dsRNA transfection; $n = 3$; mean ± standard error mean).

Dimitrijevic for technical help, Robert V. Storti for PDP1 antibodies, and the late Erminio Costa, Director of the Psychiatric Institute, for support.

References

[1] M.-C. Saleh, M. Tassetto, R. P. Van Rij et al., "Antiviral immunity in *Drosophila* requires systemic RNA interference spread," *Nature*, vol. 458, no. 7236, pp. 346–350, 2009.

[2] S. Dzitoyeva, N. Dimitrijevic, and H. Manev, "Intra-abdominal injection of double-stranded RNA into anesthetized adult *Drosophila* triggers RNA interference in the central nervous system," *Molecular Psychiatry*, vol. 6, no. 6, pp. 665–670, 2001.

[3] S. Dzitoyeva, N. Dimitrijevic, and H. Manev, "γ-aminobutyric acid B receptor 1 mediates behavior-impairing actions of alcohol in *Drosophila*: adult RNA interference and pharmacological evidence," *Proceedings of the National Academy of Sciences of the United States of America*, vol. 100, no. 9, pp. 5485–5490, 2003.

[4] A. Goto, S. Blandin, J. Royet, J.-M. Reichhart, and E. A. Levashina, "Silencing of Toll pathway components by direct injection of double-stranded RNA into *Drosophila* adult flies," *Nucleic Acids Research*, vol. 31, no. 22, pp. 6619–6623, 2003.

[5] S.-C. Lin, M.-H. Lin, P. Horváth, K. L. Reddy, and R. V. Storti, "PDP1, a novel *Drosophila* PAR domain bZIP transcription factor expressed in developing mesoderm, endoderm and ectoderm, is a transcriptional regulator of somatic muscle genes," *Development*, vol. 124, no. 22, pp. 4685–4696, 1997.

[6] S. A. Cyran, A. M. Buchsbaum, K. L. Reddy et al., "vrille, Pdp1, and dClock form a second feedback loop in the *Drosophila* circadian clock," *Cell*, vol. 112, no. 3, pp. 329–341, 2003.

[7] C. Hermann, R. Saccon, P. R. Senthilan et al., "The circadian clock network in the brain of different *Drosophila* species," *Journal of Comparative Neurology*, vol. 521, no. 2, pp. 367–388, 2013.

[8] K. L. Reddy, A. Wohlwill, S. Dzitoeva, M.-H. Lin, S. Holbrook, and R. V. Storti, "The *Drosophila* PAR domain protein 1 (Pdp1) gene encodes multiple differentially expressed mRNAs and proteins through the use of multiple enhancers and promoters," *Developmental Biology*, vol. 224, no. 2, pp. 401–414, 2000.

[9] C. Lim, J. Lee, E. Koo, and J. Choe, "Targeted inhibition of Pdp1ε abolishes the circadian behavior of *Drosophila melanogaster*,"

Biochemical and Biophysical Research Communications, vol. 364, no. 2, pp. 294–300, 2007.

[10] F. Gachon, P. Fonjallaz, F. Damiola et al., "The loss of circadian PAR bZip transcription factors results in epilepsy," *Genes and Development*, vol. 18, no. 12, pp. 1397–1412, 2004.

[11] F. Gachon, N. Leuenberger, T. Claudel et al., "Proline- and acidic amino acid-rich basic leucine zipper proteins modulate peroxisome proliferator-activated receptor α (PPARα) activity," *Proceedings of the National Academy of Sciences of the United States of America*, vol. 108, no. 12, pp. 4794–4799, 2011.

[12] L.-H. Yang, T.-M. Chen, S.-T. Yu, and Y.-H. Chen, "Olanzapine induces SREBP-1-related adipogenesis in 3T3-L1 cells," *Pharmacological Research*, vol. 56, no. 3, pp. 202–208, 2007.

[13] C. Combet, C. Blanchet, C. Geourjon, and G. Deléage, "NPS@: network protein sequence analysis," *Trends in Biochemical Sciences*, vol. 25, no. 3, pp. 147–150, 2000.

[14] M. Imbesi, S. Dzitoyeva, L. W. Ng, and H. Manev, "5-Lipoxygenase and epigenetic DNA methylation in aging cultures of cerebellar granule cells," *Neuroscience*, vol. 164, no. 4, pp. 1531–1537, 2009.

[15] G. Asher and U. Schibler, "Crosstalk between components of circadian and metabolic cycles in mammals," *Cell Metabolism*, vol. 13, no. 2, pp. 125–137, 2011.

[16] K. Xu, X. Zheng, and A. Sehgal, "Regulation of feeding and metabolism by neuronal and peripheral clocks in *Drosophila*," *Cell Metabolism*, vol. 8, no. 4, pp. 289–300, 2008.

[17] J. Benito, V. Hoxha, C. Lama et al., "The circadian output gene takeout is regulated by *Pdp1ε*," *Proceedings of the National Academy of Sciences of the United States of America*, vol. 107, no. 6, pp. 2544–2549, 2010.

[18] B. Dauwalder, S. Tsujimoto, J. Moss, and W. Mattox, "The *Drosophila* takeout gene is regulated by the somatic sex-determination pathway and affects male courtship behavior," *Genes and Development*, vol. 16, no. 22, pp. 2879–2892, 2002.

[19] M. T. Flowers, M. P. Keller, Y. Choi et al., "Liver gene expression analysis reveals endoplasmic reticulum stress and metabolic dysfunction in SCD1-deficient mice fed a very low-fat diet," *Physiological Genomics*, vol. 33, no. 3, pp. 361–372, 2008.

[20] D. B. Allison, J. W. Newcomer, A. L. Dunn et al., "Obesity among those with mental disorders: a National Institute of Mental Health meeting report," *American Journal of Preventive Medicine*, vol. 36, no. 4, pp. 341–350, 2009.

[21] R. Coccurello and A. Moles, "Potential mechanisms of atypical antipsychotic-induced metabolic derangement: clues for understanding obesity and novel drug design," *Pharmacology and Therapeutics*, vol. 127, no. 3, pp. 210–251, 2010.

[22] J. Minet-Ringuet, P. C. Even, P. Valet et al., "Alterations of lipid metabolism and gene expression in rat adipocytes during chronic olanzapine treatment," *Molecular Psychiatry*, vol. 12, no. 6, pp. 562–571, 2007.

[23] Z. Yang, J.-Y. Yin, Z.-C. Gong et al., "Evidence for an effect of clozapine on the regulation of fat-cell derived factors," *Clinica Chimica Acta*, vol. 408, no. 1-2, pp. 98–104, 2009.

10

Ezetimibe: Its Novel Effects on the Prevention and the Treatment of Cholesterol Gallstones and Nonalcoholic Fatty Liver Disease

Ornella de Bari,[1] Brent A. Neuschwander-Tetri,[1] Min Liu,[2] Piero Portincasa,[3] and David Q.-H. Wang[1]

[1] Division of Gastroenterology and Hepatology, Department of Internal Medicine, Edward Doisy Research Center,
  Saint Louis University School of Medicine, 1100 S. Grand Boulevard, Room 205, St. Louis, MO 63104, USA
[2] Department of Pathology and Laboratory Medicine, University of Cincinnati College of Medicine, Cincinnati, OH 45237, USA
[3] Department of Internal Medicine and Public Medicine, Clinica Medica "A. Murri", University of Bari Medical School,
  70124 Bari, Italy

Correspondence should be addressed to David Q.-H. Wang, dwang15@slu.edu

Academic Editor: William M. Pandak

The cholesterol absorption inhibitor ezetimibe can significantly reduce plasma cholesterol concentrations by inhibiting the Niemann-Pick C1-like 1 protein (NPC1L1), an intestinal sterol influx transporter that can actively facilitate the uptake of cholesterol for intestinal absorption. Unexpectedly, ezetimibe treatment also induces a complete resistance to cholesterol gallstone formation and nonalcoholic fatty liver disease (NAFLD) in addition to preventing hypercholesterolemia in mice on a Western diet. Because chylomicrons are the vehicles with which the enterocytes transport cholesterol and fatty acids into the body, ezetimibe could prevent these two most prevalent hepatobiliary diseases possibly through the regulation of chylomicron-derived cholesterol and fatty acid metabolism in the liver. It is highly likely that there is an intestinal and hepatic cross-talk through the chylomicron pathway. Therefore, understanding the molecular mechanisms whereby cholesterol and fatty acids are absorbed from the intestine could offer an efficacious novel approach to the prevention and the treatment of cholesterol gallstones and NAFLD.

1. Introduction

The small intestine is a unique organ providing dietary and reabsorbed biliary cholesterol to the body [1–3]. High plasma total and low-density lipoprotein (LDL) cholesterol concentrations are an important risk factor for cardiovascular diseases. The restriction of dietary calories, cholesterol, and saturated fat has been used as the primary initial therapeutic modality for the treatment of patients with dyslipidemia [4]. However, the reduction of dietary cholesterol is frequently not associated with a significant decrease in circulating LDL cholesterol levels, despite significant restrictions in dietary intake. Therefore, pharmacological intervention aimed to reduce intestinal cholesterol absorption is potentially an effective way of lowering plasma total and LDL cholesterol concentrations [2]. The use of cholesterol absorption inhibitors for treating hypercholesterolemia has a long history, and several classes of compounds such as

hydrophilic bile acid ursodeoxycholic acid (UDCA) [2], the bile acid sequestrants, specific lipase inhibitors, the intestinal acyl-CoA:cholesterol acyltransferase (ACAT) inhibitors [5, 6], and cholesterol ester transfer protein inhibitors [7] have been developed, and some of them are currently being evaluated in clinical trials. Recently, the discovery and development of ezetimibe, a novel, selective, and potent inhibitor that effectively blocks intestinal absorption of dietary and biliary cholesterol, opened a new door to the treatment of hypercholesterolemia [2, 8–11]. Ezetimibe, which can be administered either as monotherapy or in combination with statins, has been shown to be a safe and efficacious treatment for hypercholesterolemia, potentially enabling more patients to reach recommended LDL cholesterol standards set by the National Cholesterol Education Program Adult Treatment Panel III guidelines [12].

Unexpectedly, it was found that ezetimibe treatment can induce a complete resistance to cholesterol gallstone formation [13] and nonalcoholic fatty liver disease (NAFLD) in addition to its effect on hypercholesterolemia in mice on a Western diet [14]. Furthermore, ezetimibe can prevent gallstones by effectively reducing intestinal cholesterol absorption and biliary cholesterol secretion and protecting gallbladder motility function by desaturating bile in mice. Treatment with ezetimibe also promotes the dissolution of gallstones by forming an abundance of unsaturated micelles in bile. Furthermore, ezetimibe significantly reduces biliary cholesterol saturation and retards cholesterol crystallization in biles of patients with gallstones [15]. It is also found that ezetimibe could prevent fatty liver by reducing hepatic lipogenesis in mice on a high-fat diet and attenuating diet-induced insulin resistance, a state known to drive hepatic lipogenesis through elevated circulating insulin levels [16]. Therefore, it is highly likely that ezetimibe could be a novel approach to reduce biliary cholesterol content and hepatic triglyceride accumulation, and thus a promising strategy for preventing or treating cholesterol gallstones and NAFLD, by inhibiting intestinal cholesterol absorption [15].

In this paper, we will review recent progress in understanding the biochemical and physical-chemical mechanisms, whereby ezetimibe could prevent or treat cholesterol gallstones and NAFLD, the two most prevalent hepatobiliary diseases that constitute a considerable health care burden in the USA.

2. Chemistry and Pharmacology of Ezetimibe

Ezetimibe (SCH 58235), 1-(4-fluorophenyl)-(3R)-[3-(4-fluorophenyl)-(3S)-hydroxypropyl]-(4S)-(4-hydroxyphenyl)-2-azetidinone, and an analog, SCH 48461, (3R)-(3-Phenylpropyl)-1, (4S)-bis(4-methoxyphenyl)-2-azetidinone, are highly selective intestinal cholesterol absorption inhibitors. They can effectively and potently prevent the absorption of cholesterol by inhibiting the uptake and transport of dietary and biliary cholesterol across the apical membrane of enterocytes [17]. It has been found that the 2-azetidinones are able to inhibit cholesterol absorption at very low doses and induce significant reductions in plasma cholesterol concentrations in humans and in a series of different animal models [2, 18–20]. After oral administration, ezetimibe undergoes rapid monoglucuronidation in enterocytes and the liver during its first pass. Because ezetimibe and its glucuronide are enterohepatically recirculated, it is most likely that they could repeatedly produce an inhibitory action on the Niemann-Pick C1-like 1 protein (NPC1L1) on the apical membrane of enterocytes, exhibiting multiple peaks of serum drug concentrations with an elimination half-life of approximately 22 hours [21]. This may explain why ezetimibe has a longer duration of action and why its therapeutic effects persist for several days after its cessation. These observations support the notion that once-daily dosing should be sufficient for an adequate therapeutic effect. It has also been demonstrated that 12 hours after oral administration of the glucuronide (SCH-60663), more than 95% of the compound still can be found in the intestine. Because the glucuronide is more potent in

inhibiting cholesterol absorption than ezetimibe, it confirms that ezetimibe acts directly in the intestine as glucuronide [6]. Studies with [125I]-labeled ezetimibe glucuronide and [14C]-labeled cholesterol have found that the glucuronide could block the cholesterol uptake into the enterocytes [2] because it is often detected in the brush border membrane, a site predominantly associated with cholesterol uptake and transepithelial transport. Furthermore, ezetimibe and its analogs are relatively small molecules that may not be able to change the physical-chemical nature of the intraluminal environment, nor affect the enterohepatic flux of bile acids [2].

A careful analysis of 399 patients receiving either placebo- or ezetimibe-doses of 0.25, 1, 5, or 10 mg once daily found a median percentage reduction of plasma LDL cholesterol levels of 0%, 12.7%, 14.7%, 15.8%, and 19.4%, respectively [22]. Ten milligrams of ezetimibe daily reduces intestinal cholesterol absorption by 54% compared with placebo. This effect is accompanied by a decrease in plasma LDL cholesterol levels of 20%, a compensatory increase of 89% in hepatic cholesterol biosynthesis (versus placebo) and is also associated with a decrease in the absorption of plant sterols that are highly structurally related to cholesterol [22]. During ezetimibe treatment, there is a marked compensatory increase in cholesterol biosynthesis in the liver, but not in the extrahepatic organs, and an accelerated loss of cholesterol in the feces with little or no change in the rate of conversion of cholesterol to bile acids. Therefore, the combined administration of the cholesterol absorption inhibitor (ezetimibe) and the 3-hydroxy-3-methylglutaryl coenzyme A (HMG-CoA) reductase inhibitors (e.g., atorvastatin or simvastatin) produces an enhanced reduction in plasma total and LDL cholesterol levels, as well as provides a complementary treatment strategy for high-risk patients, including patients with homozygous familial hypercholesterolemia [23]. These results showed that ezetimibe in combination with HMG-CoA reductase inhibitors would be particularly effective at reducing plasma cholesterol levels in humans.

In the early studies, it was reported that ezetimibe may not affect intestinal absorption of triglycerides, fatty acids, bile acids, or fat-soluble vitamins, including vitamins A, D, E, and α- and β-carotenes [24]. More recently, intestinal fatty acid absorption was carefully reexamined by a sensitive and physiologically accurate method, the sucrose polybehenate technique in mice [25, 26]. Instead of monitoring the appearance in plasma of digestion products from an acutely delivered bolus of oil, fecal excretion of dietary fat is measured by this technique and normalized to the excretion of a nonabsorbable fat, sucrose polybehenate, incorporated into the diet [26]. It is observed that on the chow diet, dietary fatty acid absorption is significantly reduced from approximately 95% in control mice to about 87% in ezetimibe-treated mice. Moreover, ezetimibe treatment can significantly reduce intestinal absorption of saturated fatty acids in a graded manner that correlates with chain length. Thus, intestinal absorption of palmitate (16:0) and stearate (18:0) is reduced from approximately 90% and 70% in control mice to 80% and 50% in ezetimibe-treated mice, respectively.

Ezetimibe: Its Novel Effects on the Prevention and the Treatment of Cholesterol Gallstones and
Nonalcoholic Fatty Liver Disease

85

Intestinal absorption efficiency of medium-chain saturated fatty acids is more moderately affected, possibly because medium-chain fatty acid absorption is less dependent on the chylomicron pathway. Myristate (14 : 0) absorption is reduced by 7–10% and laurate (12 : 0) absorption by 4% in ezetimibe-treated mice as compared with control mice [26]. These experiments strongly indicate that ezetimibe can reduce intestinal absorption efficiency of not only cholesterol but also long-chain fatty acids in mice. Additionally, it has been found that besides plasma total and LDL cholesterol concentrations, ezetimibe reduces liver cholesteryl ester levels in a dose-dependent fashion in cholesterol-fed hamsters, rats, and monkeys by inhibiting intestinal cholesterol absorption. However, ezetimibe does not significantly affect plasma HDL cholesterol or triglyceride levels. It was found from an indirect measurement of chylomicrons from plasma, but not lymph, that cynomolgus monkeys fed a single high-cholesterol meal and treated with an ezetimibe analogue displayed a significant reduction in the chylomicron cholesterol content, but not in the triglyceride content [27]. These results suggest that it may be important to carefully investigate the absorption and lymphatic transfer of cholesterol and fatty acids in lymph-fistula animal models. Because chylomicrons and chylomicron remnants may be atherogenic [28], further investigation of this phenomenon might shed more light on the mechanism of the antiatherogenic effect of ezetimibe. It will also be important to investigate whether ezetimibe could influence the lipid and lipoprotein composition of chylomicrons and their physical structure, as well as their assembly and secretion by the enterocytes into the lymph in animals and humans. Of note is that while statins may increase the clearance of chylomicron remnants, they do not reduce the cholesterol content of chylomicrons. Therefore, the combination of a statin and ezetimibe could be highly effective in reducing the atherogenic potential of chylomicrons [29].

3. Mechanisms of Ezetimibe Action on Intestinal Absorption of Cholesterol and Fatty Acids

Although ezetimibe reduces intestinal cholesterol absorption, it does not influence intestinal gene expression levels of *Abcg5*, *Abcg8*, *Sr-b1*, and *Abca1* in mice [2, 30]. Employing a genomic-bioinformatics approach, Altmann et al. [31] identified transcripts containing expression patterns and structural characteristics anticipated in cholesterol transporters (e.g., sterol-sensing and transmembrane domains, extracellular signal peptides) and established a strong candidate for the ezetimibe-sensitive cholesterol transporter, the awkwardly named Niemann-Pick C1-like protein 1 (NPC1L1). NPC1L1 has 50% amino acid homology to NPC1 [31], which is defective in the cholesterol storage disease Niemann-Pick type C and functions in intracellular cholesterol trafficking [32]. However, in contrast to *NPC1* that is expressed in many tissues [33, 34], *NPC1L1* is expressed predominantly in the gastrointestinal tract with peak expression in the proximal jejunum. Subfractionation of brush border membranes suggests that NPC1L1 is associated with the apical membrane fraction of enterocytes. Moreover, NPC1L1 deficient mice show a ~65% reduction in intestinal cholesterol absorption (16%) compared with wild-type mice (45%). The cholesterol absorption efficiency in NPC1L1 deficient mice is unaffected by ezetimibe or cholic acid, supporting the presence of redundant alternative pathways [31]. These studies strongly suggest that NPC1L1 could be an ezetimibe-sensitive target protein and is responsible for cholesterol uptake by the enterocyte for intestinal absorption (Figure 1) [31].

Many of the details of how ezetimibe prevents cholesterol absorption have been elucidated, and recently a molecular mechanism for cholesterol uptake mediated by the NPC1L1 has been proposed. The NPC1L1 protein recycles between the plasma membrane facing the extracellular space and the endocytic recycling compartment [35]. If the cholesterol concentration in the intestinal lumen is high, it is incorporated into the plasma membrane and is sensed by NPC1L1 that is localized on the surface of apical membrane of the enterocytes [34]. Both NPC1L1 and cholesterol are then internalized together through clathrin/AP2-mediated endocytosis [36]. The clathrin-coated globular vesicles are transported along microfilaments to the endocytic recycling compartment where large quantities of cholesterol and NPC1L1 are subsequently stored [36, 37]. If the intracellular cholesterol level is low, endocytic recycling compartment-localized NPC1L1 free of cholesterol moves back to the plasma membrane along microfilaments to transfer new cholesterol as it is absorbed by the enterocytes. The key role of the NPC1L1 inhibitor ezetimibe is to prevent NPC1L1 from entering the AP2-mediated clathrin-coated vesicles. At this stage, the endocytosis of NPC1L1 is inhibited and cholesterol absorption is decreased [36].

Although it has been observed that ezetimibe can reduce intestinal fatty acid absorption in mice, the molecular mechanism of this action is still unclear. As reviewed above, the deletion of the *Npc1l1* gene also reduces intestinal absorption of fatty acids, especially long-chain fatty acids. A potential mechanism may be that inhibition of intestinal cholesterol absorption by ezetimibe could somehow influence intestinal expression of genes involved in fatty acid uptake and transport. It is well known that hydrolysis and absorption of dietary fat (mainly triglycerides) are extremely efficient processes (>90%). However, it remains a matter of debate whether intestinal fatty acid absorption occurs solely by passive diffusion or also by protein-facilitated transport. Some studies have suggested that fatty acid transporter/cluster determinant 36 (FAT/CD36) may play a role in intestinal fatty acid absorption [38, 39]. Thus, it has been hypothesized that ezetimibe may have potential inhibitory effects on "protein-facilitated" absorption of fatty acids by enterocytes [26]. As found by Western blot analysis, protein concentrations of fatty acid transport protein 4 (FATP4) in the small intestine are significantly reduced by approximately 50% in ezetimibe-treated mice compared with control mice (Figure 1), which is associated with reduced intestinal absorption of long-chain saturated fatty

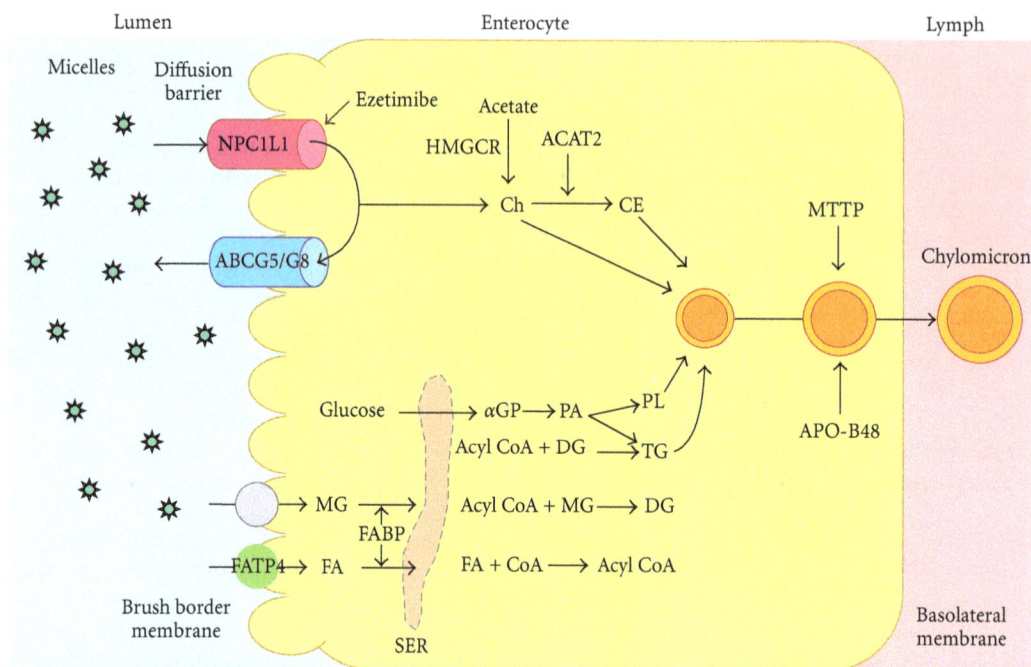

FIGURE 1: Within the intestinal lumen, the micellar solubilization of sterols facilitates movement through the diffusion barrier overlying the surface of the absorptive cells. In the presence of bile acids, large amounts of the sterol molecules are delivered to the aqueous-membrane interface so that their uptake rate is greatly increased. The Niemann-Pick C1-like 1 protein (NPC1L1), a newly identified sterol influx transporter, is located at the apical membrane of the enterocyte and may actively facilitate the uptake of cholesterol and plant sterols by promoting the passage of these molecules across the brush border membrane of the enterocyte. In contrast, ABCG5/G8 promote active efflux of cholesterol and plant sterols from the enterocyte into the intestinal lumen for excretion. The combined regulatory effects of NPC1L1 and ABCG5/G8 play a critical role in modulating the amount of cholesterol that reaches the lymph from the intestinal lumen. Ezetimibe may reduce cholesterol uptake by the enterocytes through the NPC1L1 pathway, possibly a transporter-facilitated mechanism. Absorbed cholesterol as well as some that is newly synthesized from acetate by 3-hydroxy-3-methylglutaryl-CoA reductase (HMGCR) within the enterocyte is esterified by acyl-CoA:cholesterol acyltransferase isoform 2 (ACAT2) to form cholesteryl esters. It is likely that fatty acids (FA) and monoacylglycerol (MG) could be taken up into enterocytes by facilitated transport. With the assistance of fatty acid binding protein 4 (FABP4), fatty acids and monoacylglycerol are transported into the smooth endoplasmic reticulum (SER) where they are used for the synthesis of diacylglycerol (DG) and then triacylglycerol (TG). Glucose is transported into the SER for the synthesis of phospholipids (PL) through the phosphatidic acid (PA) pathway (abbreviation: α-GP, α-glycerophosphate). All of these lipids participate in the formation of chylomicrons, a process which also requires the synthesis of apolipoprotein (APO)-B48 and the activity of microsomal triglyceride transfer protein (MTTP). As observed in lymph, the core of the secreted chylomicrons contains triglycerides and cholesteryl esters and the surface of the particles is a monolayer containing phospholipids, mainly phosphatidylcholine, unesterified cholesterol and apolipoproteins including APO-B48, APO-AI, and APO-AIV. Therefore, intestinal cholesterol absorption is a multistep process that is regulated by multiple genes. Reproduced with modifications and with permission from [50].

acids [26, 40]. It is unclear whether this inhibitory effect on intestinal FATP4 is induced by ezetimibe through a direct or indirect action pathway. Another explanation is that ezetimibe treatment significantly reduces cholesterol absorption so that the physical structure of chylomicrons may be modified and their assembly, and/or secretion into the lymphatics may be impaired. Because chylomicrons are a crucial vehicle for the transfer of cholesterol and fatty acids as triglyceride from the intestinal lumen to the lymph, impairing their formation by reducing cholesterol availability may induce a secondary action on fatty acid absorption. Because of this possible mechanism of action, it will be important to examine the physical structure of chylomicrons and their assembly and secretion into lymph to prove this hypothesis.

4. Physical-Chemistry of Bile, Physical Forms of Cholesterol Carrier, and Pathophysiology of Cholesterol Gallstones

Cholesterol, phospholipids, and bile salts are three major lipid components of bile in animals and humans [41]. Because cholesterol is virtually insoluble in an aqueous medium such as bile, specialized transport mechanisms are required to maintain it in solution and the mechanism for its solubilization in bile is complex. Similarly, phospholipids are insoluble in water and require carrier vehicles in bile. Bile salts have the property of amphiphilicity with both hydrophilic and hydrophobic areas of the molecules and are soluble in aqueous solutions to varying degrees, depending on the number and characteristics of hydroxyl groups and side chains, as well as the composition of the particular aqueous solution. Bile salt monomers can aggregate spontaneously to form simple micelles when their concentration exceeds the critical micellar concentration [42]. As defined, a micelle is a colloidal aggregation of molecules of an amphipathic compound (e.g., bile salt) in which the hydrophobic portion of each molecule faces inward and the hydrophilic groups point outward [41]. The formation of simple micelles of bile salts alone depends primarily on the concentration of bile salts. Thus, micelles are formed at, but not below, a critical micellar concentration of bile salts in bile, which is approximately 2 mmol/L [41]. The formation of micelles is also influenced by the concentrations of biliary solids and counterions, by the type of bile salt (i.e., by its degree of hydroxylation and whether it is conjugated with taurine or glycine or not), and by the temperature and pH of the bile. These simple micelles (~3 nm in diameter) are small, thermodynamically stable aggregates that are principally composed of bile salts [43]. The cholesterol can be solubilized within the hydrophobic center of the micelle. Also, simple micelles of bile salts are capable of solubilizing and incorporating phospholipids. This enables the micelles—then referred to as mixed micelles—to solubilize at least three times the amount of cholesterol solubilized by simple micelles. The solubility of cholesterol in mixed micelles is enhanced when the concentration of total lipids (bile salts, phospholipids, and cholesterol) in bile is high. Moreover, maximal solubility occurs when the molar ratio of phospholipids to bile salts is between 0.2 and 0.3 [41]. Mixed micelles (~4–8 nm in diameter) are large, thermodynamically stable aggregates that are composed of bile salts, cholesterol, and phospholipids. Their size varies depending on the relative proportion of bile salts and phospholipids. The shape of a mixed micelle is that of a lipid bilayer with the hydrophilic groups of the bile salts and phospholipids aligned on the "outside" of the bilayer, interfacing with the aqueous bile, and the hydrophobic groups on the "inside." Cholesterol molecules can, therefore, be solubilized on the inside of the bilayer away from the aqueous areas on the outside. The amount of cholesterol that can be solubilized in micelles depends on the relative proportions of bile salts and phospholipids, with additional phospholipids aiding in cholesterol solubilization [41].

Studies using techniques such as quasielastic light-scattering spectroscopy (QLS) and electron microscopy to investigate the physical-chemistry of model and native bile samples have defined more complex mechanisms of cholesterol solubilization in bile [41, 44, 45]. Beside simple and mixed micelles, biliary vesicles, nonmicellar carriers of cholesterol, do exist in bile for the solubilization of cholesterol. Vesicles are unilamellar spherical structures and contain phospholipids, cholesterol, and little, if any, bile salts. Thus, vesicles (~40 to 100 nm in diameter) are substantially larger than either simple or mixed micelles, but much smaller than liquid crystals (~500 nm in diameter) that are composed of multilamellar spherical structures. Vesicles are present in large quantities in hepatic bile and are presumably secreted by the hepatocyte [41].

Vesicles in bile have one of two distinct origins. Those formed at the canalicular membrane of hepatocytes are unilamellar and rich in phosphatidylcholines compared with cholesterol (i.e., contain one cholesterol molecule per three phosphatidylcholine molecules). Because of increasing bile salt concentrations in the biliary tree, these vesicles rapidly undergo structural rearrangements and are therefore detectable only in bile specimens analyzed immediately after collection. A second type of vesicle forms spontaneously in bile when the capacity of mixed and simple micelles to solubilize cholesterol is exceeded. These unilamellar or multilamellar vesicles are cholesterol-rich, with cholesterol content reaching as high as two cholesterol molecules per phosphatidylcholine molecule.

When concentrations of bile salts are relatively low, vesicles are relatively stable, especially in dilute hepatic bile. Moreover, vesicles may transform or convert completely to mixed micelles when bile salt concentrations in concentrated gallbladder bile are increased. When the bile salt concentration is not high enough, only some vesicles convert to micelles. Because relatively more phospholipids than cholesterol can be transferred from vesicles to mixed micelles, the residual vesicles, now remodeled, may be rich in cholesterol relative to the phospholipids. If the remaining vesicles have a relatively low cholesterol/phospholipid ratio (<1), they are relatively stable. However, if the cholesterol/phospholipid ratio in vesicles is >1, vesicles become increasingly unstable [46]. These cholesterol-rich vesicles may transfer some cholesterol to less cholesterol-rich vesicles or to micelles, or may fuse or aggregate to form larger (~500 nm in diameter) multilamellar vesicles that may now be termed liposomes or liquid crystals [41]. Liquid crystals are visible by polarizing light microscopy as lipid droplets with birefringence in the shape of a Maltese cross. Liquid crystals are inherently unstable and may form solid cholesterol monohydrate crystals, which is termed cholesterol nucleation. As a result, the nucleation of cholesterol monohydrate crystal induces a decrease in the amount of cholesterol contained in vesicles but not of cholesterol in micelles, supporting the concept that vesicles could serve as the primary source of cholesterol during cholesterol nucleation and crystallization [41].

It is well known that cholesterol cholelithiasis is a multifactorial disease influenced by a complex interaction of genetic and environmental factors [15, 42, 47]. Based on recent studies on humans and mouse models, a novel concept has been proposed that interactions of five defects could play an important role in determining the formation of cholesterol gallstones (Figure 2), which are considered in terms of *LITH* genes (genetic defect), thermodynamics (solubility defect), kinetics (nucleation defect), stasis (residence time defect), and lipid sources (metabolic defect) [48]. Furthermore, cholesterol gallstone formation represents a failure of biliary cholesterol homeostasis in which the physical-chemical balance of cholesterol solubility in bile is disturbed [41, 42, 47]. The liver is the source of cholesterol-supersaturated bile in the gallbladder with cholesterol gallstones. Thus, gallstones can be viewed in one sense as a liver disease because some metabolic defects or a combination of defects within the liver result in hypersecretion of biliary

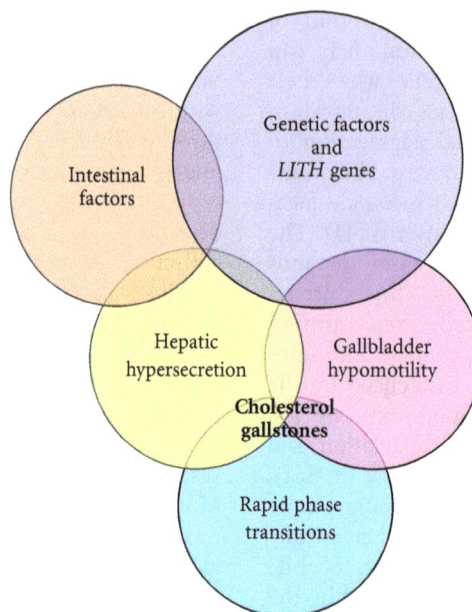

FIGURE 2: Venn diagram of five primary defects: genetic factors and *LITH* genes, hepatic hypersecretion, gallbladder hypomotility, rapid phase transitions, and intestinal factors. The hypothesis proposed is that hepatic cholesterol hypersecretion into bile is the primary defect and is the outcome in part of a complex genetic predisposition. The downstream effects include gallbladder hypomotility and rapid phase transitions. A major result of gallbladder hypomotility is alteration in the kinetics of the enterohepatic circulation of bile salts, resulting in increased cholesterol absorption and reduced bile salt absorption that lead to abnormal enterohepatic circulation of bile salts and diminished biliary bile salt pool size. Not only does gallbladder hypomotility facilitate cholesterol nucleation/crystallization, but also it allows the gallbladder to retain cholesterol monohydrate crystals. Although a large number of candidate *Lith* genes have been identified in mouse models, the identification of human *LITH* genes and their contributions to gallstones require further investigation. Reproduced with modifications and with permission from [48].

cholesterol. As noted, supersaturated bile is a prerequisite for cholesterol gallstone formation, and hypersecretion of biliary cholesterol is the primary metabolic abnormality responsible for initiating cholelithiasis. However, the gallbladder and intestine also conspire as part of a "vicious cycle" that creates physical-chemical instabilities in bile and culminates in the formation of cholesterol gallstones. Therefore, the formation of cholesterol gallstones is the final consequence of excess secretion of cholesterol from the liver into bile [42, 49]. It has been hypothesized that reducing cholesterol bioavailability in the liver for biliary secretion can prevent the formation of cholesterol gallstones and promote the dissolution of cholesterol crystals and gallstones. This information on the physical-chemistry of bile and the physical forms of cholesterol carriers can help us in understanding why ezetimibe could have a potential therapeutic effect on cholesterol gallstones.

5. Effects of Ezetimibe on the Prevention and the Treatment of Cholesterol Gallstones

Although some, but not all, studies found that high dietary cholesterol is associated with increased hepatic secretion of biliary cholesterol, epidemiological investigations have clearly demonstrated that cholesterol cholelithiasis is prevalent in cultures consuming a "Western" diet consisting of high total calories, cholesterol, saturated fatty acids, refined

carbohydrates, proteins, and salt, as well as low fiber content. Many studies have found that the gallstone incidence in North and South American as well as European populations is significantly higher than that in Asian and African populations [15, 53, 54]. Furthermore, several clinical studies have found an association between the increased incidence of cholesterol gallstones in China and a "westernization" of the traditional Chinese diet. In Japan, cholesterol cholelithiasis once was rare, but over the past 40 years with the adoption of Western-type dietary habits, the incidence has increased markedly [15, 55, 56]. Moreover, it has been observed that there is a significant and positive correlation between the efficiency of intestinal cholesterol absorption and the prevalence of cholesterol gallstone formation in mice, suggesting that high efficiency of intestinal cholesterol absorption and high dietary cholesterol are two independent risk factors in the formation of cholesterol gallstones [57]. In addition, in mouse studies, targeted deletion of the acyl-CoA:cholesterol acyltransferase gene 2 (*Acat2*) resulted in the lack of cholesterol ester synthesis in the small intestine. This causes a marked reduction in intestinal cholesterol absorption and a complete resistance to diet-induced cholesterol gallstones [13, 15]. Furthermore, the absence of expression of intestinal APO-B48, but not APO-B100, reduces biliary cholesterol secretion and cholelithogenesis, possibly by decreasing intestinal absorption and hepatic bioavailability [15, 58]. Reduced gallstone prevalence in

Ezetimibe: Its Novel Effects on the Prevention and the Treatment of Cholesterol Gallstones and
Nonalcoholic Fatty Liver Disease

89

lithogenic diet-fed apolipoprotein E knockout mice may be explained by decreased availability of chylomicron-derived cholesterol in the liver for biliary secretion [15, 59]. These studies support the notion that high dietary cholesterol through the chylomicron pathway could provide an important source of excess cholesterol molecules for secretion into bile, thereby inducing cholesterol-supersaturated bile and enhancing cholelithogenesis [15].

Indeed, because biliary cholesterol hypersecretion is an important prerequisite for cholesterol gallstone formation [15, 42, 47], inhibition of cholesterol absorption in the intestine, or hepatic uptake of chylomicron remnants has become an attractive alternative to decrease biliary cholesterol secretion and saturation [15]. Since ezetimibe significantly suppresses intestinal cholesterol absorption via the NPC1L1 pathway [15, 60], possibly a transporter-facilitated mechanism [15, 34], this should reduce the cholesterol content of the liver, which in turn decreases bioavailability of cholesterol for biliary secretion [15].

It has been found that ezetimibe induces a significant dose-dependent reduction in intestinal cholesterol absorption efficiency, coupled with a significant dose-dependent decrease in biliary cholesterol outputs and gallstone prevalence rates (Figure 3). In particular, even under high dietary cholesterol loads, cholesterol gallstones can be prevented by ezetimibe in C57L mice carrying the *Lith1* and *Lith2* genes that predispose to cholesterol stone formation [15]. Although ezetimibe substantially reduces cholesterol concentrations and to some extent phospholipid concentrations, but not bile acid concentrations in gallbladder bile, all crystallization pathways and phase boundaries on the bile phase diagram are not influenced by ezetimibe [15]. Furthermore, in company with increased doses of ezetimibe, the relative lipid compositions of pooled gallbladder bile samples are progressively shifted down and to the left of the phase diagram, entering the one-phase micellar zone where there is an abundance of unsaturated micelles, but never solid cholesterol crystals or liquid crystals. Because the micellar cholesterol solubility is dramatically increased in gallbladder bile, the cholesterol molecules can be transferred from the cholesterol monohydrate surface into unsaturated micelles. As a result, gallstones become smaller and eventually dissolved (Figure 4) [15]. This excellent physical-chemical mechanism could explain, in part, how ezetimibe treatment prevents cholesterol gallstone formation in mice.

Enlarged fasting gallbladder volume, together with impaired postprandial and interdigestive gallbladder emptying, is a frequent and distinctive feature in gallstone patients [15, 61, 62], indicating that the gallbladder is another key player in cholelithogenesis. This type of "gallbladder stasis" provides time for nucleation of cholesterol crystals and their aggregation into macroscopic stones [15, 42, 47, 62]. Under conditions of cholesterol-supersaturated bile, the gallbladder absorbs a large amount of cholesterol, thereby resulting in the accumulation of excess cholesterol in the gallbladder wall. Because gallbladder absorptive cells apparently cannot assemble lipoproteins for lipid transport into plasma, the absorbed cholesterol is converted to cholesteryl ester and stored in the mucosa and lamina propria. As a result, excess

cholesterol in smooth muscle cells could stiffen sarcolemmal membranes and decouple the G-protein-mediated signal transduction that usually occurs when CCK binds to its receptor, thereby further paralyzing gallbladder contractile function and consequently impairing gallbladder emptying function. These lithogenic effects on gallbladder motility function can be completely inhibited by ezetimibe [15, 63]. This effect of ezetimibe on protecting gallbladder motility can mostly be attributed to the desaturation of bile.

Ursodeoxycholic acid (UDCA) is currently used as a first-line pharmacological therapy to treat only a subgroup of symptomatic patients with small, radiolucent cholesterol gallstones [15, 47, 64]. Extensive clinical studies have shown that long-term administration of UDCA promotes the dissolution of cholesterol gallstones and prevents the recurrence of gallstones after extracorporeal shock wave lithotripsy [15, 65]. However, because of a failure to titrate the dose adequately, optimal use of UDCA is not always achieved in clinical practice [15]. It should be pointed out that the hydrophilic bile acid UDCA can greatly favor the formation of vesicles in bile, which can enhance the growth of liquid crystals on the cholesterol monohydrate surface and their subsequent dispersion might occur during gallstone dissolution. Consequently, liquid crystalline dissolution allows the transport of a great amount of cholesterol from stones [15]. Because the cholelitholytic mechanism of ezetimibe is totally different from that of hydrophilic bile acids such as UDCA, it has been proposed that a combined therapy of ezetimibe and UDCA could be a faster means to promote the dissolution of cholesterol gallstones, because of the two distinct mechanisms via the formation of unsaturated micelles by ezetimibe and a liquid crystalline mesophase by UDCA [15], respectively.

A clinical study has been performed to examine whether ezetimibe would reduce biliary cholesterol concentrations in gallstone patients compared to overweight subjects without gallstones [15]. It was observed that 30 days after starting the medication, ezetimibe at 20 mg/day significantly reduced cholesterol concentrations and cholesterol saturation indexes (CSIs) of gallbladder biles in gallstone patients (Table 1), similar to the results as observed in mouse studies [15]. Consequently, cholesterol crystallization was retarded and detection time of cholesterol monohydrate crystals was significantly delayed as analyzed by polarized light microscopy. Although similar results between mice and gallstone patients have been observed regarding the effect of ezetimibe on the reduction in bile cholesterol concentrations and cholesterol crystallization, a long-term human study is needed to observe whether ezetimibe can reduce gallstone prevalence and completely dissolve gallstones [15].

It should be emphasized that there is a difference in tissue distribution of NPC1L1 between mice and humans. In mice, NPC1L1 is expressed only in the intestine, while in humans, it can be detected in both the intestine and liver [34]. Because of this, it has been hypothesized that ezetimibe may have different effects on biliary cholesterol output in humans than in mice. It has been found that the secretion efficiency of biliary cholesterol is most likely determined by the net effect between efflux and influx of

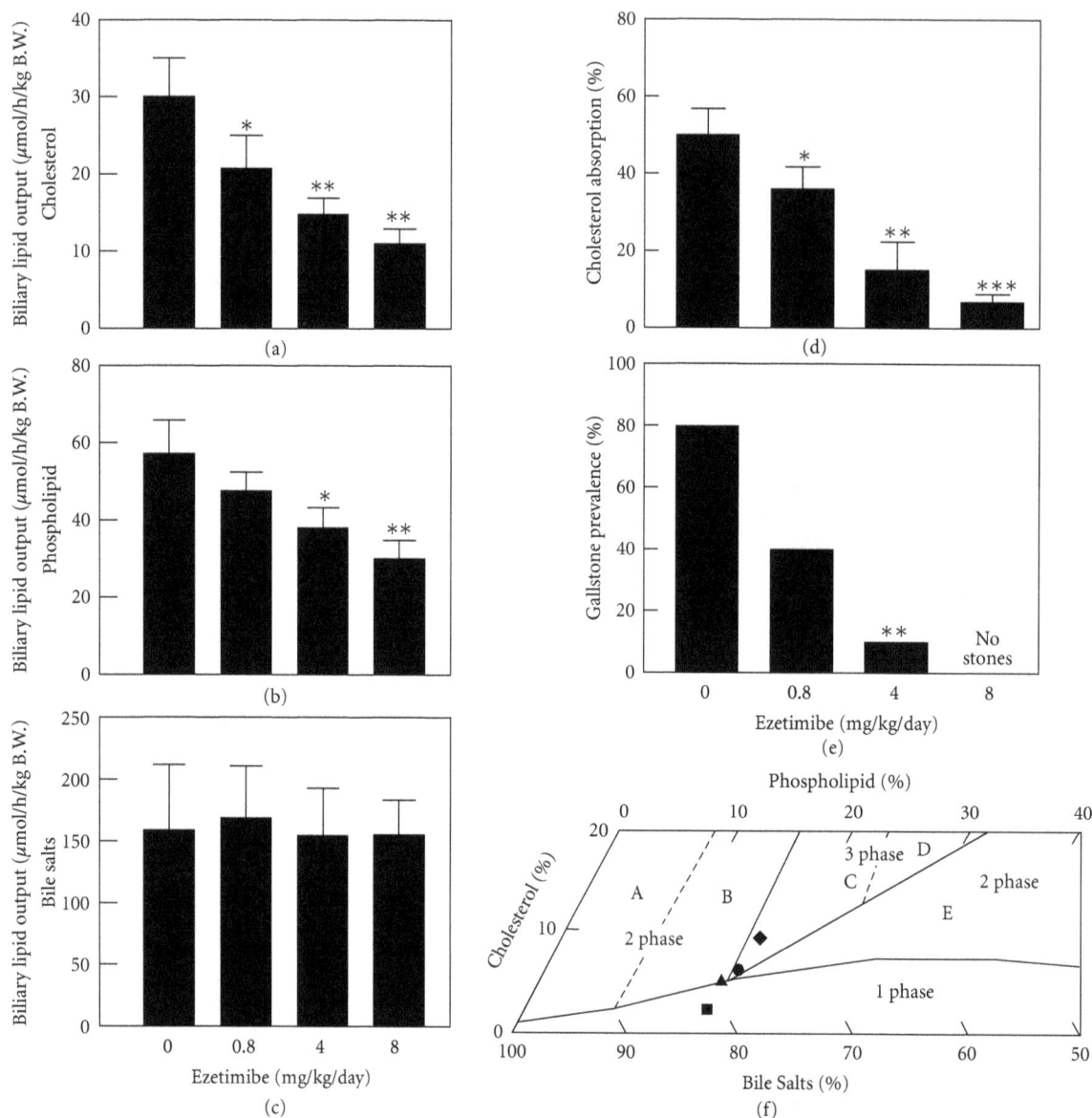

Figure 3: Effect of ezetimibe on the prevention of cholesterol gallstones. Ezetimibe significantly reduced, in a dose-dependent fashion, hepatic output of (a) biliary cholesterol and (b) phospholipid, but not (c) bile salts. *$P < 0.05$, **$P < 0.01$, and ***$P < 0.001$, compared with mice fed the lithogenic diet and receiving no ezetimibe. (d) There is a clear dose-dependent reduction in intestinal cholesterol absorption efficiency from $50 \pm 6\%$ to $4 \pm 2\%$ in chow-fed with mice, as measured by the fecal dual-isotope ratio method. (e) When doses of ezetimibe are increased from 0 to 4 mg/kg/day, gallstone prevalence rates are reduced from 80% to 10% in mice fed with the lithogenic diet for 8 weeks. No gallstones are found in mice treated with ezetimibe at 8 mg/kg/day. (f) The relative lipid composition of pooled gallbladder bile from mice fed with the lithogenic diet and receiving no ezetimibe are located in the central three-phase zone, where bile is composed of solid cholesterol monohydrate crystals, liquid crystals, and saturated micelles at equilibrium. In contrast, administration of the highest dose (8 mg/kg/day) of ezetimibe resulted in the relative biliary lipid composition of pooled gallbladder bile plotted in the one-phase micellar zone, even upon the lithogenic diet feeding for 8 weeks. By phase analysis, these bile samples are composed of unsaturated micelles at equilibrium. A symbol ♦ represents relative lipid composition of pooled gallbladder bile at 8 weeks on the lithogenic diet supplemented with ezetimibe at 0; • 0.8; ▲ 4; ■ 8 mg/kg/day. Reproduced with modifications and with permission from [15].

cholesterol molecules across the canalicular membrane of the hepatocyte, which could be regulated by the ABCG5/G8-dependent and independent pathways as well as the NPC1L1 pathway [15, 58]. Indeed, ezetimibe treatment can reduce bile cholesterol content and CSIs and prolong detection times of cholesterol monohydrate crystals in humans. One

possible reason for these results in humans is that because biliary cholesterol secretion is a unique path for excretion of cholesterol from the body in humans and animals, hepatic ABCG5/G8 may play a stronger role in the regulation of biliary cholesterol secretion than NPC1L1. Another possible explanation is that in the gut-liver axis, the intestinal

Ezetimibe: Its Novel Effects on the Prevention and the Treatment of Cholesterol Gallstones and Nonalcoholic Fatty Liver Disease

91

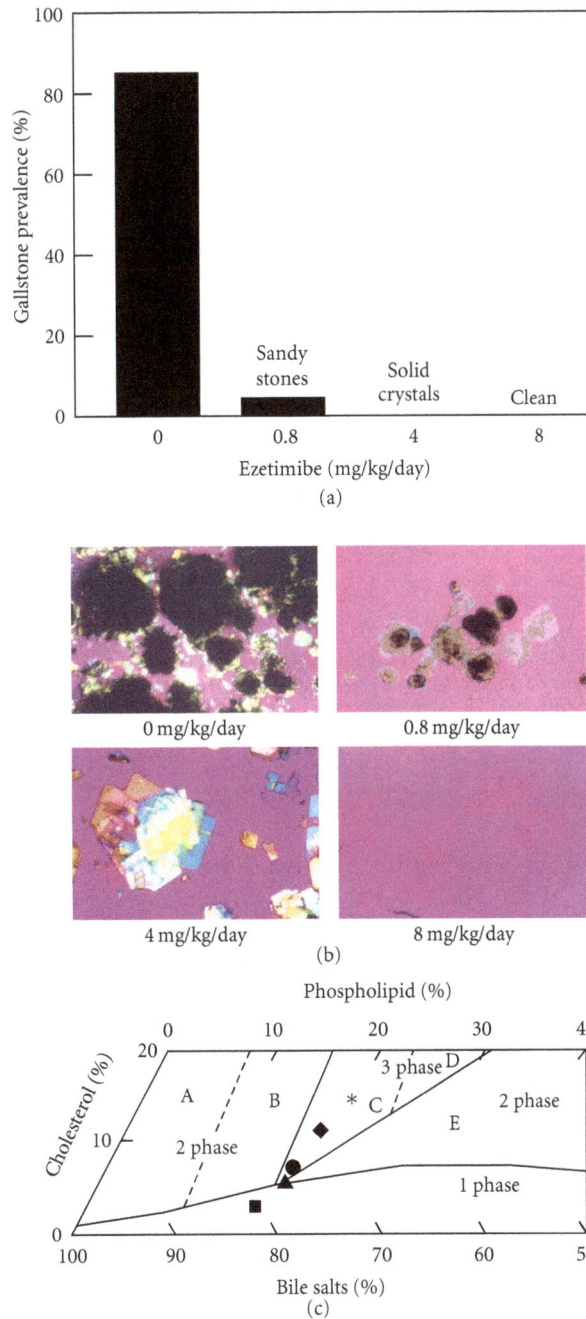

FIGURE 4: Effect of ezetimibe on the dissolution of cholesterol gallstones. (a) For gallstone dissolution experiments, mice with preexisting gallstones were fed a chow diet alone for 8 weeks, which does not result in a spontaneous dissolution of gallstones. In contrast, treatment with ezetimibe at 0.8 to 8 mg/kg/day induces rapid dissolution of gallstones. Gallstones were completely dissolved by the highest (8 mg/kg/day) dose of ezetimibe. (b) Representative photomicrographs of mucin gel, liquid crystals, cholesterol monohydrate crystals, and gallstones as observed in gallbladder biles at week 8 after ezetimibe treatment. All magnifications are ×800, except for ezetimibe treatment at 0 and 0.8 mg/kg/day, which are ×400, by polarizing light microscopy. (c) The relative lipid composition of pooled gallbladder bile from mice fed 8 weeks with the chow diet supplemented with varying doses of ezetimibe is plotted on a condensed phase diagram. Because of a 12-week feeding period of the lithogenic diet, the relative lipid composition of pooled gallbladder bile from mice that have formed cholesterol gallstones is located in the central three-phase zone. Although the lithogenic diet is replaced with the chow diet for 8 weeks, the relative biliary lipid composition of bile is still in region C, where at equilibrium the bile is composed of solid cholesterol crystals, liquid crystals, and saturated micelles. By feeding varying doses of ezetimibe, the relative lipid composition of pooled gallbladder bile gradually shifts down and, finally, enters the one-phase micellar zone. These alterations explain that gallstones are dissolved through an unsaturated micelle mechanism. A symbol ∗ represents relative lipid composition of pooled gallbladder bile from mice that have preexisting gallstones and before ezetimibe treatment; ◆ relative lipid composition of pooled gallbladder bile at the end of the gallstone dissolution study at week 8 of feeding the chow diet only (control); • 0.8; ▲ 4; ■ 8 mg/kg/day of ezetimibe. Reproduced with modifications and with permission from [15].

TABLE 1: Plasma and biliary lipids before (day 0) and at day 30 after ezetimibe treatment in humans (20 mg/day)[a].

Parameter	Overweight subjects without gallstones		Gallstone patients	
	Before	After	Before	After
BMI (kg/m^2)	31.5 ± 3.8	31.4 ± 3.4	27.0 ± 2.8	27.1 ± 2.3
Plasma lipid concentrations				
Total Ch (mg/dL)	220 ± 41	168 ± 29[b]	223 ± 32	193 ± 26
LDL Ch (mg/dL)	144 ± 53	99 ± 36	145 ± 26	115 ± 23[b]
HDL Ch (mg/dL)	44 ± 13	37 ± 13	45 ± 11	45 ± 11
TG (mg/dL)	164 ± 88	160 ± 104	166 ± 64	165 ± 76
Biliary lipid compositions of gallbladder biles				
Ch (mole%)	7.4 ± 0.7	6.8 ± 1.9	9.3 ± 1.9	7.2 ± 1.2[b]
PL (mole%)	20.2 ± 2.4	21.8 ± 2.5	19.3 ± 2.8	20.0 ± 3.5
BS (mole%)	72.4 ± 2.9	71.4 ± 3.9	71.4 ± 4.3	72.8 ± 4.2
Ch/PL ratio	0.37 ± 0.03	0.31 ± 0.08	0.48 ± 0.05	0.37 ± 0.06[c]
Ch/BS ratio	0.10 ± 0.01	0.10 ± 0.03	0.13 ± 0.03	0.10 ± 0.02
[TL] (g/dL)	5.3 ± 0.4	5.0 ± 0.9	5.5 ± 0.7	5.3 ± 0.8
CSI	1.2 ± 0.1	1.0 ± 0.2	1.6 ± 0.2	1.3 ± 0.2[b]
CDT (days)	6.4 ± 1.1	10.4 ± 1.1[c]	4.0 ± 1.2	7.0 ± 1.3[c]

[a]Values were determined from overweight subjects without gallstones (n = 5) and gallstone patients (n = 7).
[b]P < 0.05 and [c]P < 0.01, compared with before ezetimibe treatment (paired t test).
BMI: body mass index; TG: triglycerides; Ch: cholesterol; PL: phospholipids; BS: bile salts; [TL]: total lipid concentrations; CSI: cholesterol saturation index; CDT: crystal detection time.
Reproduced with slightly modifications and with permission from [15].

NPC1L1 may play a significant role in providing dietary and reabsorbed biliary cholesterol to the body, and the inhibition of its functions by ezetimibe significantly reduces cholesterol absorption. So, the bioavailability of cholesterol from intestinal sources for biliary secretion is decreased markedly. In contrast, inhibition of the hepatic NPC1L1 by ezetimibe has a weak effect on biliary cholesterol secretion and CSI values [15]. More interestingly, similar to humans, the Golden Syrian hamster displays the abundance of NPC1L1 in the small intestine that far exceeds that in other regions of the gastrointestinal tract such as liver and gallbladder [66]. The tissue distribution pattern of NPC1L1 is nearly similar between hamsters and humans. It was found that the ezetimibe-induced reduction in intestinal cholesterol absorption is coupled with a decrease in the absolute and relative cholesterol levels in bile in hamsters fed a high-cholesterol diet [66]. These results are consistent with a recent finding that ezetimibe treatment significantly reduces biliary cholesterol saturation in patients with gallstones.

Overall, ezetimibe treatment can prevent cholesterol gallstones mainly through inhibiting intestinal cholesterol absorption so that hepatic secretion of biliary cholesterol is reduced, and gallbladder motility function is preserved by desaturating bile (Figure 5). Also, ezetimibe promotes the dissolution of cholesterol gallstones through a greater capacity to form an abundance of unsaturated micelles. Therefore, ezetimibe is a novel and potential cholelitholytic agent for both preventing and treating cholesterol gallstones [15].

6. Pathophysiology of Nonalcoholic Fatty Liver Disease (NAFLD)

Nonalcoholic fatty liver disease (NAFLD) is a chronic liver disease, which includes a spectrum of hepatic pathology ranging from simple triglyceride accumulation in hepatocytes (hepatic steatosis) to hepatic steatosis with inflammation (steatohepatitis), fibrosis, and cirrhosis in the absence of alcohol abuse and other causes [67–69]. NAFLD is characterized pathologically by macrovesicular steatosis, mild diffuse lobular mixed acute and chronic inflammation, perivenular and zone 3 perisinusoidal collagen deposition, hepatocyte ballooning, poorly formed Mallory-Denk bodies, glycogen nuclei in periportal hepatocytes, lobular lipogranulomas, and PAS-diastase-resistant Kupffer cells [70, 71].

NAFLD was once proposed to be the result of two distinct but related "hits" to the hepatocyte [72, 73]. The first "hit" is the development of lipid accumulation and hepatic steatosis because of an imbalance of hepatic lipid metabolism, which leads to either excessive lipid influx, decreased lipid clearance, or both [70]. At this point, steatosis is potentially reversible and does not necessarily induce permanent hepatic injury. Although it is less common and occurs in approximately 5% of individuals with steatosis, the second "hit" is more virulent, being an inflammatory process that is induced probably by oxidative stress, lipid peroxidation, and cytokine action [74]. The resulting lobular inflammation causes ballooning degeneration and perisinusoidal fibrosis, which promote apoptosis, and hepatocellular death. These alterations eventually induce scarring and

Ezetimibe: Its Novel Effects on the Prevention and the Treatment of Cholesterol Gallstones and
Nonalcoholic Fatty Liver Disease

93

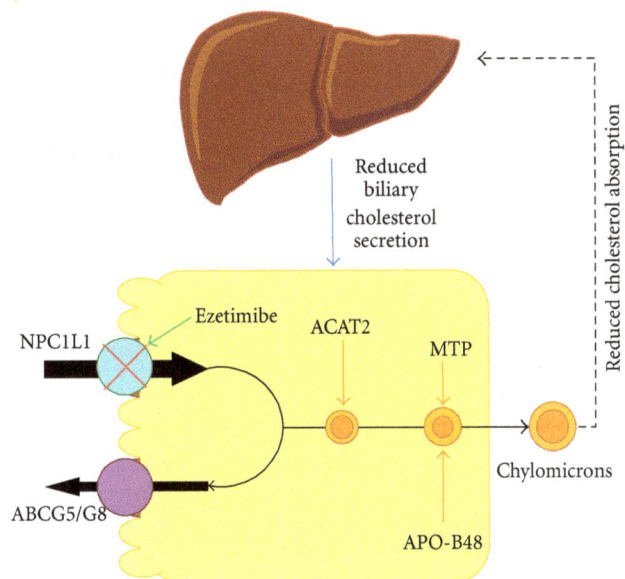

FIGURE 5: Pathways underlying the absorption of cholesterol from the intestinal lumen and its delivery to the liver. High dietary cholesterol delivery through the chylomicron pathway could provide an important source of excess cholesterol molecules for hepatic secretion into bile, thereby inducing cholesterol-supersaturated bile and enhancing cholesterol gallstone formation. Ezetimibe significantly suppresses cholesterol absorption from the small intestine via the Niemann-Pick C1-like 1 (NPC1L1) pathway, possibly by a transporter-facilitated mechanism. This effect of ezetimibe could significantly diminish the cholesterol content of the liver, which in turn remarkably decreases bioavailability of cholesterol for hepatic secretion into bile. ABCG5/G8: ATP-binding cassette (transporters) G5 and G8; ACAT2: acyl-CoA:cholesterol acyltransferase isoform 2; APO-B48: apolipoprotein B48; MTTP: microsomal triglyceride transfer protein. See text for details.

progression to nonalcoholic steatohepatitis (NASH) [75]. However, many studies have been unable to prove that either oxidant stress or lipid peroxidation is necessary for the development of steatohepatitis in humans.

Recently, the lipotoxicity model of NASH pathogenesis has emerged based on evidence showing that triglyceride often accumulates in the liver as a parallel rather than pathogenic process during lipotoxic hepatocellular injury (Figure 6) [51]. Thus, it has been hypothesized that metabolites of unesterified fatty acids play a critical role in inducing lipotoxic injury in the liver. The generation of lipotoxic metabolites of fatty acids typically occurs in parallel with the accumulation of triglyceride droplets (steatosis), resulting in a phenotype recognized as NASH, where steatosis and features of cellular injury are present together [51]. Metabolic abnormalities predisposing to lipotoxic injury include an increased supply or impaired disposal of unesterified fatty acids. More importantly, insulin resistance could play a central role in these processes by allowing unsuppressed lipolysis in adipocytes resulting in an excessive flow of fatty acids from adipose tissues and also impairing peripheral glucose disposal [51]. De novo lipogenesis in the liver using excessive dietary carbohydrate as a substrate for fatty acid synthesis is also a significant contributor to the burden of saturated fatty acids in the liver. Fatty acid disposal in the liver occurs through oxidative pathways and through the formation of triglyceride which is either stored temporarily as lipid droplets or secreted as VLDL [51]. Additional factors, including oxidative stress, mitochondrial dysfunction, gut-derived lipopolysaccharide and adipocytokines,

may promote further hepatocellular damage [76, 77]. These processes can lead to inflammation, necrosis, apoptosis and fibrogenesis, which may ultimately lead to cirrhosis, liver failure, hepatocellular carcinoma and death [78].

7. Potential Therapeutic Effects of Ezetimibe on NAFLD

Although the role of dietary fat in the pathogenesis of NAFLD continues to be investigated, evidence from animal studies supports the concept that fat overconsumption plays an important role in the etiology of hepatic steatosis [79]. It has been found that feeding a high-fat diet can induce a significant accumulation of lipids in the liver of animals such as mice and rats [80]. In humans, a large amount of dietary fat could result in the accumulation of triglyceride in the liver, but stable isotope studies found that up to only 15% of lipids accumulated in the liver are derived directly from dietary fat [81, 82]. In contrast, a low-carbohydrate diet, which is otherwise rich in protein and fat, has been used as treatment for NAFLD [83]. Furthermore, long-term overconsumption of fat could increase risk for obesity and insulin resistance, which enhances susceptible to NAFLD [84].

Indeed, mice and rats develop hepatic steatosis in response to a high-fat diet and their livers are enlarged and appear grossly pale. Histopathological studies from these livers reveal that hepatocytes are filled with multi-locular droplets of varying sizes (Figure 7) [52]. Strikingly,

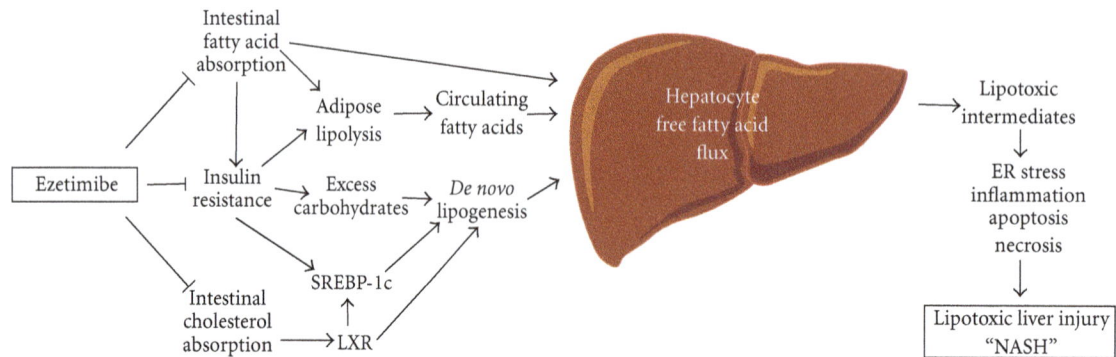

FIGURE 6: Potential therapeutic effects of ezetimibe on nonalcoholic fatty liver disease and steatohepatitis (NAFLD and NASH). On the basis of the lipotoxicity model of NAFLD and NASH [51], it has been proposed that metabolites of unesterified fatty acids may induce lipotoxic hepatocellular injury manifested as ER stress, inflammation, apoptosis, necrosis, and dysmorphic features such as ballooning and Mallory-Denk body formation. The generation of lipotoxic metabolites of fatty acids often takes place in parallel with the accumulation of triglyceride droplets (steatosis) in the liver. A high-fat diet often causes insulin resistance, a state that is associated with hyperinsulinemia and hyperglycemia. Because insulin resistance promotes an excessive flow of fatty acids from adipose tissue and also impairs peripheral glucose disposal, these alterations increase the need for fatty acid disposal in the liver through oxidative pathways and through the formation of triglyceride which is then either stored temporarily as lipid droplets or secreted as VLDL. Furthermore, elevated blood insulin and glucose activate transcription factors SREBP-1c to increase hepatic lipogenic gene expression. In addition, intestinal cholesterol absorption promotes hepatic lipogenesis via cholesterol-dependent activation of LXR. Ezetimibe treatment could block (i) intestinal fatty acid absorption, which could reduce a delivery of fatty acids from the gut to the adipose tissue through the chylomicron pathway; (ii) diet-induced insulin resistance in part by reducing intestinal fatty acid absorption; (iii) cholesterol-driven lipogenesis by inhibiting intestinal cholesterol absorption, which together may substantially reduce the burden of fatty acids on the liver. ER: endoplasmic reticulum; LXR: liver X receptor; SREBP-1c: sterol regulatory element-binding protein-1c.

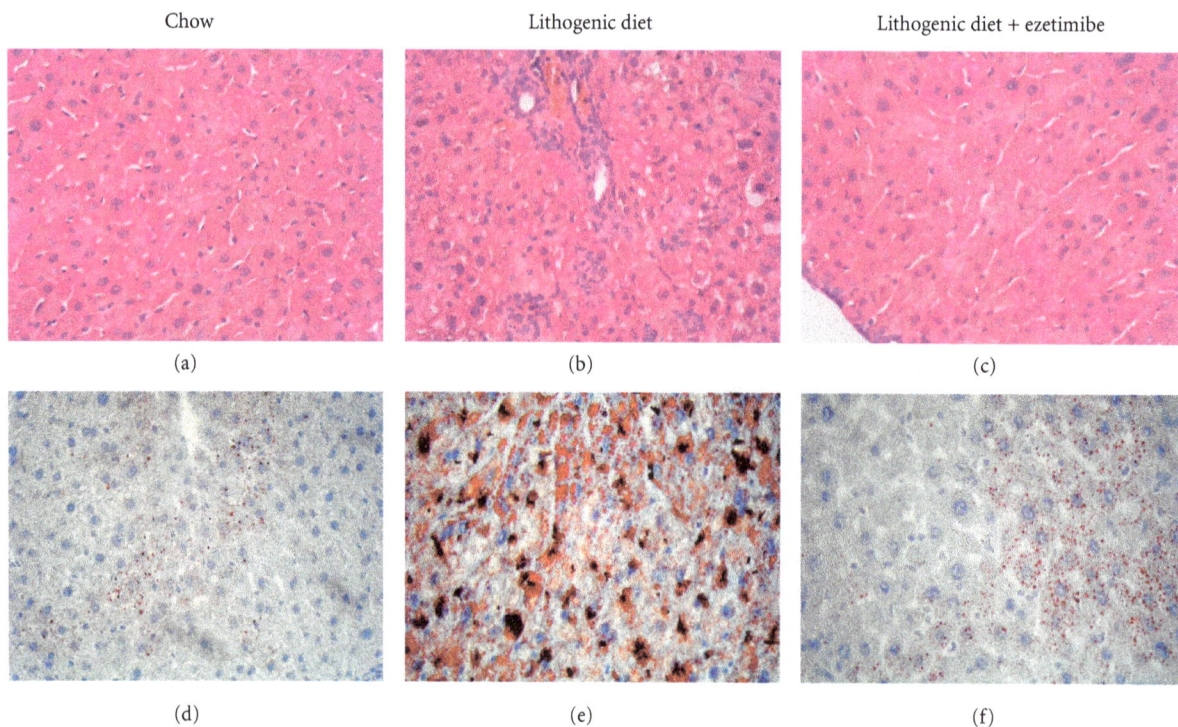

FIGURE 7: Histological characterization of the hepatic response to ezetimibe in mice fed a chow versus a lithogenic diet for 4 weeks. The liver samples from mice fed with different diets and treated with or without ezetimibe were isolated and subjected to histological analysis. Panels (a)–(c) show representative liver histology with hematoxylin-eosin staining and panels (d)–(f) show Oil Red O staining. (a) and (d) Mice fed with the chow diet. (b) and (e) Mice fed with the lithogenic diet without ezetimibe. (c) and (f) Mice fed eith the lithogenic diet with ezetimibe. The lithogenic diet induced a significant accumulation of triglyceride and cholesteryl ester in the liver as well as hepatocyte damage and inflammation. Interestingly, ezetimibe treatment markedly reduced the accumulation of lipids and prevented hepatic inflammation. Reproduced with modifications and with permission from [52].

Ezetimibe: Its Novel Effects on the Prevention and the Treatment of Cholesterol Gallstones and
Nonalcoholic Fatty Liver Disease

95

these diet-induced pathological abnormalities are completely absent in livers not only from ezetimibe-treated mice, but also from NPC1L1 deficient mice [79]. In addition, no signs of inflammatory cell infiltration are found in these livers. Hepatic concentrations of both triglyceride and cholesteryl ester are significantly reduced in ezetimibe-treated mice compared with chow-fed control mice [79].

Although a high-fat diet may promote fat accumulation in the liver by simply providing more substrate for triglyceride synthesis, an important mechanism whereby a high-fat diet may drive hepatic steatosis is by causing selective insulin resistance [79, 85]. The increased circulating insulin fails to suppress hepatic gluconeogenesis but can promote hepatic lipogenesis. In contrast, ezetimibe treatment could prevent diet-induced hepatic steatosis, weight gain, and insulin resistance [79]. These alterations are associated with reduced circulating insulin levels, hepatic *de novo* fatty acid synthesis, and hepatic levels of mRNAs for lipogenic genes including glucokinase, an enzyme critical in conversion of glucose to fat. Because elevated blood insulin increases hepatic lipogenic gene expression via transcription factors such as SREBP-1c [16, 86, 87] and glucokinase is an important mediator in this lipogenic pathway [16, 88], ezetimibe treatment may protect against diet-induced hepatic steatosis by reducing hepatic lipogenesis, mostly through preventing diet-induced insulin resistance and the associated hyperinsulinemia.

Because excessive amounts of cholesterol are lipogenic through activation of LXR by its metabolites [16, 89, 90], reduced intestinal cholesterol absorption by ezetimibe could significantly decrease cholesterol content in the liver. This may prevent diet-induced hepatic steatosis in part by reducing cholesterol-dependent LXR activation in the liver [16].

Nevertheless, ezetimibe treatment indeed plays a significant role in preventing diet-induced fatty liver in animals such as mice and rats; however, its therapeutic effect on NAFLD needs to be further investigated and proven in humans.

8. Future Research Directions and Clinical Applications

Ezetimibe is a highly potential and selective cholesterol absorption inhibitor that prevents absorption of cholesterol from dietary and biliary sources by suppressing uptake and transport of cholesterol through the enterocytes. Although there is clear evidence showing that ezetimibe can inhibit cholesterol absorption through the NPC1L1 pathway, careful and systematic studies are needed to confirm whether ezetimibe could reduce intestinal absorption of fatty acids in animal models by direct measurement of their absorption and lymphatic transport and studies need to be undertaken in humans by a balance method of intestinal fatty acid absorption. Because of significantly reduced absorption of intestinal cholesterol and fatty acids, the physical structure of chylomicrons and their metabolism in adipose tissues and liver could be influenced by ezetimibe treatment. To evaluate treatment duration, clinical response rates and the

overall cost-benefit analysis on cholesterol gallstones and NAFLD, long-term human studies are needed. Similar to atherosclerosis, the risk for cholesterol gallstone formation and NAFLD increases with dyslipidemia, hyperinsulinemia, obesity, diabetes, sedentary lifestyle, and aging. It is highly likely that the long-term administration of ezetimibe may benefit this group of subjects who could have a high predisposition to cholesterol gallstones and NAFLD.

Acknowledgments

This work was supported in part by research Grants DK54012, DK73917 (D. Q.-H. Wang), and DK70992 (M. Liu) from the National Institutes of Health (US Public Health Service), and FIRB 2003 RBAU01RANB002 (P. Portincasa) from the Italian Ministry of University and Research. P. Portincasa was a recipient of the short-term mobility grant 2005 from the Italian National Research Council (CNR). The authors are most grateful to Dr. Juan F. Miquel (Pontificia Universidad Católica de Chile) for providing them with high resolution figures for republishing their results in Figure 7 of the paper.

References

[1] D. Q. H. Wang, "Regulation of intestinal cholesterol absorption," *Annual Review of Physiology*, vol. 69, pp. 221–248, 2007.

[2] F. Lammert and D. Q. H. Wang, "New insights into the genetic regulation of intestinal cholesterol absorption," *Gastroenterology*, vol. 129, no. 2, pp. 718–734, 2005.

[3] L. P. Duan, H. H. Wang, and D. Q. H. Wang, "Cholesterol absorption is mainly regulated by the jejunal and ileal ATP-binding cassette sterol efflux transporters Abcg5 and Abcg8 in mice," *Journal of Lipid Research*, vol. 45, no. 7, pp. 1312–1323, 2004.

[4] A. Manhas and J. A. Farmer, "Hypolipidemic therapy and cholesterol absorption," *Current Atherosclerosis Reports*, vol. 6, no. 2, pp. 89–93, 2004.

[5] W. Insull Jr., M. Koren, J. Davignon et al., "Efficacy and short-term safety of a new ACAT inhibitor, avasimibe, on lipids, lipoproteins, and apolipoproteins, in patients with combined hyperlipidemia," *Atherosclerosis*, vol. 157, no. 1, pp. 137–144, 2001.

[6] S. M. Grundy, J. I. Cleeman, C. N. Merz et al., "Coordinating Committee of the National Cholesterol Education Program. Implications of recent clinical trials for the National Cholesterol Education Program Adult Treatment Panel III Guidelines," *Journal of the American College of Cardiology*, vol. 44, no. 3, pp. 720–732, 2004.

[7] M. H. Davidson, K. Maki, D. Umporowicz, A. Wheeler, C. Rittershaus, and U. Ryan, "The safety and immunogenicity of a CETP vaccine in healthy adults," *Atherosclerosis*, vol. 169, no. 1, pp. 113–120, 2003.

[8] R. H. Knopp, C. A. Dujovne, A. Le Beaut et al., "Evaluation of the efficacy, safety, and tolerability of ezetimibe in primary hypercholesterolaemia: a pooled analysis from two controlled phase III clinical studies," *International Journal of Clinical Practice*, vol. 57, no. 5, pp. 363–368, 2003.

[9] C. A. Dujovne, H. Bays, M. H. Davidson et al., "Reduction of LDL cholesterol in patients with primary hypercholesterolemia by SCH 48461: results of a multicenter dose-ranging

study," *Journal of Clinical Pharmacology*, vol. 41, no. 1, pp. 70–78, 2001.

[10] H. E. Bays, P. B. Moore, M. A. Drehobl et al., "Effectiveness and tolerability of ezetimibe in patients with primary hypercholesterolemia: pooled analysis of two phase II studies," *Clinical Therapeutics*, vol. 23, no. 8, pp. 1209–1230, 2001.

[11] C. A. Dujovne, M. P. Ettinger, J. F. McNeer et al., "Efficacy and safety of a potent new selective cholesterol absorption inhibitor, ezetimibe, in patients with primary hypercholesterolemia," *American Journal of Cardiology*, vol. 90, no. 10, pp. 1092–1097, 2002.

[12] C. M. Ballantyne, "Role of selective cholesterol absorption inhibition in the management of dyslipidemia," *Current Atherosclerosis Reports*, vol. 6, no. 1, pp. 52–59, 2004.

[13] K. K. Buhman, M. Accad, S. Novak et al., "Resistance to diet-induced hypercholesterolemia and gallstone formation in ACAT2-deficient mice," *Nature Medicine*, vol. 6, no. 12, pp. 1341–1347, 2000.

[14] L. Jia, Y. Ma, G. Liu, and L. Yu, "Dietary cholesterol reverses resistance to diet-induced weight gain in mice lacking Niemann-Pick C1-Like 1," *Journal of Lipid Research*, vol. 51, no. 10, pp. 3024–3033, 2010.

[15] H. H. Wang, P. Portincasa, N. Mendez-Sanchez, M. Uribe, and D. Q.-H. Wang, "Effect of ezetimibe on the prevention and dissolution of cholesterol gallstones," *Gastroenterology*, vol. 134, no. 7, pp. 2101–2110, 2008.

[16] L. Jia, Y. Ma, S. Rong et al., "Niemann-pick C1-like 1 deletion in mice prevents high-fat diet-induced fatty liver by reducing lipogenesis," *Journal of Lipid Research*, vol. 51, no. 11, pp. 3135–3144, 2010.

[17] M. van Heek, C. Farley, D. S. Compton, L. M. Hoos, A. Smith-Torhan, and H. R. Davis, "Ezetimibe potently inhibits cholesterol absorption but does not affect acute hepatic or intestinal cholesterol synthesis in rats," *British Journal of Pharmacology*, vol. 138, no. 8, pp. 1459–1464, 2003.

[18] S. B. Rosenblum, T. Huynh, A. Afonso et al., "Discovery of 1-(4-fluorophenyl)-(3R)-[3-(4-fluorophenyl)-(3S)-hydroxypropyl]-(4S)-(4-hydroxyphenyl)-2-azetidinone (SCH 58235): a designed, potent, orally active inhibitor of cholesterol absorption," *Journal of Medicinal Chemistry*, vol. 41, no. 6, pp. 973–980, 1998.

[19] D. A. Burnett, M. A. Caplen, H. R. Davis, R. E. Burrier, and J. W. Clader, "2-Azetidinones as inhibitors of cholesterol absorption," *Journal of Medicinal Chemistry*, vol. 37, no. 12, pp. 1733–1736, 1994.

[20] J. W. Clader, "The discovery of ezetimibe: a view from outside the receptor," *Journal of Medicinal Chemistry*, vol. 47, no. 1, pp. 1–9, 2004.

[21] T. Kosoglou, P. Statkevich, A. O. Johnson-Levonas, J. F. Paolini, A. J. Bergman, and K. B. Alton, "Ezetimibe: a review of its metabolism, pharmacokinetics and drug interactions," *Clinical Pharmacokinetics*, vol. 44, no. 5, pp. 467–494, 2005.

[22] T. Sudhop and K. von Bergmann, "Cholesterol absorption inhibitors for the treatment of hypercholesterolaemia," *Drugs*, vol. 62, no. 16, pp. 2333–2347, 2002.

[23] C. Gagné, D. Gaudet, and E. Bruckert, "Efficacy and safety of ezetimibe coadministered with atorvastatin or simvastatin in patients with homozygous familial hypercholesterolemia," *Circulation*, vol. 105, no. 21, pp. 2469–2475, 2002.

[24] M. H. Davidson, "Ezetimibe: a novel option for lowering cholesterol," *Expert Review of Cardiovascular Therapy*, vol. 1, no. 1, pp. 11–21, 2003.

[25] R. J. Jandacek, J. E. Heubi, and P. Tso, "A novel, noninvasive method for the measurement of intestinal fat absorption," *Gastroenterology*, vol. 127, no. 1, pp. 139–144, 2004.

[26] E. D. Labonté, L. M. Camarota, J. C. Rojas et al., "Reduced absorption of saturated fatty acids and resistance to diet-induced obesity and diabetes by ezetimibe-treated and Npc1l1−/− mice," *American Journal of Physiology—Gastrointestinal and Liver Physiology*, vol. 295, no. 4, pp. G776–G783, 2008.

[27] M. van Heek, D. S. Compton, and H. R. Davis, "The cholesterol absorption inhibitor, ezetimibe, decreases diet-induced hypercholesterolemia in monkeys," *European Journal of Pharmacology*, vol. 415, no. 1, pp. 79–84, 2001.

[28] R. J. Havel, "Remnant lipoproteins as therapeutic targets," *Current Opinion in Lipidology*, vol. 11, no. 6, pp. 615–620, 2000.

[29] A. L. Catapano, "Ezetimibe: a selective inhibitor of cholesterol absorption," *European Heart Journal Supplements*, vol. 3, no. 5, pp. E6–E10, 2001.

[30] J. J. Repa, J. M. Dietschy, and S. D. Turley, "Inhibition of cholesterol absorption by SCH 58053 in the mouse is not mediated via changes in the expression of mRNA for ABCA1, ABCG5, or ABCG8 in the enterocyte," *Journal of Lipid Research*, vol. 43, no. 11, pp. 1864–1874, 2002.

[31] S. W. Altmann, H. R. Davis, L. J. Zhu et al., "Niemann-Pick C1 Like 1 protein is critical for intestinal cholesterol absorption," *Science*, vol. 303, no. 5661, pp. 1201–1204, 2004.

[32] E. D. Carstea, J. A. Morris, K. G. Coleman et al., "Niemann-Pick C1 disease gene: homology to mediators of cholesterol homeostasis," *Science*, vol. 277, no. 5323, pp. 228–231, 1997.

[33] S. K. Loftus, J. A. Morris, E. D. Carstea et al., "Murine model of Niemann-Pick C disease: mutation in a cholesterol homeostasis gene," *Science*, vol. 277, no. 5323, pp. 232–235, 1997.

[34] J. P. Davies, B. Levy, and Y. A. Ioannou, "Evidence for a niemann-pick C (NPC) gene family: identification and characterization of NPC1L1," *Genomics*, vol. 65, no. 2, pp. 137–145, 2000.

[35] A. Di Ciaula, D. Q. H. Wang, H. H. Wang, L. Bonfrate, and P. Portincasa, "Targets for current pharmacologic therapy in cholesterol gallstone disease," *Gastroenterology Clinics of North America*, vol. 39, no. 2, pp. 245–264, 2010.

[36] L. Ge, J. Wang, W. Qi et al., "The cholesterol absorption inhibitor ezetimibe acts by blocking the sterol-induced internalization of NPC1L1," *Cell Metabolism*, vol. 7, no. 6, pp. 508–519, 2008.

[37] T. Y. Chang and C. Chang, "Ezetimibe blocks internalization of the NPC1L1/cholesterol complex," *Cell Metabolism*, vol. 7, no. 6, pp. 469–471, 2008.

[38] V. A. Drover, D. V. Nguyen, C. C. Bastie et al., "CD36 mediates both cellular uptake of very long chain fatty acids and their intestinal absorption in mice," *Journal of Biological Chemistry*, vol. 283, no. 19, pp. 13108–13115, 2008.

[39] F. Nassir, B. Wilson, X. Han, R. W. Gross, and N. A. Abumrad, "CD36 is important for fatty acid and cholesterol uptake by the proximal but not distal intestine," *Journal of Biological Chemistry*, vol. 282, no. 27, pp. 19493–19501, 2007.

[40] M. H. Davidson, N. Abate, C. M. Ballantyne et al., "Ezetimibe/simvastatin compared with atorvastatin or rosuvastatin in lowering to specified levels both LDL-C and each of five other emerging risk factors for coronary heart disease: non-HDL-cholesterol, TC/HDL-C, apolipoprotein B, apo-B/apo-A-I, or

Ezetimibe: Its Novel Effects on the Prevention and the Treatment of Cholesterol Gallstones and
Nonalcoholic Fatty Liver Disease

97

C-reactive protein," *Journal of Clinical Lipidology*, vol. 2, no. 6, pp. 436–446, 2008.

[41] H. H. Wang, P. Portincasa, and D. Q. H. Wang, "Molecular pathophysiology and physical chemistry of cholesterol gallstones," *Frontiers in Bioscience*, vol. 13, no. 2, pp. 401–423, 2008.

[42] D. Q. H. Wang and N. H. Afdhal, "Genetic analysis of cholesterol gallstone formation: searching for Lith (gallstone) genes," *Current Gastroenterology Reports*, vol. 6, no. 2, pp. 140–150, 2004.

[43] D. Q. H. Wang, D. E. Cohen, and M. C. Carey, "Biliary lipids and cholesterol gallstone disease," *Journal of Lipid Research*, vol. 50, pp. S406–S411, 2009.

[44] N. A. Mazer and M. C. Carey, "Quasi-elastic light-scattering studies of aqueous biliary lipid systems: cholesterol solubilization and precipitation in model bile solutions," *Biochemistry*, vol. 22, no. 2, pp. 426–442, 1983.

[45] G. J. Somjen and T. Gilat, "A non-micellar mode of cholesterol transport in human bile," *FEBS Letters*, vol. 156, no. 2, pp. 265–268, 1983.

[46] K. J. van Erpecum, "Biliary lipids, water and cholesterol gallstones," *Biology of the Cell*, vol. 97, no. 11, pp. 815–822, 2005.

[47] P. Portincasa, A. Moschetta, and G. Palasciano, "Cholesterol gallstone disease," *The Lancet*, vol. 368, no. 9531, pp. 230–239, 2006.

[48] H. H. Wang, P. Portincasa, N. H. Afdhal, and D. Q. H. Wang, "Lith genes and genetic analysis of cholesterol gallstone formation," *Gastroenterology Clinics of North America*, vol. 39, no. 2, pp. 185–207, 2010.

[49] H. H. Wang, N. H. Afdhal, and D. Q. H. Wang, "Overexpression of estrogen receptor α increases hepatic cholesterogenesis, leading to biliary hypersecretion in mice," *Journal of Lipid Research*, vol. 47, no. 4, pp. 778–786, 2006.

[50] D. Q.-H. Wang and D. E. Cohen, "Absorption and excretion of cholesterol and other sterols," in *Clinical Lipidology: A Companion to Braunwald's Heart Disease*, C. M. Ballantyne, Ed., pp. 26–44, Saunders Elsevier, Philadelphia, Pa, USA, 1st edition, 2008.

[51] B. A. Neuschwander-Tetri, "Hepatic lipotoxicity and the pathogenesis of nonalcoholic steatohepatitis: the central role of nontriglyceride fatty acid metabolites," *Hepatology*, vol. 52, no. 2, pp. 774–788, 2010.

[52] S. Zúñiga, H. Molina, L. Azocar et al., "Ezetimibe prevents cholesterol gallstone formation in mice," *Liver International*, vol. 28, no. 7, pp. 935–947, 2008.

[53] A. K. Diehl, "Epidemiology and natural history of gallstone disease," *Gastroenterology Clinics of North America*, vol. 20, no. 1, pp. 1–19, 1991.

[54] N. Méndez-Sánchez, D. Zamora-Valdés, N. C. Chávez-Tapia, and M. Uribe, "Role of diet in cholesterol gallstone formation," *Clinica Chimica Acta*, vol. 376, no. 1-2, pp. 1–8, 2007.

[55] M. Nagase, H. Tanimura, M. Setoyama, and Y. Hikasa, "Present features of gallstones in Japan: a collective review of 2,144 cases," *The American Journal of Surgery*, vol. 135, no. 6, pp. 788–790, 1978.

[56] F. Nakayama and H. Miyake, "Changing state of gallstone disease in Japan: composition of the stones and treatment of the condition," *The American Journal of Surgery*, vol. 120, no. 6, pp. 794–799, 1970.

[57] D. Q. H. Wang, S. Tazuma, D. E. Cohen, and M. C. Carey, "Feeding natural hydrophilic bile acids inhibits intestinal cholesterol absorption: studies in the gallstone-susceptible mouse," *American Journal of Physiology—Gastrointestinal and Liver Physiology*, vol. 285, no. 3, pp. G494–G502, 2003.

[58] H. H. Wang and D. Q. H. Wang, "Reduced susceptibility to cholesterol gallstone formation in mice that do not produce apolipoprotein B48 in the intestine," *Hepatology*, vol. 42, no. 4, pp. 894–904, 2005.

[59] L. Amigo, V. Quiones, P. Mardones et al., "Impaired biliary cholesterol secretion and decreased gallstone formation in apolipoprotein E-deficient mice fed a high-cholesterol diet," *Gastroenterology*, vol. 118, no. 4, pp. 772–779, 2000.

[60] H. R. Davis Jr., D. S. Compton, L. Hoos, and G. Tetzloff, "Ezetimibe, a potent cholesterol absorption inhibitor, inhibits the development of atherosclerosis in apoE knockout mice," *Arteriosclerosis, Thrombosis, and Vascular Biology*, vol. 21, no. 12, pp. 2032–2038, 2001.

[61] P. Portincasa, A. Moschetta, A. Colecchia, D. Festi, and G. Palasciano, "Measurements of gallbladder motor function by ultrasonography: towards standardization," *Digestive and Liver Disease*, vol. 35, supplement 3, pp. S56–S61, 2003.

[62] P. Portincasa, A. Di Ciaula, H. H. Wang et al., "Coordinate regulation of gallbladder motor function in the gut-liver axis," *Hepatology*, vol. 47, no. 6, pp. 2112–2126, 2008.

[63] A. Mathur, J. J. Walker, H. H. Al-Azzawi et al., "Ezetimibe ameliorates cholecystosteatosis," *Surgery*, vol. 142, no. 2, pp. 228–233, 2007.

[64] "Efficacy and indications of ursodeoxycholic acid treatment for dissolving gallstones: a multicenter double-blind trial. Tokyo Cooperative Gallstone Study Group," *Gastroenterology*, vol. 78, no. 3, pp. 542–548, 1980.

[65] M. Sackmann, H. Niller, U. Klueppelberg et al., "Gallstone recurrence after shock-wave therapy," *Gastroenterology*, vol. 106, no. 1, pp. 225–230, 1994.

[66] M. A. Valasek, J. J. Repa, G. Quan, J. M. Dietschy, and S. D. Turley, "Inhibiting intestinal NPC1L1 activity prevents diet-induced increase in biliary cholesterol in Golden Syrian hamsters," *American Journal of Physiology—Gastrointestinal and Liver Physiology*, vol. 295, no. 4, pp. G813–G822, 2008.

[67] D. J. Chiang, M. T. Pritchard, and L. E. Nagy, "Obesity, diabetes mellitus, and liver fibrosis," *American Journal of Physiology—Gastrointestinal and Liver Physiology*, vol. 300, no. 5, pp. G697–G702, 2011.

[68] E. Fassio, E. Álvarez, N. Domínguez, G. Landeira, and C. Longo, "Natural history of nonalcoholic steatohepatitis: a longitudinal study of repeat liver biopsies," *Hepatology*, vol. 40, no. 4, pp. 820–826, 2004.

[69] P. Angulo, "Obesity and nonalcoholic fatty liver disease," *Nutrition Reviews*, vol. 65, no. 6, part 2, pp. S57–S63, 2007.

[70] S. G. Hübscher, "Histological assessment of non-alcoholic fatty liver disease," *Histopathology*, vol. 49, no. 5, pp. 450–465, 2006.

[71] D. E. Kleiner, E. M. Brunt, M. van Natta et al., "Design and validation of a histological scoring system for nonalcoholic fatty liver disease," *Hepatology*, vol. 41, no. 6, pp. 1313–1321, 2005.

[72] D. G. Tiniakos, M. B. Vos, and E. M. Brunt, "Nonalcoholic fatty liver disease: pathology and pathogenesis," *Annual Review of Pathology: Mechanisms of Disease*, vol. 5, pp. 145–171, 2010.

[73] C. P. Day and O. F. W. James, "Steatohepatitis: a tale of two "hits"?" *Gastroenterology*, vol. 114, no. 4, pp. 842–845, 1998.

[74] P. B. Soeters and R. F. Grimble, "Dangers, and benefits of the cytokine mediated response to injury and infection," *Clinical Nutrition*, vol. 28, no. 6, pp. 583–596, 2009.

[75] C. Lackner, "Hepatocellular ballooning in nonalcoholic steatohepatitis: the pathologist's perspective," *Expert Review*

of Gastroenterology & Hepatology, vol. 5, no. 2, pp. 223–231, 2011.

[76] L. A. Adams and P. Angulo, "Recent concepts in non-alcoholic fatty liver disease," *Diabetic Medicine*, vol. 22, no. 9, pp. 1129–1133, 2005.

[77] P. Angulo, "Nonalcoholic fatty liver disease," *Revista de Gastroenterología de México*, vol. 70, supplement 3, pp. 52–56, 2005.

[78] S. Zheng, L. Hoos, J. Cook et al., "Ezetimibe improves high fat and cholesterol diet-induced non-alcoholic fatty liver disease in mice," *European Journal of Pharmacology*, vol. 584, no. 1, pp. 118–124, 2008.

[79] L. Jia, J. L. Betters, and L. Yu, "Niemann-pick C1-like 1 (NPC1L1) protein in intestinal and hepatic cholesterol transport," *Annual Review of Physiology*, vol. 73, pp. 239–259, 2011.

[80] M. H. Oosterveer, T. H. van Dijk, U. J. F. Tietge et al., "High fat feeding induces hepatic fatty acid elongation in mice," *PLoS One*, vol. 4, no. 6, Article ID e6066, 2009.

[81] K. M. Utzschneider and S. E. Kahn, "Review: the role of insulin resistance in nonalcoholic fatty liver disease," *Journal of Clinical Endocrinology and Metabolism*, vol. 91, no. 12, pp. 4753–4761, 2006.

[82] K. L. Donnelly, C. I. Smith, S. J. Schwarzenberg, J. Jessurun, M. D. Boldt, and E. J. Parks, "Sources of fatty acids stored in liver and secreted via lipoproteins in patients with nonalcoholic fatty liver disease," *Journal of Clinical Investigation*, vol. 115, no. 5, pp. 1343–1351, 2005.

[83] D. Tendler, S. Lin, W. S. Yancy et al., "The effect of a low-carbohydrate, ketogenic diet on nonalcoholic fatty liver disease: a pilot study," *Digestive Diseases and Sciences*, vol. 52, no. 2, pp. 589–593, 2007.

[84] H. Grønbaek, K. L. Thomsen, J. Rungby, O. Schmitz, and H. Vilstrup, "Role of nonalcoholic fatty liver disease in the development of insulin resistance and diabetes," *Expert Review of Gastroenterology & Hepatology*, vol. 2, no. 5, pp. 705–711, 2008.

[85] G. I. Shulman, "Cellular mechanisms of insulin resistance," *Journal of Clinical Investigation*, vol. 106, no. 2, pp. 171–176, 2000.

[86] G. Chen, G. Liang, J. Ou, J. L. Goldstein, and M. S. Brown, "Central role for liver X receptor in insulin-mediated activation of SREBP-1c transcription and stimulation of fatty acid synthesis in liver," *Proceedings of the National Academy of Sciences of the United States of America*, vol. 101, no. 31, pp. 11245–11250, 2004.

[87] K. Uyeda and J. J. Repa, "Carbohydrate response element binding protein, ChREBP, a transcription factor coupling hepatic glucose utilization and lipid synthesis," *Cell Metabolism*, vol. 4, no. 2, pp. 107–110, 2006.

[88] R. Dentin, J. P. Pégorier, F. Benhamed et al., "Hepatic glucokinase is required for the synergistic action of ChREBP and SREBP-1c on glycolytic and lipogenic gene expression," *Journal of Biological Chemistry*, vol. 279, no. 19, pp. 20314–20326, 2004.

[89] B. A. Janowski, P. J. Willy, T. R. Devi, J. R. Falck, and D. J. Mangelsdorf, "An oxysterol signalling pathway mediated by the nuclear receptor LXRα," *Nature*, vol. 383, no. 6602, pp. 728–731, 1996.

[90] J. J. Repa, G. Liang, J. Ou et al., "Regulation of mouse sterol regulatory element-binding protein-1c gene (SREBP-1c) by oxysterol receptors, LXRα and LXRβ," *Genes and Development*, vol. 14, no. 22, pp. 2819–2830, 2000.

11

A Pleiotropic Role for the Orphan Nuclear Receptor Small Heterodimer Partner in Lipid Homeostasis and Metabolic Pathways

Gabriella Garruti,[1,2] Helen H. Wang,[2] Leonilde Bonfrate,[3] Ornella de Bari,[2,3] David Q.-H. Wang,[2] and Piero Portincasa[3]

[1] Section of Endocrinology, Department of Emergency and Organ Transplantations, University of Bari "Aldo Moro" Medical School, Piazza G. Cesare 11, 70124 Bari, Italy
[2] Division of Gastroenterology and Hepatology, Department of Internal Medicine, Edward Doisy Research Center, Saint Louis University School of Medicine, 1100 S. Grand Boulevard, Room 205, St. Louis, MO 63104, USA
[3] Department of Biomedical Sciences and Human Oncology, Clinica Medica "A. Murri", University of Bari Medical School, Piazza G. Cesare 11, 70124 Bari, Italy

Correspondence should be addressed to Piero Portincasa, p.portincasa@semeiotica.uniba.it

Academic Editor: B. A. Neuschwander-Tetri

Nuclear receptors (NRs) comprise one of the most abundant classes of transcriptional regulators of metabolic diseases and have emerged as promising pharmaceutical targets. Small heterodimer partner (SHP; NR0B2) is a unique orphan NR lacking a DNA-binding domain but contains a putative ligand-binding domain. SHP is a transcriptional regulator affecting multiple key biological functions and metabolic processes including cholesterol, bile acid, and fatty acid metabolism, as well as reproductive biology and glucose-energy homeostasis. About half of all mammalian NRs and several transcriptional coregulators can interact with SHP. The SHP-mediated repression of target transcription factors includes at least three mechanisms including direct interference with the C-terminal activation function 2 (AF2) coactivator domains of NRs, recruitment of corepressors, or direct interaction with the surface of NR/transcription factors. Future research must focus on synthetic ligands acting on SHP as a potential therapeutic target in a series of metabolic abnormalities. Current understanding about the pleiotropic role of SHP is examined in this paper, and principal metabolic aspects connected with SHP function will be also discussed.

1. Introduction

Nuclear receptors (NRs) constitute a unique family of ligand-modulated transcription factors. NRs mediate cellular response to small lipophilic endogenous and exogenous ligands [1, 2] and are responsible for sensing a number of hormones, including steroid and thyroid hormones, and act as positive and negative regulators of the expression of specific genes [3–5]. Therefore, NRs play a central role in many aspects of mammalian development, as well as lipid homeostasis, physiology, and metabolism. NRs make up one of the most abundant classes of transcriptional regulators in the body and have emerged as promising pharmaceutical targets.

Classically, NRs consist of several functional domains, that is, a variable N-terminal ligand-independent transactivation domain (which often exhibits a constitutive transcription activation function (AF-1)), a highly conserved DNA-binding domain (DBD) that contains two zinc fingers, a hinge domain (a variable linker region), and a multifunctional C-terminal domain. Furthermore, the C-terminal domain includes the ligand binding (LBD), the dimerization interface, and the ligand-dependent transactivation domain AF-2 [1, 6].

Small heterodimer partner (SHP; NR0B2 for nuclear receptor subfamily 0, group B, member 2; MIM number 604630, 601665) is a member of the mammalian NR superfamily, due to the presence of a putative ligand-binding domain (LBD) [7]. SHP functions as a corepressor through heterodimeric interaction with a wide array of nuclear receptors and repressing their transcriptional activity. SHP achieves its goal via several members of the NR superfamily that are able to regulate SHP expression. However, SHP is also a unique and atypical NR because it lacks the classical DNA-binding domain (DBD), generally present in other NRs [8]. The *NR0B* family of NRs consists of 2 orphan receptors: SHP and DAX-1 (dosage-sensitive sex reversal adrenal hypoplasia congenita (AHC) critical region on the X chromosome, gene 1). DAX1 is a gene whose mutation causes the X-linked adrenal hypoplasia congenita [9] and is the only family member that lacks a conventional DBD. DAX-1 (NR0B1) is therefore seen as the closest relative of SHP in the NR superfamily [10–12]. Both SHP and DAX-1 appear to be specific to vertebrates. In this respect, no homologous genes have been found in *Drosophila melanogaster* or *Caenorhabditis elegans* [12]. Whereas SHP is different from other conventional NRs both structurally and functionally, it acts as a ligand-regulated receptor in metabolic pathways [13]. SHP belongs to the orphan subfamily since there is no known ligand for this receptor, except for some retinoid-related molecules [14]. SHP inhibits transcriptional activation by working on several other nuclear receptors, that is, directly modulating the activities of conventional nuclear receptors by acting as an inducible and tissue-specific corepressor [12, 15]. The discovery of SHP dates back to 1996 [10]; since then, this orphan NR has been identified as a key transcriptional regulator of signaling pathways [8, 16] involving fundamental biological functions and metabolic processes. Such processes include cholesterol, bile acid and fatty acid metabolism, glucose and energy homeostasis, and reproductive biology [17]. Experiments performed by fluorescence *in situ* hybridization (FISH) analysis of the human metaphase chromosome have shown that SHP is found at a single locus on chromosome 1 at position 1p36.1 and consists of two exons and a single intron spanning approximately 1.8 kb with 257 amino acids in humans [18]. In mice and rats, SHP resides on chromosomes 4 and 5, respectively, both consisting of 260 amino acids. SHP expression is predominantly observed in the liver [10, 18], but it is also detected at lower levels in other tissues, including the pancreas, spleen, small intestine, colon, gallbladder, kidney, adrenal gland, ovary, lung, brainstem, cerebellum, heart, and thymus (Table 1) [19–21].

The genomic structure and human SHP domain structure are depicted in Figure 1 [15]. SHP is indeed able to repress the transcriptional activities of its target NRs and transcriptional regulators through two functional Leu-Xaa-Xaa-Leu-Leu- (LXXLL-) like motifs [22–24]. Such motifs appear to be essential for the interaction with the (activation function 2) AF-2 domains of several sets of NRs [22, 23]. The human SHP is enriched by another 12 amino acids [25–36], and this region between helix 6 and 7 is also involved in the repression of the transactivation of NRs [37].

About half of all mammalian NRs and several transcriptional coregulators can interact with SHP [12]. Since SHP lacks DNA-binding domain, it exerts the inhibitory effects

TABLE 1: Small heterodimer partner (SHP) expression [10, 18–21].

LIVER (greater)*
Spleen*
Pancreas*
Central nervous system (brainstem and cerebellum)
Adrenal gland*
Intestine (duodenum*, jejunum*, ileum*, and colon)
Gallbladder, stomach*, kidney*, ovary, lung, prostate, testis, uterus, heart*, thymus, and epididymis

All organs in the mouse. Astericks indicate SHP expression in humans [18, 133].

through protein-protein interaction [10]. SHP expression seems to follow a circadian rhythm in the liver, involving the CLOCK-BMAL1 pathway and suggesting that some of the regulatory functions of SHP and deriving functions must be temporal [19, 20, 38].

Gene expression of SHP is regulated by several factors including NRs, transcription factors, and a number of additional conditions and substances, as extensively reported in Table 2. Also, the central role of SHP is clear since this NR is able to act as a coregulator for wide range of targets, namely, NRs/transcription factors/transcriptional coregulators and few different molecules, as depicted in Table 3. In general, SHP acts as a repressor of the transcriptional activity of the specific interacting partner (via LBD of the partner and NR boxes of SHP) [12, 39–43]. However, it is also demonstrated that SHP is able to upregulate gene transcription, as in the case of PPARα and PPARγ [44–46] and NF-κB [44].

Both N-terminal NR interaction domain and C-terminal domain of SHP are important for repression [47, 48]. Overall, the SHP-mediated repression of target transcription factors occurs by at least three distinct transcriptional repression mechanisms (Figure 2).

A first mechanism involves direct interference with the AF-2 coactivator domain of NRs (competition for coactivator binding, leading to the repression of NR-mediated transcriptional activity). This is the case for the inhibition of estrogen receptor α (ERα) and estrogen receptor β (ERβ) [49].

A second mechanism for the SHP-mediated repression involves the recruitment of corepressors including direct interactions among mammalian homolog of the *Saccharomyces cerevisiae* transcriptional corepressor Sin3p (mSin3A), human Brahma (Brm), SWItch/Sucrose NonFermentable (SWI/SNF) complexes leading to the repression of cholesterol 7α-hydroxylase (CYP7A1) [50].

A third mechanism of inhibition of SHP involves the direct interaction with the surface of NR or transcription factor, resulting in the blockade of DNA binding and the consequent inhibition of its transcriptional activity. This is the case for RAR-RXR heterodimers [10], PXR-RXR binding to DNA by SHP [1], interaction with hepatocyte nuclear factor (HNF4), or Jun family of the activator protein 1 (AP-1) transcription factor complex (JunD) [51, 52].

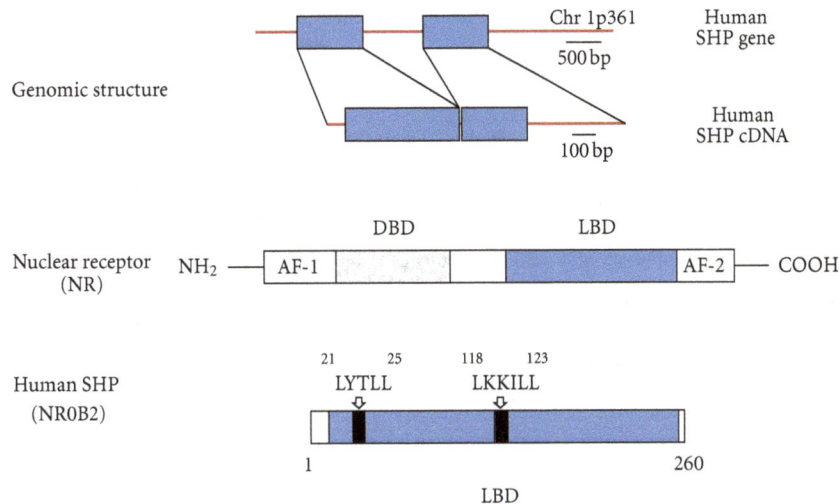

FIGURE 1: Top: the genomic structure of human SHP. Rectangles represent the two exons with a single intron spanning approximately 1.8 kilobases and located on a single locus on chromosome 1p36.1 [18]. The region 5′ includes ≈600 nucleotides from the transcription start site and is characterized by promoter activity. Bottom: typical nuclear receptor is compared with the domain structure of human SHP. The canonical structure of NR includes the N-terminal activation function 1 (AF1) domain, DNA-binding domain (DBD), ligand-binding domain (LBD), and C-terminal activation function 2 (AF2) domain. SHP lacks the DBD. Two functional LXXLL-related motifs (also named as NR boxes) are typical of the human SHP structural domains. Such motifs are located in the putative N-terminal helix 1 of the LBD and in the C-terminal region of the helix 5. While active NRs exhibit glutamic acid in AF-2, the SHP AF-2 domain is replaced with aspartic acid. Adapted from Chanda et al. [15] and Shulman and Mangelsdorf [130].

All three mechanisms might occur sequentially or alternatively according to type of cells and promoters [12].

Clearly, information on factors that increase or decrease SHP expression and that are regulated by SHP is essential for understanding the regulatory effects of this orphan NR. Few years of research have not been enough to identify a true ligand. Interestingly, it is suggested that targeting posttranslational modifications of SHP may be an effective therapeutic strategy. Selected groups of genes could be controlled to cure a vast range of metabolic and SHP-related diseases [53]. Overall, the huge amount of information on SHP function is currently available, making this NR essential in a number of functions involving cholesterol and bile acid metabolism, lipogenesis, glucose metabolism, steroid hormone biosynthesis, xenobiotic homeostasis/metabolism, and cell cycle.

In particular, the ability of SHP in interacting with different metabolic signaling pathways including bile acids and lipid homeostasis, fat mass, adipocytes, and obesity will be reviewed here.

2. Bile Acids and Lipid Homeostasis

The wide ability of SHP to target multiple genes in diverse signaling pathways points to the key role of SHP in various biological processes, including the metabolism of bile salts, glucose, and fatty acids. Both unique structure and functional properties account for the complexity of SHP signaling. Studies suggest that loss of SHP might positively affect cholesterol and bile acid homeostasis in pathophysiologically relevant conditions [54]. Bile acids (BAs) are amphipatic cholesterol metabolites which are synthesized in the liver, secreted into bile, stored in the gallbladder, and secreted postprandially into the duodenum. BAs are synthesized from cholesterol, and this pathway provides the elimination of excess cholesterol in the body [55]. Moreover, BAs should be seen as physiological detergents which, in the small intestine, are essential for the absorption, transport, and distribution of lipophilic molecules, including dietary lipids, steroids, and lipid-soluble vitamins. In the intestine, BAs undergo extensive metabolism by the intestinal microflora. A high efficient system is the enterohepatic circulation of BAs [55, 56], where more than 90–95% of BAs are returned to the liver from the terminal ileum via the portal vein. Thus, the concentration of BAs in serum, liver, and intestine is tightly regulated to prevent damage to enterohepatic tissues due to their strong detergent moiety [57–59]. The major rate-limiting step in biosynthetic pathway of BAs in humans is initiated by cholesterol 7α-hydroxylase (CYP7A1), the microsomal P450 liver enzyme, to produce two primary BAs, cholic acid, and chenodeoxycholic acid, essential in the overall balance of cholesterol homeostasis. Sterol 12α hydroxylase (CYP8B1) catalyzes the synthesis of cholic acid, a step which determines the cholic acid to CDCA ratio in the bile [60]. Secondary bile acids (deoxycholic acid and lithocholic acid) and tertiary bile acids (ursodeoxycholic acid) in humans are produced following intestinal dehydroxylation of primary bile acids by intestinal bacteria [58, 61].

Regulation of BA biosynthesis is highly coordinated and is mediated by key NRs including the orphan receptor, liver receptor homologue-1 (LRH1; NR5A2), the hepatocyte nuclear factor 4α (HNF4α), SHP, and the bile acid receptor farnesoid X receptor (FXR; NR1H4). Thus, the activation of FXR initiates a feedback regulatory loop via induction of SHP, which suppresses LRH-1- and HNF4α-dependent

TABLE 2: Regulators of the Shp gene promoter [12, 39–43].

(1) Nuclear receptors

Protein	Model(s)/putative function
ERα	Uterus, pituitary, kidney, and adrenal gland, HepG2 cell lines/biological effects of estrogens, LDL/HDL metabolism [134].
ERRα, β, γ	SHP promoter is activated by the ERRγ, while SHP inhibits ERRγ transactivation (autoregulatory loop). SHP and ERRγ coexpressed in several tissues (e.g., pancreas, kidney, and heart). Role in some forms of moderate obesity? SHP also physically interacts with ERR α and β isoforms (yeast two-hybrid and biochemical assays) [133].
FXR	Downregulation of CYP7A1-mediated bile acid biosynthesis by the FXR/SHP/LRH-1 cascade in the liver [64].
LXRα	Direct regulation of SHP and repression of CYP7A1-mediated bile acid biosynthesis (in humans not in rodents). Effect on cholesterol homeostasis [135].
LRH-1	Liver/formation of heterodimeric SHP/LRH-1 complex > inactivation of LRH-1 > SHP repression (autoregulatory negative feedback) [64, 65, 136]. Also involved in the CLOCK-BMAL1 circadian activation of SHP [38].
PPARγ	Liver/PPARγ decreases gluconeogenic gene expression by the PPARγ/RXRα heterodimer binding to the PPRE in the human SHP promoter. A mechanism explaining the SHP-mediated acute antigluconeogenic effects of PPARγ [137].
SF-1	At least five binding sites for SF-1 detected in the promoter region of SHP. Rat testis and adrenal glands, human fetal adrenal gland [136].

(2) Transcription factors

Protein	Model(s)/putative function
CLOCK-BMAL1	Liver/SHP displays a circadian expression pattern involving CLOCK-BMAL1 (core circadian clock component). Regulation of SHP promoter together with LRH-1 and SHP. Relevance for circadian liver function? [38].
E2A proteins (E47, E12, E2/5)	HepG2, HeLa, and CV-1 cells/bHLH transcription factors, the E2A proteins activate human (not mouse) hSHP promoter. E47 and SF-1 stimulate cooperatively SHP promoter. The Id protein inhibits E47 binding to hSHP promoter. A role for tissue-specific gene regulation, B-cell differentiation, tumor suppression? [138].
HNF-1α	Liver/modulation of bile acid and liver cholesterol synthesis via the FXR/SHP/LRH-1 complex and effect on CYP7A1 [69].
HNF4α	Pancreatic β-cells/decreased expression of SHP may be indirectly mediated by a downregulation of HNF4α. SHP can repress its own transcriptional activation by inhibiting HNF4 α function (feedback autoregulatory loop) and, indirectly (via HNF4 α), HNF1α function. Relevance for pancreatic islet differentiation, insulin secretion, synthesis [116].
JNK/c-Jun/AP-1	Primary rat hepatocytes/bile acid downregulation of CYP7A1-dependent bile acid biosynthesis via the JNK/cJun/AP1 pathway. SHP promoter is a direct target of activated c-Jun binding to AP-1 element [139]. Also, in HL-60 leukemia cells, c-Jun increases the transcriptional activation of the SHP promoter to activate the expression of Shp genes associated with the cascade regulation of monocytic differentiation [140].
SMILE	HEK-293T, HepG2, MCF-7, T47D, MDA-MB-435, HeLa, PC-3, C2C12, NIH 3T3, K28, Y-1, and TM4 cell lines/SMILE isoforms (SMILE-L and SMILE-S) regulate the SHP-driven inhibition of ERs transactivation in a cell-type-specific manner [25, 26, 39].
SREBP-1	Liver/effect on human (not mouse) SHP promoter. Cholesterol and bile acid homeostasis, fatty acid synthesis [27].
USF-1	HepG2, H4IIE, and AML12 cells/HGF activates AMPK signaling pathway in hepatocytes, E-box-binding transcription factor USF-1, and binding to the Shp gene promoter. SHP induction of gene expression leads to inhibition of hepatic gluconeogenesis due to SHP-repressed transcription factor HNF4α [28].

(3) Transcriptional coregulators

Protein	Model(s)/putative function
RNF31	NCI-H295R (H295R) adrenocortical carcinoma cell line, COS-7 and HeLa cells/RNF31 interacts with SHP, stabilizes DAX-1, and is required for DAX-1-mediated repression of transcription. Relevant as coregulator of steroidogenic pathways [43].
SRC-1	Murine macrophage cell line RAW 264.7, HeLa, and CV-1 cells/SHP interacts negatively with SRC-1 (a transcription coactivator of nuclear receptors and other transcription factors including NF-κB). See also oxLDL in this table [44].

TABLE 2: Continued.

(4) *Other SHP inducers*	
Factor	Model(s)/putative function
Bile acids (final intermediates)	Experiments in HepG2 cells/treatment with chenodeoxycholic acid and late intermediates in the classic pathway of bile acid synthesis: 26-OH-THC (5β-cholestane-$3\alpha,7\alpha,12\alpha,26$-tetrol), THCA ($3\alpha,7\alpha,12\alpha$-trihydroxy-5 β-cholestanoic acid), 26-OHDHC (5β-cholestane-$3\alpha,7\alpha,26$-triol), DHCA ($3\alpha,7\alpha$-dihydroxy-5β-cholestanoic acid) resulted in 2.4-6.5-fold increase in SHP mRNA expression [132]. Confirmed by Ourlin et al. with the two FXR ligands chenodeoxycholic acid and cholic acid [1].
Guggulsterone (plant sterol)	Active extract from Commiphora Mukul. FXR antagonist. In Fisher rats, guggulsterone increased transcription of bile salt export pump (BSEP) mRNA and SHP expression [29].
GW4064 (ligand)	Synthetic FXR-selective agonist [29]. In primary cultured human hepatocytes, GW4064 treatment was associated with a marked induction of *SHP* ($\approx$70-fold) and complete suppression of *CYP7A1* [64, 65]. In HepG2 cells, GW4064 (1uM) induced a 3.9-fold increase in SHP mRNA expression. Confirmed by [30].
Interleukins (various)	IL-1Ra ($-/-$) mice/high cytokine levels in IL-1Ra ($-/-$) mice reduce mRNA expression of CYP7A1 with concurrent upregulation of SHP mRNA expression [31]. SHP significantly expressed in IFN-γ/CH11-resistant HepG2 cells [32].
PGC-1α (gene expression inducer)	COS-7 cell lines/PGC-1α mediates the ligand-dependent activation of FXR and transcription of Shp gene. Relevance in mitochondrial oxidative metabolism in brown fat, skeletal muscle, and liver gluconeogenesis [33].
PMRT1 (group of protein arginine methyltransferases)	Hepatic cell lines/PRMT1 functions as FXR coactivator and has a role in chromatin remodeling. PRMT1 induces BSEP and SHP and downregulation of NTCP and CYP7A1 (targets of SHP) [30].
Procyanidins (polyphenols)	Grape seed procyanidin extract is given orally in male Wistar rats. Increase of liver mRNA levels of small heterodimer partner (SHP) (2.4-fold), cholesterol 7α-hydroxylase (CYP7A1), and cholesterol biosynthetic enzymes with improved lipidogenic profile and atherosclerotic risk [34].
(5) *Factors/conditions associated with SHP repression*	
β Klotho (type I membrane protein)	In βKlotho ($-/-$) mice: enhanced bile acid synthesis with attenuation of bile acid-mediated induction of *Shp*. βKlotho involved in CYP7A1 selective regulation [35].
IL-1β (interleukin)	SHP downregulation [36].
oxLDL (oxidized low density lipoprotein)	Murine macrophage cell line RAW 264.7, HeLa, and CV-1 cells/oxLDL decreased SHP expression. SHP transcription coactivator of NF-κB which became progressively inert in oxLDL-treated RAW 264.7 cells (see also Table 3). Relevance for differentiation mechanism of resting macrophage cells into foam cells and resulting atherogenesis [44].

AP-1: adaptor protein-1; bHLH: basic helix-loop-helix; DAX1: dosage-sensitive sex reversal adrenal hypoplasia congenita critical region on the X chromosome, gene 1; E2A: E2A2 gene products belonging to the basic helix-loop-helix (bHLH) family of transcriptor factors; ERα: estrogen receptorα; ERRγ: estrogen receptor-related receptor-γ; FXR: farnesoid X receptor; HGF: Hepatocyte growth factor; HNF-1α: hepatocyte nuclear factor-1α; HNF4α: hepatocyte nuclear factor-4α; Id: inhibitor of differentiation; IL-1Ra ($-/-$): interleukin-1 receptor antagonist; JNK: Jun N-terminal kinase; LRH-1: liver receptor homologue-1; LXRα: liver X receptorα; NFκB: nuclear factor-κB; NR: nuclear receptor; NTCP: Na$^+$-taurocholate cotransport peptide; oxLDL: oxidized low-density lipoprotein; PGC-1: PPARγ (peroxisome-proliferator-activated receptor γ) coactivator-1α; PMRT1: protein arginine methyltransferase type 1; PPRE: PPAR response element; RNF31: member of the ring-between-ring (RBR) family of E3 ubiquitin ligases; RXR α: retinoid X receptor; SF-1: steroidogenic factor-1; SHP: small (short) heterodimer partner; hSHP: human small (short) heterodimer partner; SMILE: SHP-interacting leucine zipper protein; SRC-1: steroid receptor coactivator-1; SREBP-1: sterol regulatory element binding protein-1; USF-1: upstream stimulatory factor-1.

expression of the two major pathway enzymes cholesterol 7hydroxylase (CYP7A1) and sterol 12 hydroxylase (CYP8B1).

The BA feedback regulation primarily occurs since BAs act as transcriptional regulators for the expression of the gene encoding CYP7A1. Both cholic acid and chenodeoxycholic acid function as endogenous ligands for the nuclear bile acid receptor FXR [62]. FXR expression is high in the intestine and liver, the two sites where BAs reach high concentrations to activate FXR. The transcription by FXR includes heterodimerization with retinoid X receptors (RXRs) in the cytoplasm, translocation into the nucleus, and binding to DNA response elements in the regulatory regions of target genes [63]. When the bind of BAs to FXR, SHP transcription is increased [60, 64, 65], this alteration leads to the inhibition of LRH-1 activity or HNF4α on the BA response elements

(BAREs) of CYP7A1 and CYP8B1 promoters [64, 65]. In this scenario, BA synthesis is downregulated by a precise feedback regulatory mechanism, which represents the major pathway under normal physiological conditions [64–66] (Figure 3). LRH1 is also a well-known activator of *Shp* gene transcription [64, 65], and this step leads to an autoregulatory loop of gene expression by SHP [42]. This step also includes the G protein pathway suppressor 2 (GPS2) interacting with FXR, LRH-1, and HNF4α to regulate CYP7A1 and CYP8B1 expression in human hepatocytes [67] (Table 3). A critical role in maintaining cholesterol homeostasis for CYP7A1 has been recently advocated in a model of in Cyp7a1-tg mice [68].

The hepatocyte nuclear factor-1α (HNF1α), which haploinsufficiency causes the Maturity-onset diabetes of the young type 3 (MODY3), also appears to modulate SHP expression

TABLE 3: SHP targets [12, 39–43].

(1) Nuclear receptors	
Protein	Model(s)/putative function
AR	The AR/SHP interaction leads to >95% inhibition of AR via the LXXLL motifs. Mechanisms involve inhibition of AR ligand-binding domain and AR N-terminal domain-dependent transactivation and competing with AR coactivators [23].
CAR, RAR, TR	HepG2 and JEG-3 cells/early evidence that SHP interacts with several receptor superfamily members and inhibits transactivation. CAR is an NR-inducing CYP2 and CYP3 genes involved in the metabolism of xenobiotics [10, 24].
DAX-1	Human embryonic kidney 293 cells/beside individual homodimerization of DAX1 and SHP, this is the first evidence of DAX1-SHP heterodimerization in the nucleus of mammalian cells. Involvement of the LXXLL motifs and AF-2 domain of DAX1 in this interaction. Distinct functions for SHP (different from transcriptional repressor) are anticipated [141, 142].
ER	293 human embryo kidney cells, Cos7 kidney cells/direct inhibitory binding of SHP to ERs via LXXLL-related motifs to the AF-2 domain [21]. RL95-2 human endometrial carcinoma cells/SHP inhibits the agonist activity of 4-hydroxytamoxifen displaying a potent inhibitory effect for ERα >ERβ. Direct interaction of SHP with ER and inhibition of ER transcriptional activity [143]. Prevention of tamoxifen-induced estrogen agonistic effects and neoplastic changes in the endometrium in women with breast cancer taking tamoxifen?
ERRγ	HeLa (human cervical carcinoma), CV-1 (green monkey kidney), and HEK 293 (human embryonic kidney) cell lines/SHP inhibits ERRγ transactivation by physical interaction with the 3 members of the ERR subfamily. Interaction is dependent on N-terminal receptor interaction domain of SHP and AF-2 surface of ERRγ. Part of the autoregulatory mechanism of gene expression going through ERRγ/SHP/ERRγ. A potential role in some forms of moderate human obesity during SHP mutations [133].
GR	293 human embryo kidney cells and COS-7 monkey kidney/SHP inhibits the transcriptional activity of GR via the LXXLL motif. Physiological role of SHP in glucocorticoid signaling and gluconeogenesis [22]. See also HNF4 [90] and Foxo1 [115].
HNF4	Human ANG transgenic mice and HepG2 cells treated with bile acids/evidence that bile acids negatively regulate the human ANG gene through the FXR/SHP-mediated process (inhibition of the binding of HNF4 to the ANG promoter) [90]. Mechanisms: SHP binds the AF-2 region and the N-terminal region of HNF4 and inhibits the binding of HNF4 to DNA. Also, modulation of HNF4 activity by SHP has important metabolic effects and interacts with the pathway of gluconeogenesis [47](see text and Foxo1) [115].
LRH-1	HepG2 cells/SHP interacts directly with the orphan receptor LRH-1 (AF-2 surface) and competes with other coactivators, leading to repression of LRH-1 transcriptional activity [48]. Demonstration that repression of CYP7A1 and bile acid synthesis requires coordinate interaction/transcription of FXR/LRH-1/SHP autoregulatory cascade, essential for maintenance of bile acid-induced negative feedback, and therefore hepatic cholesterol metabolism [65] (see also Figure 2).
LXRα	In vitro experiments and in vivo human colon Caco-2 cells/SHP directly inhibits the transcriptional activity of LXRα via the AF-2 domain. Relevance for direct downregulation of specific LXR target genes (controlling CYP7A1, ABCA1, ABCG1, ABCG5, ABCG8, CETP, ApoE, SREBP-1c) and therefore cholesterol-bile acid homeostasis [144].
Nur77 (NGFI-B)	HepG2 cells/Nur77 plays a key role in apoptosis of many cell types and cancer cells. Evidence that SHP functions to repress the transcriptional function of Nur77 (binding coactivator CBP, see elsewhere in this table). SHP plays a protective role in the Nur77-mediated apoptosis in liver. Mutations in SHP: a role also for affect initiation and progression of inflammatory liver diseases such as alcoholic hepatitis and hepatic viral infections? [32].
PPARα	In vitro binding assays and in vivo experiments/the promoter regions of the genes encoding the first two enzymes of the peroxisomal beta-oxidation pathway (AOx, HD), contain transcriptional regulatory sequences (PPRE) bound by the PPARα/RXRα heterodimeric complex. SHP-inhibited transcription by PPARα/RXRα heterodimers from the AOx-PPRE. SHP potentiated transcription by PPARα/RXRα heterodimers from the HD-PPRE (evidence of SHP-dependent upregulation PPARα-mediated gene transcription) [46].
PPARγ	In vitro experiments, COS-7 cells/Shp gene expressed also in adipose tissue. SHP induces PPAR activation via C terminus (direct binding to the DBD/hinge region of PPARγ) and inhibition of the repressor activity of NCoR. SHP may act as an endogenous enhancer of PPARγ by competing with NCoR [45]. Mutant SHP proteins display less enhancing activity for PPARγ compared with wild-type SHP, and a human model leading to mild obesity and insulin resistance has been described in Japanese during naturally occurring mutations [111] (see also text and Table 3).

PXR	*In vitro* experiments, human hepatocytes, mouse model on cholic acid-supplemented diet/SHP act as potent repressor of PXR transactivation. Upon sensing xenobiotics and bile acid precursors, PXR controls CYP3A gene induction and inhibits CYP7α, acting on both bile acid synthesis and catabolism. PXR function might be also inhibited in the presence of cholic acid, chenodeoxycholic acid-dependent SHP upregulation [1].
RXR	HepG2 cells/demonstration that SHP acts as a transcriptional repressor for RXR. Full inhibition by SHP requires its direct repressor activity [47].
SHP	Human embryonic kidney 293 cells/LXXLL motifs and AF-2 domain are involved in SHP homodimerization in the nucleus (similarly to DAX1-SHP heterodimerization). NR0B family members use similar mechanisms for homodimerization as well as heterodimerization. Distinct functions for SHP (different from transcriptional repressor) are anticipated [141, 142].

(2) Transcription factors

Protein	Model(s)/putative function
ARNT	RL95-2 human endometrial carcinoma cells/TCDD binds to AHR (a member of bHLH-PAS family of transcription factors). Studies on physical and functional interaction of SHP with the ligand AHR/ARNT heterodimer showed that SHP inhibits the transcriptional activity of ARNT (not AHR) *in vitro*. Consequent inhibition of binding of AHR/ARNT to XREs. [41]. Relevance for expression of several genes involved in drug and hormone metabolism [145].
BETA2/NeuroD	293T, COS-7, CV-1 cells/BETA2/NeuroD is a member of tissue-specific class B bHLH proteins and cats as a positive regulator of insulin gene expression [146] and neuronal differentiation [147]. SHP physically interacts and inhibits helix-loop-helix transcription factor BETA2/NeuroD transactivation of an E-box reporter in mouse pancreas islets. The inhibitory effect of SHP requires its C-terminal repression domain, interference with coactivator p300 for binding to BETA2/NeuroD, and direct transcriptional repression function. Relevance for development of the nervous system and the maintenance and formation of pancreatic and enteroendocrine cells [148].
C/EBPα	HepG2 hepatoma cells/SHP interacts directly with C/EBPα and represses C/EBPα-driven PEPCK gene transcription. Overall, a role for SHP in regulation of hepatic gluconeogenes is driven by C/EBPα activation in the liver [149].
Foxo1	C57BL/6J mice and HepG2 and HEK293T cells/treatment with chenodeoxycholic acid was associated with FXR-dependent SHP induction, downregulation of gluconeogenic gene expression (G6Pase, PEPCK, FBP1), interaction of the forkhead transcription factor Foxo1 with SHP, and repression of Foxo1-mediated G6Pase transcription (competition with CBP). A similar mechanism is postulated for SHP-driven HNF-4 repression of PEPCK, FBP1 transcription. A mechanism by which bile acids metabolism is linked to gluconeogenic gene expression via an SHP-dependent regulatory pathway [115].
HNF3 (Foxa)	HepG2, 293T, NIH3T3, and HeLa cells, primary hepatocytes/SHP physically interacts and inhibits the transcriptional activity of the forkhead transcription factor HNF3 (isoforms α, β, γ). Relevance for SHP-driven regulation of gluconeogenic genes encoding G6Pase, PEPCK, and bile acid synthesis (CYP7A1), via inhibition of DNA-binding of HNF3 [51].
Jun D	Two rat models of liver fibrosis and Hepatic Stellate cells (HSC)/promoting the ligand-induced FXR-SHP cascade (by the FXR ligand 6-EDCA, in rat models) and overexpressing SHP in HSC prevented fibrogenic changes in the liver. SHP binds JunD and inhibits DNA binding of adaptor protein (AP)-1 induced by thrombin. FXR ligands as therapeutic agents to treat liver fibrosis? [52].
NF-κB	Murine macrophage cell line RAW 264.7/SHP acts as a positive transcription coactivator of NF-κB and essential for NF-κB transactivation by palmitoyl lysophosphatidylcholine (one of the oxLDL constituents). Relevance for differentiation mechanism of resting macrophage cells into foam cells and resulting atherogenesis (see also [44]).
Smad	HepG2, CV-1, and HeLa cells/SHP represses Smad3-induced transcription by competing for the coactivator p300. SHP therefore represses TGF-β-induced gene expression. Relevance for TGF-β-dependent regulation of cell growth, apoptosis, carcinogenesis, and regeneration following liver injury [40]. SHP-Smad3 interaction similar to SHP-BETA2/NeuroD [148].
TRAF6, p65	Macrophages/a novel function of SHP in innate immunity involving Toll-like receptors (TLRs). SHP negatively regulates TLR signaling to NF-κB. Likely, SHP negatively regulates immune responses initiated by various pathogen-recognition receptors by forming a complex with TRAF6 and effect on TRAF6 ubiquitination. In the cytosol of LPS-stimulated cells. SHP also acts as specific transrepressor of the transcription factor p65 (part of the p50/p65 heterodimer found in NF-κB). An additional role for SHP in sepsis and inflammatory disease? [128, 129].

TABLE 3: Continued.

(3) *Transcriptional coregulators*	
Protein	Model(s)/putative function
Brm, BAF155, BAF47, mSin3A, Swi/Snf	HepG2 cells/The *CYP7A1* gene was used as a model system. SHP has direct interaction with corepressors at the level of native chromatin. SHP directly interacted and mediated the recruitment of mSin3A-Swi/Snf-Brm chromatin remodelling complex to the *CYP7A1* promoter (TATA and BARE II region of the promoter). Also, the mSinA3/HDAC1 corepressor complex is inhibiting transcription by histone deacetylation. SHP also interacted with known proteins belonging to the Swi/Snf complex (BAF155, BAF47). This mechanism explains the complex and subtle SHP-driven inhibition of hepatic bile acid synthesis [50].
CBP	HepG2 cells, CV-1 cells/SHP binds coactivator CBP and competes with Nur77. The mechanism explains the repression of the transcriptional function of Nur77, which is fundamental in apoptosis in the liver [32].
EID-1	Cos-7 cells/SHP specifically interacts with EID-1 providing inhibitory mechanisms. EID-1 (a non-HDAC cofactor) acts as inhibitor of the coregulator complex EID1–p300–CBP. Results clarify essential repression mechanisms of SHP involving coinhibitory factors (upstream targets) distinct from NRs corepressor [12, 150].
G9a, HDAC-1	Caco-2, HepG2, HeLa, Cos-1 cells/SHP localized exclusively in nuclease-sensitive euchromatin regions. SHP can functionally interact with HDAC-1 (HDAC of class I) and the euchromatic histone 3 methylase G9a, and the unmodified K9-methylated histone 3 [151]. Additional data on mechanisms involved SHP-driven repressive activity, involving also target genes regulated by G9a and SHP-mediated inhibition of hepatic bile acid synthesis via coordinated chromatin modification at target genes [152].
GPS2	Cos-7, HepG2, Huh7 cells/SHP negatively interacts with GPS2 (a stoichiometric subunit of the NR corepressor, N-Cor) complex, involved in bile acid synthesis and differential coregulation of CYP7A1 and CYP8B1 expression [67].
SIRT1	HepG2, HEK293T (293T), and HeLa cells/SIRT1 is a HDAC of class III. SHP recruits SIRT1 (activating deacetylase activity of SIRT1) to repress LRH1 transcriptional activity as well as inhibition LRH1 target gene promoter activity and mRNA levels. A novel mechanism is described for SHP repressive action and control of bile acid homeostasis. SIRT1 in working concertedly with NRs and affecting chromatin remodeling in target gene promoters [42].
SMRT/NcoR	Hepatoma cell lines/studies on the role of SHP in CAR-mediated transactivation of the CYP2B gene. SHP might interact with subunits of functionally distinct coregulator complexes, including HDAC3-N-CoR-SMRT [24, 120].
(4) *Others*	
Factor	Model(s)/putative function
miRNA-206	SHP $^{-/-}$ mice/SHP as an important transcriptional activator of miRNA-206 gene expression via a cascade dual inhibitory mechanism involving AP1 but also YY1 and ERRγ. Relevance for multiple steps involving cellular development, proliferation, and differentiation [153].
RNA Pol II	Caco-2 cells/within the pathway of SHP-LXR interaction, it is shown that SHP can interact *in vitro* with RNA polymerase II but not with TFIID and TFIIE transcription initiation factor II D (TFIID), general transcription factor II E (TFIIE) (components of the basal transcription machinery). A further mechanism by which SHP could inhibit both basal and induced transactivation [144].

ABCA1, ABCG1, ABCG5, and ABCG8: ATP-binding cassette transporters; AP1: transcription factor activator protein 1; AHR: aryl hydrocarbon receptor (AHR); ARNT: aryl hydrocarbon receptor (AHR)/AHR nuclear translocator protein; ANG: angiotensin; AOx, acyl-CoA oxidase; ApoE: apolipoprotein E; bHLH-PAS: basic helix–loop–helix-PAS; AR: androgen receptor; BAFs: Brm- or Brg-1-associated factors; BARE: bile acid response element; Brm: human Brahma; CAR: constitutive androstane receptor; CBP: CREB-binding protein; C/EBPα: CCAAT/enhancer-binding protein α; CETP: cholesteryl ester transfer protein; CREB: coactivator cAMP-response element-binding protein; CYP7A1: cholesterol-7-α-hydroxylase; DAX1: dosage-sensitive sex reversal adrenal hypoplasia congenita critical region on the X chromosome: gene 1; DBD: DNA-binding domain; 6-ECDCA, 6-ethylchenodeoxycholic acid; EID1: E1A-like inhibitor of differentiation 1; ER: estrogen receptor; ERRγ: estrogen receptor-related receptor-γ; FBP1: fructose-1,6-bisphosphatase; FXR: farnesoid X receptor; G6Pase: glucose-6-phosphate; GR: glucocorticoid receptor; GPS2: G protein pathway suppressor 2; HD: enoyl-CoA hydratase/3-hydroxyacyl-CoA dehydrogenase; HDACs: histone deacetylases; HDAC-1: histone deacetylase-1; HDAC-1: histone deacetylase-3; JunD: predominat Jun family protein; HNF3/Foxa: hepatocyte nuclear factor-3; HNF4: hepatocyte nuclear factor-4; LPS: lipopolysaccharides; LXRα: liver X receptorα; LRH-1: liver receptor homologue-1; miRNAs (miR): microRNAs; NcoR: nuclear receptor corepressor; NF-κB: nuclear factor-κB; Nur77: nuclear growth factor I-B; PEPCK: phosphoenolpyruvate carboxykinase; PPRE: peroxisome proliferator-response elements; PXR: pregnane X receptors RAR: retinoid acid receptor; RNA Pol II: RNA polymerase II; RXR: retinoid X receptor; SIRT1: sirtuin1; SREBP-1c: sterol regulatory element-binding protein-1c; TCDD, 2,3,7,8-tetrachlorodibenzo-p-dioxin; TFIID: transcription initiation factor II D (TFIID); TFIIE: transcription factor II E; TGF-β: transforming growth factor-β; TLRs: Toll-like receptors; TR: thyroid receptor; TRAF6: TNF-receptor-associated factor-6; XRE, xenobiotic response element; YY1: Ying Yang 1.

via the FXR pathway. In this respect, HNF1α ($-/-$) mice displayed a defect in bile acid transport, increased bile acid and liver cholesterol synthesis, and impaired HDL metabolism [69].

A role for SHP in mediating the recruitment of mSin3A-Swi/Snf to the CYP7A1 promoter, with chromatin remodeling and gene repression, has been described. In HepG2 cells, Kemper et al. [50] have shown that bile acid treatment

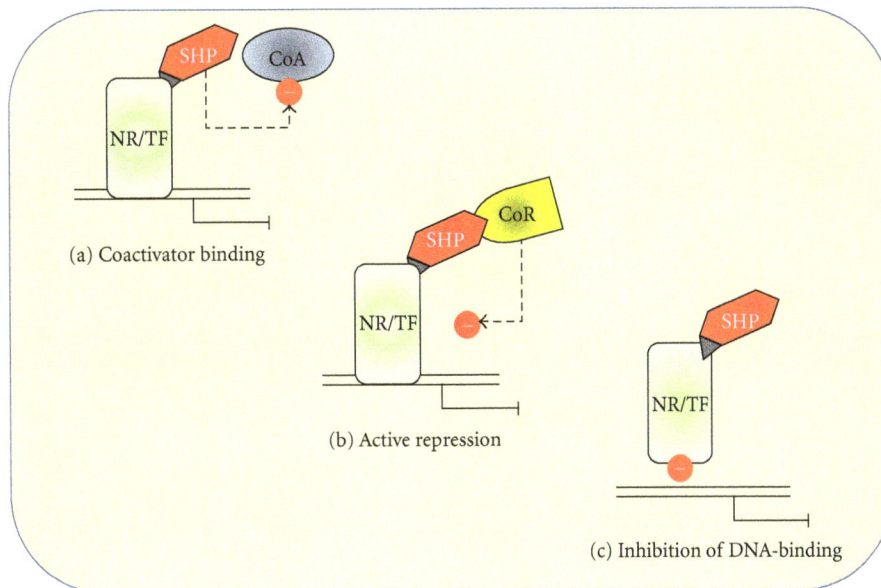

FIGURE 2: The SHP-mediated repression of target transcription factors occurs by at least three distinct transcriptional repression mechanism: (a) direct interference with the AF-2 coactivator domain of NRs (competition for coactivator binding, leading to the repression of NR-mediated transcriptional activity); (b) recruitment of corepressors, resulting in active repression; (c) direct interaction with the surface of NR or transcription factor, resulting in the blockade of DNA binding and the consequent inhibition of its transcriptional activity. See text for details. The dotted arrows and (-) symbols indicate inhibition. CoA: coactivator; CoR: corepressor; NR: nuclear receptor; SHP: small heterodimer factor; TF: transcription factor. Modified after [12, 15, 131].

resulted in SHP-mediated recruitment of transcriptional coregulators mSin3A and Swi/Snf complex to the promoter, chromatin remodeling, and gene repression (Table 3). This is an additional mechanism involving transformation of nucleosome conformation for the repression by SHP of genes activated by various NRs. In line with such results, increased synthesis and accumulation of BAs occurs in SHP $(-/-)$ mice, due to the loss of SHP repression and consequent derepression of the rate-limiting CYP7A1 and cholesterol 12α-hydroxylase (CYP8B1) (the rate-determining enzyme of the alternative but minor BA synthesis pathway) in the biosynthetic pathway [70–72].

Mechanisms independent of the FXR/SHP/LRH pathway might also exist, since BAs feeding to SHP $(-/-)$ mice reduced the levels of CYP7A1 mRNA to similar levels of control mice [70, 71]. Such SHP-independent and alternative pathways include the protein kinase C/Jun N-terminal kinase (PKC/JNK) pathway [73], the FXR/FGFR4 (FGF receptor 4) pathway [57, 74], the cytokine/JNK pathway [75], the pregnane X receptor (PXR) mediated pathway [76], and the JNK/c-Jun signaling pathway [77].

Another study demonstrated, in SHP $(-/-)$ mice on a background of 129 strain, the protection against hypercholesterolemia in three different models: an atherogenic diet, hypothyroidism, and SHP $(-/-)$ mice intercrossed with LDLR $(-/-)$ mice (to generate SHP/LDLR double $(-/-)$ mice in a mixed 129-C57BL/6 background). When fed an atherogenic diet, the latter strain was almost completely resistant to diet-mediated increases in triglyceride, very low-density lipoprotein (VLDL) cholesterol, and low-density lipoprotein (LDL) cholesterol but had an increase in high-density lipoprotein (HDL) cholesterol as compared with LDLR $(-/-)$ mice. Such results point to the protection against dyslipidemia following the inhibition of hepatic SHP expression, although no antagonist ligands have yet been identified for SHP [78]. We have recently examined biliary lipid secretion and cholesterol gallstone formation in male SHP $(-/-)$ and $(+/+)$ mice before and during the feeding of a lithogenic diet for 56 days [79]. Deletion of the Shp gene significantly increased hepatic bile salt synthesis, and doubled the increase of biliary bile salt outputs in SHP $(-/-)$ mice than in $(+/+)$ mice. The intestinal bile acid pool size was significantly greater in SHP $(-/-)$ mice than in $(+/+)$ mice. These increased BAs are efficacious ligands of FXR and can stimulate the expression of intestinal fibroblast growth factor 15 (FGF15) in mice through the FXR signaling pathway, which is consistent with the expanded bile acid pool size in SHP $(-/-)$ mice. At 14 days on the lithogenic diet, fasting gallbladder volume was significantly larger in SHP $(+/+)$ mice than in $(-/-)$ mice [80].

Indeed, FGF15/19 (mouse and human orthologs, resp.) is another FXR gene target in the intestine and appears to contribute to the fine tuning of bile acid synthesis in the liver. Thus, a model for FXR-mediated repression of bile acid synthesis should also take into account the bile acid-mediated activation of intestinal FXR and FGF15 in the small intestine (while the FXR-SHP pathway is activated in the liver). According to the most plausible view, FGF15 acts as a hormone to signal between intestine and liver. The secreted FGF15 by the intestine circulates to the liver, likely through

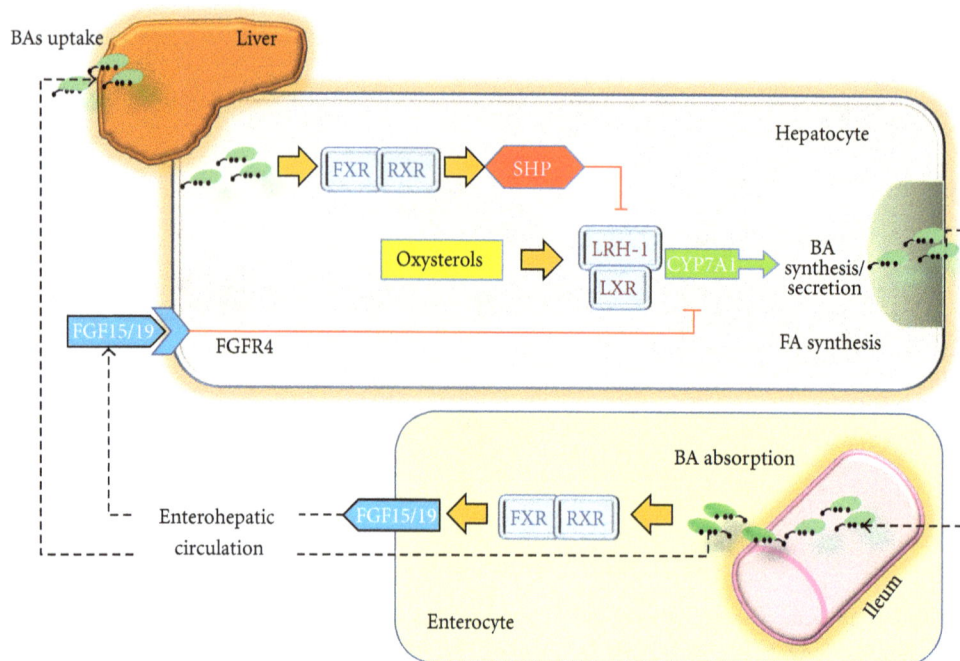

FIGURE 3: The potential molecular mechanisms of crosstalk between nuclear receptors LXR and FXR–SHP–LRH-1 regulatory cascade in the liver and intestine. Bile acids act as ligands for FXR, which regulates transcription by binding as a heterodimer with RXRs. This step results in increased SHP expression. SHP in turn inhibits LRH-1, preventing the activation of target genes that participate in bile acid and fatty acid synthesis. In the absence of bile acids, LRH-1 acts together with LXR to stimulate bile acid synthesis [64, 65, 132]. The important pathways in the intestine that contribute to modulation of bile acid synthesis are also depicted (see text for details). There is a bile-acid-mediated activation of intestinal FXR and, as a result, the release of FGF15 in the small intestine. The secreted FGF15 by the intestine circulates to the liver, likely through the portal circulation or lymph flow [81] and induces the activation of FGFR4 in the liver. The FGF15/FGFR4 pathway synergizes with SHP in vivo to repress CYP7A1 expression [57]. Bas: bile acids; FGF: fibroblast growth factor; FGFR4: FGF receptor; FXR: farnesoid X receptor; LRH-1: liver receptor homologue-1; LXR: liver X receptor; RXR: retinoid X receptors; SHP: short heterodimer partner. Adapted from Ory [66] and Inagaki et al. [57].

the portal circulation or lymph flow [81], and induces the activation of FGFR4 in the liver. As shown in Figure 3, the FGF15/FGFR4 pathway synergizes with SHP *in vivo* to repress CYP7A1 expression [57]. In humans, a similar mechanism should involve the FGF19. Of note, activation of FXR transcription in the intestine protected the liver from cholestasis in mice by inducing FGF15 expression and reducing the hepatic pool of BA. This suggests a potential approach to reverse cholestasis in patients [82]. Hepatic fatty acid homeostasis is also regulated by SHP since regulating these genes involves in fatty acid uptake, synthesis, and export [83–87]. In a study exploring global gene expression profiling combined with chromatin immunoprecipitation assays in transgenic mice constitutively expressing SHP in the liver, overexpression of *SHP* in the liver was associated with the depletion of the hepatic bile acid pool and a concomitant accumulation of triglycerides in the liver [84]. By contrast, fat accumulation induced by a high-cholesterol or high-fat diet is prevented by the deletion of *SHP* [88, 89]. The pleiotropic role of SHP can also be found in the case of nonalcoholic liver steatosis since *OB/SHP* double (−/−) mice (a model of severe obesity and insulin resistance) became resistant to liver steatosis and showed improved insulin sensitivity [86].

Another interesting role for SHP emerged after it was found that BAs negatively regulate the human angiotensinogen (ANG) gene. ANG is the precursor of vasoactive octapeptide angiotensin II, and BAs act through the SHP pathway by preventing hepatocyte nuclear factor-4 (HNF4) from binding to the human ANG promoter [90].

3. Fat Mass, Adipocytes, and Obesity

SHP appears to play a central role in obesity. Human obesity is considered a polygenic disorder characterized by partly known abnormal molecular mechanisms resulting in increased fat mass, with an imbalance between the energy acquired from nutrients that dissipated as heat (i.e., thermogenesis). In this respect, weight stability requires a balance between calories consumed and calories expended [91]. In adipose tissue depots, two main types of adipocytes exist, that is, brown adipocytes and white adipocytes. In several animal species, some adipose tissue sites mainly include brown adipocytes (BATs) and the other contains mainly white adipocytes (WATs). BAT dissipates chemical energy to produce heat either as a defense against cold [92] or as energy expenditure to compensate food intake [93, 94]. The unusual function of BAT might be better understood by considering that they share a common origin with myocytes [95, 96], and BAT was indeed considered something in between muscle and adipose tissue [95]. BAT is deemed as the major site for sympathetic (adrenergic) mediated adaptive thermogenesis; this pathway involves the uncoupling protein-1 (UCP1). WAT is mainly implicated in the regulation of lipid storage and catabolism but also in the synthesis and secretion of adipokines [97–100]. While the percentage of young men with BAT is high, the activity of BAT is reduced in men who are overweight or obese [101]. Thermogenesis unequivocally exists in both humans and animals, and BAT is the major site of thermogenesis which can be increased by environmental factors (i.e., adaptive thermogenesis). In both human and animal species, dietary composition, chronic cold exposure, and exercise may increase thermogenesis [102]. As far as adipose tissue biology is concerned, SHP seems to play a distinct regulatory function in WAT, as compared with BAT. A number of experiments have focused on animal models of obesity and subtle molecular changes. SHP-deficient mice are protected against high-fat-diet-induced obesity [89].

Peroxisome proliferator-activated receptor (PPAR) γ co-activator-1 (PGC-1) family members are multifunctional transcriptional coregulators. PGC-1 acts as a molecular switch in several metabolic pathways. In particular, PGC-1α and PGC-1β regulate mitochondrial biogenesis, adaptive thermogenesis, fatty acid and glucose metabolism, fiber-type switching in skeletal muscle, peripheral circadian clock, and development of the heart [103]. In particular, SHP functions as a negative regulator of energy production in BAT [89] because SHP is a negative regulator of PGC-1α expression in BAT. In turn, PCG-1α is a coactivator of uncoupling protein 1 (UCP1) which plays a major role in energy dissipation as heat in multilocular BAT of different animal species and humans [104–106]. Fat-specific (BAT) SHP-overexpressed transgenic mice had increased body weight and adiposity. Energy metabolism, however, was increased, and BAT cold exposure function was enhanced with activation of thermogenic genes and mitochondrial biogenesis (enhanced β1-AR gene expression and PGC1α). Compared with wild-type mice on a high-fat diet, SHP overexpression was associated with enhanced diet-induced obesity phenotype with weight gain, increased adiposity, and severe glucose intolerance. An additional feature of SHP transgenic mice was a decreased diet-induced adaptive thermogenesis, increased intake of food, and decreased physical activity [107]. This leads to the conclusion that, although expressed at low levels in fat, activation of SHP in adipocytes has a strong effect on weight gain and diet-induced obesity [107]. Moreover, if mechanisms linked to energy metabolism and the development of obesity are considered, SHP has distinct roles in WAT and BAT. As previously mentioned, while SHP deletion in obese leptin-deficient mice (ob/ob) prevented the development of nonalcoholic fatty liver and improved peripheral insulin sensitivity [86], SHP deletion did not overcome the severe obesity caused by leptin deficiency. A significant protective effect from obesity by SHP deficiency was likely associated with the low basal level of SHP expressed in fat. Adipogenesis appears to be influenced by SHP: when SHP was overexpressed in 3T3-L1 preadipocytes, cell differentiation was inhibited, as well as the accumulation of neutral lipids within the cells. Thus, SHP may act as a molecular switch governing adipogenesis. In particular, SHP appears to be a potent adipogenic suppressor, and preadipocytes are kept in an undifferentiated state through the inhibition of the adipogenic transcription factors and stimulators [108]. Further studies will address whether the loss of SHP function results in inhibition of lipid accumulation in adipocytes, similar to what is observed in hepatocytes. In a future clinical setting, treatment of obesity might also include drugs able to mimic or stimulate the effects of SHP. Mutations in the Shp gene have also been reported in patients with lipodystrophy carrying four different polymorphisms [109].

SHP mutations may not be considered a common cause of severe obesity. A number of important clinical studies have examined this issue (Table 4); however, Hung et al. [110] in UK examined the relationships between genetic variation in SHP and weight at birth, adiposity, and insulin levels in three different populations (the Genetics of Obesity Study) GOOS, the Avon Longitudinal Study of Parents and Children (ALSPAC), and the Ely studies). In the 329 cases of severe early-onset obesity (GOOS study), two novel and rare missense mutations (R34G and R36G) were identified which might in part contribute to obesity in the probands. Furthermore, two common polymorphisms, namely, G171A (12% of subjects with higher birth weight) and −195CTGAdel (16% of subjects with lower birth weight) were found. In the ALSPAC cohort of 1,079 children, the G171A variant was associated with increased body mass index and waist circumference together with higher insulin secretion 30 minutes after glucose load. Thus, whereas mutations in the Shp gene cannot be seen as a common cause of severe human obesity, genetic variation in the Shp gene locus may influence birth weight and have effects on body size. The effect might ultimately involve insulin secretion by the negative regulation between SHP and the hepatocyte nuclear factor-4α (HFN-4α), a transcription factor involved in differentiation and function of pancreatic β-cells [110].

A possibility is that decreased SHP expression or function results in increased HFN-4α activity with a cascade of events, including fetal hyperinsulinemia, and increased birth weight. At a later stage, sustained hyperinsulinemia might be responsible of insulin resistance and obesity of the adult [110].

Mutations in the Shp gene were also associated with influence on birth weight, mild obesity, and insulin levels in the study by Nishigori et al. on 274 Japanese subjects [111]. Mutations in several genes encoding transcription factors of the hepatocyte nuclear factor (HNF) cascade are associated with maturity-onset diabetes of the young (MODY). MODY is a monogenic form of early-onset diabetes mellitus (defective insulin secretion with normal body weight), and SHP is deemed as a plausible candidate MODY gene; this is because SHP is able to inhibit the transcriptional activity of the hepatocyte nuclear factor-4α (HFN-4α), a key member of the MODY regulatory network. Thus, further studies have looked for segregation of SHP mutations with MODY in a cohort of Japanese patients with early-onset diabetes. In this context, variants in SHP appeared to cosegregate with

TABLE 4: Studies on the association between SHP (NR0B2) genetic variation and birth weight, high BMI obesity, and fasting insulin diabetes.

Author	Country	Study populations/ mutation	Subjects number	Mutation(s)	Association with birth weight increase	Association with BMI/obesity	Association with increased insulin levels	Association with diabetes	Conclusions
Nishigori et al. [111]	Japan	Young-onset type 2 diabetes	274	In 7 subjects, 5 different mutations (H53fsdel10, L98fsdel9insAC, R34X, A195S, R213C) and 1 apparent polymorphism (R216H) (all in a heterozygous state)	Yes	Yes	—	No	*Shp* genetic variation: most common monogenic determinant of obesity and increased birth weight in Japanese
Hung et al. [110]	UK	GOOS (severe early-onset obesity)	329	R34G and R36C Missense mutations	Yes	Yes	Yes	—	Genetic variation in the SHP locus may influence birth weight and have effects on BMI, possibly through effects on insulin secretion
				G171A (12%)	Yes	No (selection of extreme obesity: stronger effect from other major gene?)	—	—	
				-195CTGAdel (16%) common polymorphisms	No (lower birth weight)		No (lower fasting levels)	—	Subtle effects in heterozygosity, stronger effects in homozygosity
	UK	ALSPAC (cohort of children)	1,079	G171A		Yes (higher BMI and waist circumference at 7 yrs)	Yes (higher fasting levels and 30-min response)		
				-195CTGAdel	No	No (lower BMI)	—	—	
	UK	Ely Study (Caucasian adults)	600	G171A		Yes (BMI increased)			
				-195CTGAdel	Data not available	Yes (female: higher BMI), No (male: lower BMI)	No	—	

TABLE 4: Continued.

Author	Country	Study populations/mutation	Subjects number	Mutation(s)	Association with birth weight increase	Association with BMI/obesity	Association with increased insulin levels	Association with diabetes	Conclusions
				Birth weight: the only child homozygous for the A allele had a birth weight ≥4 kg		No			Mutations in SHP < UK than in Japanese obese type 2 subjects; G171A coding polymorphism in 14.1% of UK subjects
Mitchell et al. [113]	UK	Young-onset type 2 diabetes, obesity, birth weight	1,927	Obesity: no association if G/A genotype; yes (?) (if A/A homozygotes)	No	Yes (?)	—	No	The A allele (G/A genotype) not associated with obesity or increased birth weight; Homozygous for the rare A allele: predisposed to moderate obesity and possibly increased birth weight
		Early-onset obesity (men)	750	2 silent variants c.65C4T [p. Y22Y], c.339G4A [p. P113P]; 3 missense variants c.100C4G [p. R34G], c.278G4A [p. G93D], c.415C4A [p. P139H]		Yes (only among obese)			Very low prevalence of functional SHP variants associated with obesity among Danes; A role for G171A polymorphism low penetrance SHP variants) for obesity risk in Europe? Major differences in prevalence and impact of SHP variants between Danish and Japanese obese
Echwald et al. [114]	Denmark	Nonobese controls; Functional analyses in MIN6-m9 and HepG2 cell lines	795	G171A polymorphism (8.9%); No variants G171A polymorphism (7.1)	—	No (P = 0.07 versus obese); 93D mutant protein: reduced in vitro inhibition of the HNF4α transactivation of the HNF-1α promoter expression	—	—	

Note: SHP is expressed in the liver, pancreas, spleen, small intestine, and adrenal gland in humans [18] and inhibits the transcriptional activity of hepatocyte nuclear factor-4 α (HNF4α). ALSPAC: Avon Longitudinal Study of Parents and Children; GOOS: Genetics of Obesity Study; HNF4α: hepatocyte nuclear factor-4α.

increased body mass index in families, thus contributing to obesity among Japanese subjects. Also, increased risk of morbidity was observed in another study from Japan, examining patients with type 2 diabetes and *SHP* mutations [112].

Major differences, however, might exist in the prevalence and function of *SHP* variants in different populations. Of note, the results from other Caucasian cohorts did not confirm the association between *SHP* mutation and obesity [113, 114]. Echwald et al. conducted an elegant study on the prevalence of *SHP* variants by single-strand conformational polymorphism and heteroduplex analysis among 750 Danish obese men with early-onset obesity [114]. As control, a cohort of 795 nonobese control subjects was genotyped using PCR-RFLP. Functional analyses of the identified coding region variants were performed in both MIN6-m9 and HepG2 cell lines. Five novel variants were identified (including 3 missense variants (c.100C>G [p.R34G], c.278G>A [p.G93D], and c.415C>A [p.P139H]) and 2 silent variants (c.65C>T [p.Y22Y] and c.339G>A [p.P113P])). The previously reported [111] c.512G>C [p.G171A] common polymorphism was identified; however, the prevalence of functional *SHP* variants associated with obesity was considerably lower among Danish subjects (1 out of 750 obese, none of control subjects), compared to the prevalence observed in Japan by Nishigori et al. [111]. Mitchell at al. [113] investigated SHP variants in 1927 UK subjects according to type 2 diabetes, obesity, and birth weight. Although reporting a raised body mass index among homozygous carriers of the 171A variant (<1%), this polymorphism was unlikely to be associated with all three conditions in Caucasians. Taken together, the above-mentioned studies suggest that the 171A variant might contribute only to subsets of polygenic obesity.

4. Other Functions of SHP

The existence of multiple interactions of SHP with NRs, transcription factors and transcriptional cofactors (Tables 2 and 3) points to the pleiotropic and central role of SHP in the body.

SHP has been hypothesized to act in glucose homeostasis via complex pathways involving the inhibition of glucocorticoid receptors (GR) in mammalian cells and the inhibition of PGC-1 gene, a coactivator of NRs important for gluconeogenic gene expression and the PGC-1-regulated phospho(enol)pyruvate carboxykinase (PEPCK) promoter. Such steps underscore a physiologically relevant role for SHP in modulating hepatic glucocorticoid action [22]. Following the bile acid-induced induction, SHP inhibited a number of other pathways, including the HNF4α-mediated transactivation of the PEPCK and fructose biphosphate (FBP) promoters, as well as the transactivation of the glucose-6-phosphatase (G6Pase) promoter mediated by Foxo1 [115]. The interaction between SHP inhibitory function and the 3 isoforms (α, β, and γ) of the hepatocyte nuclear factor-3 (HNF4) points to the regulatory role of SHP on gluconeogenesis [51]. A role for SHP in insulin secretion pathway has also been reported. Mutations in hepatocyte nuclear factor

1α (HNF-1α) is associated with maturity-onset diabetes of the young type 3. This condition depends on impaired insulin secretory response in pancreatic beta cells.

Indeed, loss of HNF-1α function in HNF-1α (−/−) mice resulted in altered expression of genes involved in glucose-stimulated insulin secretion, but also insulin synthesis, and beta-cell differentiation. Pancreatic islets of HNF-1α (−/−) mice showed a distinctive reduction of SHP expression and a downregulation of the HNF4α gene expression. Since SHP appears to repress its own transcriptional activation following heterodimerization with HNF4α, a feedback autoregulatory loop between SHP and HNF4α has been hypothesized [116]. Also, SHP likely functions as a negative regulator of pancreatic islet insulin secretion. SHP (−/−) mice were characterized by hypoinsulinemia, increased glucose-dependent response of islets, increased peripheral insulin sensitivity, and increased glycogen stores [117]. The role played by SHP in the regulation of hepatic gluconeogenesis has also emerged in a number of additional experiments. For example, the liver of SHP (−/−) mice showed increased glycogen stores [117], while hepatic *Shp* gene expression (induced by the antidiabetic biguanide drug metformin) was associated with inhibition of hepatic gluconeogenesis. Induction of SHP was achieved via AMP-activated protein kinase (AMPK) and associated with downregulation of essential gluconeogenic enzyme genes, that is, phosphoenolpyruvate carboxykinase (PEPCK), glucose-6-phosphatase (G6Pase) [118], and fructose-1,6-bisphosphatase (FBP1) [119].

PGC-1 gene is a coactivator of NRs, and this step is relevant for gluconeogenic gene expression. Yamagata et al. [119] showed that bile acid (chenodeoxycholic acid) was able to induce the downregulation of PGC-1 gene, and this mechanism involved forkhead transcription factors (Foxo1, Foxo3a, Foxo4) via a SHP-dependent manner.

Drug metabolism and detoxification might be regulated by SHP. This is also the case for excess BAs: the pregnane X receptor (PXR) induces CYP3A and inhibits CYP7α, both involved in biochemical pathways leading to the conversion of cholesterol into primary BAs, whereas CYP3A is also involved in the detoxification of toxic secondary bile acid derivatives. SHP acts as a potent repressor of PXR transactivation, and this finding suggests that PXR can act on both bile acid synthesis and elimination detoxification [1]. Additional mechanisms involved in the SHP-dependent control of pathways of drug metabolism have been identified. The expression of genes involved with the metabolism of xenobiotics might be regulated by SHP in the spleen acting on (aryl hydrocarbon receptor (AHR)/AHR nuclear translocator (ARNT)) AHR/ARNT heterodimers which, in turn, bind to xenobiotic response elements (XREs) at the level of specific DNA sequences [41]. A number of genes involved in hormone and drug metabolism would be expressed (i.e., UGT16, ALDH3, CYP1A1, CYP1A2, CYP1B1, etc.). SHP also appears to downregulate the constitutive-androstane-receptor- (CAR-) mediated CYP2B1 gene expression, induced by phenobarbital to form the CAR/RXR heterodimer which, in turn, binds to 2 DR-4 sites to form the phenobarbital responsive unit in the CYP2B gene [120] (Table 3). One role of SHP in steroidogenesis has been identified in the testes

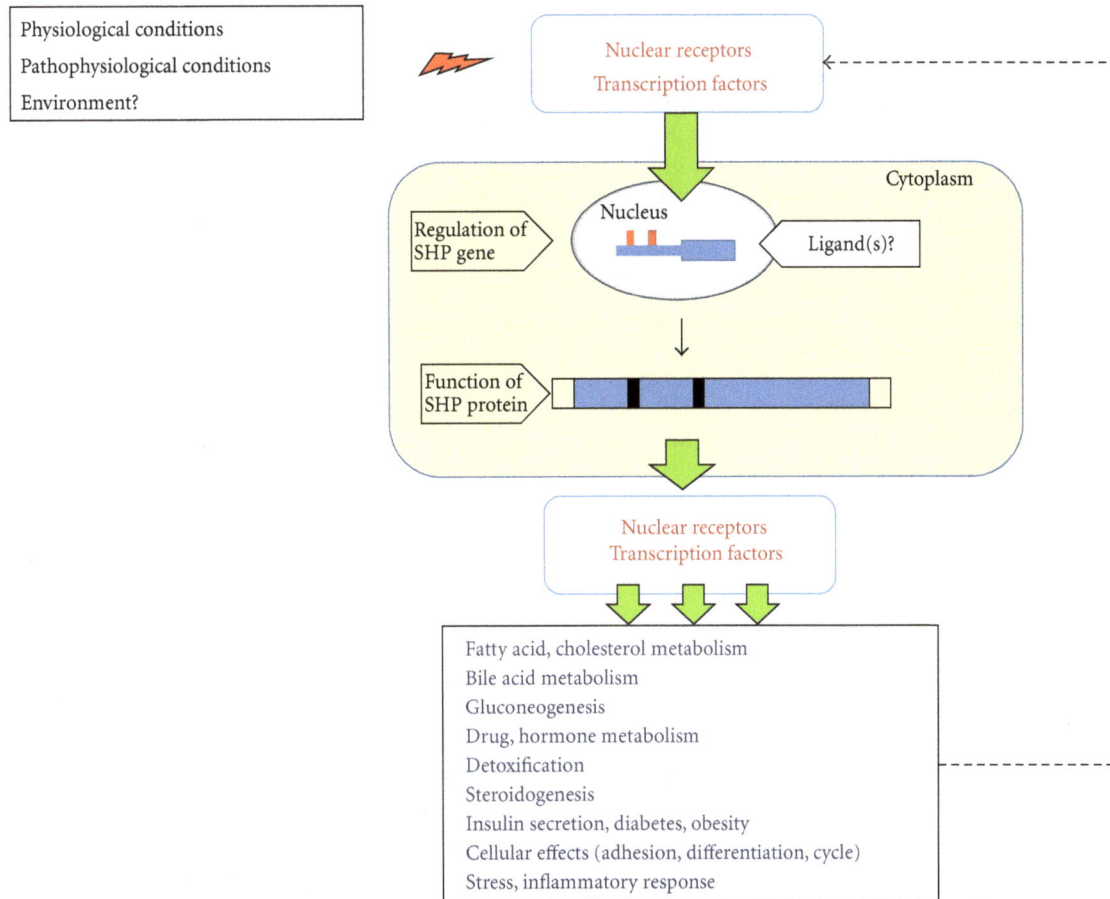

FIGURE 4: Schematic diagram of the function and gene regulation of SHP. Different conditions will lead to activation of nuclear receptors and/or transcription factors able to regulate *Shp* gene expression in the nucleus and protein synthesis in the cytoplasm. The protein acts as a transcriptional corepressor of a number of other nuclear receptors and transcription factors involved in a wide series of regulatory pathways. The potential role of a feedback mechanism and of ligand(s) is hypothesized.

with influence on testosterone synthesis and germ cell differentiation [121] and in the intestine for glucocorticoid synthesis [122].

A role for SHP in cell proliferation and apoptosis signaling is emerging. Depending on the cell type, SHP seems to have both inhibitory and stimulatory effects on apoptosis. However, the manipulation of SHP through the synthetic ligands adamantyl-substituted retinoid-related (ARR) compounds 6-[3-(1-adamantyl)-4-hydroxyphenyl]-2-naphthale-necarboxylic acid (CD437/AHPN) and 4-[3-(1-adamantyl)-4-hydroxyphenyl]-3-chlorocinnamic acid (3-Cl-AHPC) induces apoptosis of a number of malignant cells (i.e., leukemia and breast carcinoma) both *in vitro* and *in vivo* [123, 124]. The complex mechanism implies binding of ARR and 3-Cl-AHPC to SHP with formation of a corepressor complex containing Sin3A and nuclear receptor corepressor (*N*-CoR) which activate local control of mitochondrial function and apoptosis, with a limiting function on tumorigenesis [17, 123] (Table 3). SHP appears to be also involved in DNA methylation and acting as a tumor suppressor, at least in the human and mouse livers [125–127]. Whether manipulation of SHP will be helpful in the treatment of hepatic and other

gastrointestinal cancers is still a matter of research. The recent finding that SHP negatively regulates TLR signaling to NF-κB has raised the interest for the role of SHP in mechanisms governing innate immunity. SHP appears to negatively regulate the expression of genes encoding inflammatory molecules. Of note, direct binding of NF-κB seems to occur in resting cells, while binding of SHP to TRAF6 occurs in LPS-stimulated cells [128, 129].

5. Conclusions and Perspectives

A remarkable number of metabolic functions in the body appear to be regulated by the orphan unique NR, small heterodimer partner SHP, which targets a complex set of genes in multiple pathways as a transcriptional corepressor (Figure 4). Pathways include fatty acid metabolism, glucose homeostasis, and drug-hormone detoxification. When looking at complex mechanisms leading to some important *lipidopathies*, that is, obesity and liver steatosis, enlightening data about the regulatory function of SHP are provided by studies using *Shp*-deleted and *Shp*-overexpressed animal models. Most likely, a condition of *Shp* deficiency might counteract lipid accumulation and improve plasma lipoprotein profiles. Further studies are urgently needed to confirm that such an important metabolic regulatory mechanism of SHP is true and has high translational value. To date, however, no synthetic antagonists or agonists for SHP are available, and one should keep in mind that rather divergent and somewhat elusive data have been observed regarding the loss of SHP function in humans and rodents. Thus, careful examination of subtle SHP intrinsic functions is essential to dissect potential modulatory pathways of SHP for a variety of metabolic abnormalities but also in tumorigenesis. Moreover, identifying specific endogenous ligands and synthetic agonists of SHP will pave the way to for therapeutic intervention. The effect of synthetic ligands on SHP modulation in hepatocytes and adipocytes, for example, might represent therapeutic tools for the treatment of constituents of the metabolic syndrome, namely, hypercholesterolemia, overweight obesity, and liver steatosis.

Abbreviations

AF1:	Activation function 1
AF2:	Activation function 2
AHR/ARNT:	Aryl hydrocarbon receptor (AHR)/AHR nuclear translocator (ARNT)
ALDH3:	Human aldehyde dehydrogenase 3 gene
AMPK:	AMP-activated protein kinase
ARR:	Adamantyl-substituted retinoid-related molecules
$\beta1$-AR:	$\beta1$-adrenergic receptor
BAREs:	Bile acids response elements
BAT:	Brown adipose tissue
Brm:	Human Brahma (a catalytic component of the SWI/SNF-related chromatin-remodeling complex)
CAR:	Constitutive androstane receptor
CD437/AHPN:	6-[3-(1-Adamantyl)-4-hydroxyphenyl]-2-naphthalenecarboxylic acid
3-Cl-AHPC:	4-[3-(1-Adamantyl)-4-hydroxyphenyl]-3-chlorocinnamic acid
CYP3A:	Hepatic cytochrome P-4503A
CYP1A1:	Cytochrome P450, family 1, subfamily A, polypeptide 1
CYP1A2:	Cytochrome P450, family 1, subfamily A, polypeptide 2
CYP1B1:	Cytochrome P450, family 1, subfamily B, polypeptide 1
CYP2B1:	Cytochrome P450, subfamily IIB
CYP7A1:	Cholesterol 7 alpha-hydroxylase
CYP8B1:	Cholesterol 12alpha-hydroxylase
DBD:	DNA-binding domain
DAX-1:	Dosage-sensitive sex reversal adrenal hypoplasia congenita (AHC) critical region on the X chromosome, gene 1
DR-4:	Nuclear receptor half-site repeat
ER:	Estrogen receptor
FBP1:	Fructose-1,6-bisphosphatase
FGF:	Fibroblast growth factor
FGFR:	Fibroblast growth factor receptor
FISH:	Fluorescence *in situ* hybridization
Foxo:	Forkhead transcription factors
FXR:	Farnesoid X receptor
G6Pase:	Glucose-6-phosphatase
GPCR:	G-protein-coupled Receptor
GPS2:	G protein pathway suppressor 2
GR:	Glucocorticoid receptors
HDL:	High-density lipoprotein
HNF:	Hepatocyte nuclear factor
JNK:	Jun N-terminal kinase
JUN-D:	Jun family of the activator protein 1 (AP-1) transcription factor complex
LBD:	Ligand-binding domain
LDL:	Low-density lipoprotein
LDLR:	Low-density lipoprotein receptor
LRH1:	Liver receptor homologue 1
3T3-L1:	Mouse embryonic fibroblast adipose-like cell line.
LXXLL:	Leu-Xaa-Xaa-Leu-Leu
mSin3A:	Mammalian homolog of the Saccharomyces cerevisiae transcriptional corepressor Sin3p
N-CoR:	Nuclear receptor corepressor
NRs:	Nuclear receptors
PEPCK:	Phosphoenolpyruvate carboxykinase
PGC-1:	Peroxisome proliferator-activated receptor-γ (PPAR-γ) coactivator-1
PKC:	Protein kinase C
PXR:	Pregnane X receptor
RAR:	Retinoid acid receptor
RNA Pol II:	RNA polymerase II
RNF31	Member of the ring-between-ring (RBR) family of E3 ubiquitin ligases
RXR:	Retinoid X receptor
SHP:	Small heterodimer partner
SMILE:	SHP-interacting leucine zipper protein
SMRT:	Silencing mediator of retinoid and thyroid hormone receptor
SRC-1:	Steroid receptor coactivator-1
SWI/SNF:	SWItch/Sucrose NonFermentable (yeast nucleosome remodeling complex composed of several proteins which are products of the SWI and SNF genes)
UCP1:	Uncoupling protein-1
USF-1:	Upstream stimulatory factor-1
TFIID:	Transcription initiation factor II D
TFIIE:	Transcription factor II E
UGT-16:	Uridine 5′-diphospho (UDP) glucuronosyl transferase family member
VLDL:	Very-low-density lipoprotein
XREs:	Xenobiotic response elements
WAT:	White adipose tissue.

Conflict of Interests

The authors declare that there is no conflict of interests.

Acknowledgments

This work was supported in part by research grants DK54012 and DK73917 (D. Q.-H. Wang) from the National Institutes of Health (US Public Health Service), FIRB 2003 RBAU01RANB002 (P. Portincasa) from the Italian Ministry of University and Research, and ORBA10ROPA from the University of Bari. P. Portincasa was a recipient of the short-term mobility grant 2005 from the Italian National Research Council (CNR). L. Bonfrate was a recipient of the travel grant for young investigators from the European Society for Clinical Investigation, 2011. Due to space limitations, the authors apologize to those whose publications related to the discussed issues could not be cited.

References

[1] J. C. Ourlin, F. Lasserre, T. Pineau et al., "The small heterodimer partner interacts with the pregnane X receptor and represses its transcriptional activity," *Molecular Endocrinology*, vol. 17, no. 9, pp. 1693–1703, 2003.

[2] H. Gronemeyer, J. Å. Gustafsson, and V. Laudet, "Principles for modulation of the nuclear receptor superfamily," *Nature Reviews Drug Discovery*, vol. 3, no. 11, pp. 950–964, 2004.

[3] D. J. Mangelsdorf, C. Thummel, M. Beato et al., "The nuclear receptor super-family: the second decade," *Cell*, vol. 83, no. 6, pp. 835–839, 1995.

[4] D. J. Mangelsdorf and R. M. Evans, "The RXR heterodimers and orphan receptors," *Cell*, vol. 83, no. 6, pp. 841–850, 1995.

[5] K. Pardee, A. S. Necakov, and H. Krause, "Nuclear receptors: small molecule sensors that coordinate growth, metabolism and reproduction," *Subcellular Biochemistry*, vol. 52, pp. 123–153, 2011.

[6] H. Gronemeyer and D. Moras, "How to finger DNA," *Nature*, vol. 375, no. 6528, pp. 190–191, 1995.

[7] E. Lalli and P. Sassone-Corsi, "DAX-1, an unusual orphan receptor at the crossroads of steroidogenic function and sexual differentiation," *Molecular Endocrinology*, vol. 17, no. 8, pp. 1445–1453, 2003.

[8] Y. S. Lee, D. Chanda, J. Sim, Y. Y. Park, and H. S. Choi, "Structure and function of the atypical orphan nuclear receptor small heterodimer partner," *International Review of Cytology*, vol. 261, pp. 117–158, 2007.

[9] F. Muscatelli, T. M. Strom, A. P. Walker et al., "Mutations in the DAX-1 gene give rise to both X-linked adrenal hypoplasia congentia and hypogonadotropic hypogonadism," *Nature*, vol. 372, no. 6507, pp. 672–676, 1994.

[10] W. Seol, H. S. Choi, and D. D. Moore, "An orphan nuclear hormone receptor that lacks a DNA binding domain and heterodimerizes with other receptors," *Science*, vol. 272, no. 5266, pp. 1336–1339, 1996.

[11] E. Zanaria, F. Muscatelli, B. Bardoni et al., "An unusual member of the nuclear hormone receptor superfamily responsible for X-linked adrenal hypoplasia congenita," *Nature*, vol. 372, no. 6507, pp. 635–641, 1994.

[12] A. Båvner, S. Sanyal, J.-Å. Gustafsson, and E. Treuter, "Transcriptional corepression by SHP: molecular mechanisms and physiological consequences," *Trends in Endocrinology and Metabolism*, vol. 16, no. 10, pp. 478–488, 2005.

[13] J. Miao, S.-E. Choi, S. M. Seok et al., "Ligand-dependent regulation of the activity of the orphan nuclear receptor, small heterodimer partner (SHP), in the repression of bile acid biosynthetic CYP7A1 and CYP8B1 genes," *Molecular Endocrinology*, vol. 25, no. 7, pp. 1159–1169, 2011.

[14] L. Farhana, M. I. Dawson, M. Leid et al., "Adamantyl-substituted retinoid-related molecules bind small heterodimer partner and modulate the Sin3A repressor," *Cancer Research*, vol. 67, no. 1, pp. 318–325, 2007.

[15] D. Chanda, J. H. Park, and H. S. Choi, "Molecular basis of endocrine regulation by orphan nuclear receptor small heterodimer partner," *Endocrine Journal*, vol. 55, no. 2, pp. 253–268, 2008.

[16] V. Giguere, "Orphan nuclear receptors: from gene to function," *Endocrine Reviews*, vol. 20, no. 5, pp. 689–725, 1999.

[17] Y. Zhang, C. H. Hagedorn, and L. Wang, "Role of nuclear receptor SHP in metabolism and cancer," *Biochimica et Biophysica Acta*, vol. 1812, no. 8, pp. 893–908, 2011.

[18] H. K. Lee, Y. K. Lee, S. H. Park et al., "Structure and expression of the orphan nuclear receptor SHP gene," *Journal of Biological Chemistry*, vol. 273, no. 23, pp. 14398–14402, 1998.

[19] A. L. Bookout, Y. Jeong, M. Downes, R. T. Yu, R. M. Evans, and D. J. Mangelsdorf, "Anatomical profiling of nuclear receptor expression reveals a hierarchical transcriptional network," *Cell*, vol. 126, no. 4, pp. 789–799, 2006.

[20] X. Yang, M. Downes, R. T. Yu et al., "Nuclear receptor expression links the circadian clock to metabolism," *Cell*, vol. 126, no. 4, pp. 801–810, 2006.

[21] L. Johansson, A. Båvner, J. S. Thomsen, M. Färnegårdh, J. Å. Gustafsson, and E. Treuter, "The orphan nuclear receptor SHP utilizes conserved LXXLL-related motifs for interactions with ligand-activated estrogen receptors," *Molecular and Cellular Biology*, vol. 20, no. 4, pp. 1124–1133, 2000.

[22] L. J. Borgius, K. R. Steffensen, J. Å. Gustafsson, and E. Treuter, "Glucocorticoid signaling is perturbed by the atypical orphan receptor and corepressor SHP," *Journal of Biological Chemistry*, vol. 277, no. 51, pp. 49761–49766, 2002.

[23] J. Gobinet, G. Auzou, J. C. Nicolas, C. Sultan, and S. Jalaguier, "Characterization of the interaction between androgen receptor and a new transcriptional inhibitor, SHP," *Biochemistry*, vol. 40, no. 50, pp. 15369–15377, 2001.

[24] W. Seol, M. Chung, and D. D. Moore, "Novel receptor interaction and repression domains in the orphan receptor SHP," *Molecular and Cellular Biology*, vol. 17, no. 12, pp. 7126–7131, 1997.

[25] Y. B. Xie, B. Nedumaran, and H. S. Choi, "Molecular characterization of SMILE as a novel corepressor of nuclear receptors," *Nucleic Acids Research*, vol. 37, no. 12, pp. 4100–4115, 2009.

[26] Y. B. Xie, J. H. Park, D. K. Kim et al., "Transcriptional corepressor SMILE recruits SIRT1 to inhibit nuclear receptor estrogen receptor-related receptor γ transactivation," *Journal of Biological Chemistry*, vol. 284, no. 42, pp. 28762–28774, 2009.

[27] H. J. Kim, J. Y. Kim, J. Y. Kim et al., "Differential regulation of human and mouse orphan nuclear receptor small heterodimer partner promoter by sterol regulatory element binding protein-1," *Journal of Biological Chemistry*, vol. 279, no. 27, pp. 28122–28131, 2004.

[28] D. Chanda, T. Li, K.-H. Song et al., "Hepatocyte growth factor family negatively regulates hepatic gluconeogenesis via induction of orphan nuclear receptorsmall heterodimer partner in primary hepatocytes," *Journal of Biological Chemistry*, vol. 284, no. 42, pp. 28510–28521, 2009.

[29] J. Cui, L. Huang, A. Zhao et al., "Guggulsterone is a farnesoid X receptor antagonist in coactivator association assays but acts to enhance transcription of bile salt export pump,"

Journal of Biological Chemistry, vol. 278, no. 12, pp. 10214–10220, 2003.

[30] G. Rizzo, B. Renga, E. Antonelli, D. Passeri, R. Pellicciari, and S. Fiorucci, "The methyl transferase PRMT1 functions as co-activator of farnesoid X receptor (FXR)/9-cis retinoid X receptor and regulates transcription of FXR responsive genes," *Molecular Pharmacology*, vol. 68, no. 2, pp. 551–558, 2005.

[31] K. Isoda, S. Sawada, M. Ayaori et al., "Deficiency of interleukin-1 receptor antagonist deteriorates fatty liver and cholesterol metabolism in hypercholesterolemic mice," *Journal of Biological Chemistry*, vol. 280, no. 8, pp. 7002–7009, 2005.

[32] M. G. Yeo, Y. G. Yoo, H. S. Choi, Y. K. Pak, and M. O. Lee, "Negative cross-talk between Nur77 and small heterodimer partner and its role in apoptotic cell death of hepatoma cells," *Molecular Endocrinology*, vol. 19, no. 4, pp. 950–963, 2005.

[33] E. Kanaya, T. Shiraki, and H. Jingami, "The nuclear bile acid receptor FXR is activated by PGC-1α in a ligand-dependent manner," *Biochemical Journal*, vol. 382, no. 3, pp. 913–921, 2004.

[34] J. M. Del Bas, J. Fernández-Larrea, M. Blay et al., "Grape seed procyanidins improve atherosclerotic risk index and induce liver CYP7A1 and SHP expression in healthy rats," *FASEB Journal*, vol. 19, no. 3, pp. 479–481, 2005.

[35] S. Ito, T. Fujimori, A. Furuya, J. Satoh, Y. Nabeshima, and Y.-I. Nabeshima, "Impaired negative feedback suppression of bile acid synthesis in mice lacking βKlotho," *Journal of Clinical Investigation*, vol. 115, no. 8, pp. 2202–2208, 2005.

[36] M. J. Evans, K. Lai, L. J. Shaw, D. C. Harnish, and C. C. Chadwick, "Estrogen receptor α inhibits IL-1β induction of gene expression in the mouse liver," *Endocrinology*, vol. 143, no. 7, pp. 2559–2570, 2002.

[37] Y. Y. Park, H. J. Kim, J. Y. Kim et al., "Differential role of the loop region between helices H6 and H7 within the orphan nuclear receptors small heterodimer partner and DAX-1," *Molecular Endocrinology*, vol. 18, no. 5, pp. 1082–1095, 2004.

[38] A. Oiwa, T. Kakizawa, T. Miyamoto et al., "Synergistic regulation of the mouse orphan nuclear receptor SHP gene promoter by CLOCK-BMAL1 and LRH-1," *Biochemical and Biophysical Research Communications*, vol. 353, no. 4, pp. 895–901, 2007.

[39] Y. B. Xie, O. H. Lee, B. Nedumaran et al., "SMILE, a new orphan nuclear receptor SHP-interacting protein, regulates SHP-repressed estrogen receptor transactivation," *Biochemical Journal*, vol. 416, no. 3, pp. 463–473, 2008.

[40] J. H. Suh, J. Huang, Y. Y. Park et al., "Orphan nuclear receptor small heterodimer partner inhibits transforming growth factor-β signaling by repressing Smad3 transactivation," *Journal of Biological Chemistry*, vol. 281, no. 51, pp. 39169–39178, 2006.

[41] C. M. Klinge, S. C. Jernigan, K. E. Risinger et al., "Short heterodimer partner (SHP) orphan nuclear receptor inhibits the transcriptional activity of aryl hydrocarbon receptor (AHR)/AHR nuclear translocator (ARNT)," *Archives of Biochemistry and Biophysics*, vol. 390, no. 1, pp. 64–70, 2001.

[42] D. Chanda, Y. B. Xie, and H. S. Choi, "Transcriptional corepressor shp recruits sirt1 histone deacetylase to inhibit LRH-1 transactivation," *Nucleic Acids Research*, vol. 38, no. 14, pp. 4607–4619, 2010.

[43] A. Ehrlund, E. H. Anthonisen, N. Gustafsson et al., "E3 ubiquitin ligase RNF31 cooperates with DAX-1 in transcriptional repression of steroidogenesis," *Molecular and Cellular Biology*, vol. 29, no. 8, pp. 2230–2242, 2009.

[44] Y. S. Kim, C. Y. Han, S. W. Kim et al., "The orphan nuclear receptor small heterodimer partner as a novel coregulator of nuclear factor-κB in oxidized low density lipoprotein-treated macrophage cell line RAW 264.7," *Journal of Biological Chemistry*, vol. 276, no. 36, pp. 33736–33740, 2001.

[45] H. Nishizawa, K. Yamagata, I. Shimomura et al., "Small heterodimer partner, an orphan nuclear receptor, augments peroxisome proliferator-activated receptor γ transactivation," *Journal of Biological Chemistry*, vol. 277, no. 2, pp. 1586–1592, 2002.

[46] A. Kassam, J. P. Capone, and R. A. Rachubinski, "The short heterodimer partner receptor differentially modulates peroxisome proliferator-activated receptor α-mediated transcription from the peroxisome proliferator-response elements of the genes encoding the peroxisomal β-oxidation enzymes acyl-CoA oxidase and hydratase-dehydrogenase," *Molecular and Cellular Endocrinology*, vol. 176, no. 1-2, pp. 49–56, 2001.

[47] Y. K. Lee, H. Dell, D. H. Dowhan, M. Hadzopoulou-Cladaras, and D. D. Moore, "The orphan nuclear receptor SHP inhibits hepatocyte nuclear factor 4 and retinoid X receptor transactivation: two mechanisms for repression," *Molecular and Cellular Biology*, vol. 20, no. 1, pp. 187–195, 2000.

[48] Y. K. Lee and D. D. Moore, "Dual mechanisms for repression of the monomeric orphan receptor liver receptor homologous protein-1 by the orphan small heterodimer partner," *Journal of Biological Chemistry*, vol. 277, no. 4, pp. 2463–2467, 2002.

[49] L. Johansson, J. S. Thomsen, A. E. Damdimopoulos, G. Spyrou, J. A. Gustafsson, and E. Treuter, "The orphan nuclear receptor SHP inhibits agonist-dependent transcriptional activity of estrogen receptors ERα and ERβ," *Journal of Biological Chemistry*, vol. 274, no. 1, pp. 345–353, 1999.

[50] J. K. Kemper, H. Kim, J. Miao, S. Bhalla, and Y. Bae, "Role of an mSin3A-Swi/Snf chromatin remodeling complex in the feedback repression of bile acid biosynthesis by SHP," *Molecular and Cellular Biology*, vol. 24, no. 17, pp. 7707–7719, 2004.

[51] J. Y. Kim, H. J. Kim, K. T. Kim et al., "Orphan nuclear receptor small heterodimer partner represses hepatocyte nuclear factor 3/foxa transactivation via inhibition of its DNA binding," *Molecular Endocrinology*, vol. 18, no. 12, pp. 2880–2894, 2004.

[52] S. Fiorucci, E. Antonelli, G. Rizzo et al., "The nuclear receptor SHP mediates inhibition of hepatic stellate cells by FXR and protects against liver fibrosis," *Gastroenterology*, vol. 127, no. 5, pp. 1497–1512, 2004.

[53] D. Kanamaluru, Z. Xiao, S. Fang et al., "Arginine methylation by PRMT5 at a naturally occurring mutation site Is critical for liver metabolic regulation by small heterodimer partner," *Molecular and Cellular Biology*, vol. 31, no. 7, pp. 1540–1550, 2011.

[54] M. O. Hoeke, J. R. M. Plass, J. Heegsma et al., "Low retinol levels differentially modulate bile salt-induced expression of human and mouse hepatic bile salt transporters," *Hepatology*, vol. 49, no. 1, pp. 151–159, 2009.

[55] S. Sherlock and J. Dooley, *Diseases of the Liver and Biliary System*, Blackwell Science, Oxford, UK, 2002.

[56] J. Y. Chiang, "Regulation of bile acid synthesis," *Frontiers in Bioscience*, vol. 3, pp. d176–d193, 1998.

[57] T. Inagaki, M. Choi, A. Moschetta et al., "Fibroblast growth factor 15 functions as an enterohepatic signal to regulate bile acid homeostasis," *Cell Metabolism*, vol. 2, no. 4, pp. 217–225, 2005.

[58] C. V. Diogo, I. Grattagliano, P. J. Oliveira, L. Bonfrate, and P. Portincasa, "Re-wiring the circuit: mitochondria as a pharmacological target in liver disease," *Current Medicinal Chemistry*, vol. 18, no. 35, pp. 5448–5465, 2011.

[59] I. Grattagliano, S. Russmann, C. Diogo et al., "Mitochondria in chronic liver disease," *Current Drug Targets*, vol. 12, no. 6, pp. 879–893, 2011.

[60] J. Y. L. Chiang, "Regulation of bile acid synthesis: pathways, nuclear receptors, and mechanisms," *Journal of Hepatology*, vol. 40, no. 3, pp. 539–551, 2004.

[61] D. F. Jelinek, S. Andersson, C. A. Slaughter, and D. W. Russell, "Cloning and regulation of cholesterol 7α-hydroxylase, the rate-limiting enzyme in bile acid biosynthesis," *Journal of Biological Chemistry*, vol. 265, no. 14, pp. 8190–8197, 1990.

[62] D. J. Parks, S. G. Blanchard, R. K. Bledsoe et al., "Bile acids: natural ligands for an orphan nuclear receptor," *Science*, vol. 284, no. 5418, pp. 1365–1368, 1999.

[63] B. A. Laffitte, H. R. Kast, C. M. Nguyen, A. M. Zavacki, D. D. Moore, and P. A. Edwards, "Identification of the DNA binding specificity and potential target genes for the farnesoid X-activated receptor," *Journal of Biological Chemistry*, vol. 275, no. 14, pp. 10638–10647, 2000.

[64] B. Goodwin, S. A. Jones, R. R. Price et al., "A regulatory cascade of the nuclear receptors FXR, SHP-1, and LRH-1 represses bile acid biosynthesis," *Molecular Cell*, vol. 6, no. 3, pp. 517–526, 2000.

[65] T. T. Lu, M. Makishima, J. J. Repa et al., "Molecular basis for feedback regulation of bile acid synthesis by nuclear receptos," *Molecular Cell*, vol. 6, no. 3, pp. 507–515, 2000.

[66] D. S. Ory, "Nuclear receptor signaling in the control of cholesterol homeostasis: have the orphans found a home?" *Circulation Research*, vol. 95, no. 7, pp. 660–670, 2004.

[67] S. Sanyal, A. Båvner, A. Haroniti et al., "Involvement of corepressor complex subunit GPS2 in transcriptional pathways governing human bile acid biosynthesis," *Proceedings of the National Academy of Sciences of the United States of America*, vol. 104, no. 40, pp. 15665–15670, 2007.

[68] T. Li, M. Matozel, S. Boehme et al., "Overexpression of cholesterol 7α-hydroxylase promotes hepatic bile acid synthesis and secretion and maintains cholesterol homeostasis," *Hepatology*, vol. 53, no. 3, pp. 996–1006, 2011.

[69] D. Q. Shih, M. Bussen, E. Sehayek et al., "Hepatocyte nuclear factor-1α is an essential regulator of bile acid and plasma cholesterol metabolism," *Nature Genetics*, vol. 27, no. 4, pp. 375–382, 2001.

[70] L. Wang, Y. K. Lee, D. Bundman et al., "Redundant pathways for negative feedback regulation of bile acid production," *Developmental Cell*, vol. 2, no. 6, pp. 721–731, 2002.

[71] T. A. Kerr, S. Saeki, M. Schneider et al., "Loss of nuclear receptor SHP impairs but does not eliminate negative feedback regulation of bile acid synthesis," *Developmental Cell*, vol. 2, no. 6, pp. 713–720, 2002.

[72] S. Khurana, J.-P. Raufman, and T. L. Pallone, "Bile acids regulate cardiovascular function," *Clinical and Translational Science*, vol. 4, no. 3, pp. 210–218, 2011.

[73] R. T. Stravitz, Y. P. Rao, Z. R. Vlahcevic et al., "Hepatocellular protein kinase C activation by bile acids: implications for regulation of cholesterol 7α-hydroxylase," *American Journal of Physiology*, vol. 271, no. 2, pp. G293–G303, 1996.

[74] J. A. Holt, G. Luo, A. N. Billin et al., "Definition of a novel growth factor-dependent signal cascade for the suppression of bile acid biosynthesis," *Genes and Development*, vol. 17, no. 13, pp. 1581–1591, 2003.

[75] E. De Fabiani, N. Mitro, A. C. Anzulovich, A. Pinelli, G. Galli, and M. Crestani, "The negative effects of bile acids and tumor necrosis factor-α on the transcription of cholesterol 7α-hydroxylase gene (CYP7A1) converge to hepatic nuclear factor-4: a novel mechanism of feedback regulation of bile acid synthesis mediated by nuclear receptors," *Journal of Biological Chemistry*, vol. 276, no. 33, pp. 30708–30716, 2001.

[76] T. Li and J. Y. L. Chiang, "Mechanism of rifampicin and pregnane X receptor inhibition of human cholesterol 7α-hydroxylase gene transcription," *American Journal of Physiology*, vol. 288, no. 51, pp. G74–G84, 2005.

[77] T. Li, A. Jahan, and J. Y. L. Chiang, "Bile acids and cytokines inhibit the human cholesterol 7α-hydroxylase gene via the JNK/c-Jun pathway in human liver cells," *Hepatology*, vol. 43, no. 6, pp. 1202–1210, 2006.

[78] H. B. Hartman, K. D. Lai, and M. J. Evans, "Loss of small heterodimer partner expression in the liver protects against dyslipidemia," *Journal of Lipid Research*, vol. 50, no. 2, pp. 193–203, 2009.

[79] H. H. Wang, P. Portincasa, N. Mendez-Sanchez, M. Uribe, and D. Q. H. Wang, "Effect of ezetimibe on the prevention and dissolution of cholesterol gallstones," *Gastroenterology*, vol. 134, no. 7, pp. 2101–2110, 2008.

[80] H. H. Wang, P. Portincasa, M. Liu, and D. Q. H. Wang, "Rapid postprandial gallbladder refilling and increased turnover of bile prevent cholesterol crystallization in small heterodimer partner (SHP) knockout mice. DDW 2011, Chicago, USA," *Gastroenterology*, vol. 140, no. 5, pp. S67–S68, 2011.

[81] I. Bjorkhem, R. Blomstrand, A. Lewenhaupt, and L. Svensson, "Effect of lymphatic drainage on 7α-hydroxylation of cholesterol in rat liver," *Biochemical and Biophysical Research Communications*, vol. 85, no. 2, pp. 532–540, 1978.

[82] S. Modica, M. Petruzzelli, E. Bellafante et al., "Selective activation of nuclear bile acid receptor FXR in the intestine protects mice against cholestasis," *Gastroenterology*, vol. 142, no. 2, pp. 355–365, 2011.

[83] M. Watanabe, S. M. Houten, L. Wang et al., "Bile acids lower triglyceride levels via a pathway involving FXR, SHP, and SREBP-1c," *Journal of Clinical Investigation*, vol. 113, no. 10, pp. 1408–1418, 2004.

[84] K. Boulias, N. Katrakili, K. Bamberg, P. Underhill, A. Greenfield, and I. Talianidis, "Regulation of hepatic metabolic pathways by the orphan nuclear receptor SHP," *EMBO Journal*, vol. 24, no. 14, pp. 2624–2633, 2005.

[85] K. E. Matsukuma, L. Wang, M. K. Bennett, and T. F. Osborne, "A key role for orphan nuclear receptor liver receptor homologue-1 in activation of fatty acid synthase promoter by liver X receptor," *Journal of Biological Chemistry*, vol. 282, no. 28, pp. 20164–20171, 2007.

[86] J. Huang, J. Iqbal, P. K. Saha et al., "Molecular characterization of the role of orphan receptor small heterodimer partner in development of fatty liver," *Hepatology*, vol. 46, no. 1, pp. 147–157, 2007.

[87] P. Delerive, C. M. Galardi, J. E. Bisi, E. Nicodeme, and B. Goodwin, "Identification of liver receptor homolog-1 as a novel regulator of apolipoprotein AI gene transcription," *Molecular Endocrinology*, vol. 18, no. 10, pp. 2378–2387, 2004.

[88] L. Wang, Y. Han, C. S. Kim, Y. K. Lee, and D. D. Moore, "Resistance of SHP-null mice to bile acid-induced liver damage," *Journal of Biological Chemistry*, vol. 278, no. 45, pp. 44475–44481, 2003.

[89] L. Wang, J. Liu, P. Saha et al., "The orphan nuclear receptor SHP regulates PGC-1α expression and energy production in brown adipocytes," *Cell Metabolism*, vol. 2, no. 4, pp. 227–238, 2005.

[90] Y. Shimamoto, J. Ishida, K. Yamagata et al., "Inhibitory effect of the small heterodimer partner on hepatocyte nuclear factor-4 mediates bile acid-induced repression of the human angiotensinogen gene," *Journal of Biological Chemistry*, vol. 279, no. 9, pp. 7770–7776, 2004.

[91] D. Mozaffarian, T. Hao, E. B. Rimm, W. C. Willett, and F. B. Hu, "Changes in diet and lifestyle and long-term weight gain in women and men," *New England Journal of Medicine*, vol. 364, no. 25, pp. 2392–2404, 2011.

[92] S. Kajimura, P. Seale, and B. M. Spiegelman, "Transcriptional control of brown fat development," *Cell Metabolism*, vol. 11, no. 4, pp. 257–262, 2010.

[93] N. J. Rothwell and M. J. Stock, "Effect of chronic food restriction on energy balance, thermogenic capacity, and brown-adipose-tissue activity in the rat," *Bioscience Reports*, vol. 2, no. 8, pp. 543–549, 1982.

[94] N. J. Rothwell and M. J. Stock, "Effects of feeding a palatable "cafeteria" diet on energy balance in young and adult lean (+/?) Zucker rats," *British Journal of Nutrition*, vol. 47, no. 3, pp. 461–471, 1982.

[95] B. Cannon and J. Nedergaard, "Developmental biology: neither fat nor flesh," *Nature*, vol. 454, no. 7207, pp. 947–948, 2008.

[96] P. Seale, B. Bjork, W. Yang et al., "PRDM16 controls a brown fat/skeletal muscle switch," *Nature*, vol. 454, no. 7207, pp. 961–967, 2008.

[97] M. Klingenspor, "Cold-induced recruitment of brown adipose tissue thermogenesis," *Experimental Physiology*, vol. 88, no. 1, pp. 141–148, 2003.

[98] P. Trayhurn, "Adipose tissue in obesity—an inflammatory issue," *Endocrinology*, vol. 146, no. 3, pp. 1003–1005, 2005.

[99] P. Trayhurn and I. S. Wood, "Signalling role of adipose tissue: adipokines and inflammation in obesity," *Biochemical Society Transactions*, vol. 33, no. 5, pp. 1078–1081, 2005.

[100] I. S. Wood, B. Wang, J. R. Jenkins, and P. Trayhurn, "The pro-inflammatory cytokine IL-18 is expressed in human adipose tissue and strongly upregulated by TNFα in human adipocytes," *Biochemical and Biophysical Research Communications*, vol. 337, no. 2, pp. 422–429, 2005.

[101] W. D. Van Marken Lichtenbelt, J. W. Vanhommerig, N. M. Smulders et al., "Cold-activated brown adipose tissue in healthy men," *New England Journal of Medicine*, vol. 360, no. 15, pp. 1500–1508, 2009.

[102] B. B. Lowell and B. M. Spiegelman, "Towards a molecular understanding of adaptive thermogenesis," *Nature*, vol. 404, no. 6778, pp. 652–660, 2000.

[103] C. Liu and J. D. Lin, "PGC-1 coactivators in the control of energy metabolism," *Acta Biochimica et Biophysica Sinica*, vol. 43, no. 4, pp. 248–257, 2011.

[104] J. A. Stuart, S. Cadenas, M. B. Jekabsons, D. Roussel, and M. D. Brand, "Mitochondrial proton leak and the uncoupling protein 1 homologues," *Biochimica et Biophysica Acta*, vol. 1504, no. 1, pp. 144–158, 2001.

[105] G. Garruti and D. Ricquier, "Analysis of uncoupling protein and its mRNA in adipose tissue deposits of adult humans," *International Journal of Obesity*, vol. 16, no. 5, pp. 383–390, 1992.

[106] F. Bouillaud, F. Villarroya, E. Hentz, S. Raimbault, A. M. Cassard, and D. Ricquier, "Detection of brown adipose tissue uncoupling protein mRNA in adult patients by a human genomic probe," *Clinical Science*, vol. 75, no. 1, pp. 21–27, 1988.

[107] I. Tabbi-Anneni, R. Cooksey, V. Gunda et al., "Overexpression of nuclear receptor SHP in adipose tissues affects diet-induced obesity and adaptive thermogenesis," *American Journal of Physiology*, vol. 298, no. 5, pp. E961–E970, 2010.

[108] G. Song, K. Park, and L. Wang, "Gene expression profiling reveals a diverse array of pathways inhibited by nuclear receptor SHP during adipogenesis," *International Journal of Clinical and Experimental Pathology*, vol. 2, no. 3, pp. 275–285, 2009.

[109] H. Cao and R. A. Hegele, "Identification of polymorphisms in the human SHP1 gene," *Journal of Human Genetics*, vol. 47, no. 8, pp. 445–447, 2002.

[110] C. C. C. Hung, I. S. Farooqi, K. Ong et al., "Contribution of variants in the small heterodimer partner gene to birthweight, adiposity, and insulin levels: mutational analysis and association studies in multiple populations," *Diabetes*, vol. 52, no. 5, pp. 1288–1291, 2003.

[111] H. Nishigori, H. Tomura, N. Tonooka et al., "Mutations in the small heterodimer partner gene are associated with mild obesity in Japanese subjects," *Proceedings of the National Academy of Sciences of the United States of America*, vol. 98, no. 2, pp. 575–580, 2001.

[112] M. Enya, Y. Horikawa, E. Kuroda et al., "Mutations in the small heterodimer partner gene increase morbidity risk in Japanese type 2 diabetes patients," *Human Mutation*, vol. 29, no. 11, pp. E271–E277, 2008.

[113] S. M. S. Mitchell, M. N. Weedon, K. R. Owen et al., "Genetic variation in the small heterodimer partner gene and young-onset type 2 diabetes, obesity, and birth weight in U.K. subjects," *Diabetes*, vol. 52, no. 5, pp. 1276–1279, 2003.

[114] S. M. Echwald, K. L. Andersen, T. I. A. Sørensen et al., "Mutation analysis of NROB2 among 1545 danish men identifies a novel c.278G > a (p.G93D) variant with reduced functional activity," *Human Mutation*, vol. 24, no. 5, pp. 381–387, 2004.

[115] K. Yamagata, H. Daitoku, Y. Shimamoto et al., "Bile acids regulate gluconeogenic gene expression via small heterodimer partner-mediated repression of hepatocyte nuclear factor 4 and Foxo1," *Journal of Biological Chemistry*, vol. 279, no. 22, pp. 23158–23165, 2004.

[116] D. Q. Shih, S. Screenan, K. N. Munoz et al., "Loss of HNF-1α function in mice leads to abnormal expression of genes involved in pancreatic islet development and metabolism," *Diabetes*, vol. 50, no. 7-12, pp. 2472–2480, 2001.

[117] L. Wang, J. Huang, P. Saha et al., "Orphan receptor small heterodimer partner is an important mediator of glucose homeostasis," *Molecular Endocrinology*, vol. 20, no. 11, pp. 2671–2681, 2006.

[118] Y. D. Kim, K. G. Park, Y. S. Lee et al., "Metformin inhibits hepatic gluconeogenesis through AMP-activated protein kinase-dependent regulation of the orphan nuclear receptor SHP," *Diabetes*, vol. 57, no. 2, pp. 306–314, 2008.

[119] K. Yamagata, K. Yoshimochi, H. Daitoku, K. Hirota, and A. Fukamizu, "Bile acid represses the peroxisome proliferator-activated receptor-γ coactivator-1 promoter activity in a small heterodimer partner-dependent manner," *International Journal of Molecular Medicine*, vol. 19, no. 5, pp. 751–756, 2007.

[120] Y. Bae, J. K. Kemper, and B. Kemper, "Repression of CAR-mediated transactivation of CYP2B genes by the orphan nuclear receptor, short heterodimer partner (SHP)," *DNA and Cell Biology*, vol. 23, no. 2, pp. 81–91, 2004.

[121] D. H. Volle, R. Duggavathi, B. C. Magnier et al., "The small heterodimer partner is a gonadal gatekeeper of sexual maturation in male mice," *Genes and Development*, vol. 21, no. 3, pp. 303–315, 2007.

[122] M. Mueller, A. Atanasov, I. Cima, N. Corazza, K. Schoonjans, and T. Brunner, "Differential regulation of glucocorticoid synthesis in murine intestinal epithelial versus adrenocortical cell lines," *Endocrinology*, vol. 148, no. 3, pp. 1445–1453, 2007.

[123] Y. Zhang and L. Wang, "Nuclear receptor small heterodimer partner in apoptosis signaling and liver cancer," *Cancers*, vol. 3, no. 1, pp. 198–212, 2011.

[124] Y. Zhang, J. Soto, K. Park et al., "Nuclear receptor SHP, a death receptor that targets mitochondria, induces apoptosis and inhibits tumor growth," *Molecular and Cellular Biology*, vol. 30, no. 6, pp. 1341–1356, 2010.

[125] Y. Y. Park, H. S. Choi, and J. S. Lee, "Systems-level analysis of gene expression data revealed NR0B2/SHP as potential tumor suppressor in human liver cancer," *Molecules and Cells*, vol. 30, no. 5, pp. 485–491, 2010.

[126] Y. Zhang, P. Xu, K. Park, Y. Choi, D. D. Moore, and L. Wang, "Orphan receptor small heterodimer partner suppresses tumorigenesis by modulating cyclin D1 expression and cellular proliferation," *Hepatology*, vol. 48, no. 1, pp. 289–298, 2008.

[127] N. He, K. Park, Y. Zhang, J. Huang, S. Lu, and L. Wang, "Epigenetic inhibition of nuclear receptor small heterodimer partner is associated with and regulates hepatocellular carcinoma growth," *Gastroenterology*, vol. 134, no. 3, pp. 793–802, 2008.

[128] J.-M. Yuk, D.-M. Shin, H.-M. Lee et al., "The orphan nuclear receptor SHP acts as a negative regulator in inflammatory signaling triggered by Toll-like receptors," *Nature Immunology*, vol. 12, no. 8, pp. 742–751, 2011.

[129] R. Beyaert, "SHP works a double shift to control TLR signaling," *Nature Immunology*, vol. 12, no. 8, pp. 725–727, 2011.

[130] A. I. Shulman and D. J. Mangelsdorf, "Retinoid X receptor heterodimers in the metabolic syndrome," *New England Journal of Medicine*, vol. 353, no. 6, pp. 604–615, 2005.

[131] L. Wang, "Role of small heterodimer partner in lipid homeostasis and its potential as a therapeutic target for obesity," *Clinical Lipidology*, vol. 5, no. 4, pp. 445–448, 2010.

[132] T. Nishimaki-Mogami, M. Une, T. Fujino et al., "Identification of intermediates in the bile acid synthetic pathway as ligands for the farnesoid X receptor," *Journal of Lipid Research*, vol. 45, no. 8, pp. 1538–1545, 2004.

[133] S. Sanyal, J. Y. Kim, H. J. Kim et al., "Differential regulation of the orphan nuclear receptor Small Heterodimer Partner (SHP) gene promoter by orphan nuclear receptor ERR isoforms," *Journal of Biological Chemistry*, vol. 277, no. 3, pp. 1739–1748, 2002.

[134] K. Lai, D. C. Harnish, and M. J. Evans, "Estrogen receptor α regulates expression of the orphan receptor small heterodimer partner," *Journal of Biological Chemistry*, vol. 278, no. 38, pp. 36418–36429, 2003.

[135] B. Goodwin, M. A. Watson, H. Kim, J. Miao, J. K. Kemper, and S. A. Kliewer, "Differential regulation of rat and human CYP7A 1 by the nuclear oxysterol receptor liver X receptor-α," *Molecular Endocrinology*, vol. 17, no. 3, pp. 386–394, 2003.

[136] Y. K. Lee, K. L. Parker, H. S. Choi, and D. D. Moore, "Activation of the promoter of the orphan receptor SHP by orphan receptors that bind DNA as monomers," *Journal of Biological Chemistry*, vol. 274, no. 30, pp. 20869–20873, 1999.

[137] H.-i. Kim, Y.-K. Koh, T.-H. Kim et al., "Transcriptional activation of SHP by PPAR-γ in liver," *Biochemical and Biophysical Research Communications*, vol. 360, no. 2, pp. 301–306, 2007.

[138] H. J. Kim, J. Y. Kim, Y. Y. Park, and H. S. Choi, "Synergistic activation of the human orphan nuclear receptor SHP gene promoter by basic helix-loop-helix protein E2A and orphan nuclear receptor SF-1," *Nucleic Acids Research*, vol. 31, no. 23, pp. 6860–6872, 2003.

[139] S. Gupta, R. T. Stravitz, P. Dent, and P. B. Hylemon, "Down-regulation of cholesterol 7α-hydroxylase (CYP7A1) gene expression by bile acids in primary rat hepatocytes is mediated by the c-Jun N-terminal kinase pathway," *Journal of Biological Chemistry*, vol. 276, no. 19, pp. 15816–15822, 2001.

[140] Y. H. Choi, M. J. Park, K. W. Kim, H. C. Lee, Y. H. Choi, and J. Cheong, "The orphan nuclear receptor SHP is involved in monocytic differentiation, and its expression is increased by c-Jun," *Journal of Leukocyte Biology*, vol. 76, no. 5, pp. 1082–1088, 2004.

[141] A. K. Iyer, Y. H. Zhang, and E. R. B. McCabe, "Dosage-sensitive sex reversal adrenal hypoplasia congenita critical region on the X chromosome, gene 1 (DAX1) (NR0B1) and Small Heterodimer Partner (SHP) (NR0B2) form homodimers individually, as well as DAX1-SHP heterodimers," *Molecular Endocrinology*, vol. 20, no. 10, pp. 2326–2342, 2006.

[142] A. K. Iyer, Y. H. Zhang, and E. R. B. McCabe, "LXXLL motifs and AF-2 domain mediate SHP (NR0B2) homodimerization and DAX1 (NR0B1)-DAX1A heterodimerization," *Molecular Genetics and Metabolism*, vol. 92, no. 1-2, pp. 151–159, 2007.

[143] C. M. Klinge, S. C. Jernigan, and K. E. Risinger, "The agonist activity of tamoxifen is inhibited by the short heterodimer partner orphan nuclear receptor in human endometrial cancer cells," *Endocrinology*, vol. 143, no. 3, pp. 853–867, 2002.

[144] C. Brendel, K. Schoonjans, O. A. Botrugno, E. Treuter, and J. Auwerx, "The small heterodimer partner interacts with the liver X receptor α and represses its transcriptional activity," *Molecular Endocrinology*, vol. 16, no. 9, pp. 2065–2076, 2002.

[145] S. Kress and W. F. Greenlee, "Cell-specific regulation of human CYP1A1 and CYP1B1 genes," *Cancer Research*, vol. 57, no. 7, pp. 1264–1269, 1997.

[146] F. J. Naya, C. M. M. Stellrecht, and M. J. Tsai, "Tissue-specific regulation of the insulin gene by a novel basic helix-loop-helix transcription factor," *Genes and Development*, vol. 9, no. 8, pp. 1009–1019, 1995.

[147] J. E. Lee, S. M. Hollenberg, L. Snider, D. L. Turner, N. Lipnick, and H. Weintraub, "Conversion of Xenopus ectoderm into neurons by NeuroD, a basic helix-loop-helix protein," *Science*, vol. 268, no. 5212, pp. 836–844, 1995.

[148] J. Y. Kim, K. Chu, H. J. Kim et al., "Orphan nuclear receptor small heterodimer partner, a novel corepressor for a basic helix-loop-helix transcription factor BETA2/NeuroD," *Molecular Endocrinology*, vol. 18, no. 4, pp. 776–790, 2004.

[149] M. J. Park, H. J. Kong, H. Y. Kim, H. H. Kim, J. H. Kim, and J. H. Cheong, "Transcriptional repression of the gluconeogenic gene PEPCK by the orphan nuclear receptor SHP through inhibitory interaction with C/EBPα," *Biochemical Journal*, vol. 402, no. 3, pp. 567–574, 2007.

[150] A. Båvner, L. Johansson, G. Toresson, J. Å. Gustafsson, and E. Treuter, "A transcriptional inhibitor targeted by the atypical orphan nuclear receptor SHP," *EMBO Reports*, vol. 3, no. 5, pp. 478–484, 2002.

[151] K. Boulias and I. Talianidis, "Functional role of G9a-induced histone methylation in small heterodimer partner-mediated

transcriptional repression," *Nucleic Acids Research*, vol. 32, no. 20, pp. 6096–6103, 2004.

[152] S. Fang, J. Miao, L. Xiang, B. Ponugoti, E. Treuter, and J. K. Kemper, "Coordinated recruitment of histone methyltransferase G9a and other chromatin-modifying enzymes in SHP-mediated regulation of hepatic bile acid metabolism," *Molecular and Cellular Biology*, vol. 27, no. 4, pp. 1407–1424, 2007.

[153] G. Song and L. Wang, "Nuclear receptor SHP activates miR-206 expression via a cascade dual inhibitory mechanism," *PLoS ONE*, vol. 4, no. 9, Article ID e6880, 2009.

12

Protectors or Traitors: The Roles of PON2 and PON3 in Atherosclerosis and Cancer

Ines Witte,[1] Ulrich Foerstermann,[1] Asokan Devarajan,[2]
Srinivasa T. Reddy,[2,3] and Sven Horke[1]

[1] Department of Pharmacology, University Medical Center of the Johannes-Gutenberg University Mainz, Obere Zahlbacher Street 67, 55131 Mainz, Germany
[2] Department of Medicine, University of California, Los Angeles, CA 90095, USA
[3] Department of Molecular and Medical Pharmacology, University of California, Los Angeles, CA 90095, USA

Correspondence should be addressed to Sven Horke, horke@uni-mainz.de

Academic Editor: Mira Rosenblat

Cancer and atherosclerosis are major causes of death in western societies. Deregulated cell death is common to both diseases, with significant contribution of inflammatory processes and oxidative stress. These two form a vicious cycle and regulate cell death pathways in either direction. This raises interest in antioxidative systems. The human enzymes paraoxonase-2 (PON2) and PON3 are intracellular enzymes with established antioxidative effects and protective functions against atherosclerosis. Underlying molecular mechanisms, however, remained elusive until recently. Novel findings revealed that both enzymes locate to mitochondrial membranes where they interact with coenzyme Q10 and diminish oxidative stress. As a result, ROS-triggered mitochondrial apoptosis and cell death are reduced. From a cardiovascular standpoint, this is beneficial given that enhanced loss of vascular cells and macrophage death forms the basis for atherosclerotic plaque development. However, the same function has now been shown to raise chemotherapeutic resistance in several cancer cells. Intriguingly, PON2 as well as PON3 are frequently found upregulated in tumor samples. Here we review studies reporting PON2/PON3 deregulations in cancer, summarize most recent findings on their anti-oxidative and antiapoptotic mechanisms, and discuss how this could be used in putative future therapies to target atherosclerosis and cancer.

1. Introduction

Most studies in the field of paraoxonases (PONs) deal with cardiovascular diseases, such as atherosclerosis and diabetes, where PONs exert protective functions in cell culture as well as animal studies. It has been anticipated that the known antioxidative functions of PONs, including PON2 and PON3, were central to their effects although underlying molecular mechanisms remained obscure. However, recent findings caused a significant progress in this field because molecular pathways of PON2 and PON3 functions have been largely revealed. Moreover, the result of the cell-protective function were shown to play a vital role in survival and stress resistance of cancer cells, along with the finding that numerous tumors overexpressed these enzymes. There, PON2 and PON3 appear to increase chemotherapeutic resistance and favor cell survival. In this review, we summarize the most recent findings and discuss the role of PON2/PON3 in

atherosclerosis and cancer. A future perspective gives an outlook on how PONs may be targets of novel therapeutic approaches.

2. Altered Expression Levels of Paraoxonase Enzymes in Cancer

It is established that oxidative stress from mitochondria plays an important role in apoptosis and also leads to premature aging and cancer. There is growing scientific consensus that antioxidants or proteins with antioxidative functions, such as paraoxonases, can lower the incidence of, for example, cardiovascular and neurodegenerative diseases. On the other hand, recent studies have shown that various types of cancer obviously take advantage of this protection by enhanced expression of the antioxidative paraoxonase proteins. In the following section, we give an overview of studies that assessed expression of PON1, PON2, or PON3 in various cancers, with the majority of studies seemingly reporting a deregulation of these proteins.

PON1 levels and activity are lower in many inflammatory and oxidative stress-associated diseases [1]. Also, serum PON1 and arylesterase activities were reduced in patients with epithelial ovarian cancer [2] and lung cancer [3]. Uyar et al. found that Q allele of PON1 was more frequent in renal cancer patients [4], and Antognelli at al. reported that certain PON1 genotypes were prone to increased risk of prostate cancer [5]. More recently, the presence of the variant alleles of the Q192R and L55M SNPs of PON1, both of which result in an amino acid replacement that alters PON1 activity, were found associated with a 18–29% increased risk of aggressive prostate cancer [6]. These studies clearly demonstrate a link between PON1 and cancer etiology; however, PON1 is not the scope of this review. We will focus on the role of PON2 and PON3 in cancer based on recent discoveries on the mechanism of action of these proteins in proliferation and apoptosis.

Research on paraoxonases is a relatively young field, and still much of our understanding comes from findings related to PON1. Back in 1999, our knowledge about PON2 and PON3 was extremely limited although few studies emerged that reported genetic associations with metabolic diseases [7]. There are two common single nucleotide polymorphisms (SNPs) in PON2—G148A and C311S—that have been associated with disease phenotypes. In essence, an association between these SNPs and several diseases was demonstrated. For PON2-G/A148 this is true, for instance, for higher plasma glucose [8], higher plasma HDL cholesterol [9], and lower plasma LDL cholesterol [10]. With respect to S/C311, Stoltz et al. reported that this mutation determines the lactonase activity of PON2 which links to hydrolysis of important bacterial virulence factors [11]. However, a subsequent study from our lab did not confirm this finding [12]. The impact of PON2-S/C311 on lactone hydrolysis thus merits further investigation. This may similarly apply for its role in coronary heart disease, where at least one study reported an association [13] that was not found in a subsequent meta-analysis [14].

Despite the established and prevailing role of paraoxonases in cardiovascular diseases and relevant parameters, more recent studies revealed an emerging association of PONs with cancer. For example, microarray studies observed an overexpression of PON2 in some solid tumors like hepatocellular carcinoma, prostate carcinoma [15, 16], and several others, which are illustrated in Table 1. Additionally, in various leukemia gene expression profiling studies, an upregulation of PON2 could be demonstrated; an example is pediatric acute lymphoblastic leukemia (ALL) [17]. Importantly, a subsequent study identified PON2 as member of a very small group of upregulated genes that characterized pediatric ALL patients with very poor outcome prognosis [18]. In another form of leukemia, chronic myeloid leukemia (CML), PON2 was also identified in an outcome-specific gene expression signature of primary imatinib-resistant patients [19]. Moreover, a marked overexpression of PON2

was observed in lymphocytes infected with T-cell leukemia virus [20].

In contrast to PON2, there are fewer studies for PON3 with the tendency of more diverse results (see Table 2). For instance, a downregulation was demonstrated in a meta-analysis of expression profiles in hepatocellular carcinomas (HCC) [21] and in ovarian serous papillary carcinomas (OSPCs) [22] shown by oligonucleotide microarrays. However, there are various other such analyses, which showed altered expression of PON3 (up as well as downregulated) in different types of cancers. For overview, consult the Gene Expression Atlas found at http://www.ebi.ac.uk/gxa/. In general, it should be noted that these association studies show no direct proof for a physiological relevance of these proteins in cancer, nor do such studies give any clues about their functions and mechanisms.

In addition to the listed microarray data, our very recent analyses showed that the PON2 level is increased in some tumors at the protein level (Table 1). We showed a moderate PON2 overexpression in pancreas, liver, kidney, and lung tumors and an over 10-fold upregulation of PON2 in thymus tumors and non-Hodgkins lymphomas [23]. Assessment of PON2 protein levels is not feasible in hundreds of cancer samples. Therefore, we previously used cDNA arrays, developed for differential gene expression analysis and validation of hundreds of different human tissues. We showed that PON2 is ~2–4-fold overexpressed in the tumors from urinary bladder, liver, kidney, lymphoid tissues, and endometrium/uterus in comparison to normal tissue [23], which are in accordance with western blot analyses. Despite some other tissues, where no increase in the expression level was observed, human tumors of the thyroid gland, testis, prostate, and pancreas showed a slight upregulation of PON2 (Table 1).

Using the same cDNA arrays as for PON2, our group showed a considerably increased PON3 expression in all tested cancer types, except cervix [24]. Remarkably, the intensity of PON3 overexpression was markedly enhanced compared to that of PON2. In this array over 10-fold upregulation of PON3 in tumors from endometrium/uterus and stomach was shown and over 3-fold induction in samples from pancreas, urinary bladder, thyroid, prostate, pancreas, liver testis, and lung cancers. These results could be verified with another matched array particularly for lung cancer (normal versus diseased samples from the same patient). But in contrast to PON2, PON3 expression appears to be largely restricted to cells derived from solid tumors [24]. One reason for the high expression level of PON3 in cancer tissue is certainly the low basal expression level of PON3 in healthy tissues but may nevertheless suggest a role for PON3 in cell death escape.

An interesting phenomenon is obvious upon closer inspection of the array data. A tumor subtype and stage-specific analysis revealed that both PON2 and PON3 are upregulated rather in the early stages and some subtypes of cancer, whereas the expression in the late stages of the tumor seems to be declining (see Figure 1). This could indicate that, especially in the early stages of tumor formation, the antioxidative and antiapoptotic function of PON2 and

TABLE 1: Expression levels of PON2 in various tumor tissues and/or cancer cell cultures. Microarray experiment (array express) listings are according to the Gene Atlas Database. Protein and cDNA levels according to [23]. Cell culture expression levels were roughly estimated as relative level comparing to A549, grouped into *low*, *medium*, or *high*.

Tissue (cancer)	Protein level (fold of normal tissue)	cDNA array (fold of normal tissue)	Microarray studies	Cell culture (expression level in cell line)
Kidney	2	2.2	Upregulated in renal carcinoma (E-MTAB-37)	Medium (HEK293)
Liver	1.7	2.2	Overexpressed (Li et al. [15])	High (Huh7/HepG2)
Lung	1.3	1	Upregulated in lung adenocarcinoma (E-MEXP-231/E-MTAB-37) Downregulated in small cell lung carcinoma (E-GEOD-4127)	High (A549; H661; H1299)
Spleen	0.5	n/a		
Pancreas	1.4	1.6	Upregulated in pancreatic carcinoma (E-MTAB-37)	
Thymus	11.5	n/a		
Urinary bladder	n/a	4.1		High (HT1367/RT112)
Esophagus	n/a	0.6	Upregulated in esophageal cancer (E-MTAB-62)	
Stomach	n/a	1	Upregulated in gastric carcinoma (E-GEOD-2685)	
Ovar	n/a	1	Upregulated (E-MTAB-62)	
Cervix	n/a	1	Upregulated (E-MTAB-37/E-MTAB-62)	Medium (HeLa)
Adrenal gland	n/a	1	Downregulated in adrenocortical carcinoma (E-TABM-311)	
Thyroid gland	n/a	1.4	Upregulated (E-GEOD-3467/E-GEOD-3678)	
Prostate	n/a	1.6	Overexpressed (Ribarska, T. et al. [16] E-MTAB-62)	
Testis	n/a	1.7		Low (SuSa/GCT27/833K)
Uterus/endometrium	n/a	2.1		
Lymphoid tissue	n/a	2.5		
Leukemias (various)	n/a		Upregulated in pediatric ALL (Ross et al. and Kang et al. [17, 18])	Low in AML-like Nalm6/EOL; Jurkat Tcells; PML-like HL60/HCW2; CML-like KCL Medium in blast crisis line K562; CML-like lama; AML-like THP1/MonoMac6/HEL
Non-Hodgkin	11.9	n/a	Downregulated (E-MTAB-37)	

PON3 is important and beneficial as it helps generating the platform for malignant transformation. This could represent a potential approach of innovative therapies trying to normalize the otherwise overexpressed PONs.

A first direct hint to this theory came from our recent study demonstrating that PON2 increased chemoresistance in leukemic cells [23], which is in line with genetic association studies where PON2 upregulation was associated with imatinib resistance in CML patients [19] and poor prognosis in cohorts of pediatric ALL [17, 18]. In support of the hypothesis, the same study [23] revealed that knockdown of endogenous PON2 caused spontaneous apoptosis of several human cancer cell lines—an intriguing but somewhat unexpected finding given the viability of PON2-deficient mice (the residual PON2 expression in these mice [25] may be comparable to efficient cell culture RNAi experiments).

An exciting question is how tumors achieve an increase in PON2 and/or PON3 expression, and this should be a major goal of future studies. Certainly there is no general answer to this question. Most likely, underlying mechanisms are individual for each given tumor. One simple explanation could be that, in some tissues, for example, papillary renal cell kidney carcinoma or prostate adenocarcinoma, chromosome 7, which contains the PON cluster, is amplified [16]. Another reason might be that the regulation depends on several signaling pathways, which are linked to reactive oxygen species and cancer, for example, PPAR-γ, AP-1, β-catenin/Wnt, NF-κB, HIF-1α, PI3K, and Nrf2 [26]. In accordance, earlier studies showed that PON2 expression is enhanced by oxidative stress [27], PI3K/PDGFR, PPARγ, and NADPH oxidase activation as well as by AP-1 activation [28, 29]. The urokinase plasminogen activator (uPA) system

TABLE 2: Expression levels of PON3 in various tumor tissues and/or cancer cell cultures. Microarray experiment (array express) listings are according to the Gene Atlas Database. cDNA levels according to [24]. Cell culture expression levels were roughly estimated as relative level comparing to A549, grouped into *low*, *medium*, or *high*.

Tissue (cancer)	cDNA array (fold of normal tissue)	Microarray studies	Cell culture (expression level in cell line)
Kidney	2.2	Downregulated in clear cell sarcoma of the kidney (E-GEOD-2712/E-TABM-282)	Not detectable (HEK293)
Liver	4.9	Downregulated in hepatocellular carcinoma (HCC) (Choi et al. [21])	High (Huh7) Medium (HepG2)
Lung	3.4	Upregulated in lung adenocarcinoma (E-MTAB-37/E-MTAB-62)	Medium (A549)
Pancreas	3.2	Upregulated in pancreatic carcinoma (E-MTAB-37)	
Urinary bladder	3.8		Not detectable (HT1367/RT112)
Esophagus	1.8		
Stomach	9.5		
Ovar	2.1	Downregulated in ovarian serous papillary carcinomas (OSPCs) (Santin et al. [22])	
Cervix	0.5	Downregulated in cervical carcinoma (E-MTAB-62) Upregulated in cervical carcinoma (E-MTAB-37)	Not detectable (HeLa)
Adrenal gland	1.5		
Thyroid gland	2.6	Downregulated in papillary thyroid carcinoma (E-GEOD-3467)	
Prostate	4.5	Downregulated in prostate carcinoma (E-MTAB-62)	
Testis	5.3		Not detectable (SuSa/GCT27/833K)
Uterus/endometrium	16.2		
Lymphoid tissue	2.3		
Leukemias (various)	n/a		Not detectable in AML-like Nalm6; Jurkat Tcells; PML-like HL60/HCW2 blast crisis line K562; CML-like lama; AML-like THP1 high in CML-like KCL
Non-Hodgkin	n/a	Downregulated (E-MTAB-37)	

may also be relevant, as this is increased in numerous cancers and upregulates PON2 [29].

A point of interest is why some tumors upregulate PON2 or PON3. One of the hallmarks of cancer is resistance to cell death [30]. It has been found that paraoxonases 2 and 3 provide a protection against mitochondrial cell death signaling [23, 24]. Their overexpression lowered susceptibility to different chemotherapeutics (e.g., imatinib, doxorubicin, and staurosporine) in cell culture models via diminishing proapoptotic mitochondrial O_2^- formation. It is established that oxidative stress and chronic inflammation are closely linked to cell death and cancer [26]. Therefore, it appears conceivable that tumors take advantage of the antioxidative function of PON2/PON3 to escape cell death.

3. The Antioxidative Mechanisms of PON2/PON3

Inflammation and oxidative stress contribute to the etiology of almost every known disease. Reactive oxygen species generated by enzymatic and nonenzymatic systems modify lipids and sterols, producing oxidized lipids and oxidized sterols that, if unchecked, produce a vicious cycle of undesirable inflammation and more oxidative stress. Atherosclerosis is a chronic inflammatory disease characterized by the focal accumulation of numerous cells, lipids, and extracellular matrices in the intima of arteries. Although reduced levels of high density lipoprotein (HDL) and elevated levels of low density lipoprotein (LDL) cholesterol are accepted

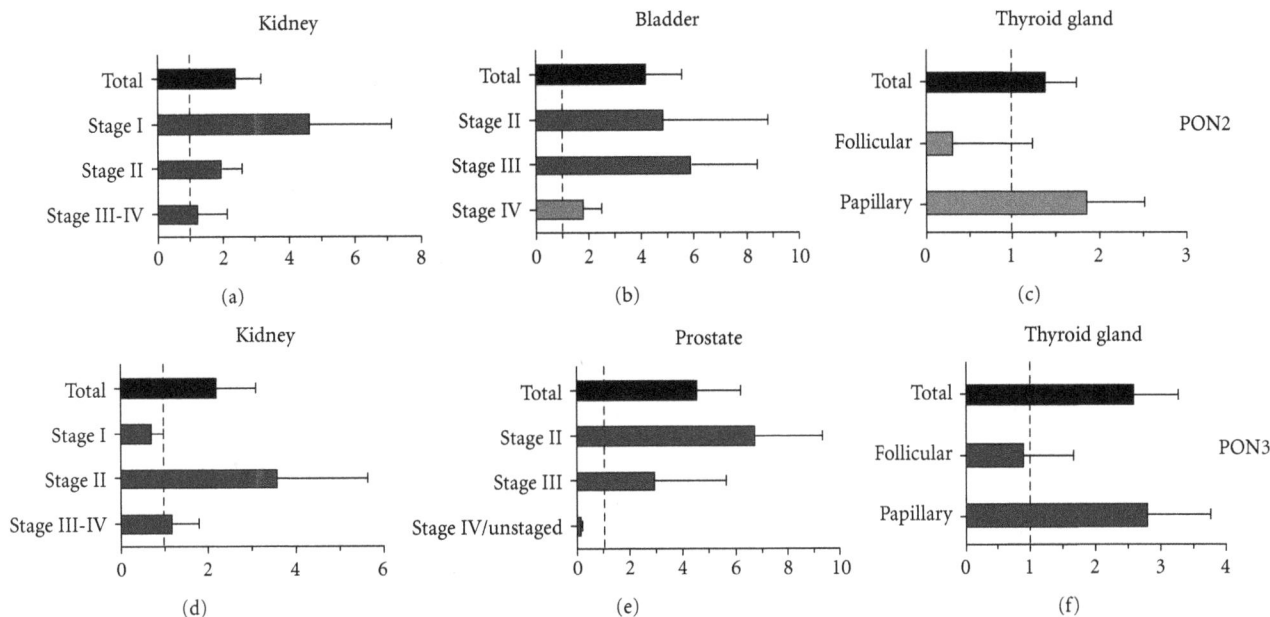

FIGURE 1: PON2 and PON3 are found overexpressed in early rather than late stages of tumors. Indicated cancer tissues were analyzed for PON2/PON3 cDNA levels (normalized to GAPDH) relative to healthy controls. Values were taken from recently performed arrays [23, 24].

risk factors for this disease, atherogenesis cannot solely be explained by cholesterol or lipid deposition in the arterial wall. Accumulating evidence suggests that oxidative stress plays a fundamental role in atherosclerosis. In particular, the oxidation theory for atherosclerosis proposes that LDL is a major target of oxidation and is involved in both the initiation as well as progression of atherosclerosis [31].

Although there has been a focus on PON1 due to its association with HDL, a number of studies demonstrated that PON2 and PON3 protect cells and tissues from oxidative stress by reducing reactive oxygen species [1, 25, 32–37]. PON2 and PON3 can inhibit LDL oxidation and enhance the antioxidant properties and cholesterol efflux capacity of HDL even though they are not readily found on the lipoproteins [1, 25, 32–37]. Moreover, in animal models, both PON2 and PON3 have been shown to abrogate the development of atherosclerosis [25, 35, 38]. These preclinical studies clearly demonstrated that PON2 and PON3 (similar to PON1) are (a) anti-atherogenic and (b) targets for therapy. However, to date, the physiological substrates and roles for PON2 and PON3 have not been elucidated, which similarly applies to PON1.

Recent studies suggest that PON2 [12, 38, 39] and PON3 [24] modulate the levels of reactive species in cells and in animal models demonstrating for the first time a physiological molecular link between PON proteins and oxidative stress. Based on the earlier result that PON2 was found in subcellular mitochondrial fractions [40], Altenhöfer et al. demonstrated that PON2 prevents the ubisemiquinone-mediated mitochondrial superoxide generation and apoptosis independent of its lactonase activity [12]. During Q cycle, unstable intermediate ubisemiquinone (coenzyme Q_{10} [CoQ_{10}^-]) can donate electron to molecular oxygen (instead of cytochrome c) leading to superoxide production and

reduced ETC activity [41–43]. Devarajan et al. reported that (a) PON2 is present in the inner mitochondrial membrane (IMM), and (b) binds with high affinity to coenzyme Q10 (CoQ10), an important component of the ETC [38]. Steady-state concentrations of ubisemiquinone are increased in the IMM resulting in superoxide formation when treated with inhibitors of ETC, antimycin, or rotenone [43]. Devarajan et al. demonstrated that overexpression of PON2 reduces superoxide levels induced by either antimycin or rotenone suggesting that PON2 sequesters ubisemiquinone. Moreover, PON2-deficient mice harbour reduced ETC complex I + III activities, oxygen consumption, ATP levels, and enhanced mitochondrial oxidative stress further suggesting that PON2 maintains the respiratory chain by promoting the sequestration of the unstable reactive intermediate ubisemiquinone, thereby preventing the superoxide production. Supporting our hypothesis, previously, it has been shown that mitochondrial superoxide is inversely related to the amount of CoQ10 bound to membrane proteins [44]. Similar to PON2, Schweikert et al. have demonstrated that PON3 is also localized to mitochondria, protects against mitochondrial oxidative stress, and demonstrated that Q10 is associated with purified PON3-GFP protein [24]. This illustrates that the antiatherogenic effects of PON2/3 are, in part, mediated by their role in mitochondrial function (Figure 2). Since increased production of reactive oxygen species (ROS) as a result of mitochondrial dysfunction play a role in the development of many inflammatory diseases including atherosclerosis, the recent data on PON2 and PON3 provide a mechanistic direction for the scores of epidemiological studies that show a link between PON proteins and numerous inflammatory diseases including Type II diabetics and cancer.

Atherosclerosis and insulin resistance are multifactorial diseases that are commonly associated with dyslipidemia, oxidative stress, obesity, hypertension, and chronic inflammation. The liver is not only the primary site of lipid metabolism, but is a major site for glucose uptake, production, and storage. Its role in glucose metabolism is strongly influenced by systemic as well as local oxidative and inflammatory stimuli [45, 46], which in turn influences whole-body insulin responsiveness [47]. Hepatic glucose metabolism is strongly influenced by oxidative stress and proinflammatory stimuli. Given the elevated oxidative stress levels and abnormal lipid metabolism reported previously in PON2-deficient mice [25, 38], Bourquard et al. hypothesized that atherosclerosis may be accompanied by impaired hepatic insulin signaling and showed that PON2 deficiency is associated with inhibitory insulin-mediated phosphorylation of hepatic insulin receptor substrate-1 (IRS-1) [39]. Factors secreted from activated macrophage cultures derived from PON2-deficient mice are sufficient to modulate insulin signaling in cultured hepatocytes in a manner similar to that observed *in vivo* [39]. It was further demonstrated that modulation of hepatic insulin sensitivity by PON2 is mediated by a shift in the balance of NO and ONOO$^-$ (peroxynitrite) formation. These studies show that PON2 plays an important role in insulin sensitivity by its ability to modulate reactive species most likely as a result of PON2's association with mitochondrial function.

Oxidative stress has long been associated with the pathophysiology of cancer. In particular, enhanced ROS formation increases DNA damage, genome instability, and cell proliferation especially during cancer initiation. On the other hand, oxidative stress also counteracts tumorigenesis, as it induces senescence and drives apoptosis and other cell death pathways [48]. The precise spatiotemporal control of ROS generation is therefore a critical regulator of cell survival and death, for instance since overwhelming mitochondrial oxidative stress exerts apoptotic rather than protumorigenic functions. Nevertheless, reactive oxygen species may be conducive to the vitality of cancer cells and drive signaling transduction pathways, which lead to activation of redox-sensitive transcription factors and genes involved in cancer cell growth, proliferation, and survival [26]. In conclusion, PON2 and PON3 reduce oxidative stress and inflammation and thus act as central regulators of diseases, including cancer and atherosclerosis.

4. Paraoxonases and the Regulation of Cell Death

The antioxidative effects of PON2 and PON3 were reported long ago, but underlying mechanisms were uncovered just recently [12, 24, 38]. This similarly applies to the cell death-reducing activity of PON2, where discovery [40] and mechanistic realization [23] were separated by years. Based on the latest knowledge, these enzymes modulate execution of the apoptotic program. In this chapter, we review their involvement in apoptosis and discuss their putative functions in other cell death pathways.

Tumor cells evolve a plethora of strategies to resist cell death with the intrinsic apoptotic program being implicated as a major barrier to cancer formation. Execution of intrinsic cell death is mainly controlled by the balance of pro- and antiapoptotic Bcl-2 protein family members [30, 49], because they regulate mitochondrial pore opening and cytochrome C release. Importantly, it also requires intramitochondrial redox signaling to liberate cytochrome C from its membrane attaching molecule, cardiolipin [50, 51]. In fact, this is a two-step process because neither mitochondrial membrane permeabilization alone nor redox-triggered disruption of the cytochrome C/cardiolipin interaction sufficiently activates the cascade. Recent studies revealed that PON2 and PON3, due to interaction with coenzyme Q10, diminish O_2^- release on either side of the inner mitochondrial membrane [23, 24]. This results in both lowered cardiolipin peroxidation and cytochrome C release, providing a marked resistance against apoptosis. Thus, if a cancer cell needs to escape from mitochondrial redox-dependent cell death, it appears beneficial to increase PON2 or PON3 expression. In accordance, both enzymes protected against a range of chemotherapeutics when overexpressed [23, 24]. In contrast, receptor-mediated apoptosis was unchanged, at least in type-I cells, where stimulation with TRAIL or TNF-α directly activated caspases 8 and 3. This may be different for type-II cells, which involve mitochondrial actions.

Another important stress and cell death pathway is the unfolded protein response (UPR) as a result of insurmountable ER stress [52]. Both PON2 and PON3 protected against UPR-mediated apoptosis in a similar manner, that is, by negative modulation of JNK signaling, CHOP induction, and subsequent caspase activation [12, 23, 24]. Canonical UPR signaling (via ATF6, XBP1, or p-eIF2a) was unchanged, at least by PON2, so their precise mechanisms of protection remain uncertain. Future studies must reveal if PON2/PON3 act just through their mitochondrial effects or if they modulate signaling from IRE1 to TRAF2/ASK1, from the ER to mitochondria or local ROS/Ca^{2+} responses and how they reduce JNK phosphorylation. Interestingly, PON2 overexpression was induced by ER stress and protected against UPR-triggered cell death, but this was lost upon major disturbances of Ca^{2+}—homeostasis, presumably by calpain-dependent PON2 degradation [53]. Our current studies suggest that this similarly applies to PON3 ([54] this issue and data not shown). The postulated functions of PON2 and PON3 in apoptosis and ER stress-induced cell death are summarized in Figure 3.

A vital physiologic response that regulates cellular metabolism and survival is autophagy. This pathway operates at low basal levels but can be markedly increased under specific stress conditions. It enables breakdown of macromolecular structures and organelles to allow recycling of catabolites. Therefore, autophagy may alleviate nutrient limitation as experienced by many cancer cells. However, autophagy has opposing effects on different tumor cells and may cause survival of one but death of the other [55]. Whether paraoxonases modulate this pathway is unknown and no interaction with Bcl-2 family/autophagy-related

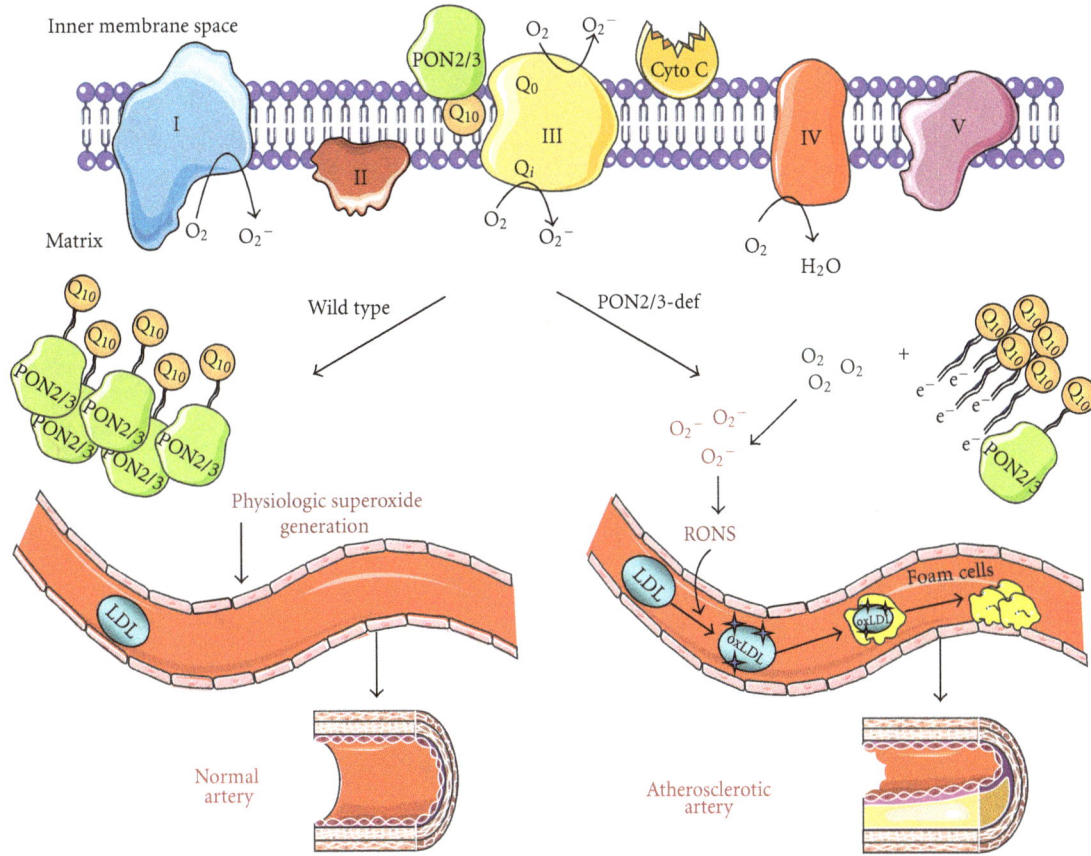

FIGURE 2: Schematic presentation of the suggested antioxidative mechanism of PON2 and PON3. A current model for the role of PON2/3 in the development of atherosclerosis. Ubisemiquinone is released from ETC in the mitochondria during Q cycle. *Right*. In the absence of PON2/3, ubisemiquinone donates electron to molecular oxygen to form superoxide; superoxide generates other reactive oxygen/nitrogen species (RONS), which oxidize LDL to form oxLDL; macrophages engulf oxLDL to form foam cells; foam cells attach to the arterial wall and subsequently develop into atherosclerotic lesions. *Left*. In the presence of PON2/3 (in wild type mice), ubisemiquinone binds to PON2/3. The binding of PON2/3 and ubisemiquinone prevents superoxide generation thereby preventing the development of atherosclerosis. Note: it is currently unknown if PON2/3 face the matrix side of the inner mitochondrial membrane or the one directed towards the innermembrane space; also, the stoichiometry of PON2/3 versus Q10 is unknown. Abbreviations: I-NADH: ubiquinone oxidoreductase, II: succinate coenzyme Q reductase, III: ubiquinol cytochrome coxidoreductase, IV: cytochrome c oxidase, V-ATP synthase, Cyto c cytochrome c, Q10-coenzyme Q10.

proteins has been reported. On the other hand, oxidative stress is mutually linked with autophagy, and there it plays an important role in cancer therapy resistance and tumor progression. The connection between ROS and autophagy is illustrated, for instance, by TNFα-induced signaling in sarcoma cells [56], or by the autophagy-relevant factor Atg4 whose delipidating activity is sensitive to mitochondrial H_2O_2 production [57] (see [58] for a detailed overview of this topic). Paraoxonases hence could have a profound impact on autophagy due to their central redox effects. Because autophagy by ROS can serve as rescue pathway but may also initiate autophagic cell death, it requires more in-depth evaluation including the origin and targets of ROS. In a similar manner, this may also be true for necroptosis (or necrotic cell death), which contrasts with the chemical- or injury-triggered necrosis and represents another, RIP1 kinase-dependent programmed cell death pathway. Necrop-tosis is of relevance, for example, for damages resulting from ischemia-reperfusion, such as stroke or myocardial

infarction. Moreover, necrotic cell death may paradoxically be even beneficial to neoplasias as this form of cell death attracts tumor-promoting inflammatory cells [30]. TNF-induced necroptosis has been shown to generate complex-I-mediated ROS in mitochondria, which is crucial to this process and accounts for ultrastructural changes observed in such cells [59]. Because PON2 as well as PON3 were able to reduce superoxide released from mitochondrial complex-I [12, 24, 38], it would hence be a promising endeavor to test PONs in TNF-induced necroptosis.

Another hallmark of tumor cells is the reprogramming of glucose metabolism in order to provide efficient fueling of the high energy demand associated with rapid cancer growth. For the most part, this is manifested as a switch to (aerobic) glycolysis but also involves two different cancer cell subpopulations—one using glucose and a second set consuming lactate produced by the former (see [30] and references therein). How overexpressed PONs could play a role in this system has not been explored, and speculations can

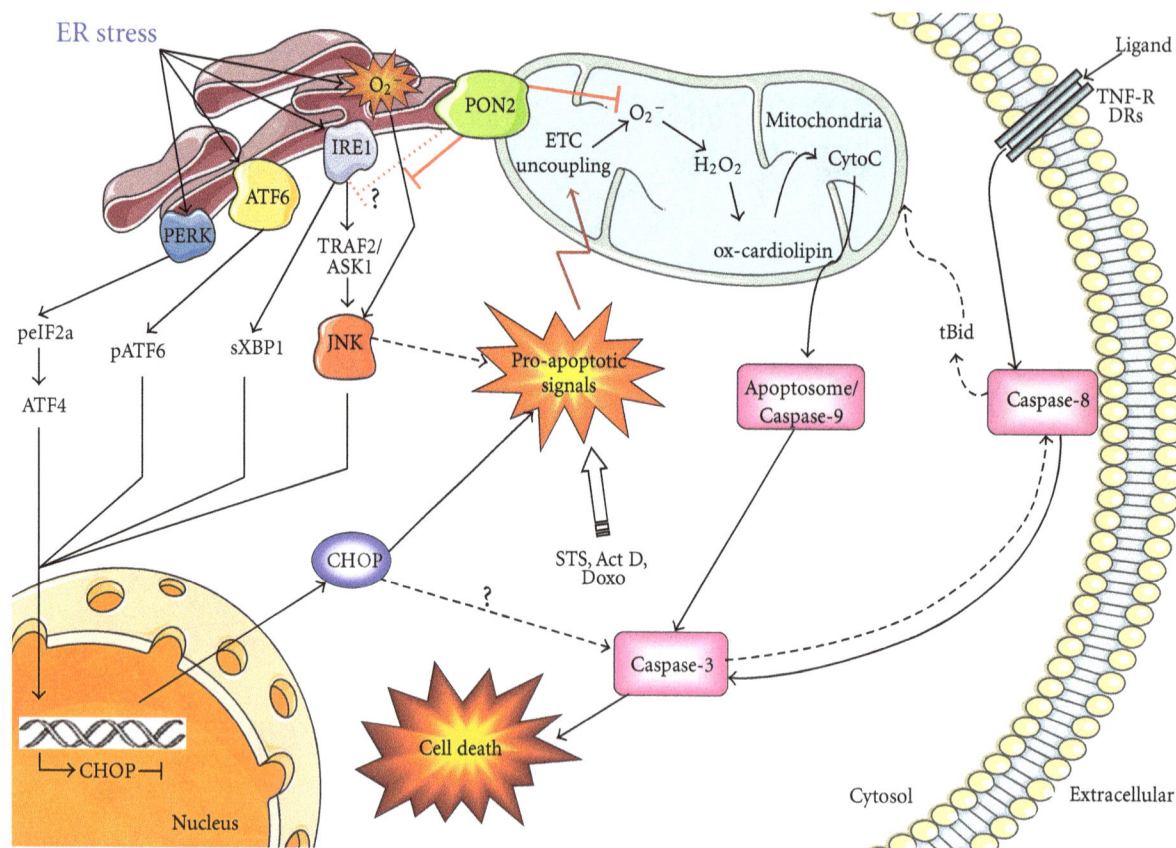

FIGURE 3: Schematic presentation of the suggested antiapoptotic mechanism of PON2 and PON3. Its ability to prevent mitochondrial O_2^- formation impacts on both ER stress-induced pathways (via acting on JNK and CHOP) as well as mitochondrial proapoptotic signaling such as cardiolipin peroxidation and cytochrome C release. See text for details. From our current understanding, PON2 is functionally interchangeable with PON3.

only be extrapolated from the PON2-deficient mice. Intriguingly, mitochondria from PON2 deficient mice produced less ATP, had impaired complex-I and -III activities, and showed enhanced oxidative stress and consumed less oxygen, resulting in an overall exhausted mitochondrial function [38]. Thus, if the opposite was true for PON2 overexpressing cells, this would ensure mitochondrial functionality and could support the energy efficacy of tumor cells.

Despite a role of paraoxonases directly in cancer cells, it could also be interesting to scan for other near-by functions. Cancer progression is determined by intracellular changes in the malignant cell itself, but also modulated by surrounding stromal cells in the tumor microenvironment. It is composed of leukocyte infiltrates consisting, for example, of endothelial cells, mast cells, T cells, and tumor-associated macrophages (TAMs). The protumorigenic TAMs are involved in critical features of neoplastic cells (such as migration & metastasis), in the inflammatory tumor microenvironment, angiogenesis, survival under hypoxia, and immune evasion [60]. Although most research groups that work on PONs employ macrophages, no study addressed TAMs to our knowledge. PON2 expression is enhanced during monocyte to macrophage differentiation in a ROS-dependent manner [28], but it is uncertain how the established anti-inflammatory and antioxidative effect of PON2 could fit

particularly into TAM functions. Thus, it may be worthwhile to assess PON2 levels in M1 versus M2 macrophages. One may speculate that TAMs have low levels of PON2, which would favor ROS formation and inflammatory responses [25, 38] and may also increase production of the metastasis-augmenting IL-1β (at least, the latter has been shown for PON2 knockdown in endothelial cells [61]). Alternatively, given that TAMs represent an interesting therapeutic target [60], monitoring their recruitment in tumors of wild-type mice compared to those deficient in PON2 or PON3 mice may also uncover new aspects of both paraoxonases and TAM infiltration with potential therapeutic implications. A direct role of PON3 in TAMs may be unlikely considering that PON3 appeared undetectable in human macrophages [27].

5. Future Perspectives

5.1. Manipulation of PON Expression: A Double-Edged Sword? The relevance of PON2 and PON3 to the cancer field has been demonstrated only recently. Few studies addressed their direct role, and our current knowledge appears somewhat fragmented. However, much of their antiapoptotic mechanism has been revealed. This allows tentative evaluation of their pharmacological usefulness. We presented multiple lines of evidence demonstrating that PON2 and PON3 are frequently found upregulated in cancer samples. Specific regulatory mechanisms are mostly unknown. The altered expression appears similar but also distinct for each of the two enzymes, varying with the tissue itself, with the specific kind of tumor and its stage, progression, differentiation, and/or metastasizing potency (see above). Up to now, two studies directly analyzed PON2 and PON3 levels in a variety of different tumors or representative cell lines [23, 24]. Addressing this question from the opposite direction, many independent laboratories used microarrays to investigate gene expressions in different tumors and often (though not always) found enhanced levels of these enzymes (see above).

Our current understanding allows to conclude that overexpression of PON2/PON3 diminishes the execution of the apoptotic program. Most likely, the antioxidative function of these enzymes represents the antiapoptotic trigger, whereas the contribution of their enzymatic activities remains unknown, if at all significant. How this works in the ER and relates to the UPR that is unknown. In the simplest model, however, this refers to electron transport within the inner mitochondrial membrane, which is in close proximity to and control of powerful apoptotic modifiers. From such perspective, upregulation of PON2 or PON3 in cells destined for apoptotic evasion appears consequential. As a logical deduction, the controlled reduction of overexpressed PON2 and/or PON3 in a given tumor may represent a novel approach to enhance its susceptibility to chemotherapeutics and to improve the therapy's effectiveness. Such hypothesis is encouraged by the observation that PON2 knockdown induced spontaneous apoptosis of several human tumor cell lines and because overexpression of either PON2 or PON3 granted robust chemotherapeutic resistance [23, 24]. PON2/PON3 expression varied substantially between different cell types, and high levels did not automatically correlate with cellular responsiveness to its knockdown. This outlines that individual approaches must be identified because PON2 and PON3, similar to already established targets, are unlikely to be beneficial in every setting. Therefore, future studies need to identify a rapid, reliable, and simple read-out system to monitor if a given tumor relies on high PON levels. This should be worthwhile, for example, in leukemic transformation in pediatric B-precursor ALL, where PON2 was among a very small group of factors highly expressed in patients with worst outcome, high risk, and affected relapse-free survival [18]. Other rewarding projects may be deduced from Table 1, where we summarized the combination (if available) of studies reporting PON2 overexpression in a given tumor. There is limited evidence for PON3 since we are just beginning to understand its function and because PON3, compared to PON2, is expressed to much lower levels and in fewer tissues. Given our recent data [24], we nevertheless conclude PON3 also represents a molecule actively involved in cell death regulation.

Can we then simply strive for a systemic downregulation of PON2/PON3 in selected cancer therapies and if so, which specific risks could be expected? PON2-deficient mice were in a pronounced inflammatory status [25] and suffered from a series of other defects linked to severe malfunctions (Witte & Horke; unpublished). Furthermore, (i) reduced PON2 levels enhanced atherogenesis in mice [25, 38], modulated monocyte chemotaxis and cell-mediated LDL oxidation [25, 32], and correlated with atherosclerosis progression in humans [62]; (ii) PON2 may have a neuroprotective role [63]; (iii) genetic associations linked PON2 with amyotrophic lateral sclerosis [64], Alzheimer disease [65], microvascular complications in diabetes [66], coronary heart disease [67], or perhaps also obesity [68]; (iv) PON2 plays a dominant role in the hydrolysis of bacterial virulence regulators [69–71] such that its knockdown may favor certain infections. In a similar manner, human PON3 also has a protective role against atherosclerosis and obesity [1, 33–35], but interpretation is complicated by the fact that there are conflicting reports on its expression pattern, which also varies with the species. Marsillach et al. found PON3 by immunohistochemistry in human aortic walls and macrophages [72], while we did not detect human PON3 message or protein in immortalized EA.hy 926 macrovascular endothelial cells, in primary HUVECs (human umbilical vein endothelial cells), SMCs (human coronary artery smooth muscle cells), or AoAFs (aortic adventitial fibroblasts; [24] and data not shown); this makes it difficult to reveal the mechanistic site of action. Moreover, human PON3 is present on HDL particles and absent in macrophages while the opposite is true for mice [1, 27]. In general, it has been postulated that human PON3 exerts its (antiatherogenic) function rather inside than outside the cells [35, 73], similar to PON2 and likely different from PON1. Studies performed by Shih et al. also revealed a role for PON3 in lipid metabolism that links to adiposity; intriguingly, this was gender-specific for yet unknown reasons [35]. Collectively, PON2 and PON3 have protective functions in cardiovascular diseases, and PON2 plays a dominant role in antibacterial defense, such that an untargeted knockdown may favor these illnesses. As a consequence, a systemic downregulation of PON2 or PON3 does not seem advantageous, as it likely causes a range of serious side effects.

Given that PON2/PON3 protect against atherosclerosis and stabilize atherosclerotic plaques (see above and [33, 36, 74]), may their upregulation then be beneficial to combat atherosclerosis? This is a relevant aspect given the overall number of deaths caused by cardiovascular diseases, which outnumber all cancers [75]. We would first need to determine if enforced PON2/PON3 expression blocks progression or, under optimal conditions, causes regression of established atherosclerotic plaques. Overexpression has been shown to prevent atherogenesis in murine models [25, 33, 35, 36, 74]; however, in clinical reality, patients show up with fully established plaques and need alleviating care, as it is too late for prophylactic approaches. Yet, there is little [33] or no evidence if PONs block progression of established plaques or even cause regression, perhaps due to their anti-inflammatory effects. Should such studies be positive, how can we exploit the beneficial effects of PON2/PON3 against atherosclerosis while concomitantly avoiding their pro-oncogenic function? The first step is the identification of pathways regulating PON expression and the identification of lead substances increasing or decreasing endogenous levels. Then, one solution may come from drug-eluting stents implanted into the atherosclerotic vascular wall—an already established clinical application. This would allow an upregulation of PON2/PON3 directly in the diseased vessel without promoting tumor formation in distant organs. Another solution may come from the specific targeting of effector molecules or pathways (once they are identified) for example, via surface receptors—a likely realistic mission given the accessibility of the vascular wall. In turn, similar approaches could be useful to downregulate PONs in cancer tissues. It would also be valuable to inhibit the interaction of PONs with coQ10 as this could block their antioxidative effect and render these enzymes useless for cancer cells. Finally, the time line may be advantageous in consideration of slow-progressing atherosclerosis and fast-progressing tumors; in some cases this may allow transient downregulation of paraoxonases to boost efficacy of anticancer therapies while not immediately causing plaque formation.

In summary, there exists a remarkable twist in the paraoxonase field since we know that PON2 and PON3 protect against cardiovascular diseases but favor tumor formation. It will be exciting to await further developments and the usefulness of these enzymes in the fight against two of the most significant human diseases.

Abbreviations

ALL: Acute lymphoblastic leukemia
CHOP: C/EBP homologous protein (growth arrest/DNA-damage inducible gene 153, GADD153)
CML: Chronic myeloid leukemia
ER: Endoplasmic reticulum
HDL: High-density lipoprotein particle
LDL: Low-density lipoprotein particle
PON: Paraoxonase
ROS: Reactive oxygen species
SOD: Superoxide dismutase
TNF-α: Tumor necrosis factor-α
TRAIL: Tumor necrosis factor-related apoptosis inducing ligand
UPR: Unfolded protein response.

Acknowledgment

Work in the lab of S. Horke has been financially supported by intramural funds of the Johannes-Gutenberg University Mainz or the University Medical Center of the JGU Mainz (MAIFOR) and by the German Research Foundation, Deutsche Forschungsgemeinschaft, Project HO-3924/4-1. Work in the lab of S.T. Reddy has been financially supported by the National Heart, Lung and Blood Institute (NHLBI; Grant no. 1RO1HL71776). Figures were produced using Servier Medical Art (http://www.servier.com/servier-medical-art/). All authors declare that they have no competing financial interests.

References

[1] C. J. Ng, D. M. Shih, S. Y. Hama, N. Villa, M. Navab, and S. T. Reddy, "The paraoxonase gene family and atherosclerosis," *Free Radical Biology and Medicine*, vol. 38, no. 2, pp. 153–163, 2005.

[2] H. Camuzcuoglu, D. T. Arioz, H. Toy, S. Kurt, H. Celik, and O. Erel, "Serum paraoxonase and arylesterase activities in patients with epithelial ovarian cancer," *Gynecologic Oncology*, vol. 112, no. 3, pp. 481–485, 2009.

[3] E. T. Elkiran, N. Mar, B. Aygen, F. Gursu, A. Karaoglu, and S. Koca, "Serum paraoxonase and arylesterase activities in patients with lung cancer in a Turkish population," *BMC Cancer*, vol. 7, article 48, 2007.

[4] O. A. Uyar, M. Kara, D. Erol, A. Ardicoglu, and H. Yuce, "Investigating paraoxonase-1 gene Q192R and L55M polymorphism in patients with renal cell cancer," *Genetics and Molecular Research*, vol. 10, no. 1, pp. 133–139, 2011.

[5] C. Antognelli, L. Mearini, V. N. Talesa, A. Giannantoni, and E. Mearini, "Association of CYP17, GSTP1, and PON1 polymorphisms with the risk of prostate cancer," *Prostate*, vol. 63, no. 3, pp. 240–251, 2005.

[6] V. L. Stevens, C. Rodriguez, J. T. Talbot, A. L. Pavluck, M. J. Thun, and E. E. Calle, "Paraoxonase I (PONI) polymorphisms and prostate cancer in the CPS-II nutrition cohort," *Prostate*, vol. 68, no. 12, pp. 1336–1340, 2008.

[7] R. A. Hegele, "Paraoxonase genes and disease," *Annals of Medicine*, vol. 31, no. 3, pp. 217–224, 1999.

[8] R. A. Hegele, P. W. Connelly, S. W. Scherer et al., "Paraoxonase-2 gene (PON2) G148 variant associated with elevated fasting plasma glucose in noninsulin-dependent diabetes mellitus," *The Journal of Clinical Endocrinology and Metabolism*, vol. 82, no. 10, pp. 3373–3377, 1997.

[9] A. P. Boright, P. W. Connelly, J. H. Brunt, S. W. Scherer, L. C. Tsui, and R. A. Hegele, "Genetic variation in paraoxonase-1 and paraoxonase-2 is associated with variation in plasma lipoproteins in Alberta Hutterites," *Atherosclerosis*, vol. 139, no. 1, pp. 131–136, 1998.

[10] R. A. Hegele, S. B. Harris, P. W. Connelly et al., "Genetic variation in paraoxonase-2 is associated with variation in plasma lipoproteins in Canadian Oji-Cree," *Clinical Genetics*, vol. 54, no. 5, pp. 394–399, 1998.

[11] D. A. Stoltz, E. A. Ozer, T. J. Recker et al., "A common mutation in paraoxonase-2 results in impaired lactonase activity," *The Journal of Biological Chemistry*, vol. 284, no. 51, pp. 35564–35571, 2009.

[12] S. Altenhöfer, I. Witte, J. F. Teiber et al., "One enzyme, two functions: PON2 prevents mitochondrial superoxide formation and apoptosis independent from its lactonase activity," *The Journal of Biological Chemistry*, vol. 285, no. 32, pp. 24398–24403, 2010.

[13] D. K. Sanghera, C. E. Aston, N. Saha, and M. I. Kamboh, "DNA polymorphisms in two paraoxonase genes (PON1 and PON2) are associated with the risk of coronary heart disease," *American Journal of Human Genetics*, vol. 62, no. 1, pp. 36–44, 1998.

[14] J. G. Wheeler, B. D. Keavney, H. Watkins, R. Collins, and J. Danesh, "Four paraoxonase gene polymorphisms in 11 212 cases of coronary heart disease and 12 786 controls: meta-analysis of 43 studies," *The Lancet*, vol. 363, no. 9410, pp. 689–695, 2004.

[15] Y. Li, Y. Li, R. Tang et al., "Discovery and analysis of hepatocellular carcinoma genes using cDNA microarrays,"

Journal of Cancer Research and Clinical Oncology, vol. 128, no. 7, pp. 369–379, 2002.

[16] T. Ribarska, M. Ingenwerth, W. Goering, R. Engers, and W. A. Schulz, "Epigenetic inactivation of the placentally imprinted tumor suppressor gene TFPI2 in prostate carcinoma," *Cancer Genomics and Proteomics*, vol. 7, no. 2, pp. 51–60, 2010.

[17] M. E. Ross, X. Zhou, G. Song et al., "Classification of pediatric acute lymphoblastic leukemia by gene expression profiling," *Blood*, vol. 102, no. 8, pp. 2951–2959, 2003.

[18] H. Kang, I. M. Chen, C. S. Wilson et al., "Gene expression classifiers for relapse-free survival and minimal residual disease improve risk classification and outcome prediction in pediatric B-precursor acute lymphoblastic leukemia," *Blood*, vol. 115, no. 7, pp. 1394–1405, 2010.

[19] O. Frank, B. Brors, A. Fabarius et al., "Gene expression signature of primary imatinib-resistant chronic myeloid leukemia patients," *Leukemia*, vol. 20, no. 8, pp. 1400–1407, 2006.

[20] C. A. Pise-Masison, M. Radonovich, R. Mahieux et al., "Transcription profile of cells infected with human T-cell leukemia virus type I compared with activated lymphocytes," *Cancer Research*, vol. 62, no. 12, pp. 3562–3571, 2002.

[21] J. K. Choi, J. Y. Choi, D. G. Kim et al., "Integrative analysis of multiple gene expression profiles applied to liver cancer study," *FEBS Letters*, vol. 565, no. 1–3, pp. 93–100, 2004.

[22] A. D. Santin, F. Zhan, S. Bellone et al., "Gene expression profiles in primary ovarian serous papillary tumors and normal ovarian epithelium: identification of candidate molecular markers for ovarian cancer diagnosis and therapy," *International Journal of Cancer*, vol. 112, no. 1, pp. 14–25, 2004.

[23] I. Witte, S. Altenhöfer, P. Wilgenbus et al., "Beyond reduction of atherosclerosis: PON2 provides apoptosis resistance and stabilizes tumor cells," *Cell Death and Disease*, vol. 2, no. 1, Article ID e112, 2011.

[24] E. M. Schweikert, A. Devarajan, I. Witte et al., "PON3 is upregulated in cancer tissues and protects against mitochondrial superoxide-mediated cell death," *Cell Death Differentiation*. In press.

[25] C. J. Ng, N. Bourquard, V. Grijalva et al., "Paraoxonase-2 deficiency aggravates atherosclerosis in mice despite lower apolipoprotein-B-containing lipoproteins: anti-atherogenic role for paraoxonase-2," *The Journal of Biological Chemistry*, vol. 281, no. 40, pp. 29491–29500, 2006.

[26] S. Reuter, S. C. Gupta, M. M. Chaturvedi, and B. B. Aggarwal, "Oxidative stress, inflammation, and cancer: how are they linked?" *Free Radical Biology and Medicine*, vol. 49, no. 11, pp. 1603–1616, 2010.

[27] M. Rosenblat, D. Draganov, C. E. Watson, C. L. Bisgaier, B. N. La Du, and M. Aviram, "Mouse macrophage paraoxonase 2 activity is increased whereas cellular paraoxonase 3 activity is decreased under oxidative stress," *Arteriosclerosis, Thrombosis, and Vascular Biology*, vol. 23, no. 3, pp. 468–474, 2003.

[28] M. Shiner, B. Fuhrman, and M. Aviram, "Paraoxonase 2 (PON2) expression is upregulated via a reduced-nicotinamide- adenine-dinucleotide-phosphate (NADPH)-oxidase-dependent mechanism during monocytes differentiation into macrophages," *Free Radical Biology and Medicine*, vol. 37, no. 12, pp. 2052–2063, 2004.

[29] B. Fuhrman, A. Gantman, J. Khateeb et al., "Urokinase activates macrophage PON2 gene transcription via the PI3K/ROS/MEK/SREBP-2 signalling cascade mediated by the PDGFR-β," *Cardiovascular Research*, vol. 84, no. 1, pp. 145–154, 2009.

[30] D. Hanahan and R. A. Weinberg, "The hallmarks of cancer," *Cell*, vol. 100, no. 1, pp. 57–70, 2000.

[31] F. Azizi, M. Rahmani, F. Raiszadeh, M. Solati, and M. Navab, "Association of lipids, lipoproteins, apolipoproteins and paraoxonase enzyme activity with premature coronary artery disease," *Coronary Artery Disease*, vol. 13, no. 1, pp. 9–16, 2002.

[32] C. J. Ng, D. J. Wadleigh, A. Gangopadhyay et al., "Paraoxonase-2 is a ubiquitously expressed protein with antioxidant properties and is capable of preventing cell-mediated oxidative modification of low density lipoprotein," *The Journal of Biological Chemistry*, vol. 276, no. 48, pp. 44444–44449, 2001.

[33] C. J. Ng, N. Bourquard, S. Y. Hama et al., "Adenovirus-mediated expression of human paraoxonase 3 protects against the progression of atherosclerosis in apolipoprotein E-deficient mice," *Arteriosclerosis, Thrombosis, and Vascular Biology*, vol. 27, no. 6, pp. 1368–1374, 2007.

[34] S. T. Reddy, A. Devarajan, N. Bourquard, D. Shih, and A. M. Fogelman, "Is it just paraoxonase 1 or are other members of the paraoxonase gene family implicated in atherosclerosis?" *Current Opinion in Lipidology*, vol. 19, no. 4, pp. 405–408, 2008.

[35] D. M. Shih, Y. R. Xia, X. P. Wang et al., "Decreased obesity and atherosclerosis in human paraoxonase 3 transgenic mice," *Circulation Research*, vol. 100, no. 8, pp. 1200–1207, 2007.

[36] C. J. Ng, S. Y. Hama, N. Bourquard, M. Navab, and S. T. Reddy, "Adenovirus mediated expression of human paraoxonase 2 protects against the development of atherosclerosis in apolipoprotein E-deficient mice," *Molecular Genetics and Metabolism*, vol. 89, no. 4, pp. 368–373, 2006.

[37] S. T. Reddy, D. J. Wadleigh, V. Grijalva et al., "Human paraoxonase-3 is an HDL-associated enzyme with biological activity similar to paraoxonase-1 protein but is not regulated by oxidized lipids," *Arteriosclerosis, Thrombosis, and Vascular Biology*, vol. 21, no. 4, pp. 542–547, 2001.

[38] A. Devarajan, N. Bourquard, S. Hama et al., "Paraoxonase 2 deficiency alters mitochondrial function and exacerbates the development of atherosclerosis," *Antioxidants and Redox Signaling*, vol. 14, no. 3, pp. 341–351, 2011.

[39] N. Bourquard, C. J. Ng, and S. T. Reddy, "Impaired hepatic insulin signalling in PON2-deficient mice: a novel role for the PON2/apoE axis on the macrophage inflammatory response," *Biochemical Journal*, vol. 436, no. 1, pp. 91–100, 2011.

[40] S. Horke, I. Witte, P. Wilgenbus, M. Krüger, D. Strand, and U. Förstermann, "Paraoxonase-2 reduces oxidative stress in vascular cells and decreases endoplasmic reticulum stress-induced caspase activation," *Circulation*, vol. 115, no. 15, pp. 2055–2064, 2007.

[41] T. Ohnishi and B. L. Trumpower, "Differential effects of antimycin on ubisemiquinone bound in different environments in isolated succinate·cytochrome c reductase complex," *The Journal of Biological Chemistry*, vol. 255, no. 8, pp. 3278–3284, 1980.

[42] A. W. Linnane and H. Eastwood, "Cellular redox regulation and prooxidant signaling systems: a new perspective on the free radical theory of aging," *Annals of the New York Academy of Sciences*, vol. 1067, no. 1, pp. 47–55, 2006.

[43] J. F. Turrens, A. Alexandre, and A. L. Lehninger, "Ubi-semiquinone is the electron donor for superoxide formation by complex III of heart mitochondria," *Archives of Biochemistry and Biophysics*, vol. 237, no. 2, pp. 408–414, 1985.

[44] A. Lass and R. S. Sohal, "Comparisons of coenzyme Q bound to mitochondrial membrane proteins among different

mammalian species," *Free Radical Biology and Medicine*, vol. 27, no. 1-2, pp. 220–226, 1999.

[45] P. J. Klover, T. A. Zimmers, L. G. Koniaris, and R. A. Mooney, "Chronic exposure to interleukin-6 causes hepatic insulin resistance in mice," *Diabetes*, vol. 52, no. 11, pp. 2784–2789, 2003.

[46] S. E. Shoelson, J. Lee, and A. B. Goldfine, "Inflammation and insulin resistance," *The Journal of Clinical Investigation*, vol. 116, no. 7, pp. 1793–1801, 2006.

[47] D. Cai, M. Yuan, D. F. Frantz et al., "Local and systemic insulin resistance resulting from hepatic activation of IKK-β and NF-κB," *Nature Medicine*, vol. 11, no. 2, pp. 183–190, 2005.

[48] R. Visconti and D. Grieco, "New insights on oxidative stress in cancer," *Current Opinion in Drug Discovery and Development*, vol. 12, no. 2, pp. 240–245, 2009.

[49] J. M. Adams and S. Cory, "The Bcl-2 apoptotic switch in cancer development and therapy," *Oncogene*, vol. 26, no. 9, pp. 1324–1337, 2007.

[50] V. E. Kagan, H. A. Bayir, N. A. Belikova et al., "Cytochrome c/cardiolipin relations in mitochondria: a kiss of death," *Free Radical Biology and Medicine*, vol. 46, no. 11, pp. 1439–1453, 2009.

[51] M. Ott, J. D. Robertson, V. Gogvadze, B. Zhivotovsky, and S. Orrenius, "Cytochrome c release from mitochondria proceeds by a two-step process," *Proceedings of the National Academy of Sciences of the United States of America*, vol. 99, no. 3, pp. 1259–1263, 2002.

[52] M. Boyce and J. Yuan, "Cellular response to endoplasmic reticulum stress: a matter of life or death," *Cell Death and Differentiation*, vol. 13, no. 3, pp. 363–373, 2006.

[53] S. Horke, I. Witte, P. Wilgenbus et al., "Protective effect of paraoxonase-2 against endoplasmic reticulum stress-induced apoptosis is lost upon disturbance of calcium homoeostasis," *Biochemical Journal*, vol. 416, no. 3, pp. 395–405, 2008.

[54] E. M. Schweikert, J. Amort, P. Wilgenbus, U. Foerstermann, J. Teiber, and S. Horke, "Paraoxonases-2 and 3 are important defense enzymes against Pseudomonas aeruginosa virulence factors due to their anti-oxidative and anti-inflammatory properties," *Journal of Lipids*, vol. 2012, Article ID 352857, 9 pages, 2012.

[55] E. White and R. S. DiPaola, "The double-edged sword of autophagy modulation in cancer," *Clinical Cancer Research*, vol. 15, no. 17, pp. 5308–5316, 2009.

[56] M. Djavaheri-Mergny, M. Amelotti, J. Mathieu et al., "NF-κB activation represses tumor necrosis factor-α-induced autophagy," *The Journal of Biological Chemistry*, vol. 281, no. 41, pp. 30373–30382, 2006.

[57] R. Scherz-Shouval, E. Shvets, E. Fass, H. Shorer, L. Gil, and Z. Elazar, "Reactive oxygen species are essential for autophagy and specifically regulate the activity of Atg4," *The EMBO Journal*, vol. 26, no. 7, pp. 1749–1760, 2007.

[58] M. Dewaele, H. Maes, and P. Agostinis, "ROS-mediated mechanisms of autophagy stimulation and their relevance in cancer therapy," *Autophagy*, vol. 6, no. 7, pp. 838–854, 2010.

[59] P. Vandenabeele, W. Declercq, F. Van Herreweghe, and T. V. Berghe, "The role of the kinases RIP1 and RIP3 in TNF-induced necrosis," *Science Signaling*, vol. 3, no. 115, article re4, 2010.

[60] A. Sica, P. Allavena, and A. Mantovani, "Cancer related inflammation: the macrophage connection," *Cancer Letters*, vol. 267, no. 2, pp. 204–215, 2008.

[61] J. B. Kim, Y.-R. Xia, C. E. Romanoski et al., "Paraoxonase-2 modulates stress response of endothelial cells to oxidized phospholipids and a bacterial quorum-sensing molecule," *Arteriosclerosis, Thrombosis, and Vascular Biology*, vol. 31, no. 11, pp. 2624–2633, 2011.

[62] G. Fortunato, M. D. Di Taranto, U. M. Bracale et al., "Decreased paraoxonase-2 expression in human carotids during the progression of atherosclerosis," *Arteriosclerosis, Thrombosis, and Vascular Biology*, vol. 28, no. 3, pp. 594–600, 2008.

[63] G. Giordano, T. B. Cole, C. E. Furlong, and L. G. Costa, "Paraoxonase 2 (PON2) in the mouse central nervous system: a neuroprotective role?" *Toxicology and Applied Pharmacology*, vol. 256, no. 3, pp. 369–378, 2011.

[64] N. Ticozzi, A. L. LeClerc, P. J. Keagle et al., "Paraoxonase gene mutations in amyotrophic lateral sclerosis," *Annals of Neurology*, vol. 68, no. 1, pp. 102–107, 2010.

[65] P. M. Erlich, K. L. Lunetta, L. A. Cupples et al., "Polymorphisms in the PON gene cluster are associated with Alzheimer disease," *Human Molecular Genetics*, vol. 15, no. 1, pp. 77–85, 2006.

[66] B. Mackness, P. Mcelduff, and M. I. Mackness, "The paraoxonase-2-310 polymorphism is associated with the presence of microvascular complications in diabetes mellitus," *Journal of Internal Medicine*, vol. 258, no. 4, pp. 363–368, 2005.

[67] H. L. Li, D. P. Liu, and C. C. Liang, "Paraoxonase gene polymorphisms, oxidative stress, and diseases," *Journal of Molecular Medicine*, vol. 81, no. 12, pp. 766–779, 2003.

[68] J. I. Joo, T. S. Oh, D. H. Kim et al., "Differential expression of adipose tissue proteins between obesity-susceptible and -resistant rats fed a high-fat diet," *Proteomics*, vol. 11, no. 8, pp. 1429–1448, 2011.

[69] S. Horke, I. Witte, S. Altenhöfer et al., "Paraoxonase 2 is down-regulated by the Pseudomonas aeruginosa quorum-sensing signal N-(3-oxododecanoyl)-L-homoserine lactone and attenuates oxidative stress induced by pyocyanin," *Biochemical Journal*, vol. 426, no. 1, pp. 73–83, 2010.

[70] J. F. Teiber, S. Horke, D. C. Haines et al., "Dominant role of paraoxonases in inactivation of the Pseudomonas aeruginosa quorum-sensing signal N-(3-oxododecanoyl)-L-homoserine lactone," *Infection and Immunity*, vol. 76, no. 6, pp. 2512–2519, 2008.

[71] D. A. Stoltz, E. A. Ozer, C. J. Ng et al., "Paraoxonase-2 deficiency enhances Pseudomonas aeruginosa quorum sensing in murine tracheal epithelia," *American Journal of Physiology*, vol. 292, no. 4, pp. L852–L860, 2007.

[72] J. Marsillach, J. Camps, R. Beltran-Debón et al., "Immunohistochemical analysis of paraoxonases-1 and 3 in human atheromatous plaques," *European Journal of Clinical Investigation*, vol. 41, no. 3, pp. 308–314, 2011.

[73] D. I. Draganov, "Human PON3, effects beyond the HDL: clues from human PON3 transgenic mice," *Circulation Research*, vol. 100, no. 8, pp. 1104–1105, 2007.

[74] Z. G. She, W. Zheng, Y. S. Wei et al., "Human paraoxonase gene cluster transgenic overexpression represses atherogenesis and promotes atherosclerotic plaque stability in ApoE-Null Mice," *Circulation Research*, vol. 104, no. 10, pp. 1160–1168, 2009.

[75] V. L. Roger, A. S. Go, D. M. Lloyd-Jones et al., "Heart disease and stroke statistics-2011 update: a report from the American Heart Association," *Circulation*, vol. 123, pp. e18–e209, 2011.

13

Paraoxonases-2 and -3 Are Important Defense Enzymes against *Pseudomonas aeruginosa* Virulence Factors due to Their Anti-Oxidative and Anti-Inflammatory Properties

Eva-Maria Schweikert,[1] Julianna Amort,[1] Petra Wilgenbus,[1] Ulrich Förstermann,[1] John F. Teiber,[2] and Sven Horke[1]

[1] *Institute of Pharmacology, University Medical Center of the Johannes Gutenberg-University Mainz, Obere Zahlbacher Straße 67, 55131 Mainz, Germany*
[2] *Division of Epidemiology, Department of Internal Medicine, The University of Texas Southwestern Medical Center, 5323 Harry Hines Boulevard, Dallas, TX 75390, USA*

Correspondence should be addressed to Sven Horke, horke@uni-mainz.de

Academic Editor: Alejandro Gugliucci

The pathogen *Pseudomonas aeruginosa* causes serious damage in immunocompromised patients by secretion of various virulence factors, among them the quorum sensing N-(3-oxododecanoyl)-L-homoserine lactone (3OC12) and the redox-active pyocyanin (PCN). Paraoxonase-2 (PON2) may protect against *P. aeruginosa* infections, as it efficiently inactivates 3OC12 and diminishes PCN-induced oxidative stress. This defense could be circumvented because 3OC12 mediates intracellular Ca^{2+}-rise in host cells, which causes rapid inactivation and degradation of PON2. Importantly, we recently found that the PON2 paralogue PON3 prevents mitochondrial radical formation. Here we investigated its role as additional potential defense mechanism against *P. aeruginosa* infections. Our studies demonstrate that PON3 diminished PCN-induced oxidative stress. Moreover, it showed clear anti-inflammatory potential by protecting against NF-κB activation and IL-8 release. The latter similarly applied to PON2. Furthermore, we observed a Ca^{2+}-mediated inactivation and degradation of PON3, again in accordance with previous findings for PON2. Our results suggest that the anti-oxidative and anti-inflammatory functions of PON2 and PON3 are an important part of our innate defense system against *P. aeruginosa* infections. Furthermore, we conclude that *P. aeruginosa* circumvents PON3 protection by the same pathway as for PON2. This may help identifying underlying mechanisms in order to sustain the protection afforded by these enzymes.

1. Introduction

The bacterium *Pseudomonas aeruginosa* is an opportunistic nosocomial pathogen, which infects the pulmonary tract of, for example, immunocompromised patients or those suffering from cystic fibrosis, pneumonia, burn wounds, HIV, or cancer chemotherapy [1]. The infection causes serious damage in the host, complicated by an often hindered antibiotic treatment due to multiresistances and biofilm formation that provides physical protection. Furthermore, *P. aeruginosa* secrets a variety of virulence factors to regulate bacterial communication and weaken the defense mechanisms of the infected host. Two important factors are the quorum sensing signal N-(3-oxododecanoyl)-L-homoserine lactone (3OC12) and the redox-active pyocyanin (PCN). 3OC12 is a mediator of the cell-density-dependent signaling system known as quorum sensing, by which the bacteria coordinate their gene expression. If bacterial density and 3OC12 concentration exceed a certain threshold, the bacteria become virulent by expression of virulence factors (immunogenic exoenzymes and toxins) and by inducing inflammation. The lactone 3OC12 has numerous immunomodulatory and inflammatory properties, such as an inhibitory effect on dendritic cells and T-cell activation [2], proinflammatory induction of IL-6 and IL-8 in airway epithelial cells and lung fibroblasts [3], and promoting apoptosis [4, 5]. Studies in mice have shown that 3OC12 is a critical determinant for bacterial colocalization and the establishment of chronic lung infections [6]. Therefore, the development of quorum sensing inhibitors would be a major advance in the ability to combat *P. aeruginosa* infections [7, 8].

The production of the virulence factor PCN is positively regulated by quorum sensing signals including 3OC12 [9]. PCN causes oxidative stress and has a broad range of effects on airway epithelial cells such as cellular senescence and ciliary dyskinesia, induction of IL-8 secretion, decrease of glutathione levels, and inhibition of catalase activity [9, 10]. The redox-activity of PCN is central to the damage observed in exposed host cells. The zwitterionic PCN transfers electrons from reduced NADH or NADPH in the cytosol to molecular oxygen leading to production of superoxide (O_2^-), which is converted to H_2O_2 [11]. Additionally, PCN causes a disturbance of the antimicrobial Duox/SCN^-/LPO-system, by consuming the same substrates (molecular oxygen and NADPH) [12]. The requirement for PCN in lung infection was demonstrated in an acute pneumonia mice model: *P. aeruginosa* strains lacking the ability to produce PCN are much more rapidly cleared from lungs and showed less virulence than the wild-type strain [13].

The paraoxonase family consists of the three members PON1, PON2, and PON3, which exhibit about 70% similarity at the amino acid level [14]. PON1 is associated with HDL in serum, whereas PON2 and PON3 are intracellular proteins. In contrast to PON2, which is ubiquitously expressed, PON3 appears restricted to fewer tissues/cells; its expression in cells relevant to cardiovascular diseases is contradictory, because Marsillach et al. [15] found PON3 by immunohistochemistry in human vascular walls and macrophages, while our studies revealed absence in macrophages, endothelial, smooth muscle, and many other cell types [16]. All three PONs share a lactonase activity with distinct and overlapping substrate specificities [17–19]. PON2 dominantly hydrolyzes 3OC12 presumably resulting in the ability to interfere with quorum sensing, which may significantly attenuate bacterial virulence of *P. aeruginosa*. In support of this concept, epithelial tracheal cells from PON2 deficient mice showed a reduced ability to inactivate 3OC12 [20]. In addition to its lactonase activity, PON2 is a major anti-oxidative protein that diminishes mitochondrial superoxide production and thus considerably determines cell survival [21–23]. In particular, it has also been shown that PON2 diminishes PCN-induced ROS production in human epithelial cells [24]. Intriguingly, 3OC12 causes a rapid Ca^{2+}-mediated PON2 inactivation and degradation in cultured cells by a yet unknown mechanism, which enables 3OC12 to protect itself from its hydrolysis by PON2 [25]. As a consequence, 3OC12 potentiates the formation of ROS induced by PCN, revealing a potential mechanism by which the bacterium may circumvent the protection afforded by PON2 [25].

Our recent studies demonstrated that PON3, much like PON2, reduced the generation of mitochondrial superoxide. We also revealed that PON3 was frequently found overexpressed in tumors. There, it reduced susceptibility to chemotherapeutics and reduced apoptosis, in line with the central involvement of mitochondrial ROS to cell death [16]. Hence, we hypothesized that PON3 attenuated PCN-induced oxidative stress, which was tested here for the first time. Our studies also included inflammatory pathways subsequent to PCN stimulation, that is, NF-κB activation and secretion of various cytokines. Our results support the concept of marked anti-inflammatory roles for PON2 and PON3. Finally, we found that both enzymes, PON2 and PON3, were inactivated and degraded in response to Ca^{2+}-disturbances caused by 3OC12. Identifying the underlying pathway(s) by which the PONs are downregulated may reveal a therapeutic target, which could be exploited to help sustain or enhance the host's PON2 and PON3 activities. Such a clinical intervention could be of great benefit in the defense against *P. aeruginosa* infections.

2. Material and Methods

2.1. Cell Culture and Material. Human endothelial EA.hy 926 cells obtained from the ATCC were cultured in Dulbecco's modified Eagle's medium without Phenol Red (Sigma, St. Louis, MO, USA) containing sodium pyruvate (PAA Laboratories, Pasching, Austria), antibiotics penicillin/streptomycin, hypoxanthine/aminopterin/thymidine supplement, L-Glutamine (Invitrogen, Carlsbad, CA, USA), and 10% (v/v) fetal calf serum (PAA). Human PON2 or PON3 cDNA was subcloned into pDsRed-Express-N1 or pEGFP-N1 plasmids (Clontech). Stable cell lines, plasmids, and transfection procedures were described before [25, 26]. HEK293 and A549 cells were from the German Collection of Microorganisms and Cell Cultures. HEK293 received the same medium as EA.hy 926, but without hypoxanthine/aminopterin/thymidine supplement. A549 cells received the same medium as HEK293, but 5% serum. Cells were cultured at 37°C in a humidified atmosphere with 5% CO_2 (10% for EA.hy 926). Pyocyanin was purchased from Cayman Chemical Company (Ann Arbor, MI, USA); Mito-HE and thapsigargin were from Molecular Probes; L-012 was from Wako Chemicals (Neuss, Germany), and all other reagents were from Sigma.

2.2. ELISA. EA.hy 926 cells were seeded one day prior to stimulation in 6-well dishes at 80% confluency (5×10^5/well). Treatment occurred in 1.5 mL medium without FCS for 16 h with or without pyocyanin (10 μM). Cell supernatants were taken for Multi-Analyte ELISArray Kit for human inflammatory cytokines (SA Biosciences, Frederick, MD, USA) according to the supplier's instructions. Absorbance was determined using a FluoStar Optima microplate reader (BMG Labtechnologies). Corrected OD was calculated by subtracting the A450 reading by the A570 reading to clear any minor optical imperfections in the ELISA plate.

2.3. Reporter Gene Assays. Cells at 80% confluency were cotransfected with pcDNA3-HA or pcDNA3-PON2-HA or pEGFP-N1 or pEGFP-N1-PON3 and a plasmid allowing for constitutive renilla luciferase expression (a kind gift of H. Kleinert, University Medical Centre, Mainz) and the NF-κB reporter plasmid (pGL4.32[luc2P/NF-κB-RE/Hygro] from Promega, Madison, WI, USA). We used Nanofect-inTM (PAA) for transfection according to the supplier's instructions. Cells were treated 24 h after transfection with pyocyanin (100 μM) for 4 h. Subsequently the NF-κB activity was measured by Dual-Luciferase Reporter Assay System (Promega) according to the supplier's instructions.

2.4. qRT-PCR. RNA isolation and cDNA generation was performed as reported previously [21]. PON3 expression level was determined by quantitative real-time PCR normalized to GAPDH as described before [21]. The following Taqman primers (Eurofins, MWG Operon) were used: PON3: sense 5'-TGGGATCACAGTCTCAGCAG-3'; antisense 5'-TCC-ACTAAGGTGCCCAACTG-3'; probe 5'-TGGAAAAAC-ATGATAACTGGGA-3'; GAPDH: sense 5'-CAACAGCCT-CAAGATCATCAGC-3'; antisense 5'-TGGCATGGACTG-TGGTCATGAG-3'; probe 5'-CCTGGCCAAGGTCATCCA-TGACAAC-3'.

2.5. Western Blotting. Preparation of lysates, SDS-PAGE, and Western blotting was performed as reported previously [25]. Rabbit-anti-PON3 polyclonal antibody was used at 1 : 750 (Sigma, St. Louis, MO, USA), rabbit-anti-PON2 [26] used at 1 : 2000. Mouse-anti-GAPDH 6C5 (Santa Cruz, Santa Cruz, CA, USA) and HRP-conjugated secondary antibodies were from Sigma or Cell Signaling Technology. Immunodetected proteins were visualized and quantitatively evaluated as described before [24].

ROS detection and determination of lactonase and lovastatinase activities were performed as described before [17, 24, 27].

2.6. Software, Statistics, and Image Acquisition. GraphPad Prism-5 was used for calculations, statistical evaluation using 1-/2-way ANOVA with Bonferroni's multiple comparisons posttest (see Figures 1–6). $P < 0.05$ was considered significant. Adobe Photoshop software was used for image acquisition. If necessary, only brightness and/or contrast were changed simultaneously for all areas of any blot.

3. Results

3.1. PON2 and PON3 Protect Cells from PCN-Induced ROS and Inflammatory Responses. PCN is essential for *P. aeruginosa* infections *in vivo* and causes oxidative stress, which leads to serious damage of airway epithelial cells [12, 13]. We recently found that PON2 decreases ROS production in EA.hy 926 cells in response to treatment with PCN [24]. Given the high degree of homology between the paraoxonases and previous descriptions of PON3 anti-oxidative effects, we wanted to test whether PON3 also attenuated PCN-induced ROS production. This was measured by loading naïve or PON3 overexpressing cells with the fluorescent ROS indicator carboxy-H_2DCFDA followed by stimulation with PCN (2.4 μM); PON2 overexpressing cells were used in these (and subsequent) studies for

comparison. As with PON2, PON3 overexpression afforded a marked protection against PCN-induced oxidative stress (Figure 1(a)). To control for a potential direct oxidation of carboxy-H_2DCFDA by PCN, an unwanted effect by the GFP tag and for effects specific to HEK293 cells, we also used the luminol derivate L-012 to report ROS in EA.hy 926 cells overexpressing a PON3-dsRed construct. As with HEK293 PON3-GFP cells, the EA.hy 926 PON3-GFP and PON3-dsRed cells show decreased ROS production after PCN treatment (Figure 1(b)).

PCN also leads to a proinflammatory response by causing the release of interleukins, which might be triggered by activation of NF-κB [12, 28]. To test for anti-inflammatory effects of PON2 and PON3, we next addressed activation of the NF-κB pathway in lung epithelial A549 cells in response to PCN treatment. Employing gene reporter studies, we cotransfected A549 cells with (i) an NF-κB firefly luciferase reporter plasmid, (ii) a renilla luciferase expression vector for normalization purposes, and (iii) a PON2 or PON3 expression plasmid. Controls received the same vectors, but without PON2/PON3 inserts. As expected, NF-κB promoter activity was significantly increased 4- to 6-fold above control cells following PCN treatment (Figure 2). Importantly, PON2 and PON3 overexpression caused a dramatic decrease in activation of this major pathway involved in inflammatory response.

Finally, because NF-κB has both pro- and anti-inflammatory properties, we wished to determine which cytokine is secreted from endothelial cells exposed to PCN and if this was altered by paraoxonase overexpression. For this purpose, we performed an ELISA with a panel of inflammatory cytokines by using cell supernatants from PCN-treated naïve EA.hy 926 cells or cells overexpressing PON2-GFP or PON3-GFP. By using this approach, we covered several major inflammatory mediators, like IL-1A, IL-1B, IL-2, IL-4, IL-6, IL-8, IL-10, IL-12, IL-17A, IFN-γ, TNF-α, and GM-CSF. Remarkably, only one single cytokine was induced by PCN in endothelial cells, namely, IL-8 (>1000 pg/mL); the IL-8 level was about 6-fold higher in PCN treated compared to untreated EA.hy 926 cells (Figure 3(a)). IL-8 release was markedly lowered by PON2 or by PON3 overexpression (about 2-fold to 4-fold, respectively; Figure 3(b)). Taken together, PON2 and PON3 act as potent anti-inflammatory enzymes, which is shown by their reducing effects on ROS production, NF-κB activation and IL-8 release in response to *P. aeruginosa* virulence factor PCN.

3.2. 3OC12 Downregulates PON3 Hydrolytic Activity and Protein. Besides the virulence factor PCN, *P. aeruginosa* secrets the quorum sensing signal 3OC12, which can be hydrolyzed and thus inactivated by PON2. In fact, it appears that PON2 has a dominant role in 3OC12 hydrolysis [27]. Our further studies showed that PON2 hydrolytic activity, mRNA, and protein are actively downregulated by 3OC12, which disrupts PON2's protection against both 3OC12 levels and PCN-induced ROS production [24]. Given that PON3 diminished PCN-triggered ROS, it is worthwhile to address effects of 3OC12 on PON3 in a similar manner. To this end, we addressed the effect of 3OC12 on PON3 hydrolytic

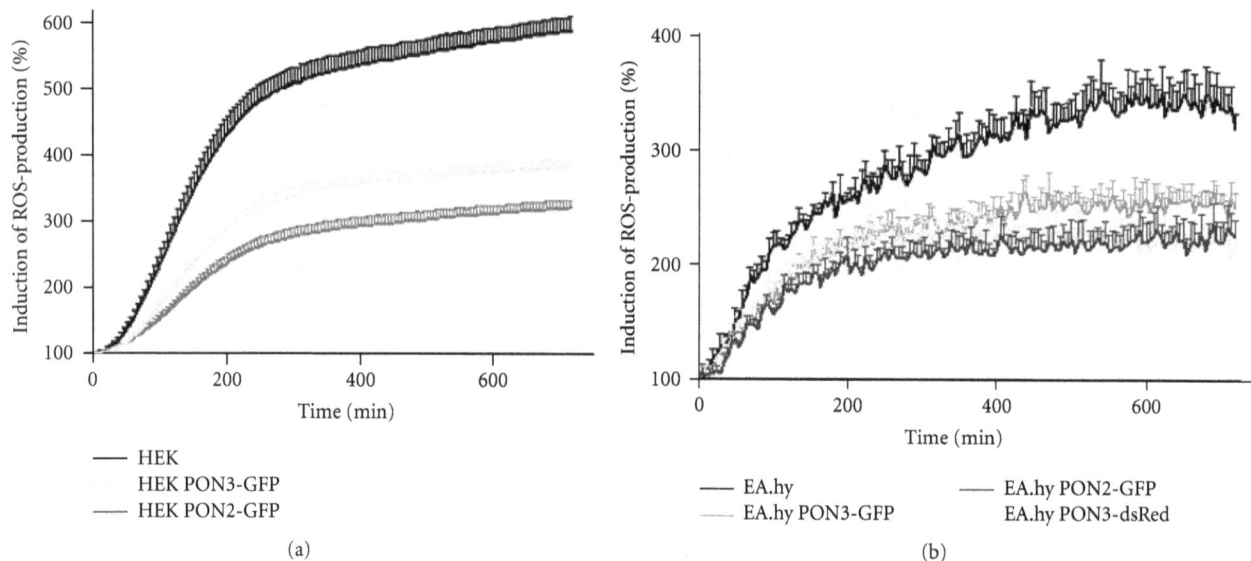

FIGURE 1: PON2 or PON3 overexpression diminishes ROS production induced by *P. aeruginosa* signaling molecule pyocyanin (PCN). (a) Naïve, PON2-GFP, or PON3-GFP overexpressing HEK293 cells were loaded with carboxy-H_2DCFDA and stimulated with PCN (2.4 μM). Carboxy-H_2DCFDA fluorescence as means of ROS was recorded over several hours. (b) Similar to A. Naïve, PON2-GFP, PON3-GFP, or PON3-dsRed overexpressing EA.hy 926 cells were loaded with L-012 and stimulated with PCN (2.4 μM). Curve maxima calculated by nonlinear regression showed statistically significant differences ($P < 0.001$) between naïve and PON2 or PON3 overexpressing cells.

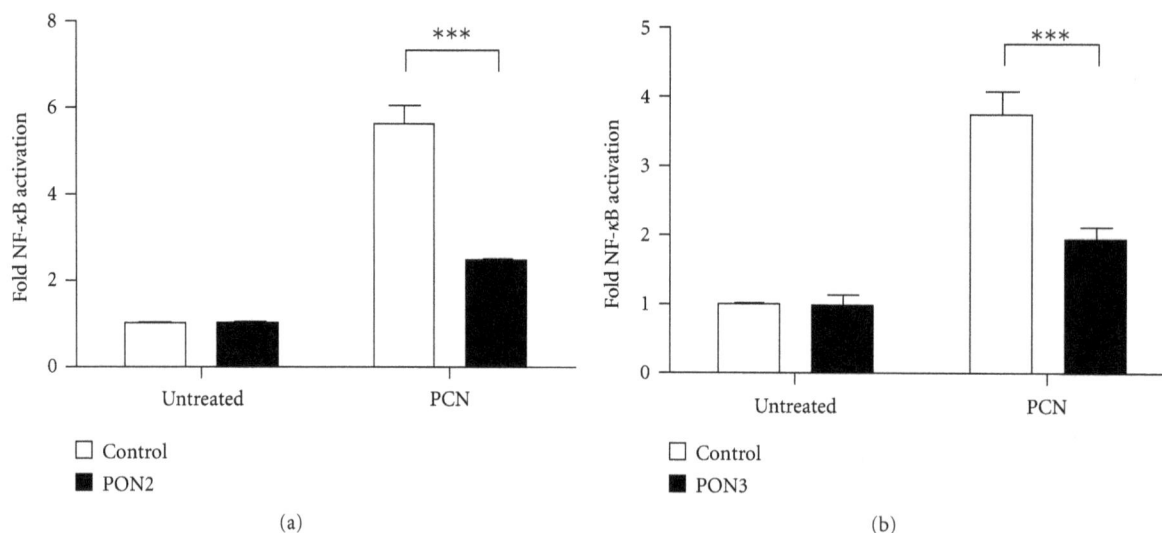

FIGURE 2: PON2 or PON3 overexpression diminishes NF-κB activation induced by *P. aeruginosa* signaling molecule PCN. A549 cells transiently overexpressing HA or PON2-HA (a) and GFP or PON3-GFP (b) were stimulated with PCN (100 μM, 4 h) and analyzed for NF-κB activation. Symbols represent $\pm$ S.E.M. $n = 6$–9; ***$P < 0.001$.

activity, mRNA, and protein. We used PON2-GFP or PON3-GFP overexpressing HEK293 cells for measuring lactonase or lovastatinase activity (for PON2 and PON3, resp.). In agreement with previous observations in other cell lines, PON2 activity was rapidly decreased after 3OC12 treatment. The PON2 hydrolytic activity was reduced by ~50% in 10 min and >80% in 30 min (Figure 4(a)). Our previous studies revealed that PON2 activity was inversely related to intracellular Ca^{2+}-homeostasis, with 3OC12 hydrolysis being depleted in response to Ca^{2+}-release as it was inhibitable by

Ca^{2+}-chelator BAPTA [24]. Interestingly, PON3 lovastatinase activity was also rapidly inactivated by 3OC12 to an almost identical level as observed for PON2 (Figure 4(b)).

Our previous data demonstrated that 3OC12 caused a pronounced calcium influx in A549 and EA.hy 926 cells in a very short time-interval. This formed the basis for an active, calcium-dependent inactivation and subsequent degradation of PON2. Therefore, we wanted to explore if PON3 is also degraded in a calcium-sensitive manner, as this could point to regulatory pathways shared by these two different

Paraoxonases-2 and -3 Are Important Defense Enzymes against Pseudomonas aeruginosa Virulence Factors due to
Their Anti-Oxidative and Anti-Inflammatory Properties

137

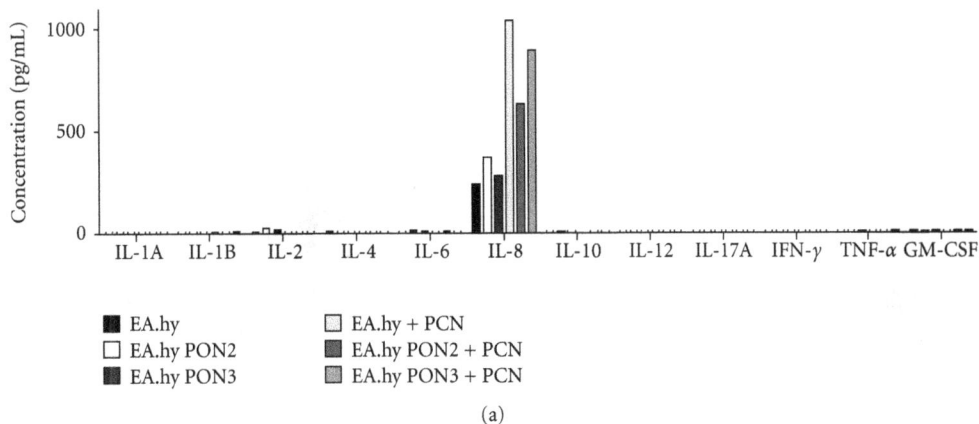

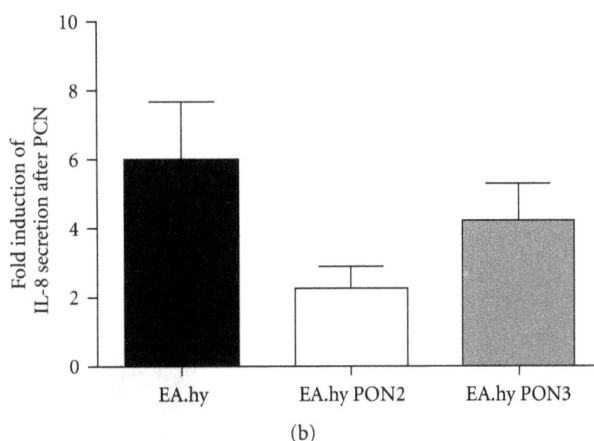

FIGURE 3: PCN induces IL-8 secretion, which can be lowered by PON2 or PON3 overexpression. (a) Naïve, PON2-GFP or PON3-GFP
overexpressing EA.hy 926 cells were treated with PCN ($10\,\mu$M, 16 h). Cell supernatants were analyzed for the secretion of the listed cytokines
and chemokines by ELISA. (b) Quantitative evaluation of results from panel (a) Fold induction of IL-8 release was calculated between
untreated and PCN-treated samples.

enzymes [24]. To this end, we treated A549 cells with 3OC12
and analyzed PON3 mRNA levels by qRT-PCR at different
time-points. Figure 5(a) shows a decrease of PON3 mRNA
after 16 h 3OC12 treatment to ~70%. Next we verified, if
PON3 mRNA is actively degraded or if the decrease results
from a discontinued transcription or reflection of normal
mRNA turnover. We treated A549 cells with 3OC12 or
the RNA synthesis inhibitor 5,6-dichlorobenzimidazole 1-β-
D-ribofuranoside (DRB), or combinations thereof. Unlike
PON2, PON3 mRNA was not actively degraded in response
to 3OC12 (Figure 5(b)).

Finally we performed Western blot analyses to monitor
PON3 protein levels after major Ca^{2+}-disturbances caused
by treatment with 3OC12. For reasons of comparison,
we used the SERCA inhibitor thapsigargin, which also
causes serious Ca^{2+}-disturbances. Treatment of A549 cells
with 3OC12 for different durations showed that PON3
protein level decreased time dependently and vanished nearly
completely after 16 h; while PON2 protein was also degraded
by ~50% after 16 h (Figure 6(a)). In accordance, A549 cells

treated with different thapsigargin concentrations showed a
significant dose-dependent degradation of PON3 and PON2
(Figure 6(b)).

4. Discussion

P. aeruginosa infections are difficult to treat since the bacteria
often develop multiple antibiotic resistance and form a
biofilm, which hinders the access of the antibiotics to the
bacteria. Additionally, *P. aeruginosa* secrets different virulence factors, which regulate the bacterial communication
and damage the infected host. Therefore, it is important
to understand *P. aeruginosa* host-pathogen interactions to
identify new potential therapeutic targets. Combined with
the knowledge gained in previous studies, our results suggest
that human paraoxonases PON2 and PON3 comprise a
major defense against virulence factor-induced oxidative
stress, inflammatory response, and cytokine release.

In this study, we revealed, for the first time, the protective
effect of PON3 against PCN-induced host cell damage. By

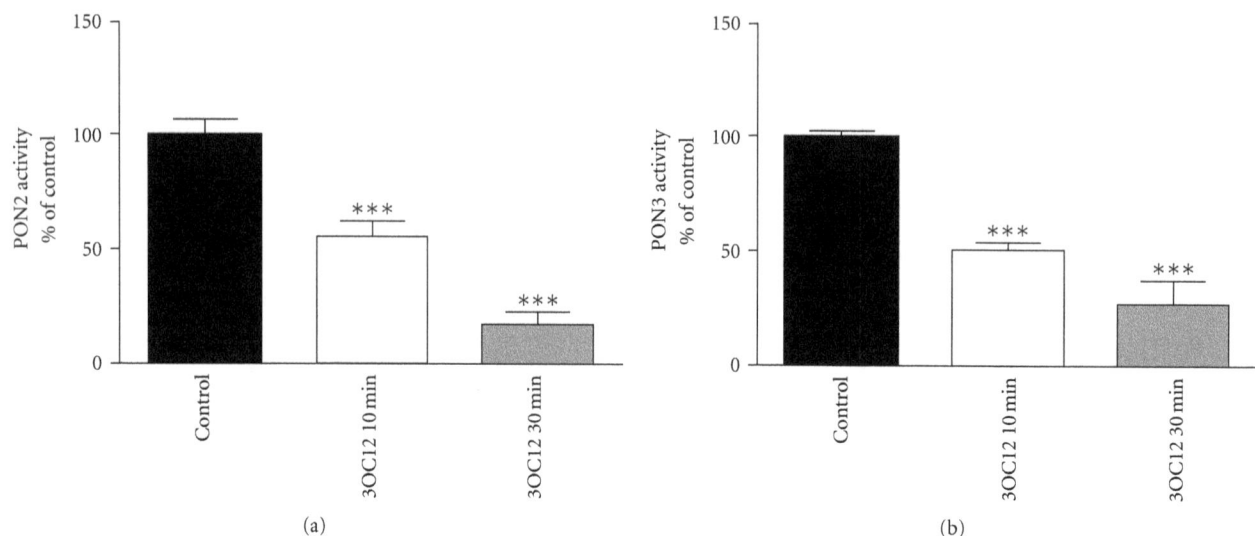

FIGURE 4: PON2 and PON3 activity decrease after 3OC12 treatment. (a) PON2-GFP overexpressing HEK293 cells were treated with 3OC12 (100 μM) for the indicated durations and tested for 3OC12-HSL hydrolytic activity. (b) PON3-GFP overexpressing HEK293 cells were treated with 3OC12 (100 μM) for the indicated durations and tested for lovastatinase hydrolytic activity. Symbols represent $\pm$ S.E.M. $n = 3$; ***$P < 0.001$.

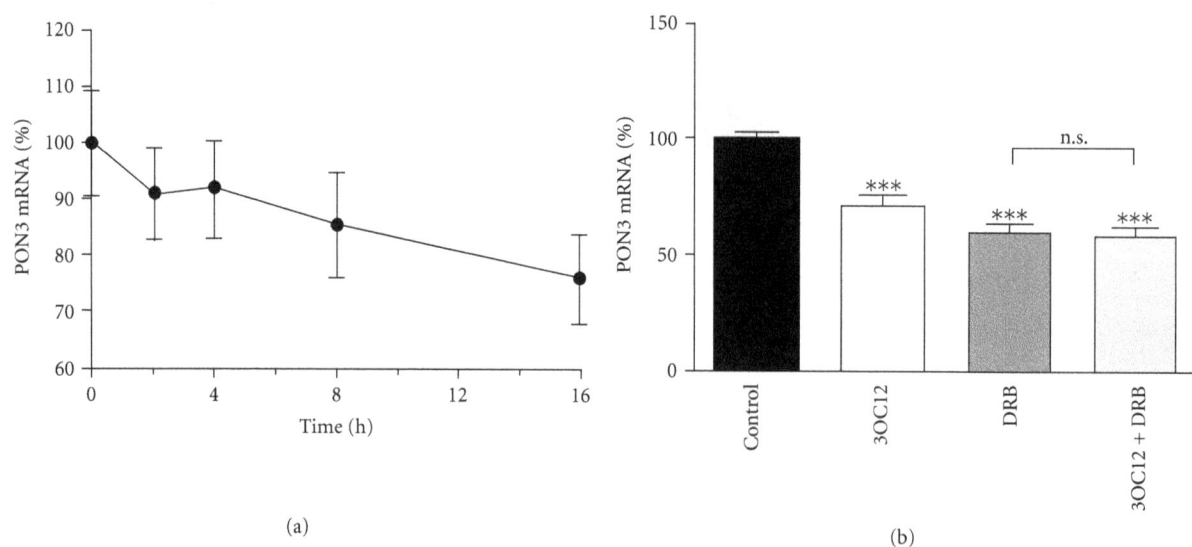

FIGURE 5: PON3 mRNA is not actively degraded in response to 3OC12 treatment. (a) A549 cells were treated with 3OC12 (100 μM) for the indicated durations and analyzed for PON3 mRNA levels by qRT-PCR. (b) A549 cells were treated with 3OC12 (100 μM, 24 h) or with DRB (100 μM, 24 h) or combinations thereof. There was no statistically significant difference in PON3 mRNA levels after DRB or DRB/3OC12 treatment. Symbols represent $\pm$ S.E.M. $n = 3$; ***$P < 0.001$ versus control.

various technical approaches and using different cell lines, we showed that PCN leads to ROS production, NF-κB-activation, and IL-8 secretion, which can be prevented by PON2 or PON3 overexpression. PCN induces the production of O_2^- and H_2O_2, which causes damage in various cell types [10]. Overexpression of PON2 or PON3 in relevant systems leads to a significant reduction of ROS production, reflecting the protective effect of both enzymes against PCN-induced oxidative damage. Importantly, PON2 and PON3 differ in their substrate specificities, as PON2 has a dominant lactonase activity, whereas PON3 has much better activity

with some large lactones or arylesters (i.e., statins; estradiol acetates). Thus, we conclude that PON2 and PON3 act by a common anti-oxidative mechanism. Intriguingly, our data also suggest that the anti-oxidative effect of PON3 is independent from its enzymatic activity. This would be in agreement with previous results demonstrating an independent anti-oxidative and hydrolytic activity of PON2 [22]. Addressing the underlying anti-oxidative mechanism of PON2 and PON3, we and others recently showed that PON2 and PON3 localize to the inner mitochondrial membrane where they interact with coenzyme Q10 (coQ10) resulting in

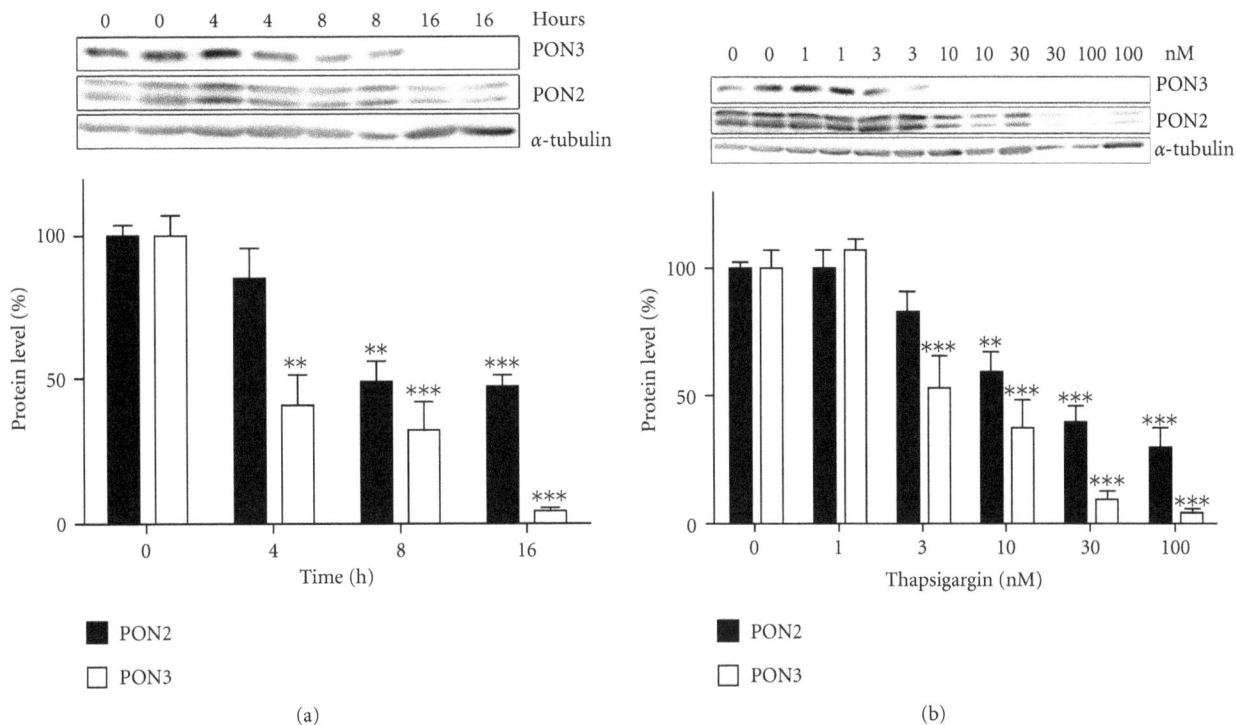

FIGURE 6: Both PON2 and PON3 are degraded after 3OC12 treatment. (a) A549 cells were treated with 3OC12 (100 μM) for the indicated durations or (b) with thapsigargin (24 h) with the indicated concentrations. Lysates (50 μg of protein) were analyzed by Western blotting using anti-PON2, anti-PON3, or anti-α-tubulin antibodies. One representative blot is shown. Results (right) are the means ± S.E.M. of three replicate analyses; $**P < 0.01$; $***P < 0.001$.

abrogated superoxide production [16, 22, 23]. According to the current model (reviewed in this issue; [29]), it is assumed that PON2 and PON3 protect against ROS formation by acting as an insulator for coQ10 to prevent coincidental superoxide production at the mitochondrial membrane.

It has also been reported that PCN increases IL-8 expression in human airway epithelial cells [28]. Similarly, we observed an increase in IL-8 secretion by PCN in endothelial EA.hy 926 cells. Interestingly, this can be lowered by PON2 or PON3. Additionally, we monitored the release of numerous proinflammatory cytokines and chemokines after PCN treatment. None of the tested factors was induced except for IL-8, suggesting that IL-8 acts as a central mediator of endothelial inflammation triggered by PCN. It is known that IL-8 promoter activity and expression can be induced by NF-κB [30, 31]. Our data also imply a role for NF-κB as mediator between PCN-induced effects on ROS formation and proinflammatory immune response by release of IL-8. In particular, PCN was able to activate NF-κB, which could be reduced by PON2 or PON3. It is fully established that NF-κB is regulated by redox signaling. Taken together, our data suggest that PON2 and PON3 anti-inflammatory activities result from their ability to prevent ROS formation, as less oxidative stress likely diminishes NF-κB activation and subsequent IL-8 release.

Furthermore, we revealed the effect of 3OC12 on PON3 hydrolytic activity, mRNA, and protein. PON3 mRNA, in contrast to PON2 mRNA [22], was not actively degraded,

indicative of independent mechanisms that regulate stability of these two mRNAs. PON3 protein levels were dramatically decreased by 3OC12 treatment. A severely altered calcium homeostasis as underlying mechanism appears highly likely, as this has previously been demonstrated for PON2 [24]. Our data may also suggest that PON2 and PON3 degradation occurs through the same Ca^{2+}-mediated pathway, as it acts in a similar time (in case of 3OC12) or dose-dependent (in case of thapsigargin) manner. Similar to PON2, PON3 hydrolytic activity was decreased much more extensively and rapidly than the protein, indicating a likely posttranslational event blocking PON2's and PON3's enzymatic function.

Our findings emphasize roles for PON2 and PON3 in the defense against *P. aeruginosa* virulence but show also that the bacterium may circumvent the protection by PON2 and PON3. Identification of the posttranslational modification of PON2, which causes its inactivation and induces the signaling pathway that mediates PON2 and PON3 downregulation, may lead to the identification of an important therapeutic approach. It may be beneficial to block the 3OC12-mediated decrease of PON2 activity, sustaining PON2's protective effect of inactivating 3OC12. Given the high similarity of PON2 and PON3 and considering the fact that both proteins are inactivated by 3OC12-mediated Ca^{2+}-disturbances, the same regulatory posttranslational modification may reside in conserved position(s). The distinct modification remains to be determined, as protein function can be altered by many modifications, like methylation, acetylation, ubiquitinylation, and glyco-sylation/deglycosylation. Our previous data showed that PON2 inactivation is rapid and reversible, which would be consistent with phosphorylation/dephosphorylation by protein kinases/phosphatases [24]. If true, about 50 potential serine, threonine, and tyrosine phosphorylation residues could be responsible for the (in-)activation of PON2. Hence, future studies are needed to identify the precise mechanism that regulates enzymatic activity of PON2 and PON3. Given that inactivation occurs within minutes and the knowledge that activity does not require cofactors (except for calcium), the presence of a Ca^{2+}-triggered, regulatory posttranslational modification appears highly likely. Revealing this mechanism is of great interest, not only for protection against *P. aeruginosa* virulence factors but also for activities of paraoxonases beyond this specific interaction.

Abbreviations

carboxy-H$_2$DCFDA:	5-(and-6)-carboxy-2′, 7′-dichlorodihydrofluorescein diacetate
coQ10:	coenzyme Q10
IL-8:	Interleukin-8
PCN:	Pyocyanin
PON:	Paraoxonase
ROS:	Reactive oxygen species
3OC12:	*N*-(3-oxododecanoyl)-L-homoserine lactone.

Acknowledgments

Work in the lab of S. Horke has been financially supported by intramural funds (MAIFOR) of the University Medical Center Mainz, by the Johannes-Gutenberg University Mainz, and by the German Research Foundation, Deutsche Forschungsgemeinschaft, Project HO-3924/4-1. All authors declare that they have no competing financial interests to declare.

References

[1] J. A. Driscoll, S. L. Brody, and M. H. Kollef, "The epidemiology, pathogenesis and treatment of *Pseudomonas aeruginosa* infections," *Drugs*, vol. 67, no. 3, pp. 351–368, 2007.

[2] P. Boontham, A. Robins, P. Chandran et al., "Significant immunomodulatory effects of *Pseudomonas aeruginosa* quorum-sensing signal molecules: possible link in human sepsis," *Clinical Science*, vol. 115, no. 11, pp. 343–351, 2008.

[3] R. S. Smith, E. R. Fedyk, T. A. Springer, N. Mukaida, B. H. Iglewski, and R. P. Phipps, "IL-8 production in human lung fibroblasts and epithelial cells activated by the Pseudomonas autoinducer N-3-oxododecanoyl homoserine lactone is transcriptionally regulated by NF-κB and activator protein-2," *Journal of Immunology*, vol. 167, no. 1, pp. 366–374, 2001.

[4] H. Li, L. Wang, L. Ye et al., "Influence of *Pseudomonas aeruginosa* quorum sensing signal molecule N-(3-oxododecanoyl) homoserine lactone on mast cells," *Medical Microbiology and Immunology*, vol. 198, no. 2, pp. 113–121, 2009.

[5] C. A. Jacobi, F. Schiffner, M. Henkel et al., "Effects of bacterial N-acyl homoserine lactones on human Jurkat T lymphocytes-OdDHL induces apoptosis via the mitochondrial pathway," *International Journal of Medical Microbiology*, vol. 299, no. 7, pp. 509–519, 2009.

[6] R. S. Smith, S. G. Harris, R. Phipps, and B. Iglewski, "The *Pseudomonas aeruginosa* quorum-sensing molecule N-(3-oxododecanoyl)homoserine lactone contributes to virulence and induces inflammation in vivo," *Journal of Bacteriology*, vol. 184, no. 4, pp. 1132–1139, 2002.

[7] T. Bjarnsholt, M. van Gennip, T. H. Jakobsen, L. D. Christensen, P. Ø. Jensen, and M. Givskov, "In vitro screens for quorum sensing inhibitors and in vivo confirmation of their effect," *Nature Protocols*, vol. 5, no. 2, pp. 282–293, 2010.

[8] M. Hentzer, H. Wu, J. B. Andersen et al., "Attenuation of *Pseudomonas aeruginosa* virulence by quorum sensing inhibitors," *The EMBO Journal*, vol. 22, no. 15, pp. 3803–3815, 2003.

[9] G. W. Lau, D. J. Hassett, H. Ran, and F. Kong, "The role of pyocyanin in *Pseudomonas aeruginosa* infection," *Trends in Molecular Medicine*, vol. 10, no. 12, pp. 599–606, 2004.

[10] H. M. Hassan and I. Fridovich, "Mechanism of the antibiotic action of pyocyanine," *Journal of Bacteriology*, vol. 141, no. 1, pp. 156–163, 1980.

[11] K. J. Reszka, Y. O'Malley, M. L. McCormick, G. M. Denning, and B. E. Britigan, "Oxidation of pyocyanin, a cytotoxic product from *Pseudomonas aeruginosa*, by microperoxidase 11 and hydrogen peroxide," *Free Radical Biology and Medicine*, vol. 36, no. 11, pp. 1448–1459, 2004.

[12] B. Rada, K. Lekstrom, S. Damian, C. Dupuy, and T. L. Leto, "The Pseudomonas toxin pyocyanin inhibits the dual oxidase-based antimicrobial system as it imposes oxidative stress on airway epithelial cells," *Journal of Immunology*, vol. 181, no. 7, pp. 4883–4893, 2008.

[13] G. W. Lau, H. Ran, F. Kong, D. J. Hassett, and D. Mavrodi, "*Pseudomonas aeruginosa* pyocyanin is critical for lung infection in mice," *Infection and Immunity*, vol. 72, no. 7, pp. 4275–4278, 2004.

[14] S. L. Primo-Parmo, R. C. Sorenson, J. Teiber, and B. N. La Du, "The human serum paraoxonase/arylesterase gene (PON1) is one member of a multigene family," *Genomics*, vol. 33, no. 3, pp. 498–507, 1996.

[15] J. Marsillach, B. Mackness, M. Mackness et al., "Immunohistochemical analysis of paraoxonases-1, 2, and 3 expression in normal mouse tissues," *Free Radical Biology & Medicine*, vol. 45, pp. 146–157, 2008.

[16] E.-M. Schweikert, A. Devarajan, I. Witte et al., "PON3 is upregulated in cancer tissues and protects against mitochondrial superoxide-mediated cell death," *Cell Death & Differentiation*. In press.

[17] D. I. Draganov, J. F. Teiber, A. Speelman, Y. Osawa, R. Sunahara, and B. N. La Du, "Human paraoxonases (PON1, PON2, and PON3) are lactonases with overlapping and distinct substrate specificities," *Journal of Lipid Research*, vol. 46, no. 6, pp. 1239–1247, 2005.

[18] J. F. Teiber, S. S. Billecke, B. N. La Du, and D. I. Draganov, "Estrogen esters as substrates for human paraoxonases," *Archives of Biochemistry and Biophysics*, vol. 461, no. 1, pp. 24–29, 2007.

[19] D. I. Draganov, "Lactonases with oragnophosphatase activity: structural and evolutionary perspectives," *Chemico-Biological Interactions*, vol. 187, no. 1–3, pp. 370–372, 2010.

[20] D. A. Stoltz, E. A. Ozer, C. J. Ng et al., "Paraoxonase-2 deficiency enhances *Pseudomonas aeruginosa* quorum sensing

in murine tracheal epithelia," *American Journal of Physiology*,
vol. 292, no. 4, pp. L852–L860, 2007.

[21] I. Witte, S. Altenhöfer, P. Wilgenbus et al., "Beyond reduction
of atherosclerosis: PON2 provides apoptosis resistance and
stabilizes tumor cells," *Cell Death and Disease*, vol. 2 , article
e112, 2011.

[22] S. Altenhöfer, I. Witte, J. F. Teiber et al., "One enzyme, two
functions: PON2 prevents mitochondrial superoxide formation and apoptosis independent from its lactonase activity,"
Journal of Biological Chemistry, vol. 285, no. 32, pp. 24398–
24403, 2010.

[23] A. Devarajan, N. Bourquard, S. Hama et al., "Paraoxonase 2
deficiency alters mitochondrial function and exacerbates the
development of atherosclerosis," *Antioxidants and Redox Signaling*, vol. 14, no. 3, pp. 341–351, 2011.

[24] S. Horke, I. Witte, S. Altenhöfer et al., "Paraoxonase 2 is downregulated by the *Pseudomonas aeruginosa* quorum-sensing
signal N-(3-oxododecanoyl)-L-homoserine lactone and attenuates oxidative stress induced by pyocyanin," *Biochemical
Journal*, vol. 426, no. 1, pp. 73–83, 2010.

[25] S. Horke, I. Witte, P. Wilgenbus et al., "Protective effect of
paraoxonase-2 against endoplasmic reticulum stress-induced
apoptosis is lost upon disturbance of calcium homoeostasis,"
Biochemical Journal, vol. 416, no. 3, pp. 395–405, 2008.

[26] S. Horke, I. Witte, P. Wilgenbus, M. Krüger, D. Strand, and
U. Förstermann, "Paraoxonase-2 reduces oxidative stress in
vascular cells and decreases endoplasmic reticulum stress-
induced caspase activation," *Circulation*, vol. 115, no. 15, pp.
2055–2064, 2007.

[27] J. F. Teiber, S. Horke, D. C. Haines et al., "Dominant role of
paraoxonases in inactivation of the *Pseudomonas aeruginosa*
quorum-sensing signal N-(3-oxododecanoyl)-L-homoserine
lactone," *Infection and Immunity*, vol. 76, no. 6, pp. 2512–2519,
2008.

[28] G. M. Denning, L. A. Wollenwebber, M. A. Railsback, C. D.
Cox, L. L. Stoll, and B. E. Britigan, "Pseudomonas pyocyanin
increases interleukin-8 expression by human airway epithelial
cells," *Infection and Immunity*, vol. 66, no. 12, pp. 5777–5784,
1998.

[29] I. Witte, U. Foerstermann, A. Devarajan, S. Reddy, and S.
Horke, "Protectors and traitors—the roles of PON2 and
PON3 in atherosclerosis and cancer," *Journal of Lipids*. In
press.

[30] C. Kunsch and C. A. Rosen, "NF-κB subunit-specific regulation of the interleukin-8 promoter," *Molecular and Cellular
Biology*, vol. 13, no. 10, pp. 6137–6146, 1993.

[31] H. B. Kang, Y. E. Kim, H. J. Kwon, D. E. Sok, and Y. Lee,
"Enhancement of NF-κB expression and activity upon differ-
entiation of human embryonic stem cell line SNUhES3," *Stem
Cells and Development*, vol. 16, no. 4, pp. 615–623, 2007.

14

Fatty Acid Oxidation and Cardiovascular Risk during Menopause: A Mitochondrial Connection?

Paulo J. Oliveira,[1] Rui A. Carvalho,[1,2] Piero Portincasa,[3] Leonilde Bonfrate,[3] and Vilma A. Sardao[1]

[1] CNC—Center for Neuroscience and Cell Biology, University of Coimbra, 3004-517 Coimbra, Portugal
[2] Department of Life Sciences, University of Coimbra, 3004-517 Coimbra, Portugal
[3] Department of Internal Medicine and Public Medicine, Clinica Medica "A. Murri", University of Bari Medical School, 70124 Bari, Italy

Correspondence should be addressed to Vilma A. Sardao, vimarisa@ci.uc.pt

Academic Editor: B. A. Neuschwander-Tetri

Menopause is a consequence of the normal aging process in women. This fact implies that the physiological and biochemical alterations resulting from menopause often blur with those from the aging process. It is thought that menopause in women presents a higher risk for cardiovascular disease although the precise mechanism is still under discussion. The postmenopause lipid profile is clearly altered, which can present a risk factor for cardiovascular disease. Due to the role of mitochondria in fatty acid oxidation, alterations of the lipid profile in the menopausal women will also influence mitochondrial fatty acid oxidation fluxes in several organs. In this paper, we propose that alterations of mitochondrial bioenergetics in the heart, consequence from normal aging and/or from the menopausal process, result in decreased fatty acid oxidation and accumulation of fatty acid intermediates in the cardiomyocyte cytosol, resulting in lipotoxicity and increasing the cardiovascular risk in the menopausal women.

1. Menopause: A Burden for Aging Women

Menopause is one of the most critical periods in women's life. Although being a natural biological process that occurs with aging, physiological alterations observed during this period can be challenging. Caused by a reduced secretion of ovarian hormones estrogen and progesterone after depletion of the storage of ovarian follicles, menopause defines the end of women menstrual cycle and their natural fertility. On average, spontaneous or natural menopause occurs around the early 50s and is confirmed after 12 months of nonpathological amenorrhoea. However, when premature ovarian failure (POF) occurs before the 40s due to pathological causes, an early or premature menopause can be induced, which is thus disconnected from the aging process properly said. When a bilateral oophorectomy is necessary, menopause occurs immediately without women experiencing the gradual transition of perimenopause. Chemotherapy can also provoke a permanent damage in ovaries and induces menopause *per se*

[1]. Women who experience an early menopause are more susceptible to certain health problems, such as osteoporosis and heart diseases, since they spend more time in their lives without the benefits of estrogens. POF can also be temporary (temporary menopause) induced by high levels of stress, excessive exercising and/or dieting, and by medications used to treat fibroids [2] and endometriosis [3]. However, as soon as women adopt a healthier life style or stop medication, the ovaries may resume normal production of hormones. Normally, menopausal transition or perimenopause starts around mid-to-late 40s and persists several years before the last menstrual period, normally for 4-5 years (Figure 1). Smoking and genetic background are two factors that can influence the timing of spontaneous menopause. Normally, smokers can reach menopause earlier than nonsmokers [4]. During perimenopause, levels of estrogen and progesterone start gradually to decline and menstrual periods become irregular. Since sex hormones are physiologically important to maintain the health and normal functioning of several

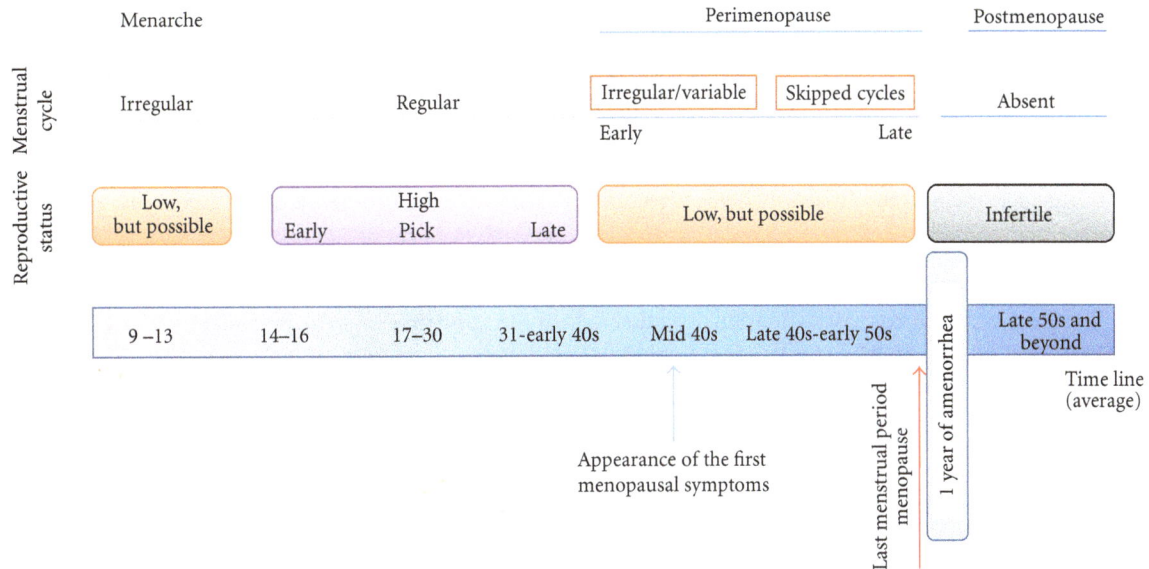

FIGURE 1: Women reproductive stages during aging: from menarche to postmenopausal. Time line represents only an average for the normal age. More details can be found in the text.

organs, such as the heart, liver, brain, and bone, hormonal changes observed during this menopausal transition may induce several chronic medical conditions [5]. All women experience menopause, but different women may cope with different symptoms. The variation of menopause phenotypes around the world and in different ethnic groups suggests both cultural and genetic influences [6, 7]. Menstrual irregularities, vaginal atrophy, and vasomotor instability are the most frequent menopausal symptoms that have been directly related with the decreased levels of female sex hormones [8].

Menopause-associated vasomotor symptoms (also known as hot flashes) include spontaneous feeling of warmth, usually on face, neck, and chest and are usually associated with perspiration, palpitations, and anxiety, being variable in frequency, duration, and severity, and can be the cause for fatigue, difficulty concentrating, and memory lapses, symptoms that have also been observed during menopause transition. The cause for menopause-associated vasomotor symptoms is not completely understood, although some theories have been proposed [8, 9].

Vaginal atrophy is also a common symptom during menopause transition. Due to loss of estrogens, vagina lining may become thinner and dryer, and the pH also changes, making the vagina more susceptible to infections. Those alterations can affect sexual function and quality of life [10].

Others menopause-associated complications include increased cardiovascular risk (see below), osteoporosis [11] and body weight gain, which can all be a combination of changes in hormone levels and aging.

Increase in body weight is another characteristic associated with menopause. Although it is known that the metabolic rate decreases with aging, the increase in body weight and visceral adipose tissue accumulation after menopause have been associated with ovarian hormone withdrawal [12]. It has been shown that, in abdominal adipocytes, estrogen regulates the expression of lipoprotein lipase (LPL) and hormone-sensitive lipase (HSL) [13]. In hepatocytes, estrogen regulates the synthesis of structural apolipoproteins for very low-density lipoproteins (VLDLs) and high-density lipoproteins (HDLs) and decreases the synthesis of hepatic lipases [14]. By regulating lipidogenesis in adipocytes and hepatocytes, estrogen modulates lipid concentration in plasma. The withdrawal of estrogens during induced or natural menopause leads to several lipid metabolism disorders. For example, dyslipidemia was also observed in bilateral oophorectomized in women [15]. Abdominal accumulation of adipose tissue and associated dyslipidemia are important components of a group of metabolic irregularities strongly related with increased cardiovascular risk in the menopausal woman.

2. Cardiovascular Disease in Women during Menopause: The Role of Hormone Replacement Therapy

2.1. Clinical Data: What Do We Know? Cardiovascular disease (CVD) is a multifactorial disease. Both bad lifestyle including inappropriate diet, sedentary life, smoking and drinking, and determined factors (e.g., aging, sex, genotype, and menopause) influence CVD [16, 17]. The impact of CVD on overall mortality in westernized countries is enormous, accounting for up to 30% of all deaths worldwide. The definition of CVD includes four major groups of diseases: coronary heart disease (CHD) disclosed by angina pectoris, myocardial infarction, heart failure, and coronary death, cerebrovascular disease such as stroke or transient ischemic attack, clinically evident peripheral artery disease, aortic atherosclerosis, and thoracic or abdominal aortic aneurysm. What is less known is that CVD is the leading cause of death in women, with more deaths than all other causes combined yearly [18]. Various studies showed a growing risk for CVD in menopausal women due to negative changes in metabolism and hemodynamic parameters [16]. According to the guidelines of the National Cholesterol Education Program (NCEP) [19], the American Heart Association (AHA), and the American College of Cardiology (ACC) [18, 20], evaluation of CVD risk factors in women must include a personal CHD history, age over 55, family history of premature CHD, diabetes mellitus, dyslipidemia, hypertension, personal history of peripheral artery disease, and smoking.

Guidelines for prevention of CVD in women were first published in 1999 by the American Heart Association (AHA) [21]. One consequence of such increased attention to gender-related health problems, is awareness of CVD as the leading cause of death among women has nearly doubled since 1997 [22]. The impact of menopause should be taken into account when discussing CVD, and this aspect has been the matter of debate [23].

Premenopausal women have a lower incidence of CVD when compared to men with the same age-range. Whereas CHD is sporadic in premenopausal women [24], the incidence of myocardial infarction increases with age in both sexes, but occurs later and after menopause [24]. Estrogen loss during menopause causes negative effects on metabolism and cardiovascular function [25], and the progression to menopause with the changes in estrogen levels decreases or cancels the women advantage versus men [26–29].

Postmenopausal women have a higher risk of coronary artery disease, atherosclerosis, and all causes of mortality [29]. A consequence of this gender-related trend is that the postmenopausal state is acknowledged as a risk factor for CHD, with a weight similar to that of male sex [30]. Furthermore, an early natural menopause appears to be associated with increased risk of CVD [31, 32], even in non-smokers.

Indeed, menopause is associated with increased total serum cholesterol, triglycerides, and fibrinogen, as well as with a decrease in high-density lipoprotein (HDL) cholesterol. A plausible explanation is that menopause is believed to be a result of fluctuations in hormonal status, primarily a deficiency in estrogen [33]. Whether other contributing factors may have a role on CVD after menopause, is less clear and difficult to demonstrate. The transition from pre-menopausal phase to menopause, for example, may induce a weight gain responsible for increased in blood pressure, total cholesterol, low-density lipoprotein (LDL), triglycerides, and fasting insulin [33]. What should be mentioned is that aging *per se* can be more important than menopause itself for a number of CHD risk factors. In the SWAN study (Study of Women's Health Across the Nation) [34], changes in traditional risk markers of CHD were evaluated in three different stages: before, within a year, and after the final menstrual period within a multiethnic group (African, American, Hispanic, Japanese, or Chinese and Caucasian women). Changes due to menopause were only represented by total cholesterol, low-density lipoprotein cholesterol, and apolipoprotein B. By contrast, chronological aging was responsible for changes in the other risk factors with a linear model. Many other potential factors might be also implicated in the sex differences in coronary heart disease [35]. The possibility that heart disease risk determines menopausal age rather than the inverse has already been proposed [36].

Oxidative stress plays a role in hypertension, hypercholesterolemia, diabetes, and promoting CVD [37]. The formation of free radicals leads to cellular oxidative stress with a contribution to the first step of endothelial damage and the progression to atherosclerotic lesion. The perpetuation of the process induces the final events of CVD, which appears to be linked to some oxidative stress biomarkers [38, 39]. Oxidative stress appears to be an emerging factor also in the pathophysiology of CVD in menopausal women. Studies have shown that during menopause the risk of CVD increases at the same time of a rise in oxidative status [40, 41].

It is still unclear if the type of menopause (surgical or natural) can have a role on cardiovascular risk. The Nurses' Health Study (1987) demonstrated that the risk of CHD was higher in patients undergoing bilateral oophorectomy compared with natural menopause. An estrogen-replacement therapy could prevent this effect [42]. In a later study, carotid artery intima-media thickness showed a positively association with years elapsed since menopause; however, according to this marker of subclinical atherosclerosis, women with natural menopause presented no difference compared with those who had surgical menopause [43]. Indeed, men with the common estrogen receptor alpha (ESR1) c.454-397CC genotype have a major risk of myocardial infarction, suggesting the potential linkage between estrogen receptors and CVD susceptibility. In this respect, a variation in estrogen receptor could clarify the contrasting results of hormone therapy on CVD susceptibility in women [44]. The apparent protective effect of hormone replacement therapy (HRT) has been a matter of debate for several years [45–47]. Prevention of CHD and osteoporosis in menopausal women was originally achieved by exogenous estrogen plus progestin, assuming a protective effect of estrogen on the heart. Additional effects included a protective effect on the bone and on colon cancer [48–52], despite increasing incidence of breast cancer [53, 54]. Two landmark studies, however, changed this view. The Women's Health Initiative (WHI) Estrogen plus Progestin (E+P) trial in 2002 showed no protection for CHD and confirmed the increased risk in breast cancer and thromboembolic disease [55].

Two years later the WHI Estrogen Alone trial confirmed the lack of effect on CHD while suggesting a trend for decreased breast cancer, with a rise in stroke and venous thromboembolic disease. A nonsignificant protective effect on CHD was seen in the younger women (ages 50 to 59) [56]. The public consequence was that hormone therapy was abandoned or was conducted with lower doses [57].

The possibility that CHD risk is lowered by earlier hormone therapy after menopause should also be considered, although results are not conclusive [58]. Whether hormone replacement therapy results in either increased or unchanged risk for stroke, is also a matter of debate [56]. Of note, recent guidelines do not identify estrogen therapy for the primary or secondary prevention of CHD [59, 60].

2.2. Animal Models: Helping to Define the Role of Estrogens. Although the WHI and the Heart and Estrogen/progestin Replacement Studies (HERS) showed no CVD protection resulting from HRT, several animal studies have suggested an important cardioprotective role for estrogens against heart failure [61], mediated by a genomic or a nongenomic estrogen-receptor-mediated signaling pathway (see [62] for a review).

Tumor necrosis factor-alpha (TNF-α) has been reported as an important factor during I/R injury and ischemia preconditioning. In a Langendorff-perfused rat heart model, estrogen reversed the deterioration of heart hemodynamics induced by TNF-α treatment [63]. Several evidences have been demonstrated that stromal cell-derived factor 1 (SDF-1) is increased in ischemic hearts and induced cardioprotection [64]. A higher expression of myocardial SDF-1 was observed in female rats in response to I/R and the increased myocardial SDF-1 production in female hearts was due to estrogen-estrogen Receptor α (ERα) interactions [65]. In C57BL/6J male mice, estrogen also induced cardioprotection after acute myocardial infarction through a decreased activity of matrix metalloproteinase-9 and increased Akt-Bcl-2 antiapoptotic signaling [66]. In a Langendorff isolated perfused rat heart model, estrogen increased the perfusion pressure and coronary resistance through activation of L-type calcium channels [67].

Estrogen-related receptor alpha (ERRα) is a transcription factor for some myocardial mitochondrial enzymes, essential to maintain cardiac energy reserves. A decrease in myocardial ERRα, regulated by the metabolic sensor AMP-activated protein kinase alpha 2 (AMPKα2), was recently reported during congestive heart failure [68]. Proteins from the intracellular lipin family are also involved in metabolism regulation. It was reported that lipin 1 is the principal protein of this family in myocardium and is also regulated by ERRα [69].

The lack of CVD protection observed during HRT has been proposed to be related with alterations in sex hormone synthesis and metabolism that can occur during aging, and can affect the hormone environment in postmenopausal women. Also age-related changes in vascular estrogen receptors (ERs) subtype, structure, expression, distribution, and the signaling pathway in the endothelium and vascular smooth muscle, preexisting CVD conditions, and structural changes in blood vessels architecture have been suggested as possible causes for the failure of HRT in CVD [70]. It also should be noticed that HRT is not only composed by estrogens, but also by a combination of estrogen and progesterone. A recent study demonstrated that a combination of 17-α-estradiol and medroxyprogesterone acetate aggravates chronic heart failure after experimental myocardial infarction, which can also explain the results from previous studies including WHI and HERS [71].

3. Cardiac Mitochondrial Fatty Acid Beta-Oxidation in Health and Disease: Where Does Menopause Stand?

The heart is one of the organs with the highest energy demand in the body, which is hardly surprising due to high energetic input required by the contractile apparatus. Although the heart is considered an omnivorous organ due to the fact that it can use several substrates for energy generation, including glucose, amino acids, lactate, and ketone bodies, fatty acids are the favored fuel for the cardiac muscle [72, 73]. In fact, the adult heart generates between 50–70% of its ATP from fatty acid beta-oxidation, which occurs mainly in mitochondria [72], possesses an elaborate system to import and process fatty acids of different lengths [72, 74]. In fact, in itself, mitochondrial function is one among different factors that impact the flux of fatty acid beta-oxidation. Others include the fatty acid supply itself, which is modulated among other factors by diet, competing substrates for the cardiac tissue, the energy demand and oxygen availability, and the regulation at a nuclear or allosteric level of enzymes which are involved in all steps of fatty acid uptake, esterification, and metabolism [72].

Fatty acids can be transported in the plasma as free fatty acids (FFAs) conjugated with albumin or as part of triacylglycerol (TAG) contained in chylomicrons or very-low density lipoproteins (VLDLs) [75, 76]. FFA concentration in the plasma is highly variable, depending not only on the diet, but also on the developmental state of the organism and if any pathology is present. For example, the amount of FFA in the plasma is known to greatly increase during myocardial infarction [77] and diabetes [78], which leads to an augmented cardiomyocyte FFA uptake and accumulation, since the concentration of FFA in the plasma is a major determinant for these two events [72]. Regardless of the mechanism underlying an acute or chronic accumulation of FFA in the plasma (reviewed in [72]), the end result of cardiomyocyte cytosolic accumulation of fatty acids can differ, depending on a wide range of factors.

The first step after entering the cardiomyocyte is conversion to CoA esters, through the action of fatty acyl CoA synthase (FACS). Fatty acid uptake by cells is made by membrane proteins with high affinity for fatty acids [79, 80], namely, the fatty acid translocase (FAT/CD36), the fatty acid binding protein (FABPpm) and a variety of fatty acid transport proteins (FATPs), as well as by simple diffusion of fatty acids through either the phospholipid bilayer or a pore or channel formed by one or more of the referred fatty acid transporter proteins [81]. Upon entering the cell, the rate of utilization is governed by a variety of factors, including malonyl-CoA, the ratio acetyl-CoA/CoA and the availability of other substrates, namely, glucose, lactate, and ketone bodies that can compete with free fatty acids as a source of acetyl-CoA [79]. Long-term regulation of uptake and utilization requires alterations in expression rates of genes encoding for fatty acid handling proteins [82]. Free fatty acids can also by themselves modulate the expression of such genes via nuclear transcription factors such as peroxisome proliferator-activated receptors (PPARs) [83].

Mitochondrial beta-oxidation of long-chain fatty acids starts with its association with CoA, forming acyl-CoA esters that are transported into mitochondria by carnitine palmitoyl transferase I (CPT-I). Beta-oxidation produces in each round one NADH, one FADH$_2$ (as part of an enzymatic complex), and one acetyl-CoA, which is further oxidized in the Krebs cycle to CO$_2$, with the concomitant further generation of three NADH, reduced FAD co-factor in succinate dehydrogenase complex, and one GTP. NADH, via NADH

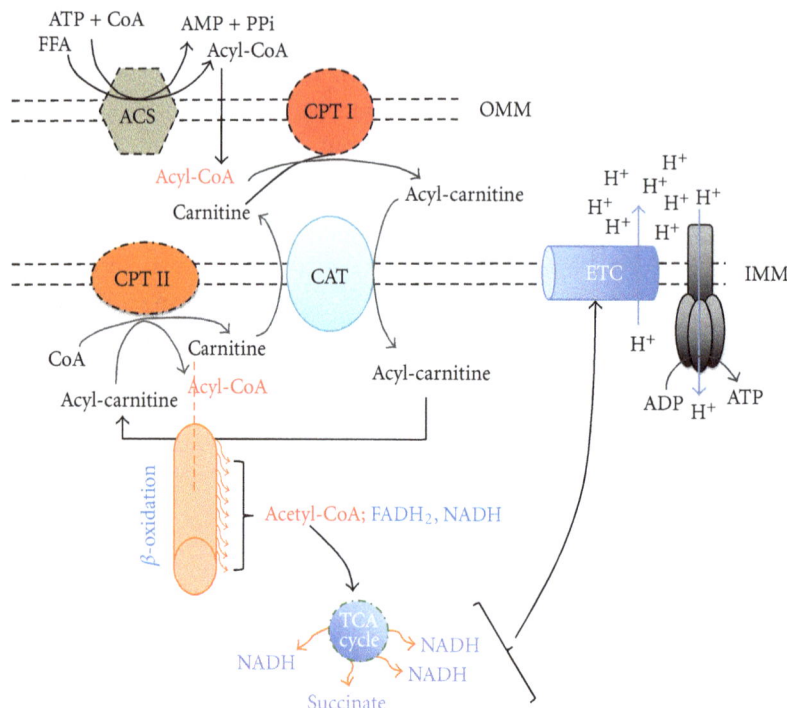

FIGURE 2: Transport of fatty acids from the cytoplasm to the mitochondrial matrix for oxidation. Following activation to acyl-CoA, CoA is exchanged for carnitine by carnitine palmityl transferase (CPT-I), which is then transported to the inside of the mitochondria where a reversal exchange takes place through the action of carnitine acylcarnitine translocase (CPT-II), and beta-oxidation machinery initiates its activity, producing reducing equivalents that feed the electron transport chain. More details are available in the text. CAT: Carnitine Acylcarnitine Translocase, FFA: free fatty acid, ACS: Acyl-coA synthase, ETC: electron transport chain, IMM: inner mitochondrial membrane, OMM: outer mitochondrial membrane, coA: coenzyme A, ATP: adenosine triphosphate, ADP: adenosine diphosphate, and AMP: adenosine monophosphate.

dehydrogenase, and succinate dehydrogenase deliver electrons to the remaining electron transport chain complexes which contribute to the generation of a proton gradient used to synthesize ATP (Figure 2). Throughout this whole process, several regulation mechanisms can operate, starting with the transport of the acyl chain to the mitochondrial matrix and ending at the accumulation of end products of the oxidation process, namely, reducing equivalents and ultimately ATP levels. The transport process is considered a major player in the control of the flux through beta-oxidation [84], mostly in intact muscle, since levels of malonyl-CoA are kept considerably high. With this type of control, it is possible for the tissues to rapidly adapt to different metabolic demands, such as in muscles [84]. An inhibition of fatty acid beta-oxidation, which as mentioned can occur at several stages, will ultimately result in free fatty acid intracellular accumulation which subsequently will be responsible for poor removal of fatty acids from plasma in any of their forms of transportation. In fact, a possible role has been attributed to female sex hormones in the development of fatty liver pregnancy on the basis of their effect in the reduction of mitochondrial fatty acid oxidation [85] and in regulating cellular energy balance in vivo by regulating the expression of the medium chain acyl coenzyme A dehydrogenase (MCAD) gene [86].

Besides mitochondrial oxidation, long-chain fatty acyl coA can also be used for the synthesis of intermediates, including TAG, diacylglycerol (DAG), and ceramide [72, 87].

Under normal intracellular concentrations, these intermediates are stored and/or channeled to different biosynthetic pathways, including biomembrane synthesis. If alterations in normal fatty acid homeostasis occur, which can originate from excessive plasma FFA content or from enhanced FACS expression and/or activity, long-chain fatty acyl coA derivatives can accumulate in cells. Depending on the tissue, accumulation of some of these intermediates can have distinct effects. For example, it is known that excessive accumulation of TAG in nonadipocyte tissues can result in different negative outcomes including impaired insulin signaling in the liver and skeletal muscle [88] and apoptosis and other metabolic disturbances in the heart [87, 89, 90]. DAG has also been determined to cause similar effects in the same tissues [88], including increased insulin resistance observed in a model of rodent high-fat diet [91]. It is interesting to note that both increases in TAG and ceramide intracardiac content did not correlate with the increased insulin resistance [91].

Ceramide, by its turn, has been demonstrated in different biological models to increase apoptotic signaling in several tissues [92–94], although evidence is scarcer for the heart [95]. It is interesting to note that ceramide derivatives have been involved in the triggering of the mitochondrial permeability transition pore (MPT pore) and outer-membrane permeabilization [96, 97], conditions closely linked with mitochondrial dysfunction and cell death [98]. In opposition, long-chain ceramide species have been shown to inhibit the MPT pore [99]. The discrepancy of results regarding ceramide implicates this lipid species in the control of mitochondrial cell death pathways.

From the short description above, it is clear that a balance between FFA cell uptake and metabolism must be reached in order to avoid the accumulation of undesired fatty acid metabolites. Also, increased reliance of fatty acids as fuel for cardiac cells has undesired effects, one of them being decreased ATP synthesis, resulting from increased ATP hydrolysis for noncontractile purposes, increased mitochondrial uncoupling due to increased activity/expression of uncoupling proteins and greater proton futile cycling, creating the so-called oxygen wasting and resulting in several physiological complications [100–102]. Interestingly, inhibition of fatty acid metabolism is proposed to be beneficial for some forms of heart failure [103].

The important question is now where the menopausal heart stands. As described above, menopause is a normal consequence of the aging process in women and is accompanied of important physiological and biochemical alterations. There are several evidences in the literature that the content in FFA in the plasma tends to increase during menopause. One particular study performed with 4-vinylcyclohexene-diepoxide- (VCD-) treated rats indicated that progressive loss of ovarian function induced by VCD results in an increase of plasma FFA, which initiated several alterations leading to the development of the metabolic syndrome [104]. This important piece of evidence mimics what is observed in the menopausal women, where an increase in circulating FFA was measured [105]. It is also known that women experience a characteristic increase in circulating lipids at the time of the final menstruation period [34], although it is difficult to evaluate the component resulting from hormonal alterations and what is the result of the normal aging process [34, 106]. The increased FFA was partly reverted by hormone-replacement therapy, showing that, at least in part, it is a hormone-dependent effect [105]. The role of estrogens in fatty acid metabolism is well described and involves different mechanisms [107–109]. One important effect is that estradiol promotes the channeling of FFA toward oxidation and away from triglyceride storage (Figure 3) by upregulating the expression of peroxisome proliferation activator receptor delta and its targets and also by directly and rapidly activating AMP-activated protein kinase (AMPK). AMPK acts as a fuel sensor that increased fatty acid beta-oxidation during higher metabolic demands [110].

The data, although still scarce and largely spread out, indicates that during menopause, fatty acid metabolism is altered. The decrease in estradiol levels may result in decreased fatty acid oxidation and increased accumulation in the adipose tissue, with hormone replacement therapies recovering the pre-menopausal fatty acid status quo. But is this so straightforward? Maybe not, one important player in fatty acid metabolism is, as described, the mitochondrion. A proper channeling of fatty acyl-CoA and subsequent beta-oxidation is necessary for the energy-generating process. It is clear that a failure of mitochondrial bioenergetics causes an unbalance in fatty acid metabolism, which may result

in the accumulation of fatty acyl-CoA esters in the cytosol of cardiomyocytes. This phenomenon could result in a larger channeling of fatty acyl-CoA esters to the synthesis of the intermediates described above, including TAG, DAG, and ceramide. It is interesting to recapitulate here that ceramide has been involved in the induction of apoptosis in a variety of biological models [92–94]. Although the relationship between increased ceramide intracellular levels in the menopausal heart and increased apoptotic signaling is still to be determined, several endpoints for increased cardiac Fas-dependent and mitochondrial-dependent apoptosis were identified in the hearts of bilateral ovariectomized Wistar rats [111, 112]. A logical question would be if there is a possible relationship between intracellular lipid metabolism alterations resulting from ovariectomy and enhanced apoptotic signaling in the heart.

Decreased fatty acid oxidation by mitochondria occurs in a variety of situations, ranging from xenobiotic-induced toxicity to several pathologies. There are many fatty acid oxidation disorders identified in humans, and which affect organs as different as muscle [113] and brain [114], which result in altered fat deposition and mitochondrial beta-oxidation. Defects are commonly present in the mitochondrial machinery that shuttles long-chain fatty acid metabolites to mitochondria, resulting in decreased beta-oxidation [113]. Several xenobiotics also alter fatty acid metabolism in different organs [115], examples are fluorochemicals [116] and the antibiotic tetracycline [117] in the liver. As for the heart, it is now becoming increasingly recognized that alterations in fatty acid uptake and/or beta-oxidation can result in the so-called fatty heart, a largely unrecognized entity for a long time, and which, as described has important cardiovascular complications [89, 118]. This subject will deserve more attention in the future.

It has been proposed that mitochondrial function in the heart decreases with the progression of aging. Alterations include loss or oxidation of cardiolipin, a tetra-acyl phospholipid involved in the activity of many oxidative phosphorylation enzymes including complex I [119–121]. This presents a clear determinant of loss of mitochondrial function and also represents a phenotype of mitochondrial membrane aging which impacts both the bioenergetics and several signaling pathways to and from mitochondria.

It is also known that aging-dependent cardiac mitochondrial effects are more specific to interfibrillar mitochondria, which is the subpopulation responsible for the majority of energy supply to the myocardium [122, 123]. Such alterations include decrease respiratory complex activity and increased oxidative stress, while a decreased capacity for beta-oxidation has also been demonstrated in an animal model for aging due to alterations in carnitine palmitoyltransferase I which were suspected to originate from a decrease in cardiolipin content [123]. Mitochondrial "power" in the heart is thus affected with aging [124], which is further illustrated by a decrease in the nuclear control of mitochondrial biogenesis and function [125] and by increased mtDNA deletions frequency found in the aged heart [126].

Adding to mitochondrial aging, *per se*, one has to have in mind that other factors may be operating in the menopausal

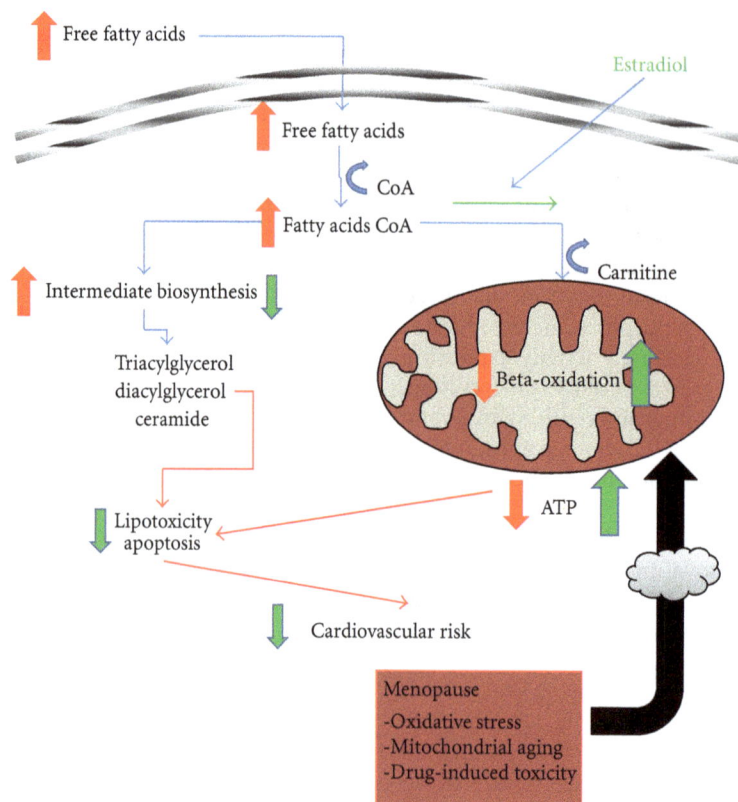

FIGURE 3: General scheme of the hypothesis raised by the present paper. It is proposed that menopause, as a condition natural to the normal aging process, is accompanied by specific mitochondrial alterations (bottom red box, arrow with a dark cloud) which decrease their ability to cope with an increased flux of long-chain fatty acyl CoA, resulting from augmented plasma levels. Inability to process fatty acyl CoA may result in accumulation of fatty acid intermediates including tri- and diacylglycerol, as well as ceramide, which causes myocardial lipotoxicity and may even result into activation of apoptotic signaling. The cardiovascular risk increases under these circumstances, which is fueled by other coexisting pathological conditions or by pharmacological interventions that present toxicity to the cardiovascular system. Estradiol (represented by green arrows) has been proposed to increase fatty acid oxidation by mitochondria, decreasing the flux through other biosynthetic pathways, preventing the potential accumulation of deleterious metabolites and increasing fatty acid-derived mitochondrial ATP production.

woman that can contribute to altered mitochondrial function and result in disrupted fatty acid metabolism. For example, the incidence of diabetes, and obesity increases during menopause [127], which also contributes to accelerate mitochondrial dysfunction [128–130]. By its turn, the menopausal woman may be under treatment with different medications which may also affect the bioenergetic efficacy of cardiac mitochondria [131, 132], especially if other conditions occur at the same time.

To summarize, ageing results into a progressive degradation of mitochondrial capacity in the heart, which, in combination with hormonal alterations resulting from menopause and its associated alterations in lipid profile, may result into a progressive decrease in lipid oxidation in mitochondria and increased lipid storage in adipocytes and formation of fatty acyl intermediates in the cytosol of cardiomyocytes (Figure 3). The development of insulin resistance, diabetes and obesity can be several faces of the same coin, the increased lipotoxicity in the cardiomyocyte of the menopausal woman. This is a clear avenue for research that still is largely unexplored and deserves attention since menopause is a condition

that affects an increasingly number of women, as the general population is progressively aging.

If the hypothesis put together in this paper is correct, then prophylactic measures that improve mitochondrial capacity in menopausal women would contribute to decrease cardiovascular risk. In fact, besides hormone replacement therapy, which replenishes estrogens and reequilibrates lipid homeostasis, other cotherapies may help improve the lipid profile in the menopausal woman through different mechanisms. For example, endurance exercise has been demonstrated to increase mitochondrial capacity in the heart [133, 134]. In a menopausal setting, twelve weeks of endurance exercise have been demonstrated to provide some benefits in increasing lipid oxidation, besides improving other cardiorespiratory parameters [135, 136]. Carnitine, which is essential to long-chain fatty acid beta-oxidation, has been shown to recover some of skeletal muscle function and inhibit alterations in ovariectomized rats [137]. Nevertheless, to the best of our knowledge, no work on the impact of carnitine on lipid profile and oxidation in the menopausal heart has been provided.

Cardiac oxidative stress after ovariectomy has also been observed in animal models [138] although evidence for increased oxidative stress in the cardiovascular system is scarce. Estrogens *per se* act as antioxidants, although it is still unclear if estrogen supplementation during menopause is completely without risks for the cardiovascular system [139, 140]. Also, it is unclear so far if antioxidant supplementations would improve mitochondrial fitness in menopausal women. Finally, an interesting alternative was proposed by Zern et al. [141]. Lyophilized grape powder was given to a group of postmenopausal women for 4 weeks. The powder was enriched in phytochemicals such as flavans, anthocyanins, quercetin, myricetin, kaempferol, and resveratrol. The results showed alterations in lipoprotein metabolism, oxidative stress, and inflammatory markers, which were all decreased in the treated group. Although the heart was not specifically targeted in the study, the results may suggest a positive impact in this organ as well. Interestingly, resveratrol is considered an activator of mitochondrial biogenesis in different model systems, acting through sirtuin-1-dependent and independent mechanisms [142–144]. The future will tell if this is a trail worth exploring.

4. Concluding Remarks

Although there are many loose ends in the story, it appears logical to consider that progressive deterioration of mitochondrial function in the aging woman with menopause contributes to the metabolic alterations observed in the heart, including a decreased capacity for lipid oxidation. A decreased mitochondrial flux of fatty acid beta-oxidation, can result in most cases in the accumulation of toxic intermediates in the cytosol and also of nonmetabolized fatty acids in mitochondria, which leads to further deterioration of mitochondrial function and progressive metabolic changes that can increase cardiovascular risk. Not only this line of thought needs to be demonstrated in animal models and humans, but if true, pharmacological, or nonpharmacological strategies must be devised to counteract this metabolic remodeling.

Acknowledgments

V. A. Sardao is supported by the Foundation for Science and Technology (FCT, Portugal), Post-doctoral Fellowship SFRH/BPD/31549/2006. Work in the authors' laboratory is funded by the FCT (PTDC/SAU-OSM/104731/2008 to P. J. Oliveira and PTDC/AGR-ALI/108326/2008 to V. A. Sardao) and by the Italian Ministry of University (FIRB 2003 RBAU01RANB002 to P. Portincasa). P. Portincasa was a recipient of the short-term mobility grant 2005 from the Italian National Research Council (CNR).

References

[1] A. M. Gordon, S. Hurwitz, C. L. Shapiro, and M. S. Leboff, "Premature ovarian failure and body composition changes with adjuvant chemotherapy for breast cancer," *Menopause*, vol. 18, no. 11, pp. 1244–1248, 2011.

[2] M. Shozu, K. Murakami, and M. Inoue, "Aromatase and leiomyoma of the uterus," *Seminars in Reproductive Medicine*, vol. 22, no. 1, pp. 51–60, 2004.

[3] A. E. Schindler, "Dienogest in long-term treatment of endometriosis," *International Journal of Women's Health*, vol. 3, pp. 175–184, 2011.

[4] D. Kaleta, B. Usidame, and K. Polańska, "Tobacco advertisements targeted on women: creating an awareness among women," *Central European Journal of Public Health*, vol. 19, no. 2, pp. 73–78, 2011.

[5] J. C. Stevenson, "A woman's journey through the reproductive, transitional and postmenopausal periods of life: impact on cardiovascular and musculo-skeletal risk and the role of estrogen replacement," *Maturitas*, vol. 70, no. 2, pp. 197–205, 2011.

[6] F. Kronenberg, "Menopausal hot flashes: a review of physiology and biosociocultural perspective on methods of assessment," *Journal of Nutrition*, vol. 140, no. 7, pp. 1380S–1385S, 2010.

[7] R. Green and N. Santoro, "Menopausal symptoms and ethnicity: the study of Women's Health Across the Nation," *Women's Health*, vol. 5, no. 2, pp. 127–133, 2009.

[8] H. D. Nelson, "Menopause," *The Lancet*, vol. 371, no. 9614, pp. 760–770, 2008.

[9] S. L. Dormire, "The potential role of glucose transport changes in hot flash physiology: a hypothesis," *Biological Research for Nursing*, vol. 10, no. 3, pp. 241–247, 2009.

[10] M. Panjari and S. R. Davis, "Vaginal DHEA to treat menopause related atrophy: a review of the evidence," *Maturitas*, vol. 70, no. 1, pp. 22–25, 2011.

[11] B. Frenkel, A. Hong, S. K. Baniwal et al., "Regulation of adult bone turnover by sex steroids," *Journal of Cellular Physiology*, vol. 224, no. 2, pp. 305–310, 2010.

[12] P. Babaei, R. Mehdizadeh, M. M. Ansar, and A. Damirchi, "Effects of ovariectomy and estrogen replacement therapy on visceral adipose tissue and serum adiponectin levels in rats," *Menopause International*, vol. 16, no. 3, pp. 100–104, 2010.

[13] S. L. Palin, P. G. McTernan, L. A. Anderson, D. W. Sturdee, A. H. Barnett, and S. Kumar, "17β-Estradiol and anti-estrogen ICI: compound 182,780 regulate expression of lipoprotein lipase and hormone-sensitive lipase in isolated subcutaneous abdominal adipocytes," *Metabolism*, vol. 52, no. 4, pp. 383–388, 2003.

[14] H. Szafran and W. Smielak-Korombel, "The role of estrogens in hormonal regulation of lipid metabolism in women," *Przegląd lekarski*, vol. 55, no. 5, pp. 266–270, 1998.

[15] T. Yoshida, K. Takahashi, H. Yamatani, K. Takata, and H. Kurachi, "Impact of surgical menopause on lipid and bone metabolism," *Climacteric*, vol. 14, no. 4, pp. 445–452, 2011.

[16] C. Vassalle, A. Mercuri, and S. Maffei, "Oxidative status and cardiovascular risk in women: keeping pink at heart," *World Journal of Cardiology*, vol. 1, no. 1, pp. 26–30, 2009.

[17] I. M. Fearon and S. P. Faux, "Oxidative stress and cardiovascular disease: novel tools give (free) radical insight," *Journal of Molecular and Cellular Cardiology*, vol. 47, no. 3, pp. 748–381, 2009.

[18] K. Tolfrey, "American Heart Association guidelines for preventing heart disease in women: 2007 Update," *Physician and Sportsmedicine*, vol. 38, no. 1, pp. 162–164, 2010.

[19] "Third report of the national cholesterol education program (NCEP) expert panel on detection, evaluation, and treatment of high blood cholesterol in adults (Adult Treatment Panel III) final report," *Circulation*, vol. 106, no. 25, pp. 3143–3421, 2002.

[20] L. Mosca, S. M. Grundy, D. Judelson et al., "AHA/ACC scientific statement: consensus panel statement. Guide to preventive cardiology for women. American Heart Association/American College of Cardiology," *Journal of American College of Cardiology*, vol. 33, no. 6, pp. 1751–1755, 1999.

[21] L. Mosca, S. M. Grundy, D. Judelson et al., "Guide to preventive cardiology for women. AHA/ACC scientific statement consensus panel statement," *Circulation*, vol. 99, no. 18, pp. 2480–2484, 1999.

[22] L. Mosca, H. Mochari-Greenberger, R. J. Dolor, L. K. Newby, and K. J. Robb, "Twelve-year follow-up of American women's awareness of cardiovascular disease risk and barriers to heart health," *Circulation: Cardiovascular Quality and Outcomes*, vol. 3, no. 2, pp. 120–127, 2010.

[23] H. Tunstall-Pedoe, "Myth and paradox of coronary risk and the menopause," *Lancet*, vol. 351, no. 9113, pp. 1425–1427, 1998.

[24] D. J. Lerner and W. B. Kannel, "Patterns of coronary heart disease morbidity and mortality in the sexes: a 26-year follow-up of the Framingham population," *American Heart Journal*, vol. 111, no. 2, pp. 383–390, 1986.

[25] G. M. Rosano, C. Vitale, G. Marazzi, and M. Volterrani, "Menopause and cardiovascular disease: the evidence," *Climacteric*, vol. 10, no. 1, pp. 19–24, 2007.

[26] J. F. Reckelhoff and C. Maric, "Editorial: sex and gender differences in cardiovascular-renal physiology and pathophysiology," *Steroids*, vol. 75, no. 11, pp. 745–746, 2010.

[27] V. Bittner, "Menopause, age, and cardiovascular risk: a complex relationship," *Journal of the American College of Cardiology*, vol. 54, no. 25, pp. 2374–2375, 2009.

[28] E. S. Kim and V. Menon, "Status of women in cardiovascular clinical trials," *Arteriosclerosis, Thrombosis, and Vascular Biology*, vol. 29, no. 3, pp. 279–283, 2009.

[29] M. Coylewright, J. F. Reckelhoff, and P. Ouyang, "Menopause and hypertension: an age-old debate," *Hypertension*, vol. 51, no. 4, pp. 952–959, 2008.

[30] S. M. Grundy, "Guidelines for cholesterol management: recommendations of the National Cholesterol Education Program's Adult Treatment Panel II," *Heart Disease and Stroke*, vol. 3, no. 3, pp. 123–127, 1994.

[31] J. S. Hong, S. W. Yi, H. C. Kang et al., "Age at menopause and cause-specific mortality in South Korean women: kangwha Cohort Study," *Maturitas*, vol. 56, no. 4, pp. 411–419, 2007.

[32] F. B. Hu, F. Grodstein, C. H. Hennekens et al., "Age at natural menopause and risk of cardiovascular disease," *Archives of Internal Medicine*, vol. 159, no. 10, pp. 1061–1066, 1999.

[33] B. L. Haddock, H. P. Hopp Marshak, J. J. Mason, and G. Blix, "The effect of hormone replacement therapy and exercise on cardiovascular disease risk factors in postmenopausal women," *Sports Medicine*, vol. 29, no. 1, pp. 39–49, 2000.

[34] K. A. Matthews, S. L. Crawford, C. U. Chae et al., "Are changes in cardiovascular disease risk factors in midlife women due to chronological aging or to the menopausal transition?" *Journal of the American College of Cardiology*, vol. 54, no. 25, pp. 2366–2373, 2009.

[35] E. Barrett-Connor, "Sex differences in coronary heart disease: why are women so superior? The 1995 Ancel Keys Lecture," *Circulation*, vol. 95, no. 1, pp. 252–264, 1997.

[36] H. S. Kok, K. M. van Asselt, Y. T. van der Schouw et al., "Heart disease risk determines menopausal age rather than the reverse," *Journal of the American College of Cardiology*, vol. 47, no. 10, pp. 1976–1983, 2006.

[37] L. Mosca, C. L. Banka, E. J. Benjamin et al., "Evidence-based guidelines for cardiovascular disease prevention in women: 2007 Update," *Journal of the American College of Cardiology*, vol. 49, no. 11, pp. 1230–1250, 2007.

[38] C. Vassalle, L. Petrozzi, N. Botto, M. G. Andreassi, and G. C. Zucchelli, "Oxidative stress and its association with coronary artery disease and different atherogenic risk factors," *Journal of Internal Medicine*, vol. 256, no. 4, pp. 308–315, 2004.

[39] E. Schwedhelm, A. Bartling, H. Lenzen et al., "Urinary 8-isoprostaglandin F2 α as a risk marker in patients with coronary heart disease: a matched case-control study," *Circulation*, vol. 109, no. 7, pp. 843–848, 2004.

[40] L. Baker, K. K. Meldrum, M. Wang et al., "The role of estrogen in cardiovascular disease," *Journal of Surgical Research*, vol. 115, no. 2, pp. 325–344, 2003.

[41] O. C. Gebara, M. A. Mittleman, P. Sutherland et al., "Association between increased estrogen status and increased fibrinolytic potential in the Framingham Offspring Study," *Circulation*, vol. 91, no. 7, pp. 1952–1958, 1995.

[42] G. A. Colditz, W. C. Willett, M. J. Stampfer et al., "Menopause and the risk of coronary heart disease in women," *New England Journal of Medicine*, vol. 316, no. 18, pp. 1105–1110, 1987.

[43] W. J. Mack, C. C. Slater, M. Xiang, D. Shoupe, R. A. Lobo, and H. N. Hodis, "Elevated subclinical atherosclerosis associated with oophorectomy is related to time since menopause rather than type of menopause," *Fertility and Sterility*, vol. 82, no. 2, pp. 391–397, 2004.

[44] A. M. Shearman, L. A. Cupples, S. Demissie et al., "Association between estrogen receptor α gene variation and cardiovascular disease," *Journal of the American Medical Association*, vol. 290, no. 17, pp. 2263–2270, 2003.

[45] F. Grodstein, M. J. Stampfer, J. E. Manson et al., "Postmenopausal estrogen and progestin use and the risk of cardiovascular disease," *New England Journal of Medicine*, vol. 335, no. 7, pp. 453–461, 1996.

[46] D. Grady, S. M. Rubin, D. B. Petitti et al., "Hormone therapy to prevent disease and prolong life in postmenopausal women," *Annals of Internal Medicine*, vol. 117, no. 12, pp. 1016–1037, 1992.

[47] T. W. Meade and A. Berra, "Hormone replacement therapy and cardiovascular disease," *British Medical Bulletin*, vol. 48, no. 2, pp. 276–308, 1992.

[48] K. M. Randell, R. J. Honkanen, H. Kröger, and S. Saarikoski, "Does hormone-replacement therapy prevent fractures in early postmenopausal women?" *Journal of Bone and Mineral Research*, vol. 17, no. 3, pp. 528–533, 2002.

[49] B. Ettinger, D. M. Black, B. H. Mitlak et al., "Reduction of vertebral fracture risk in postmenopausal women with osteoporosis treated with raloxifene: results from a 3-year randomized clinical trial. Multiple Outcomes of Raloxifene Evaluation (MORE) investigators," *Journal of the American Medical Association*, vol. 282, no. 7, pp. 637–645, 1999.

[50] B. E. Henderson, A. Paganini-Hill, and R. K. Ross, "Decreased mortality in users of estrogen replacement therapy," *Archives of Internal Medicine*, vol. 151, no. 1, pp. 75–78, 1991.

[51] T. L. Bush, E. Barrett-Connor, L. D. Cowan et al., "Cardiovascular mortality and noncontraceptive use of estrogen in women: results from the Lipid Research Clinics Program Follow-up Study," *Circulation*, vol. 75, no. 6, pp. 1102–1109, 1987.

[52] T. L. Bush, L. D. Cowan, E. Barrett Connor et al., "Estrogen use and all-cause mortality. Preliminary results from the Lipid Research Clinics Program Follow-up study," *Journal of the American Medical Association*, vol. 249, no. 7, pp. 903–906, 1983.

[53] C. Schairer, J. Lubin, R. Troisi, S. Sturgeon, L. Brinton, and R. Hoover, "Menopausal estrogen and estrogen-progestin replacement therapy and breast cancer risk," *Journal of the American Medical Association*, vol. 283, no. 4, pp. 485–491, 2000.

[54] L. Bergkvist, H. O. Adami, I. Persson, R. Hoover, and C. Schairer, "The risk of breast cancer after estrogen and estrogen-progestin replacement," *New England Journal of Medicine*, vol. 321, no. 5, pp. 293–297, 1989.

[55] J. E. Rossouw, G. L. Anderson, R. L. Prentice et al., "Risks and benefits of estrogen plus progestin in healthy postmenopausal women: principal results from the women's health initiative randomized controlled trial," *Journal of the American Medical Association*, vol. 288, no. 3, pp. 321–333, 2002.

[56] G. L. Anderson, M. Limacher, A. R. Assaf et al., "Effects of conjugated equine estrogen in postmenopausal women with hysterectomy: the Women's Health Initiative randomized controlled trial," *Journal of the American Medical Association*, vol. 291, no. 14, pp. 1701–1712, 2004.

[57] A. L. Hersh, M. L. Stefanick, and R. S. Stafford, "National use of postmenopausal hormone therapy: annual trends and response to recent evidence," *Journal of the American Medical Association*, vol. 291, no. 1, pp. 47–53, 2004.

[58] J. E. Rossouw, R. L. Prentice, J. E. Manson et al., "Postmenopausal hormone therapy and risk of cardiovascular disease by age and years since menopause," *Journal of the American Medical Association*, vol. 297, no. 13, pp. 1465–1477, 2007.

[59] S. M. Harman, E. Vittinghoff, E. A. Brinton et al., "Timing and duration of menopausal hormone treatment may affect cardiovascular outcomes," *American Journal of Medicine*, vol. 124, no. 3, pp. 199–205, 2011.

[60] L. Mosca, E. J. Benjamin, K. Berra et al., "Effectiveness-based guidelines for the prevention of cardiovascular disease in women-2011 update: a Guideline from the American Heart Association," *Circulation*, vol. 123, no. 11, pp. 1243–1262, 2011.

[61] M. Pierdominici, E. Ortona, F. Franconi, M. Caprio, E. Straface, and W. Malorni, "Gender specific aspects of cell death in the cardiovascular system," *Current Pharmaceutical Design*, vol. 17, no. 11, pp. 1046–1055, 2011.

[62] A. M. Deschamps, E. Murphy, and J. Sun, "Estrogen receptor activation and cardioprotection in ischemia reperfusion injury," *Trends in Cardiovascular Medicine*, vol. 20, no. 3, pp. 73–78, 2010.

[63] J. S. Juggi, L. J. Hoteit, F. A. Babiker, S. Joseph, and A. S. Mustafa, "Protective role of normothermic, hyperthermic and estrogen preconditioning and pretreatment on tumour necrosis factor-α-induced damage," *Experimental and Clinical Cardiology*, vol. 16, no. 2, pp. e5–e10, 2011.

[64] S. Kanki, V. F. Segers, W. Wu et al., "Stromal cell-derived factor-1 retention and cardioprotection for ischemic myocardium," *Circulation*, vol. 4, no. 4, pp. 509–518, 2011.

[65] C. Huang, H. Gu, Y. Wang, and M. Wang, "Estrogen-induced SDF-1 production is mediated by estrogen receptor-α in female hearts after acute ischemia and reperfusion," *Surgery*, vol. 150, no. 2, pp. 197–203, 2011.

[66] J. Cao, T. Zhu, L. Lu et al., "Estrogen induces cardioprotection in male C57BL/6J mice after acute myocardial infarction via decreased activity of matrix metalloproteinase-9 and increased Akt-Bcl-2 anti-apoptotic signaling," *International Journal of Molecular Medicine*, vol. 28, no. 2, pp. 231–237, 2011.

[67] L. F. Valverdea, F. D. Cedillob, M. L. Ramosa, E. G. Cerveraa, K. Quijanoa, and J. Cordobaa, "Changes induced by estradiol-ethylenediamine derivative on perfusion pressure and coronary resistance in isolated rat heart: l-type calcium channel," *Biomedical Papers*, vol. 155, no. 1, pp. 27–32, 2011.

[68] X. Hu, X. Xu, Z. Lu et al., "AMP activated protein kinase-α2 regulates expression of estrogen-related receptor-α, a metabolic transcription factor related to heart failure development," *Hypertension*, vol. 58, no. 4, pp. 696–703, 2011.

[69] M. S. Mitra, J. D. Schilling, X. Wang et al., "Cardiac lipin 1 expression is regulated by the peroxisome proliferator activated receptor γ coactivator 1α/estrogen related receptor axis," *Journal of Molecular and Cellular Cardiology*, vol. 51, no. 1, pp. 120–128, 2011.

[70] D. E. Masood, E. C. Roach, K. G. Beauregard, and R. A. Khalil, "Impact of sex hormone metabolism on the vascular effects of menopausal hormone therapy in cardiovascular disease," *Current Drug Metabolism*, vol. 11, no. 8, pp. 693–714, 2010.

[71] P. A. Arias-Loza, K. Hu, S. Frantz et al., "Medroxyprogesterone acetate aggravates oxidative stress and left ventricular dysfunction in rats with chronic myocardial infarction," *Toxicologic Pathology*, vol. 39, no. 5, pp. 867–878, 2011.

[72] G. D. Lopaschuk, J. R. Ussher, C. D. Folmes, J. S. Jaswal, and W. C. Stanley, "Myocardial fatty acid metabolism in health and disease," *Physiological Reviews*, vol. 90, no. 1, pp. 207–258, 2010.

[73] R. M. Beadle and M. Frenneaux, "Modification of myocardial substrate utilisation: a new therapeutic paradigm in cardiovascular disease," *Heart*, vol. 96, no. 11, pp. 824–830, 2010.

[74] J. Kerner and C. Hoppel, "Fatty acid import into mitochondria," *Biochimica et Biophysica Acta*, vol. 1486, no. 1, pp. 1–17, 2000.

[75] Y. G. Niu and R. D. Evans, "Very-low-density lipoprotein: complex particles in cardiac energy metabolism," *Journal of Lipid Research*, vol. 2011, Article ID 189876, 9 pages, 2011.

[76] Y. G. Niu, D. Hauton, and R. D. Evans, "Utilization of triacylglycerol-rich lipoproteins by the working rat heart: routes of uptake and metabolic fates," *Journal of Physiology*, vol. 558, no. 1, pp. 225–237, 2004.

[77] M. F. Oliver, "Control of free fatty acids during acute myocardial ischaemia," *Heart*, vol. 96, no. 23, pp. 1883–1884, 2010.

[78] A. Barsotti, A. Giannoni, P. di Napoli, and M. Emdin, "Energy metabolism in the normal and in the diabetic heart," *Current Pharmaceutical Design*, vol. 15, no. 8, pp. 836–840, 2009.

[79] G. J. van der Vusse, M. van Bilsen, J. F. Glatz, D. M. Hasselbaink, and J. J. Luiken, "Critical steps in cellular fatty acid uptake and utilization," *Molecular and Cellular Biochemistry*, vol. 239, no. 1-2, pp. 9–15, 2002.

[80] J. F. Glatz, J. J. Luiken, and A. Bonen, "Involvement of membrane-associated proteins in the acute regulation of cellular fatty acid uptake," *Journal of Molecular Neuroscience*, vol. 16, no. 2-3, pp. 123–132, 2001.

[81] J. F. Glatz, J. J. Luiken, F. A. van Nieuwenhoven, and G. J. van der Vusse, "Molecular mechanism of cellular uptake and intracellular translocation of fatty acids," *Prostaglandins Leukotrienes and Essential Fatty Acids*, vol. 57, no. 1, pp. 3–9, 1997.

[82] A. T. Turer, C. R. Malloy, C. B. Newgard, and M. V. Podgoreanu, "Energetics and metabolism in the failing heart: important but poorly understood," *Current Opinion*

in Clinical Nutrition and Metabolic Care, vol. 13, no. 4, pp. 458–465, 2010.

[83] M. E. Young, G. W. Goodwin, J. Ying et al., "Regulation of cardiac and skeletal muscle malonyl-CoA decarboxylase by fatty acids," *American Journal of Physiology: Endocrinology and Metabolism*, vol. 280, no. 3, pp. E471–E479, 2001.

[84] S. Eaton, "Control of mitochondrial β-oxidation flux," *Progress in Lipid Research*, vol. 41, no. 3, pp. 197–239, 2002.

[85] S. Grimbert, C. Fisch, D. Deschamps et al., "Effects of female sex hormones on mitochondria: possible role in acute fatty liver of pregnancy," *American Journal of Physiology*, vol. 268, no. 1, pp. G107–G115, 1995.

[86] R. Sladek, J. A. Bader, and V. Giguère, "The orphan nuclear receptor estrogen-related receptor or α is a transcriptional regulator of the human medium-cha n Acyl coenzyme A dehydrogenase gene," *Molecular and Cellular Biology*, vol. 17, no. 9, pp. 5400–5409, 1997.

[87] L. O. Li, E. L. Klett, and R. A. Coleman, "Acyl-CoA synthesis, lipid metabolism and lipotoxicity," *Biochimica et Biophysica Acta*, vol. 1801, no. 3, pp. 246–251, 2010.

[88] N. A. van Herpen and V. B. Schrauwen-Hinderling, "Lipid accumulation in non-adipose tissue and lipotoxicity," *Physiology and Behavior*, vol. 94, no. 2, pp. 231–241, 2008.

[89] L. S. Szczepaniak, R. G. Victor, L. Orci, and R. H. Unger, "Forgotten but not gone: the rediscovery of fatty heart, the most common unrecognized disease in America," *Circulation Research*, vol. 101, no. 8, pp. 759–767, 2007.

[90] N. M. Borradaile and J. E. Schaffer, "Lipotoxicity in the heart," *Current Hypertension Reports*, vol. 7, no. 6, pp. 412–417, 2005.

[91] L. Zhang, J. R. Ussher, T. Oka, V. J. Cadete, C. Wagg, and G. D. Lopaschuk, "Cardiac diacylglycerol accumulation in high fat-fed mice is associated with impaired insulin-stimulated glucose oxidation," *Cardiovascular Research*, vol. 89, no. 1, pp. 148–156, 2011.

[92] I. Chowdhury, A. Branch, M. Olatinwo, K. Thomas, R. Matthews, and W. E. Thompson, "Prohibitin (PHB) acts as a potent survival factor against ceramide induced apoptosis in rat granulosa cells," *Life Sciences*, vol. 89, no. 9-10, pp. 295–303, 2011.

[93] T. D. Mullen and L. M. Obeid, "Ceramide and apoptosis: exploring the enigmatic connections between sphingolipid metabolism and programmed cell death," *Anti-Cancer Agents in Medicinal Chemistry*. In press.

[94] H. Lee, J. A. Rotolo, J. Mesicek et al., "Mitochondrial ceramide-rich macrodomains functionalize bax upon irradiation," *PLoS ONE*, vol. 6, no. 6, article e19783, 2011.

[95] E. Usta, M. Mustafi, F. Artunc et al., "The challenge to verify ceramide's role of apoptosis induction in human cardiomyocytes—a pilot study," *Journal of Cardiothoracic Surgery*, vol. 6, no. 1, article 38, 2011.

[96] S. A. Novgorodov, Z. M. Szulc, C. Luberto et al., "Positively charged ceramide is a potent inducer of mitochondrial permeabilization," *Journal of Biological Chemistry*, vol. 280, no. 16, pp. 16096–16105, 2005.

[97] M. di Paola, P. Zaccagnino, G. Montedoro, T. Cocco, and M. Lorusso, "Ceramide induces release of pro-apoptotic proteins from mitochondria by either a Ca^{2+}-dependent or a Ca^{2+}-independent mechanism," *Journal of Bioenergetics and Biomembranes*, vol. 36, no. 2, pp. 165–170, 2004.

[98] K. W. Kinnally, P. M. Peixoto, S.-Y. Ryu, and L. M. Dejean, "Is mPTP the gatekeeper for necrosis, apoptosis, or both?" *Biochimica et Biophysica Acta*, vol. 1813, no. 4, pp. 616–622, 2011.

[99] S. A. Novgorodov, T. I. Gudz, and L. M. Obeid, "Long-chain ceramide is a potent inhibitor of the mitochondrial permeability transition pore," *Journal of Biological Chemistry*, vol. 283, no. 36, pp. 24707–24717, 2008.

[100] M. A. Cole, A. J. Murray, L. E. Cochlin et al., "A high fat diet increases mitochondrial fatty acid oxidation and uncoupling to decrease efficiency in rat heart," *Basic Research in Cardiology*, vol. 106, no. 3, pp. 447–457, 2011.

[101] N. Li, J. Wang, F. Gao, Y. Tian, R. Song, and S.-J. Zhu, "The role of uncoupling protein 2 in the apoptosis induced by free fatty acid in rat cardiomyocytes," *Journal of Cardiovascular Pharmacology*, vol. 55, no. 2, pp. 161–167, 2010.

[102] L. H. Opie and J. Knuuti, "The adrenergic-fatty acid load in heart failure," *Journal of the American College of Cardiology*, vol. 54, no. 18, pp. 1637–1646, 2009.

[103] J. S. Jaswal, W. Keung, W. Wang, J. R. Ussher, and G. D. Lopaschuk, "Targeting fatty acid and carbohydrate oxidation—a novel therapeutic intervention in the ischemic and failing heart," *Biochimica et Biophysica Acta*, vol. 1813, no. 7, pp. 1333–1350, 2011.

[104] M. J. Romero-Aleshire, M. K. Diamond-Stanic, A. H. Hasty, P. B. Hoyer, and H. L. Brooks, "Loss of ovarian function in the VCD mouse-model of menopause leads to insulin resistance and a rapid progression into the metabolic syndrome," *American Journal of Physiology: Regulatory Integrative and Comparative Physiology*, vol. 297, no. 3, pp. R587–R592, 2009.

[105] F. Pansini, G. Bonaccorsi, F. Genovesi et al., "Influence of estrogens on serum free fatty acid levels in women," *Journal of Clinical Endocrinology and Metabolism*, vol. 71, no. 5, pp. 1387–1389, 1990.

[106] C. A. Derby, S. L. Crawford, R. C. Pasternak et al., "Lipid changes during the menopause transition in relation to age and weight: the Study of Women's Health Across the Nation," *American Journal of Epidemiology*, vol. 169, no. 11, pp. 1352–1361, 2009.

[107] M. L. Power and J. Schulkin, "Sex differences in fat storage, fat metabolism, and the health risks from obesity: possible evolutionary origins," *British Journal of Nutrition*, vol. 99, no. 5, pp. 931–940, 2008.

[108] T. M. D'Eon, S. C. Souza, M. Aronovitz, M. S. Obin, S. K. Fried, and A. S. Greenberg, "Estrogen regulation of adiposity and fuel partitioning: evidence of genomic and non-genomic regulation of lipogenic and oxidative pathways," *Journal of Biological Chemistry*, vol. 280, no. 43, pp. 35983–35991, 2005.

[109] C. M. Williams, "Lipid metabolism in women," *Proceedings of the Nutrition Society*, vol. 63, no. 1, pp. 153–160, 2004.

[110] A. K. Wong, J. Howie, J. R. Petrie, and C. C. Lang, "AMP-activated protein kinase pathway: a potential therapeutic target in cardiometabolic disease," *Clinical Science*, vol. 116, no. 8, pp. 607–620, 2009.

[111] C.-M. Liou, A.-L. Yang, C.-H. Kuo, H. Tin, C.-Y. Huang, and S.-D. Lee, "Effects of 17β-estradiol on cardiac apoptosis in ovariectomized rats," *Cell Biochemistry and Function*, vol. 28, no. 6, pp. 521–528, 2010.

[112] S. D. Lee, W. W. Kuo, Y. J. Ho et al., "Cardiac Fas-dependent and mitochondria-dependent apoptosis in ovariectomized rats," *Maturitas*, vol. 61, no. 3, pp. 268–277, 2008.

[113] W.-C. Liang and I. Nishino, "State of the art in muscle lipid diseases," *Acta Myologica*, vol. 29, no. 2, pp. 351–356, 2010.

[114] M. J. Bennett, "Pathophysiology of fatty acid oxidation disorders," *Journal of Inherited Metabolic Disease*, vol. 33, no. 5, pp. 533–537, 2010.

[115] K. Begriche, J. Massart, M.-A. Robin, A. Borgne-Sanchez, and B. Fromenty, "Drug-induced toxicity on mitochondria

and lipid metabolism: mechanistic diversity and deleterious consequences for the liver," *Journal of Hepatology*, vol. 54, no. 4, pp. 773–794, 2011.

[116] J. A. Bjork, J. L. Butenhoff, and K. B. Wallace, "Multiplicity of nuclear receptor activation by PFOA and PFOS in primary human and rodent hepatocytes," *Toxicology*, vol. 288, no. 1–3, pp. 8–17, 2011.

[117] E. Freneaux, G. Labbe, P. Letteron et al., "Inhibition of the mitochondrial oxidation of fatty acids by tetracycline in mice and in man: possible role in microvesicular steatosis induced by this antibiotic," *Hepatology*, vol. 8, no. 5, pp. 1056–1062, 1988.

[118] D. J. Glenn, F. Wang, M. Nishimoto et al., "A murine model of isolated cardiac steatosis leads to cardiomyopathy," *Hypertension*, vol. 57, no. 2, pp. 216–222, 2011.

[119] G. Petrosillo, M. Matera, N. Moro, F. M. Ruggiero, and G. Paradies, "Mitochondrial complex I dysfunction in rat heart with aging: critical role of reactive oxygen species and cardiolipin," *Free Radical Biology and Medicine*, vol. 46, no. 1, pp. 88–94, 2009.

[120] E. J. Lesnefsky and C. L. Hoppel, "Cardiolipin as an oxidative target in cardiac mitochondria in the aged rat," *Biochimica et Biophysica Acta*, vol. 1777, no. 7-8, pp. 1020–1027, 2008.

[121] H. J. Lee, J. Mayette, S. I. Rapoport, and R. P. Bazinet, "Selective remodeling of cardiolipin fatty acids in the aged rat heart," *Lipids in Health and Disease*, vol. 5, article 2, 2006.

[122] S. Judge, Y. M. Jang, A. Smith, T. Hagen, and C. Leeuwenburgh, "Age-associated increases in oxidative stress and antioxidant enzyme activities in cardiac interfibrillar mitochondria: implications for the mitochondrial theory of aging," *FASEB Journal*, vol. 19, no. 3, pp. 419–421, 2005.

[123] S. W. Fannin, E. J. Lesnefsky, T. J. Slabe, M. O. Hassan, and C. L. Hoppel, "Aging selectively decreases oxidative capacity in rat heart interfibrillar mitochondria," *Archives of Biochemistry and Biophysics*, vol. 372, no. 2, pp. 399–407, 1999.

[124] C. C. Preston, A. S. Oberlin, E. L. Holmuhamedov et al., "Aging-induced alterations in gene transcripts and functional activity of mitochondrial oxidative phosphorylation complexes in the heart," *Mechanisms of Ageing and Development*, vol. 129, no. 6, pp. 304–312, 2008.

[125] J. Marín-García, Y. Pi, and M. J. Goldenthal, "Mitochondrial-nuclear cross-talk in the aging and failing heart," *Cardiovascular Drugs and Therapy*, vol. 20, no. 6, pp. 477–491, 2006.

[126] S. A. Mohamed, T. Hanke, A. W. Erasmi et al., "Mitochondrial DNA deletions and the aging heart," *Experimental Gerontology*, vol. 41, no. 5, pp. 508–517, 2006.

[127] M. R. Meyer, D. J. Clegg, E. R. Prossnitz, and M. Barton, "Obesity, insulin resistance and diabetes: sex differences and role of oestrogen receptors," *Acta Physiologica*, vol. 203, no. 1, pp. 259–269, 2011.

[128] B. Niemann, Y. Chen, M. Teschner, L. Li, R.-E. Silber, and S. Rohrbach, "Obesity induces signs of premature cardiac aging in younger patients: the role of mitochondria," *Journal of the American College of Cardiology*, vol. 57, no. 5, pp. 577–585, 2011.

[129] J. G. Duncan, "Mitochondrial dysfunction in diabetic cardiomyopathy," *Biochimica et Biophysica Acta*, vol. 1813, no. 7, pp. 1351–1359, 2011.

[130] P. J. Oliveira, "Cardiac mitochondrial alterations observed in hyperglycaemic rats—what can we learn from cell biology?" *Current Diabetes Reviews*, vol. 1, no. 1, pp. 11–21, 2005.

[131] J. Suski, M. Lebiedzinska, N. G. Machado et al., "Mitochondrial tolerance to drugs and toxic agents in ageing and

disease," *Current Drug Targets*, vol. 12, no. 6, pp. 827–849, 2011.

[132] V. A. Sardão, S. L. Pereira, and P. J. Oliveira, "Drug-induced mitochondrial dysfunction in cardiac and skeletal muscle injury," *Expert Opinion on Drug Safety*, vol. 7, no. 2, pp. 129–146, 2008.

[133] A. Ascensão, J. Lumini-Oliveira, P. J. Oliveira, and J. Magalhães, "Mitochondria as a target for exercise-induced cardioprotection," *Current Drug Targets*, vol. 12, no. 6, pp. 860–871, 2011.

[134] J. Lumini-Oliveira, J. Magalhães, C. V. Pereira, A. C. Moreira, P. J. Oliveira, and A. Ascensão, "Endurance training reverts heart mitochondrial dysfunction, permeability transition and apoptotic signaling in long-term severe hyperglycemia," *Mitochondrion*, vol. 11, no. 1, pp. 54–63, 2011.

[135] M. L. Johnson, Z. Zarins, J. A. Fattor et al., "Twelve weeks of endurance training increases FFA mobilization and reesterification in postmenopausal women," *Journal of Applied Physiology*, vol. 109, no. 6, pp. 1573–1581, 2010.

[136] Z. A. Zarins, G. A. Wallis, N. Faghihnia et al., "Effects of endurance training on cardiorespiratory fitness and substrate partitioning in postmenopausal women," *Metabolism*, vol. 58, no. 9, pp. 1338–1346, 2009.

[137] A. M. Moustafa and V. Boshra, "The possible role of L-carnitine on the skeletal muscle of ovariectomized rats," *Journal of Molecular Histology*, vol. 42, no. 3, pp. 217–225, 2011.

[138] I. Baeza, J. Fdez-Tresguerres, C. Ariznavarreta, and M. De La Fuente, "Effects of growth hormone, melatonin, oestrogens and phytoestrogens on the oxidized glutathione (GSSG)/reduced glutathione (GSH) ratio and lipid peroxidation in aged ovariectomized rats," *Biogerontology*, vol. 11, no. 6, pp. 687–701, 2010.

[139] R. E. White, R. Gerrity, S. A. Barman, and G. Han, "Estrogen and oxidative stress: a novel mechanism that may increase the risk for cardiovascular disease in women," *Steroids*, vol. 75, no. 11, pp. 788–793, 2010.

[140] J.-F. Arnal, P.-Y. Scarabin, F. Trémollières, H. Laurell, and P. Gourdy, "Estrogens in vascular biology and disease: where do we stand today?" *Current Opinion in Lipidology*, vol. 18, no. 5, pp. 554–560, 2007.

[141] T. L. Zern, R. J. Wood, C. Greene et al., "Grape polyphenols exert a cardioprotective effect in pre- and postmenopausal women by lowering plasma lipids and reducing oxidative stress," *Journal of Nutrition*, vol. 135, no. 8, pp. 1911–1917, 2005.

[142] M. Sun, F. Qian, W. Shen et al., "Mitochondrial nutrients stimulate performance and mitochondrial biogenesis in exhaustively exercised rats," *Scandinavian Journal of Medicine and Science in Sports*. In press.

[143] A. Biala, E. Tauriainen, A. Siltanen et al., "Resveratrol induces mitochondrial biogenesis and ameliorates Ang II-induced cardiac remodeling in transgenic rats harboring human renin and angiotensinogen genes," *Blood Pressure*, vol. 19, no. 3, pp. 196–205, 2010.

[144] G. Szabó, "A glass of red wine to improve mitochondrial biogenesis? Novel mechanisms of resveratrol," *American Journal of Physiology: Heart and Circulatory Physiology*, vol. 297, no. 1, pp. H8–H9, 2009.

15

A Nonradioactive Fluorimetric SPE-Based Ceramide Kinase Assay Using NBD-C$_6$-Ceramide

Helena Van Overloop, Gerd Van der Hoeven, and Paul P. Van Veldhoven

Department Cellular and Molecular Medicine, Katholieke Universiteit Leuven, Campus Gasthuisberg O&N1, LIPIT, Herestraat, Box 601, 3000 Leuven, Belgium

Correspondence should be addressed to Paul P. Van Veldhoven, Paul.VanVeldhoven@med.kuleuven.be

Academic Editor: Philip W. Wertz

Ceramide kinase (CERK) has been implicated in important cellular processes such as inflammation and apoptosis. Its activity is usually measured using radiolabeled ceramide or [γ-^{32}P]-ATP, followed by extraction, thin-layer chromatography, and detection of the formed labeled ceramide-1-phosphate. To eliminate the use of radioactivity, we developed similarly but independently from the approach by Don and Rosen (2008), a fluorescence-based ceramide kinase assay, using N-[7-(4-nitrobenz-2-oxa-1,3-diazole)]-6-aminohexanoyl-sphingenine (NBD-C$_6$-ceramide) as substrate. Its K_m value (4 μM) was comparable to that of N-hexanoyl-sphingenine (C$_6$-ceramide). The produced fluorescent NBD-C$_6$-ceramide-1-phosphate was captured by means of solid-phase extraction on an aminopropyl phase, resulting in a fast and sensitive CERK measurement. By performing this assay in a 96-well format, it is also suitable for high-throughput screening (HTS) to search for CERK modulators. A limited screen revealed that some protein kinase inhibitors (e.g., U-0126; IC$_{50}$ 4 μM) and ceramide analogues (e.g., fenretinide, AMG-9810; IC$_{50}$ 1.1 μM) affect CERK *in vitro*.

1. Introduction

Ceramide, the initial product of the sphingomyelin cycle, functions as a key component in the regulation of various cellular functions like differentiation, proliferation, apoptosis, and inflammation [1, 2]. In the past years, ceramide-1-phosphate (Cer-1-P), a metabolite of ceramide and sphingomyelin, also gained more attention and turned out to be a powerful bioactive sphingolipid. Cer-1-P has been implicated as a regulator of different cellular processes, like mitosis, apoptosis, phagocytosis, and inflammation [3, 4]. In mice lacking CERK, neutrophil homeostasis is defective leading to more severe disease upon pulmonary infection [5]. Until now, CERK is the only mammalian enzyme known to phosphorylate ceramide, but the residual Cer-1-P levels in tissues of the CERK-deficient mice [5], indicate the existence of alternative pathways to generate Cer-1-P. CERK was first described in 1989 as a Ca^{2+}-dependent lipid kinase [6] and cloned in 2002 by Sugiura et al. [7], based on similarity to sphingosine kinase. A similar cloning strategy was followed by others [8]. CERK appears to associate with (endo)membranes via a pleckstrin domain [9, 10] and is highly selective for the D-*erythro* configuration of ceramides [8, 11].

CERK activity is commonly determined using a radioactivity-based assay [7–11] based on [γ-^{32}P]ATP. After extraction of lipids, the amount of radiolabeled ceramide-1-phosphate in the organic phase is determined by directly counting [12] or by TLC followed by autoradiography and quantitation [7, 8]. To avoid the use of radioactivity, we intended to develop a fluorescence-based CERK assay, sensitive enough to be employed for cellular work and whose format would be suitable or could be adapted to search for CERK inhibitors via HTS. Based on previous work, showing that truncated ceramides such as C$_2$-ceramide (N-acetyl-sphingenine) and C$_6$-ceramide (N-hexanoyl-sphingenine), when presented bound to albumin, are well recognized by human CERK [8, 13], we tested a fluorescent analogue, namely, N-[7-(4-nitrobenz-2-oxa-1,3-diazole)]-6-aminohexanoyl-sphingenine (NBD-C$_6$-ceramide, NBD-C$_6$-Cer), in which NBD is coupled to sphingenine via an 6-aminohexanoic acid linker (see Figure 1).

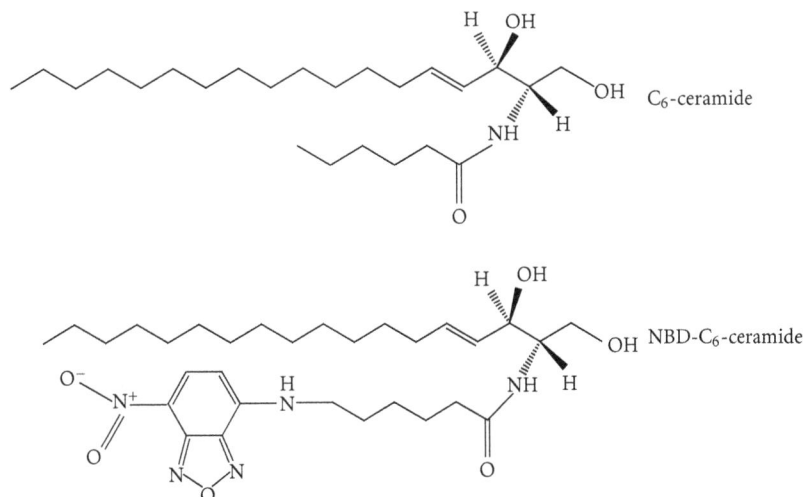

FIGURE 1: Structure of C$_6$-ceramide and NBD-C$_6$-ceramide.

This lipid was introduced several years ago by Pagano and coworkers for the study of sphingolipid metabolism and shown to be metabolized in a similar way as ceramide, being incorporated into NBD-sphingomyelin and NBD-cerebrosides [14]. During these and subsequent studies its phosphorylation, as far as we are aware of, was never detected or described, likely due to the very low activity of CERK compared to the other ceramide utilizing pathways. Also when using ceramides containing a shorter N-acyl chain (truncated ceramides), phosphorylation by intact cells is difficult to reveal, requiring ^{32}P-uploading of the cells as reported for neutrophils [15, 16], cerebellar granule cells, [17] and Hela cells [18]. Upon overexpression of CERK, detection of intracellular formed truncated ceramide-1-^{32}P is facilitated [8, 13].

Here, we studied the kinetics, revealing that NBD-C$_6$-Cer is a good substrate for CERK, both *in vitro* and *in vivo*, and developed a simple solid phase extraction scheme to measure CERK activity (this work was presented in a preliminary form at the LKI Oncoforum meeting, February 15, 2008, Leuven (Belgium) (H. Van Overloop and P. P. Van Veldhoven, Development of HTS-assays for enzymes acting on bioactive sphingolipids, new players in chemotherapy resistance. Part I. ceramide kinase)). At the start of this work, Graf et al. [19] reported that NBD-C$_6$-Cer is phosphorylated when given to CERK-expressing Cos-1 cells, as revealed by TLC of cellular extracts and scanning. Independently, Don and Rosen [20] have described in the meantime a CERK assay using the same substrate but based on liquid/liquid extraction, phase separation, and transfer of the upper phase for analysis.

2. Materials and Methods

2.1. Expression of Lipid Kinases. Recombinant *Hs*CERK was expressed in Top10F' *E. coli* cells transformed with plasmid pPVV072, coding for a (His)$_6$-tagged fusion of *Hs*CERK, as described before [8]. The harvested bacteria, resuspended in

PBS (25 mL/100 mL culture) containing a mix of protease inhibitors, were sonicated on ice (Branson Sonifier B115, microtip), followed by a clearing step (10,000 g for 10 min). Aliquots of the supernatant were frozen in liquid nitrogen, stored at −80°C and diluted 1/15 in PBS containing protease inhibitors before use. Compared to the pelleted fraction [8], the specific activity of the soluble lysate fraction is three fold lower (9.3 nmol/min·mg protein at 100 μM C$_6$-Cer/40 μM BSA for the batch used in these experiments), but it was considered to more compatible with SPE work up.

Recombinant human sphingosine kinase 1 (*Hs*SphK1) was obtained from Top10F' cells transformed with plasmid pSG003 as described before [21].

2.2. Synthesis of NBD-Hexanoyl Derivatives. N-[7-(4-nitro-benz-2-oxa-1,3-diazole)]-6-aminohexanoic acid (33 μmol, Molecular Probes), dissolved in 3 mL anhydrous dimethylformamide and activated with carbonyldiimidazole (40 μmol, Fluka), was mixed with *D,erythro*-sphingenine (33 μmol; Acros Organics), dissolved in 2 mL dimethylformamide, and stirred overnight at room temperature. After drying the reaction mixture, the amides were phase-separated in chloroform/methanol/water (1/1/0.9), the lower phase was dried, dissolved in 0.5 mL 33% methylamine in ethanol/water (7/3), and heated to 70°C for 90 min [22] to remove any formed O-acylated products. After evaporating the hydrolysis mixture, NBD-C$_6$-Cer was extracted and further purified by preparative TLC (silica G60; Merck) in solvent A (chloroform/methanol/acetic acid, 93/7/1, v/v). Stock solutions were standardized by nitrogen determination (yield 63%) and purity, based on fluoroscanning (Storm 840 with blue LED (450 nm), GE Healthcare) after TLC separation in solvent B (chloroform/acetone/methanol/acetic acid/water, 10/4/3/2/1, v/v) was 89%, substantially better than that of commercially obtained NBD-C$_6$-Cer (Sigma; Avanti Polar Lipids).

NBD-C_6-ceramide-1-phosphate (NBD-C_6-Cer-1-P) was prepared by phosphorylation of homemade NBD-Cer using bacterially expressed HsCERK under the conditions described before [8]. Briefly, the reaction mixture containing 100 μM NBD-C_6-Cer, solubilized with ethanol/BSA, was incubated with excess recombinant HsCERK for 1 h at 37°C. After acidic phase separation, the lower phase was dried and the phosphorylated NBD-C_6-Cer was further purified by preparative TLC (silica G60; Merck) in solvent B, followed by elution with chloroform/methanol/water (5/5/1, v/v). Stock solutions were standardized by organic phosphate content (yield 85%). Based on fluoroscanning of the phosphate ester, TLC-separated in solvent B, purity was estimated at >82% (based on main spot; due some streaking, actual purity is higher), being slightly better compared to the phosphate ester prepared from commercially obtained NBD-C_6-ceramide (Avanti Polar Lipids, >75%).

2.3. Separation of NBD-C_6-Cer and NBD-C_6-Cer-1-P via Solid-Phase Extraction (SPE). To document the separation of NBD-labeled CERK substrate and product, 100 μL CERK assay mixture containing NBD-C_6-Cer and NBD-C_6-Cer-1-P (both at 5 μM final concentration) was mixed with 300 μL methanol and applied on a 25 mg NH$_2$-SPE column (Varian), which had been conditioned with methanol and water. SPE devices were washed and bound lipids were eluted as described further. Total fluorescence of the flow-through and eluted fractions was measured by fluorimetry (λ_{ex} 465 nm; λ_{em} 535 nm; Tecan Infinite 200). The amount of NBD-C_6-Cer and NBD-C_6-Cer-1-P in the different fractions was estimated by drying them, redissolving the residu in chloroform/methanol (1/1, v/v), followed by separation on silica G TLC (solvent B), and scanning of the fluorescent spots.

2.4. Ceramide Kinase Measurements. To prepare the reaction mixture for the fluorescent CERK assay, largely based on previous work [8], NBD-C_6-Cer was dissolved in ethanol and mixed with 4 volumes of BSA (resulting in a molar ceramide/BSA ratio of 2.5), followed by addition of reaction mixture up to 75 μL and a 25 μL aliquot of recombinant CERK or cell lysate. Final concentrations were 5 μM NBD-C_6-Cer–1 mM ATP–50 mM Mops/NaOH pH 7.2–3 mM MgCl$_2$–40 mM NaF–1 mM dithriothreitol–100 μM orthovanadate. After 10 min at 37°C, the reaction was stopped by addition of 300 μL methanol. The mixture was applied to an NH$_2$-column (25 mg, Chromabond Multi-96, Macherey-Nagel), which had been activated with 0.5 mL methanol followed by 0.5 mL water. After washing the column with 800 μL methanol containing 2% formic acid, followed by 100 μL methanol containing 0.5 M trifluoroacetic acid (TFA), NBD-C_6-Cer-1-P was eluted into a black FIA 96-well plate (Greiner) with 250 μL methanol containing 3 M TFA. Fluorescence was measured in a multi-reader (λ_{ex} 465 nm, λ_{em} 535 nm; Tecan Infinite 200). The amount of NBD-C_6-Cer-1-P in the eluate was calculated based on a calibration curve with NBD-C_6-Cer-1-P at concentrations of 0 to 5 μM treated simultaneously and equivalently with the samples.

Fluorescence measurements from calibration curves fitted to a linear equation ($R^2 = 0.998$; $n = 12$).

CERK activity based on radioactivity was measured as described before [8], but using 1 mM [γ-^{32}P]-ATP (GE Healthcare) and reducing the volumes to obtain similar conditions as in the fluorescent assay described above. As substrates C_6-ceramide (40 μM) or NBD-C_6-Cer (5 μM), both bound to BSA (molar ceramide/BSA ratio = 2.5), were used.

3. Results and Discussion

The substrate specificity of CERK, documented by different groups [7, 8, 11], reveals that the N-acyl chain can be shortened up to two carbons [7, 8, 13] and that the presence of a bulky group in this chain is tolerated (unpublished data). Ceramide analogues with a shortened base are also phosphorylated [8]. Previously, we documented that natural ceramides (with a long N-acyl chain) are better recognized by CERK when presented in a micellar form, whereas less hydrophobic ceramide analogues, either with a shortened base or a truncated acyl chain, display better activity in the presence of BSA [8, 13]. In agreement with this observation, NBD-C_6-Cer, containing the polar NBD moiety, was well phosphorylated when bound to BSA, but substantially less when incorporated in octylglucoside/cardiolipin micelles, certainly at low substrate concentration (30 fold, less at 25 μM). Similarly, use of Triton X-100 [11] or CHAPS micelles resulted in low activities (data not shown). Using bacterially expressed human CERK in a radiometric assay and separation of the products by TLC, a K_m of 6 μM was obtained for NBD-C_6-Cer, bound to BSA (data not shown). Despite the bulky NBD-group, this value is about 2.5-fold lower than the K_m for the nonfluorescent C_6-ceramide (14 μM), obtained under the same assay conditions. A similar K_m, but based on fluorimetry, was obtained by Rosen and Don (1 μM in presence of 7 μM BSA), who also documented the inhibitory action of Triton X-100 [20]. The K_m for ATP was 168 μM (Figure 2(b)).

Given the low K_m, it appeared justified to evaluate some procedures to separate NBD-C_6-Cer from NBD-C_6-Cer-1-P in order to develop a nonradioactive CERK assay. In addition, we attempted to avoid tedious liquid extraction steps and aimed for a procedure compatible with HTS. Hence, to separate the fluorescent substrate and product, after some trials on reversed phase (NBD-C_6-Cer and its phosphate ester were both retained on C18-SPE (Varian) at 50% methanol, without or with addition of acid (0.5% (v/v) TFA) or base (0.5% (v/v) triethylamine), but coeluted with pure methanol; similar findings were obtained with Oasis-HLB (Waters) columns, except that NBD-C_6-Cer-1-P was recovered in the flow through under alkaline conditions.) and ion exchange-SPE, we focused finally on NH$_2$-SPE columns. These systems have been employed to separate phospholipids, and lipids containing a primary phosphate group, such as phosphatidate and phosphorylated phosphoinositides, do bind strongly [16, 17]. Elution of these phosphate esters is achieved by increasing the polarity of

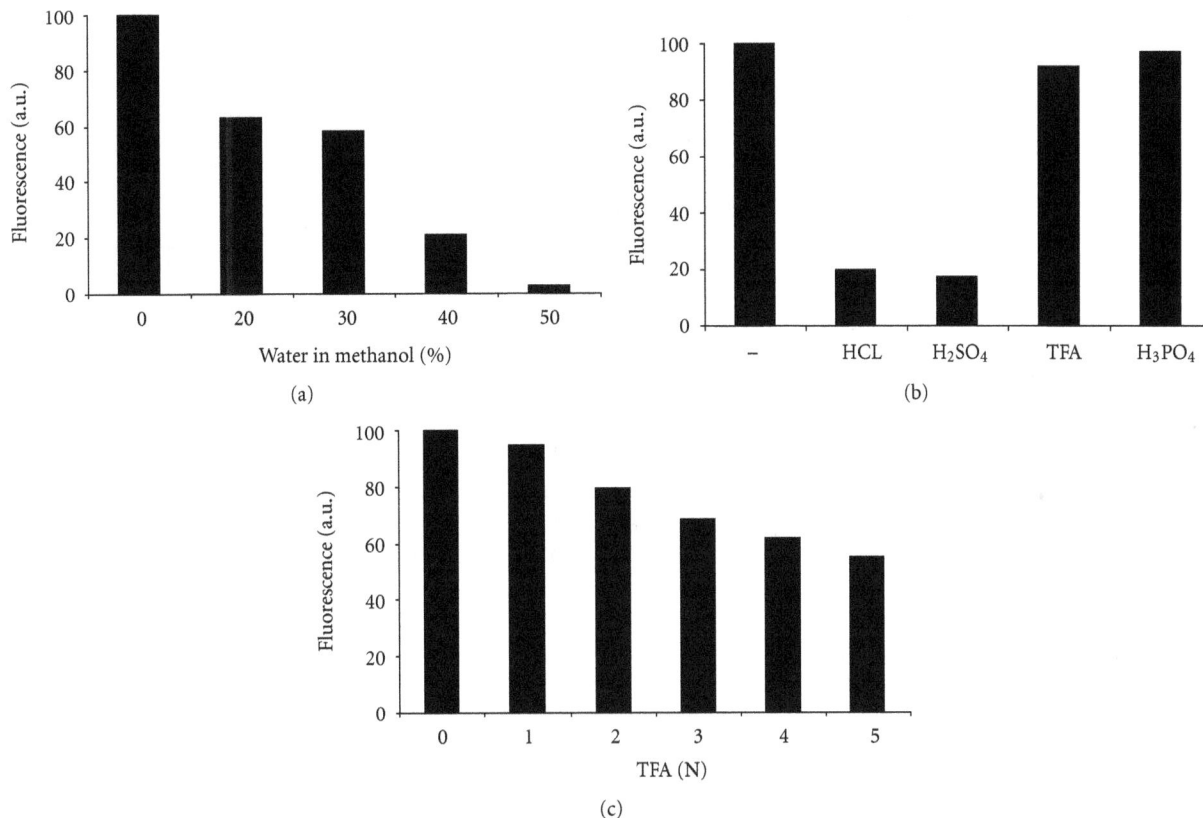

FIGURE 2: Influence of water and acidity on fluorescence of NBD-derivatives. Fluorescence of methanolic NBD-C_6-Cer-1-P solutions (1 μM, 200 μL) containing increasing amounts of water (a) different acids at 0.5 N final concentration (b) or increasing TFA concentration (c) (λ_{ex} 465 nm, λ_{em} 535 nm; Tecan Infinite 200).

the eluting solvent and addition of strong acids such as phosphoric acid [23] or HCl [24].

Given the reported dependence of the fluorescence of NBD-derivatives with regard to the lipid environment [25], solvent composition, polarity and pH [25–27] and low quantum yield in water [25], in a first step the fluorescence of NBD-C_6-Cer and its phosphate ester was evaluated in solvents and in the presence of acids. For solvent we focused on the use of methanol, being suitable for dissolution of these lipids, compatible with plastic (e.g., polystyrene), and having a low toxicity and moderate cost. As shown in Figure 2(a), increasing the amount of water in a methanolic NBD-C_6-Cer-1-P solution, strongly reduced the fluorescence. Compared to pure methanol as solvent, fluorescence dropped to 3% in a solution containing 50% of water. The influence of different acids, present at 0.5 N final concentration, on the fluorescence of NBD-C_6-Cer-1-P, dissolved in methanol, is shown in Figure 2(b). Both HCl and H_2SO_4 caused a severe drop in fluorescence (less than 20% compared to the neutral methanolic solution). The influence of TFA and H_3PO_4 was less drastic, since more than 90% of fluorescent signal remained (Figure 2(b)). This effect seems to be related to the strength of the acid, although it is not strictly linearly related to the acid dissociation constant.

When applied in pure methanol or methanol containing up to 25% water, NBD-C_6-Cer-1-P was quantitatively retained on NH_2-SPE systems. Both TFA and H_3PO_4 displaced NBD-C_6-Cer-1-P from the NH_2-phase, but TFA was chosen for further optimisation. Less volume was required to elute the phosphate ester compared to methanol containing H_3PO_4 in equal normality, and the decrease in NBD fluorescence at increasing TFA concentrations was rather small (Figure 2(c)), both factors improving the sensitivity of the assay. By increasing the TFA concentration to 3 N, it was possible to elute the bound lipid in a small volume, 250 μL when using 25 mg SPE.

Finally, we analyzed how well NBD-C_6-Cer and its phosphate ester could be separated using small SPE systems (25 mg NH_2-SPE, column format or contained in a 96-well format). Hereto, a mixture containing NBD-C_6-Cer and NBD-C_6-Cer-1-P and with a similar composition as the CERK assay mixture, was diluted with methanol (75% final concentration), and applied to the NH_2-SPE column, which had been conditioned with methanol and water. After washing the column with methanol containing 2% formic acid (FA) (800 μL) and 0.5 N TFA (100 μL), NBD-C_6-Cer-1-P was eluted from the column using methanol containing 3 N TFA (250 μL). This resulted in a good separation between NBD-C_6-Cer and NBD-C_6-Cer-1-P (If employing a larger SPE format, adjust volumes accordingly; e.g. for 100 mg NH_2-SPE columns, NBD-C_6-Cer-P is eluted with a similar yield with 1 ml 3 N TFA in methanol (data not shown)).

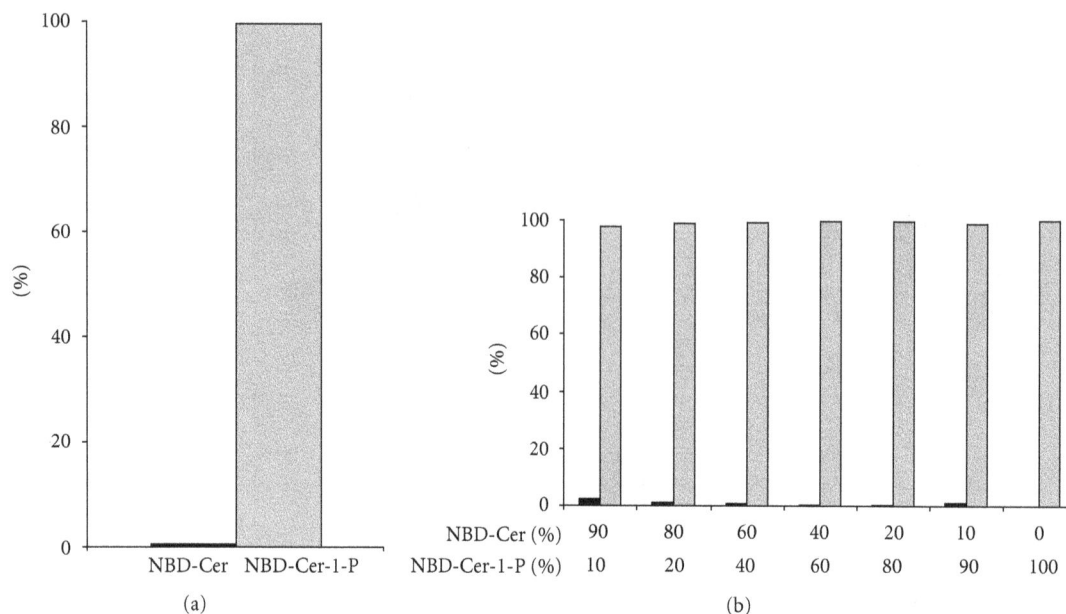

FIGURE 3: Separation of NBD-C$_6$-Cer and NBD-C$_6$-Cer-1-P via NH$_2$-SPE. A mixture containing NBD-C$_6$-Cer and NBD-C$_6$-Cer-1-P, both at 5 μM (a) or at a total concentration of 5 μM but with a variable ratio (b) in the assay medium was separated via NH$_2$-SPE, as described in Section 2. The eluted fractions were dried, resolubilized in chloroform/methanol (1/2, v/v) and the lipids were separated on silica G TLC plates (chloroform/acetone/methanol/acetic acid/water, 10/4/3/2/1, v/v), followed by fluorescence scanning of the spots (NBD-C$_6$-Cer (black bars); NBD-C$_6$-Cer-1-P (grey bars).) The result is expressed as percentage of total fluorescence in the elution fraction (a) mean $\pm$ SEM; n = 5; (b) single experiment).

When loading equal amounts of both lipids, more than 99% of NBD-C$_6$-Cer was present in the flow-through and wash fractions (99.6 $\pm$ 0.01%, mean $\pm$ SEM, n = 5), and less than 0.4% was found in the acidic eluate (0.36 $\pm$ 0.02%), based upon fluorescence-scanning of the TLC-separated fractions (data not shown). Fluorescence in the TFA-eluate was almost completely (99.4 $\pm$ 0.08%) associated with NBD-C$_6$-Cer-1-P (Figure 3(a)). Apparently, these numbers do not change when varying the relative amounts of both fluorescent lipids initially present (Figure 3(b)): about 96% of the total amount of NBD-C$_6$-Cer-1-P is present in the eluted fraction (95.6 $\pm$ 0.42%). The use of TFA at lower normality than 3 N resulted in a lower recovery for NBD-C$_6$-Cer-1-P (data not shown). Higher normalities did not improve the recoveries, but resulted in lower sensitivity because of increased quenching.

Having established optimum SPE-separation conditions, the kinetics of CERK were reevaluated with the new assay. The assay conditions were similar to the traditional radio-metric assay [8], but assay volume was reduced to 100 μL and ATP concentration was fixed at 1 mM. In addition, soluble CERK was used to avoid potential SPE clogging. To halt the reaction, 3 volumes of methanol were added, followed by transfer of the mixture to the 96-well SPE-plate. A K_m value of 4 μM was obtained (Figure 4(a)), comparable to the K_m of 6 μM obtained for NBD-C$_6$-Cer with the radiometric assay (see above). For economical use of the substrate, its concentration was fixed at 5 μM (standard assay conditions). At this concentration, although close to the K_m, phosphorylation continued at a linear rate till about 50%

of the substrate was converted (Figure 4(b)). Likewise, when varying the amounts of CERK, production of NBD-C$_6$-Cer-1-P proceeded linearly till about 50% of NBD- C$_6$-Cer was phosphorylated (data not shown). Hence, the amount of NBD-C$_6$-Cer converted under the standard conditions is a good measure for CERK activity.

As documented in Figure 5, the fluorescent assay is very convenient to document CERK activity in cultured cells. Upon overexpression of CERK, a $\pm$50-fold increase in kinase activity (1.40 versus 0.029 nmol/min·mg protein) was measured. Similar values were obtained using the radiometric assay (1.57 versus 0.030 nmol/min·mg protein), supporting the use of the fluorescence assay as a valuable alternative. The detection limit of the assay is estimated at 10 pmol NBD-C$_6$-Cer-1-P, meaning that CERK activities corresponding to 1 pmol/min can be measured (or less if incubation time is prolonged). When relying on a [γ-^{32}P]-ATP based assay, and labelling of the produced ceramide-1-^{32}P to 100–1000 dpm, this would require an input of 0.5–5 μCi/assay at 1 mM ATP and 100 μL assay volume. Since CERK activity in most tissues and cells is quite low (<40 pmol/min·mg protein) [8], our assay will facilitate further work on CERK and its regulation. In addition, the low K_m implies that NBD-C$_6$-Cer might be a handy substrate for *in vivo* CERK measurements. Indeed, when added to CERK-expressing cultured cells, formation of NBD-C$_6$-Cer-1-P can be followed by TLC analysis of the cellular lipid extracts (data not shown), in full agreement with data reported by Bornancin and coworkers [19]. By comparing

(a)

(b)

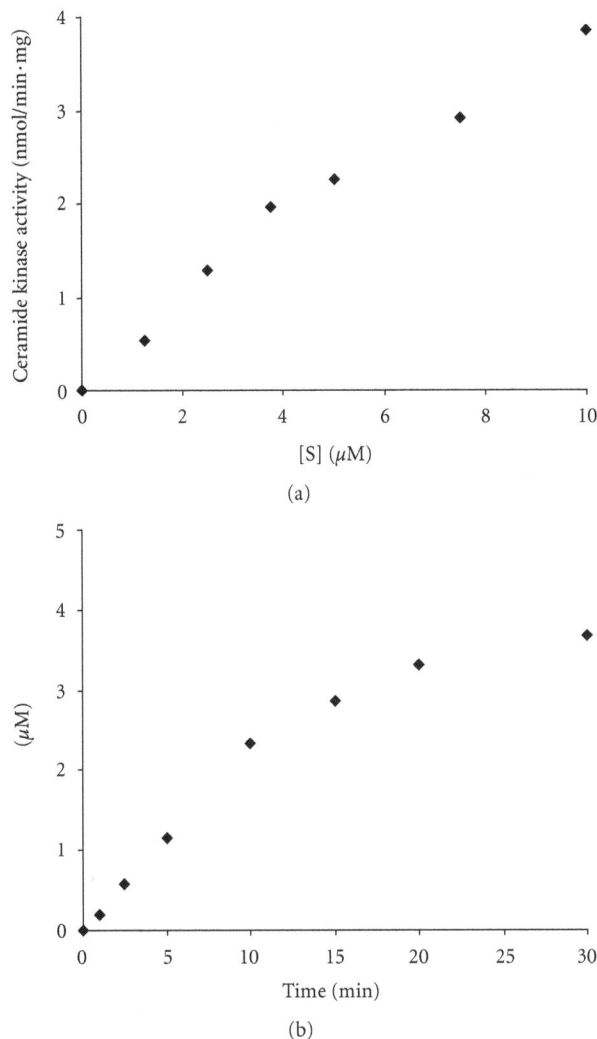

FIGURE 4: Substrate and time dependence of the kinase reaction. Recombinant bacterially expressed *Hs*CERK was incubated with the indicated concentration of NBD-C$_6$-Cer (a) or 5 μM (b) in the presence of 1 mM ATP at 37°C. The reaction was stopped at 10 min (a) or the indicated time periods (b) by addition of methanol and the mixture was applied to an NH$_2$-SPE column. The reaction product NBD-C$_6$-Cer-1-P was quantified based on fluorescence intensity of the column eluate and converted into nmol or μM, based on a calibration curve.

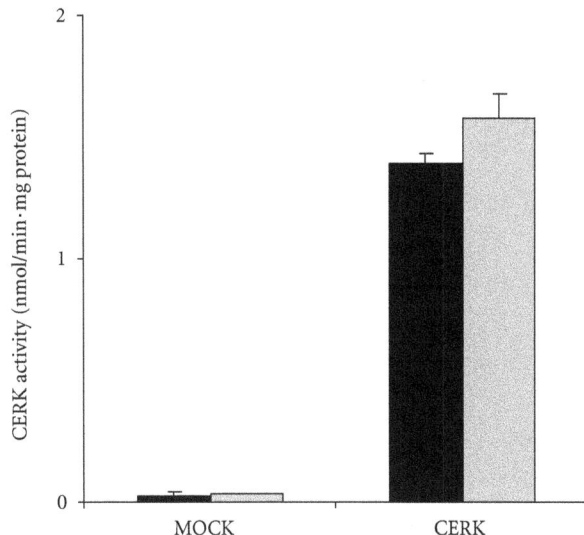

FIGURE 5: CERK activity in cultured cells. CERK activity towards NBD-C$_6$-Cer was determined in lysates from CHO cells, transfected with pCMV-Tag2B (mock) or pHVO001, coding for a Flag-HsCERK fusion [8], using the fluorescence assay (black bars) or the radiometric assay (grey bars, mean $\pm$ SEM, $n = 3$). CERK activity is expressed as nmol per mg protein per min (nmol/min·mg protein).

the scanned intensities against the fluorescence of NBD-C$_6$-Cer-1-P standards, TLC analysis of cell extracts is another means to calculate CERK activity (data not shown). During our attempts to publish this work, Don and Rosen [20] reported on the same ceramide analogue as a substrate but their assay was based on either TLC spotting for the micellar assay or for the BSA-based assay, extraction, followed by phase separation and transfer of the upper phase for analysis; the latter was done in a 96-well format. The solvent influence on the NBD-fluorescence was apparently not considered.

Omission of a liquid-liquid extraction step clearly speeds up the assay and allows for other formats like multiwell plates used in HTS. To simulate an HTS, a commercial library was tested in a 96 well format. To increase the chance to get some positive hits, we selected hereto a protein kinase inhibitor library given that their targets rely on the same cofactor as CERK. To show specificity, the same library was also tested on another lipid kinase, human sphingosine kinase 1. Various established protein kinase inhibitors appear to affect CERK (Figure 6). CERK activity was blocked (more than 95% inhibition) by AG-494, AG-825, BAY11-7082, 2-hydroxy-5-(2,5-dihydroxybenzylamino)benzoic acid, hypericin, indirubin-3'-monoxime (and its 6-bromo-derivative), piceatannol, quercetin, Ro31-8220, rottlerin, *D-erythro*-sphingosine (and *D-threo*-sphinganine; not shown), U-0126, staurosporine, and ZM449829 (at 500 μM). SPHK1 was clearly inhibited by fewer compounds, the most potent being AG-494, piceatannol, and quercetin. For the most potent CERK-inhibitors, IC$_{50}$ values were determined: U-0126 (4 μM), followed by 6-bromo-indirubin-3'-oxime (9 μM) and hypericin and rottlerin (both 19 μM). U-0126 is considered to be a selective MAP kinase kinase inhibitor and IC$_{50}$ values are indeed lower (72 nM for MEK1; 58 nM for MEK2) [28].

A few ceramide analogues and lipophilic amides, partly commercially obtained, partly homemade, were also tested as substrate and/or inhibitor (full list available upon request). Compounds that were not phosphorylated but strong inhibitory were further evaluated. From this screen, we retained fenretinide {(N-4-hydroxyphenyl)retinamide); IC$_{50}$ 1.1 μM} en AMG-9810 {(E)-3-(4-t-butylphenyl)-N-(2,3-dihydrobenzo [b][1,4]dioxin-6-yl)acrylamide; IC$_{50}$ 1.4 μM}. These compounds are known to influence other biological processes. Fenretinide binds f.i. the retinoic acid receptor, slows the growth of transformed cells, and induces apoptosis

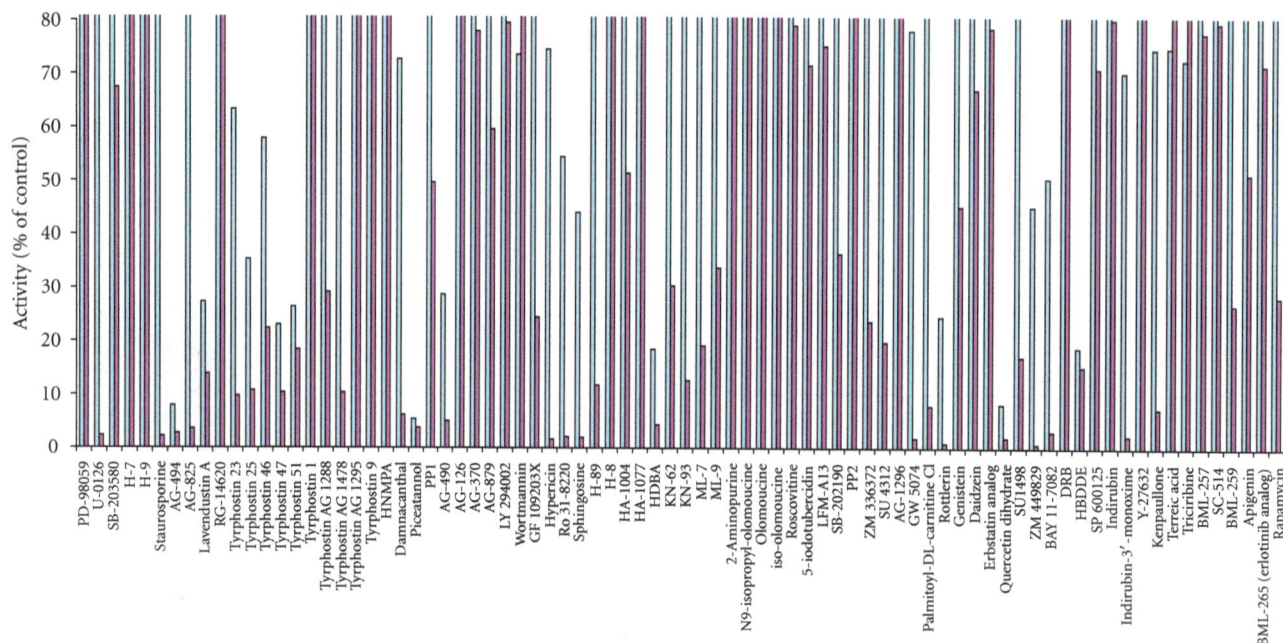

FIGURE 6: Influence of proteine kinase inhibitors on CERK via 96 well HTS. CERK was measured with $NBD-C_6$-Cer in the presence of 0.5 mM of established protein kinase inhibitors (kinase inhibitor library, Biomol), using the 96 well fluorescent SPE assay. Activity is expressed as % of the control containing 10% DMSO ($100 \pm 11\%$; mean $\pm$ SD; $n = 3$, magenta bars). CERK was not influenced by DMSO, up to 20% (data not shown). For comparison, effect of the inhibitors on sphingosine kinase is displayed in blue bars as % of control ($100 \pm 7\%$; mean $\pm$ SD; $n = 3$).

in cultured cells (effective concentrations $1–10\,\mu M$) [29], the latter likely via increasing dihydroceramide levels [30]. AMG-9810 is known as an antagonist of the vanilloid/TPRV1 receptor [31]; its endogenous ligand, anandamide, is also a fatty amide.

Summarizing, by further analyzing the substrate spectrum of CERK, it was shown that $NBD-C_6$-ceramide is a suitable substrate, allowing for a fluorescence based CERK measurement. By combining this substrate with the use of NH_2-SPE to isolate the product, a straightforward assay has been developed, useful for basic research (100 mg SPE) and adaptable to HTS for CERK inhibitors/activators (25 mg SPE-96 well format). Recently, a HTS-CERK assay was described by Munagala et al. [32], which can be miniaturized to 1,536 well plates. However, this assay is based on chemi-luminescent detection of the disappearing ATP and C_{12}-ceramide as substrate. Hence, an extra control is required for the effect of compounds on the coupling reaction/enzymes. Moreover, this assay is not applicable to crude cell/tissue lysates given the interfering presence of ATPases and other phosphatase activities [33] and the low CERK activity.

Abbreviations

C6-: Hexanoyl-
Cer-1-P: Ceramide-1-phosphate
CERK: Ceramide kinase
HTS: High throughput screening
NBD: 7-nitrobenz-2-oxa-1,3-diazole

SPE: Solid phase extraction
NBD-C-Cer: N-[NBD]-6-aminohexanoyl-sphingenine or NBD-C-ceramide
NBD-C6-Cer-1-P: N-[NBD]-6-aminohexanoyl-sphingenine-1-phosphate or NBD-C-ceramide-1-phosphate
TFA: Trifluoroacetic acid.

Acknowledgments

This work was supported by grants from the Flemish 'Fonds voor Wetenschappelijk Onderzoek' (G.0405.02), the Belgian Ministry of "Federaal Wetenschapsbeleid" (Interuniversitaire Attractiepolen IAP-P5/05), and the "Industrieel Onderzoeks-fonds KU Leuven." H. Van Overloop was paid by the latter grant. The authors like to thank Dr. P. Augustijns (Laboratory for Farmacotechnology and Biopharmacy, KU Leuven) and Dr. H. Desmedt (Laboratory of Molecular and Cellular Signaling, KU Leuven) for access to different fluorimeters and fluoscanners, as well as Dr. P. Chaltin (LRD, Leuven) for his interest in this work.

References

[1] Y. A. Hannun and L. M. Obeid, "The ceramide-centric universe of lipid-mediated cell regulation: stress encounters of the lipid kind," *Journal of Biological Chemistry*, vol. 277, no. 29, pp. 25847–25850, 2002.

[2] K. Thevissen, I. E. J. A. François, J. Winderickx, C. Pannecouque, and B. P. A. Cammue, "Ceramide involvement in apoptosis and apoptotic diseases," *Mini-Reviews in Medicinal Chemistry*, vol. 6, no. 6, pp. 699–709, 2006.

[3] N. F. Lamour and C. E. Chalfant, "Ceramide-1-phosphate: the "missing" link in eicosanoid biosynthesis and inflammation," *Molecular Interventions*, vol. 5, no. 6, pp. 358–367, 2005.

[4] A. Gómez-Muñoz, "Ceramide 1-phosphate/ceramide, a switch between life and death," *Biochimica et Biophysica Acta*, vol. 1758, no. 12, pp. 2049–2056, 2006.

[5] C. Graf, B. Zemann, P. Rovina et al., "Neutropenia with impaired immune response to *Streptococcus pneumoniae* in ceramide kinase-deficient mice," *The Journal of Immunology*, vol. 180, no. 5, pp. 3457–3466, 2008.

[6] S. M. Bajjalieh, T. F. J. Martin, and E. Floor, "Synaptic vesicle ceramide kinase. A calcium-stimulated lipid kinase that co-purifies with brain synaptic vesicles," *Journal of Biological Chemistry*, vol. 264, no. 24, pp. 14354–14360, 1989.

[7] M. Sugiura, K. Kono, H. Liu et al., "Ceramide kinase, a novel lipid kinase: molecular cloning and functional characterization," *Journal of Biological Chemistry*, vol. 277, no. 26, pp. 23294–23300, 2002.

[8] H. Van Overloop, S. Gijsbers, and P. P. Van Veldhoven, "Further characterization of mammalian ceramide kinase: substrate delivery and (stereo)specificity, tissue distribution, and subcellular localization studies," *Journal of Lipid Research*, vol. 47, no. 2, pp. 268–283, 2006.

[9] A. Carré, C. Graf, S. Stora et al., "Ceramide kinase targeting and activity determined by its N-terminal pleckstrin homology domain," *Biochemical and Biophysical Research Communications*, vol. 324, no. 4, pp. 1215–1219, 2004.

[10] P. Rovina, M. Jaritz, S. Höfinger et al., "A critical β6-β7 loop in the pleckstrin homology domain of ceramide kinase," *Biochemical Journal*, vol. 400, no. 2, pp. 255–265, 2006.

[11] D. S. Wijesinghe, A. Massiello, P. Subramanian, Z. Szulc, A. Bielawska, and C. E. Chalfant, "Substrate specificity of human ceramide kinase," *Journal of Lipid Research*, vol. 46, no. 12, pp. 2706–2716, 2005.

[12] M. Bektas, P. S. Jolly, S. Milstien, and S. Spiegel, "A specific ceramide kinase assay to measure cellular levels of ceramide," *Analytical Biochemistry*, vol. 320, no. 2, pp. 259–265, 2003.

[13] H. Van Overloop, Y. Denizot, M. Baes, and P. P. Van Veldhoven, "On the presence of C2-ceramide in mammalian tissues: possible relationship to etherphospholipids and phosphorylation by ceramide kinase," *Biological Chemistry*, vol. 388, no. 3, pp. 315–324, 2007.

[14] N. G. Lipsky and R. E. Pagano, "Sphingolipid metabolism in cultured fibroblasts: microscopic and biochemical studies employing a fluorescent ceramide analogue," *Proceedings of the National Academy of Sciences of the United States of America*, vol. 80, no. 9, pp. 2608–2612, 1983.

[15] V. T. Hinkovska-Galcheva, L. A. Boxer, P. J. Mansfield, D. Harsh, A. Blackwood, and J. A. Shayman, "The formation of ceramide-1-phosphate during neutrophil phagocytosis and its role in liposome fusion," *Journal of Biological Chemistry*, vol. 273, no. 50, pp. 33203–33209, 1998.

[16] G. Rile, Y. Yatomi, T. Takafuta, and Y. Ozaki, "Ceramide 1-phosphate formation in neutrophils," *Acta Haematologica*, vol. 109, no. 2, pp. 76–83, 2003.

[17] L. Riboni, R. Bassi, V. Anelli, and P. Viani, "Metabolic formation of ceramide-1-phosphate in cerebellar granule cells: evidence for the phosphorylation of ceramide by different metabolic pathways," *Neurochemical Research*, vol. 27, no. 7-8, pp. 711–716, 2002.

[18] B. J. Pettus, A. Bielawska, S. Spiegel, P. Roddy, Y. A. Hannun, and C. E. Chalfant, "Ceramide kinase mediates cytokine- and calcium ionophore-induced arachidonic acid release," *Journal of Biological Chemistry*, vol. 278, no. 40, pp. 38206–38213, 2003.

[19] C. Graf, P. Rovina, L. Tauzin, A. Schanzer, and F. Bornancin, "Enhanced ceramide-induced apoptosis in ceramide kinase overexpressing cells," *Biochemical and Biophysical Research Communications*, vol. 354, no. 1, pp. 309–314, 2007.

[20] A. S. Don and H. Rosen, "A fluorescent plate reader assay for ceramide kinase," *Analytical Biochemistry*, vol. 375, no. 2, pp. 265–271, 2008.

[21] S. Gijsbers, S. Asselberghs, P. Herdewijn, and P. P. Van Veldhoven, "1-O-Hexadecyl-2-desoxy-2-amino-sn-glycerol, a substrate for human sphingosine kinase," *Biochimica et Biophysica Acta*, vol. 1580, no. 1, pp. 1–8, 2002.

[22] P. P. Van Veldhoven, T. J. Matthews, D. P. Bolognesi, and R. M. Bell, "Changes in bioactive lipids, alkylacylglycerol and ceramide, occur in HIV-infected cells," *Biochemical and Biophysical Research Communications*, vol. 187, no. 1, pp. 209–216, 1992.

[23] H.-Y. Kim and N. Salem, "Separation of lipid classes by solid phase extraction," *Journal of Lipid Research*, vol. 31, no. 12, pp. 2285–2289, 1990.

[24] A. Pietsch and R. L. Lorenz, "Rapid separation of the major phospholipid classes on a single aminopropyl cartridge," *Lipids*, vol. 28, no. 10, pp. 945–947, 1993.

[25] S. Mazères, V. Schram, J. F. Tocanne, and A. Lopez, "7-nitrobenz-2-oxa-1,3-diazole-4-yl-labeled phospholipids in lipid membranes: differences in fluorescence behavior," *Biophysical Journal*, vol. 71, no. 1, pp. 327–335, 1996.

[26] J. A. Monti, S. T. Christian, and W. A. Shaw, "Synthesis and properties of a highly fluorescent derivative of phosphatidylethanolamine," *Journal of Lipid Research*, vol. 19, no. 2, pp. 222–228, 1978.

[27] A. Chattopadhyay and E. London, "Spectroscopic and ionization properties of N-(7-nitrobenz-2-oxa-1,3-diazol-4-yl)-labeled lipids in model membranes," *Biomembranes*, vol. 938, no. 1, pp. 24–34, 1988.

[28] M. F. Favata, K. Y. Horiuchi, E. J. Manos et al., "Identification of a novel inhibitor of mitogen-activated protein kinase kinase," *Journal of Biological Chemistry*, vol. 273, no. 29, pp. 18623–18632, 1998.

[29] P. H. O'Donnell, W. X. Guo, C. P. Reynolds, and B. J. Maurer, "N-(4-hydroxyphenyl)retinamide increases ceramide and is cytotoxic to acute lymphoblastic leukemia cell lines, but not to non-malignant lymphocytes," *Leukemia*, vol. 16, no. 5, pp. 902–910, 2002.

[30] H. Wang, B. J. Maurer, Y. Y. Liu et al., "N-(4-Hydroxyphenyl)retinamide increases dihydroceramide and synergizes with dimethylsphingosine to enhance cancer cell killing," *Molecular Cancer Therapeutics*, vol. 7, no. 9, pp. 2967–2976, 2008.

[31] N. R. Gavva, R. Tamir, Y. Qu et al., "AMG 9810 [(E)-3-(4-t-butylphenyl)-N-(2,3-dihydrobenzo[b][1,4] dioxin-6-yl)acrylamide], a novel vanilloid receptor 1 (TRPV1) antagonist with antihyperalgesic properties," *Journal of Pharmacology and Experimental Therapeutics*, vol. 313, no. 1, pp. 474–484, 2005.

[32] N. Munagala, S. Nguyen, W. Lam et al., "Identification of small molecule ceramide kinase inhibitors using a homogeneous

chemiluminescence high throughput assay," *Assay and Drug Development Technologies*, vol. 5, no. 1, pp. 65–73, 2007.

[33] S. Gijsbers, G. Van der Hoeven, and P. P. Van Veldhoven, "Subcellular study of sphingoid base phosphorylation in rat tissues: evidence for multiple sphingosine kinases," *Biochimica et Biophysica Acta*, vol. 1532, no. 1-2, pp. 37–50, 2001.

16

Roles of Fatty Acid Oversupply and Impaired Oxidation in Lipid Accumulation in Tissues of Obese Rats

Nicholas D. Oakes,[1] Ann Kjellstedt,[1] Pia Thalén,[1] Bengt Ljung,[1] and Nigel Turner[2,3]

[1] *AstraZeneca R&D Mölndal, 431 83 Mölndal, Sweden*
[2] *Diabetes and Obesity Program, Garvan Institute of Medical Research, Darlinghurst, NSW 2010, Australia*
[3] *School of Medical Sciences, University of New South Wales, Sydney, NSW 2052, Australia*

Correspondence should be addressed to Nicholas D. Oakes; nick.oakes@astrazeneca.com

Academic Editor: Philip W. Wertz

To test the roles of lipid oversupply versus oxidation in causing tissue lipid accumulation associated with insulin resistance/obesity, we studied *in vivo* fatty acid (FA) metabolism in obese (Obese) and lean (Lean) Zucker rats. Indices of local FA utilization and storage were calculated using the partially metabolizable [9,10-^{3}H]-(R)-2-bromopalmitate (^{3}H-R-BrP) and [U-^{14}C]-palmitate (^{14}C-P) FA tracers, respectively. Whole-body FA appearance (R_a) was estimated from plasma ^{14}C-P kinetics. Whole-body FA oxidation rate (R_{ox}) was assessed using ^{3}H$_2$O production from ^{3}H-palmitate infusion, and tissue FA oxidative capacity was evaluated *ex vivo*. In the basal fasting state Obese had markedly elevated FA levels and R_a, associated with elevated FA utilization and storage in most tissues. Estimated rates of muscle FA oxidation were not lower in obese rats and were similarly enhanced by contraction in both lean and obese groups. At comparable levels of FA availability, achieved by nicotinic acid, R_{ox} was lower in Obese than Lean. In Obese rats, FA oxidative capacity was 35% higher than that in Lean in skeletal muscle, 67% lower in brown fat and comparable in other organs. In conclusion, lipid accumulation in non-adipose tissues of obese Zucker rats appears to result largely from systemic FA oversupply.

1. Introduction

Disturbances in fatty acid metabolism may cause several key features of the insulin resistance syndrome, including impaired glucose regulation, dyslipidemia, and obesity. Thus systemic free fatty acid (FA) oversupply can decrease insulin-stimulated glucose uptake in skeletal muscle [1, 2], reduce insulin suppression of hepatic glucose production [3], and alter glucose-stimulated insulin secretion [4]. Furthermore, an oversupply of FA to liver may cause dyslipidemia, including hypertriglyceridemia and the atherogenic lipoprotein profile [5]. With regard to FA utilization, reductions in mitochondrial content and diminished fatty acid oxidation capacity in skeletal muscle and adipose tissue have been linked with obesity and insulin resistance [6–14]. Additionally, in conditions of impaired insulin action there is also a reduced ability to appropriately switch between glucose and lipid fuels (i.e., metabolic inflexibility), postulated to play an important role in the development of obesity [15].

The obese Zucker rat is an animal model possessing major metabolic features seen in conditions of human insulin resistance, including glucose metabolic insulin resistance, hypertriglyceridemia, and elevated nonadipose tissue lipid levels [16, 17], which have been implicated in both the development of insulin resistance and lipoapoptosis in tissues including the pancreas and heart [18]. Inappropriate deposition of triglycerides and other bioactive fatty acid metabolites in a tissue may result from a systemic oversupply of fatty acid or from a local defect in fatty acid oxidation. The obese Zucker rat has a loss-of-function mutation in the leptin receptor, and although leptin has been shown to enhance local rates of lipid oxidation [19, 20], the relative contribution of reduced FA utilization to the lipotoxic state observed in tissues of these animals *in vivo* is still not well established.

We were interested in elucidating the mechanisms of lipid accumulation in non-adipose tissues of obese Zucker rats: oversupply or underutilization? The aim of this study was therefore to determine how the fluxes and metabolic fate of FA are altered in obese versus lean Zucker rats. This was done in three independent experimental series. Series 1 quantified FA uptake and metabolic fate *in vivo* at the individual tissue level based on the simultaneous combined use of [9,10-^{3}H]-(R)-2-bromopalmitate (^{3}H-R-BrP) and [U-^{14}C]-palmitate (^{14}C-P). The ^{3}H-R-BrP tracer is used to estimate local FA (oxidative + nonoxidative) utilization while ^{14}C-P is used to assess non-oxidative FA disposal into lipid stores [21]. Series 2 examined FA oxidation and its dependence on FA levels, using nicotinic acid as an antilipolytic agent. Finally, Series 3 was used to assess FA oxidation capacity in a comprehensive range of metabolically important tissues *ex vivo*. Our results give strong support for the hypothesis that the major factor responsible for tissue lipid accumulation in obese Zucker rats is increased plasma FA availability.

2. Materials and Methods

2.1. Animals. Experimental procedures were approved by the local ethics review committee on animal experiments (Göteborg region). Male 8-week old Lean (FA/FA) and obese (fa/fa) Zucker rats (Charles River Wiga GmbH, Suffield, FRG) were maintained in a temperature controlled (20–22°C) room with a 12 h light-dark cycle (lights on at 06:00) and free access to rodent chow (R3 Laktamin AB, Stockholm, Sweden) and tap water.

2.2. Acute Study Preparation for Series 1 and 2. At 07:00 on the morning of the study food was withdrawn. Then at 09:00 the rats were anaesthetized with Na-thiobutabarbital (Inactin, RBI, Natick, MA), with the lean and obese rats receiving 120 and 180 mg kg^{-1} (I.P.), respectively. Body temperature was monitored using a rectal probe and maintained at 37.5°C throughout the experiment. Animals were tracheotomized and catheters were placed in the right jugular vein for tracer administration and left carotid artery for continuous monitoring of arterial blood pressure and heart rate and for blood sampling via a device allowing minimum sample volumes. Arterial catheter patency was maintained by continuous infusion (10 μL min^{-1}) of a sterile saline solution containing sodium citrate (20.6 mmol L^{-1}).

2.3. Series 1: Tissue-Specific FA Metabolism In Vivo

2.3.1. Unilateral Hindlimb Muscle Contraction. Immediately following catheterization (described above), the sciatic nerves were exposed and cut bilaterally at the gluteal level. Unilateral electrical sciatic nerve stimulation was applied with ring electrodes at 0.5 Hz to induce sustainable twitch contractions in muscles of one hind leg as described previously [22].

2.3.2. Tracer Preparation. ^{3}H-R-BrP tracer was synthesized and purified using methods described in [23]. Tracer infusates were prepared freshly each day. For each rat ~5 × 10^7 dpm

^{3}H-R-BrP and ~2.5 × 10^7 dpm ^{14}C-P (Amersham, Solna, Sweden), as well as 152 nmol Na-palmitate (Sigma, St. Louis, MO), were complexed to essentially fatty acid-free bovine serum albumin (BSA) (Sigma) as detailed in [22]. *Protocol.* Unilateral sciatic nerve stimulation was commenced 70 min after completion of surgical preparation and 20 min prior to commencing tracer administration. All blood samples were collected via the carotid catheter into K-EDTA containing tubes (Microvette CB300, Sarstedt, Nümbrecht, Germany) via a device designed to reduce sample volume. Immediately before tracer infusion a 200 μL basal blood sample was collected for determination of plasma insulin and substrate levels. Tracer administration and blood sampling were performed according to previously described methods [22]. Briefly, the albumin-palmitate-tracer complex was infused through the jugular catheter at 230 μL min^{-1} for 4 min. Blood samples were centrifuged immediately at 4°C and a 25 μL plasma aliquot placed directly into lipid extraction mixture (described in [22]) for determination of plasma ^{3}H-R-BrP and ^{14}C-P concentrations. After collection of the final blood sample, 16 min after commencing the tracer infusion, rats were killed with an overdose of thiobutabarbitol (120 mg kg^{-1}). Tissues were collected and samples (~100 mg) were combusted for determination of total ^{3}H and ^{14}C content [22]. *Calculations.* The clearance rate of ^{3}H-R-BrP by an individual tissue (K_f^*), an index of the ability of the tissue to utilize FAs, was calculated as previously described [22], as

$$K_f^* = \frac{m_B}{\int_0^T c_B(t)\, dt}, \tag{1}$$

where T is the time of tissue collection (16 min), m_B is the total tissue ^{3}H content (at $t = T$), and c_B is the arterial plasma concentration of ^{3}H-R-BrP. An index of FA utilization rate (R_f^*) was calculated as

$$R_f^* = C_P \times K_f^*, \tag{2}$$

where C_P is the arterial plasma FA concentration.

An index of the clearance of ^{14}C-P into storage products (K_{fs}) was calculated as

$$K_{fs} = \frac{m_P}{\int_0^T c_P(t)\, dt}, \tag{3}$$

where m_P is the total tissue ^{14}C content (at $t = T$) and c_P is the arterial plasma ^{14}C-P concentration. This assumes that all of the ^{14}C label originating from locally activated ^{14}C-P directed into oxidative metabolism would be lost from the tissue (largely as ^{14}CO$_2$) by the time of tissue sampling. An index of the rate of FA incorporation into storage (R_{fs}) was calculated as

$$R_{fs} = C_P \times K_{fs}. \tag{4}$$

Assuming certain conditions are met [22], R_f^* is proportional to the genuine rate of FA utilization (R_f), that is,

$$R_f^* = LC^* \times R_f \tag{5}$$

with a constant of proportionality ("lumped constant") LC^*. Since R_f is the sum of oxidative disposal ($R_{f_{ox}}$) and non-oxidative disposal (R_{fs}),

$$R_{f_{ox}} = LC^{*-1} \times R_f^* - R_{fs}. \tag{6}$$

Note that when $R_{f_{ox}} = 0$, $LC^{*-1} = R_{fs}/R_f^*$. We have previously suppressed fatty acid oxidation using pharmacological β-oxidation inhibition to obtain crude estimates of LC^* for different tissues in the rat [22]. LC^* and the reliability of the $R_{f_{ox}}$ values derived vary in a tissue-specific manner [22] with the following values assumed for the present work: skeletal muscle 0.27 and heart 0.19.

Estimates of whole-body plasma ^{14}C-P clearance (K_P) were calculated according to [22].

2.3.3. Determination of Plasma ^{3}H-R-BrP and ^{14}C-P Concentrations, as well as Tissue ^{3}H and ^{14}C-Levels.

Plasma ^{3}H-R-BrP and ^{14}C-P were resolved using an acid lipid extraction procedure. Total tissue ^{3}H and ^{14}C levels were determined by combusting tissue samples using a Packard System 387 Automated Sample Preparation Unit (Packard Instruments Co., Inc., Meriden, CT). These methods are described in detail in [22].

2.4. Series 2: Whole-Body FA Oxidation Rate and Its Dependence on Plasma FA Level

2.4.1. Groups.
Four groups (n = 2-3 per group) of both lean and obese Zucker rats were studied in order to generate a range of plasma FA levels: a vehicle control group receiving a normal saline infusion and three groups receiving intravenous nicotinic acid infusion at the doses of 10, 100 and 1000 nmol kg^{-1} min^{-1}, respectively.

2.4.2. Tracer Preparation.
Tracer infusates, ~$2 \cdot 10^8$ dpm per rat [9,10 ^{3}H] palmitic acid (^{3}H-P, Amersham, Solna, Sweden) and 305 nmol Na-palmitate (Sigma, St. Louis, MO), were freshly prepared daily. The tracer and Na-palmitate were prepared in 150 μL ethanol and added dropwise to 0.6 mL of continuously stirred 4% (w/v) essentially fatty acid-free bovine serum albumin (BSA, Sigma, St. Louis, MO) in normal saline. The infusate was made up to a final volume of ~3 mL per rat by addition of normal saline.

2.4.3. Protocol.
After a 2 h postsurgery recovery period, two basal blood samples (~150 μL) were collected 15 min apart for analysis of plasma FA, TG, glucose, and insulin. Immediately following collection of the second blood sample, intravenous infusions of nicotinic acid (or vehicle) and tracer were started. The albumin-palmitate-^{3}H-P complex was infused at a constant rate (~1×10^6 dpm min^{-1}, 17 μL min^{-1}). Arterial blood samples (~75 μL) were collected 10, 20, 40, 60, 80, 100, and 120 min after the start of tracer infusion. For each sample, plasma was separated as quickly as possible in a refrigerated centrifuge. One 25 μL aliquot was placed into 2 mL lipid extraction mixture, for determination of ^{3}H-P and ^{3}H$_2$O; the

remainder was used for analysis of FA level. After collection of the final blood sample, rats were killed with an overdose of thiobutabarbitol (120 mg kg^{-1}).

2.4.4. Measurement of Plasma Levels of ^{3}H-P and ^{3}H$_2$O.
To discriminate ^{3}H-P from total plasma ^{3}H activity, a lipid extraction and separation procedure was performed on plasma samples. This involved an initial acid lipid extraction using a mixture of isopropanol-heptane-1 mol/L acetic acid (40 : 10 : 1 vol) followed by solid phase separation of free fatty acids (including ^{3}H-P) from neutral lipids. ^{3}H$_2$O was estimated as the ^{3}H-activity in the lower (isopropanol-water) phase of the lipid extraction procedure.

2.4.5. Rates of Plasma FA Appearance (R_a) and Oxidation (R_{ox}).
Plasma FA mobilization was assessed using a constant infusion of ^{3}H-palmitate (^{3}H-P). After attainment of isotopic steady states (<40 min after the start of tracer infusion), the plasma clearance rate of ^{3}H-P (K) was calculated as

$$K = \frac{i_p}{c_p(\infty)}, \tag{7}$$

where i_p is the tracer infusion rate (dpm min^{-1}) and $c_p(\infty)$ is the steady state arterial concentration of ^{3}H-P (dpm mL^{-1}). The rate of appearance of plasma FAs (R_a) was calculated as

$$R_a = C_p \times K, \tag{8}$$

where C_p is the arterial plasma FA concentration (μmol mL^{-1}).

Appearance of ^{3}H$_2$O in the plasma was linear from the earliest time point throughout the study consistent with rapid attainment of steady state of the labeled oxidation precursor pool. The fraction of plasma FA undergoing oxidation (f_{ox}) was estimated using the relationship

$$f_{ox} = \frac{V_w \times dc_w/dt}{i_p}, \tag{9}$$

where V_w is the total water space of the rat (estimated from body weight, BW in g, using separate regression equations: for obese animals % water = $59.3 - 0.027 \times$ BW and for lean animals % water = $73.4 - 0.030 \times$ BW obtained from in-house water content analyses of obese and lean Zucker rats, resp.), c_w is the plasma concentration of ^{3}H$_2$O, and t is the time from commencement of tracer infusion. The derivative above was estimated from the slope obtained from linear regression analysis of the ^{3}H$_2$O plasma versus time data for the period $t = 10$ to $t = 120$ min. An estimate of the whole-body rate of FA clearance into oxidation (K_{ox}) was calculated as

$$K_{ox} = f_{ox} \times K. \tag{10}$$

The rate of whole-body FA oxidation (R_{ox}) was calculated as

$$R_{ox} = f_{ox} \times R_a. \tag{11}$$

2.5. Series 3: Ex Vivo Fatty Acid Oxidation. Fatty acid oxidation was measured in tissue homogenates using a modified version of a previously published method [24]. Briefly, tissues were homogenized in either 9 volumes (epididymal white adipose tissue; WAT), 19 volumes (cerebellum), or 39 volumes (heart, brown adipose tissue (BAT), liver, red gastrocnemius, and white quadriceps) of ice-cold $250 \, mmol \, L^{-1}$ sucrose, $10 \, mmol \, L^{-1}$ Tris-HCl, $1 \, mmol \, L^{-1}$ EDTA. For assessment of palmitate oxidation, $50 \, \mu L$ of tissue homogenate was then incubated with $450 \, \mu L$ reaction mixture (pH 7.4). Final concentrations of the reaction mixture were (in $mmol \, L^{-1}$) 100 sucrose, 10 Tris-HCl, 5 potassium phosphate, 80 potassium chloride, 1 magnesium chloride, 2 malate, 2 ATP, 1 dithiothreitol, 0.2 EDTA, 2 L-carnitine, 0.05 coenzyme A (CoA), 0.2 palmitate [+0.5 μCi 1-^{14}C-palmitate], and 0.3% (w/v) fatty acid-free BSA. After 90 min of incubation at 25°C, the reaction was stopped by the addition of $100 \, \mu L$ of ice-cold $1 \, mol \, L^{-1}$ perchloric acid. CO_2 produced during the 90 min incubation was collected in $100 \, \mu L$ of $1 \, mol \, L^{-1}$ sodium hydroxide. ^{14}C counts present in the acid-soluble fraction were also measured and combined with the CO_2 values to give the total palmitate oxidation rate.

2.6. Plasma Insulin and Substrate Concentrations. Insulin concentrations were determined using radioimmunoassay (rat insulin RIA kit; Linco Research, St. Charles, MO). Colorimetric kit methods were used for the measurement of plasma FA (NEFA C; Wako, Richmond, VA), triglycerides (Triglycerides/GB; Boehringer Mannheim, Indianapolis, IN), and glucose (Glucose HK; Roche, Stockholm).

2.7. Statistics. Differences between Lean and Obese groups were assessed using Student's t-tests, assuming equal group variance. Systematic between-group differences in muscle parameters were assessed using 2-way analysis of variance (ANOVA) using the program SPSS (SPSS, Chicago, IL). Linear regression analysis was performed using GraphPad Prism (GraphPad Software Inc., La Jolla, CA). Results are reported as mean ± SE. $P < 0.05$ was considered statistically significant.

3. Results

3.1. Series 1: Tissue-Specific FA Metabolism In Vivo. Body weights and general plasma factors for Series 1 animals are summarized in Table 1. Estimates of body composition, lean and fat mass, have also been made as previously described [25]. As expected, obese Zucker rats weighed approximately 50% more than age-matched lean Zucker rats, due to increased fat mass, and displayed hyperinsulinemia, hypertriglyceridemia, and a mild hyperglycemia.

Plasma FA level and rate of appearance of FA (R_a), calculated from the ^{14}C-palmitate kinetics, are presented in Figure 1. Obese animals had substantially elevated systemic FA availability, compared to lean Zuckers, due to an elevated rate of entry of FA into the plasma as shown by the R_a data. Metabolic clearance rates of ^{14}C-palmitate (K_P, expressed

TABLE 1: Body weights and plasma factors in lean and obese Zucker rats, Series 1.

	Lean	Obese
Body weight (g)	337 ± 21	512 ± 35[***]
Lean body mass (g)	291 ± 16	314 ± 18
Fat body mass (g)	47 ± 5	197 ± 18[***]
Plasma glucose (mM)	8.3 ± 0.4	12.4 ± 1.7[*]
Plasma insulin (nM)	0.4 ± 0.1	3.7 ± 0.9[***]
Plasma TG (mM)	1.2 ± 0.3	4.0 ± 0.6[***]

Data are mean ± SE ($n = 6$ per group).
[*] $P < 0.05$, [***] $P < 0.001$ versus Lean.

TABLE 2: ^{3}H-R-BrP clearance (K_f^*) and ^{14}C-P clearance into storage products (K_{fs}) in individual tissues of lean and obese Zucker rats *in vivo*.

	K_f^* (mL 100 g^{-1} min^{-1})		K_{fs} (mL 100 g^{-1} min^{-1})	
	Lean	Obese	Lean	Obese
Cerebellum	0.46 ± 0.05	0.48 ± 0.04	0.87 ± 0.07	0.79 ± 0.05
Liver	37.5 ± 2.8	32.3 ± 2.3	45.9 ± 4.4	38.9 ± 2.8
WAT	0.69 ± 0.05	0.70 ± 0.06	1.6 ± 0.1	1.4 ± 0.2
BAT	7.2 ± 0.6	3.6 ± 0.5[***]	9.4 ± 1.0	7.8 ± 1.4
Heart	29.3 ± 2.8	16.5 ± 1.4[**]	15.5 ± 1.2	17.0 ± 3.3

K_f^* indexes the ability of the tissue to take up and utilize FA (for both oxidation and storage); K_{fs} indexes the ability to store FA. Data are mean ± SE ($n = 6$ per group).
[*] $P < 0.05$, [**] $P < 0.01$, [***] $P < 0.001$ versus Lean.

per rat) were similar in lean and obese animals, despite the much greater total tissue mass of the obese animals (data not shown).

The rate of ^{3}H-R-BrP clearance from plasma into a tissue (K_f^*) provides an index of the local ability to utilize FA for both oxidative and non-oxidative metabolism (storage), independent of the direct influence of plasma FA level. Tissue-specific clearance of ^{14}C-palmitate into storage (K_{fs}) indexes the local ability to store plasma FA. Independent of group, K_f^* and K_{fs} had a range of 2 orders of magnitude across the different tissues sampled, with cerebellum having the lowest and liver the highest values (Table 2). Also independent of group is the large difference between adipose tissue types with BAT having a much greater ability to take up and store FA than WAT. Comparing results for obese and lean animals, there were no differences in K_f^* or K_{fs} values for liver, cerebellum, or WAT. In BAT and heart K_f^* was lower in the obese compared to the lean animals, while K_{fs} was similar in the two groups. This indicates a reduced ability to metabolically sequester available FA and a preferential diversion towards non-oxidative disposal in these tissues of obese compared to lean Zuckers.

Hindlimb muscle K_f^* and K_{fs} results are summarized in Table 3. Results for five corresponding muscles from both hind legs are presented: one leg subject to repetitive efferent electrical stimulation of the sciatic nerve (Stim-Leg) versus the unstimulated control leg (Con-Leg). Examining first the results in the quiescent control leg muscles, it is apparent that

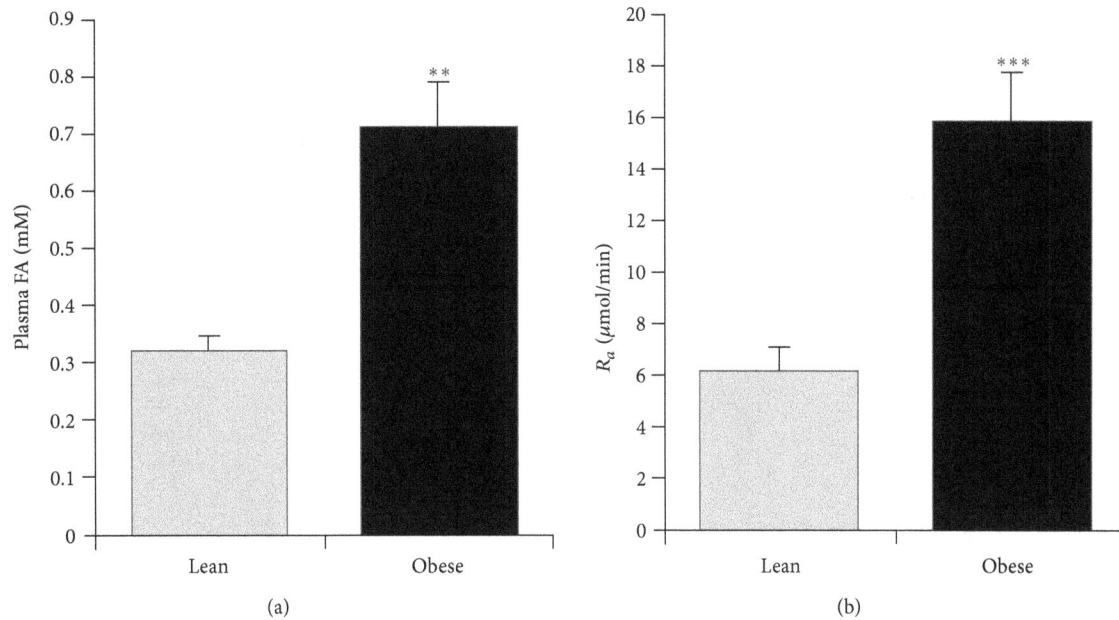

FIGURE 1: Whole-body FA metabolism in lean (Lean) and obese (Obese) Zucker rats. R_a plasma FA appearance rate. Results are expressed as mean ± SE ($n = 6$ rats per group). $^{**}P < 0.01$, $^{***}P < 0.001$ versus Lean.

TABLE 3: K_f^* and K_{fs} in individual hind leg muscles *in vivo* in lean and obese Zucker rats.

	K_f^* (mL 100 g^{-1} min^{-1})				K_{fs} (mL 100 g^{-1} min^{-1})			
	Lean		Obese		Lean		Obese	
	Con-Leg	Stim-Leg	Con-Leg	Stim-Leg	Con-Leg	Stim-Leg	Con-Leg	Stim-Leg
WG	0.76 ± 0.06	$2.03 \pm 0.23^\dagger$	0.69 ± 0.04	$1.86 \pm 0.10^\dagger$	1.87 ± 0.16	$2.54 \pm 0.12^\dagger$	1.96 ± 0.18	2.35 ± 0.14
EDL	1.08 ± 0.08	$3.38 \pm 0.21^\dagger$	1.06 ± 0.08	$2.54 \pm 0.21^{\dagger *}$	3.32 ± 0.31	$4.26 \pm 0.21^\dagger$	4.26 ± 0.44	4.65 ± 0.23
RG	1.17 ± 0.07	$2.58 \pm 0.47^\dagger$	1.39 ± 0.11	$2.10 \pm 0.17^\dagger$	3.75 ± 0.23	$5.08 \pm 0.46^\dagger$	4.48 ± 0.44	$6.37 \pm 0.54^\dagger$
WQ	0.61 ± 0.05	0.66 ± 0.07	0.51 ± 0.02	0.52 ± 0.03	1.39 ± 0.10	1.42 ± 0.12	1.17 ± 0.06	1.16 ± 0.06
RQ	1.63 ± 0.13	1.54 ± 0.13	1.86 ± 0.13	1.76 ± 0.12	4.28 ± 0.38	4.06 ± 0.30	4.56 ± 0.21	4.14 ± 0.23

Sustained twitch contractions were induced in the lower leg muscles (WG white gastrocnemius, EDL extensor digitorum longus, and RG red gastrocnemius) but not in the upper leg muscles (WQ white quadriceps and RQ red quadriceps) by repetitive electrical stimulation of the sciatic nerve of one leg (Stim-Leg). The contralateral, control leg (Con-Leg) was not subjected to sciatic nerve stimulation. Results represent mean ± SEM ($n = 6$ per group). $^\dagger P < 0.05$ versus Con-Leg (paired t-test, effect of sciatic nerve stimulation), $^* P < 0.05$ versus Lean (unpaired t-test, effect of obesity status).

independent of group, K_f^* and K_{fs} roughly rank according to expected oxidative capacity with glycolytic muscle (WQ and WG) < intermediate mixed fiber type muscle (EDL) < highly oxidative muscle (RG and RQ). There was no systematic difference in K_f^* in the quiescent muscles between lean and obese groups. K_{fs} did however tend to be modestly higher (by 15%, $P < 0.01$, ANOVA) in the quiescent muscles of the obese compared to the lean animals. These results suggest a similar ability to metabolically sequester available FA in the two groups but that in the obese animals there was a slight preference for disposal of FA into non-oxidative metabolism compared to the lean animals.

Electrical stimulation of the sciatic nerve at the gluteal level induced twitch contractions in the lower leg muscles (WG, EDL, and RG) but not in the thigh muscles (WQ and RQ). Correspondingly, for each lower leg muscle the K_f^* value was significantly higher in the Stim-Leg versus Con-Leg, while there were no differences in K_f^* between the Stim-Leg and Con-Leg for the 2 noncontracting thigh muscles (Table 3). The average contraction-induced increase in lower leg muscle K_f^* was similar in both lean and obese groups (group effect, $P > 0.05$, ANOVA). Sciatic nerve stimulation tended to induce a small increase in K_{fs} (3 out of 3 muscles in lean and 1 out of 3 muscles in obese, Table 3) which when averaged over all muscles was not significantly different in lean versus obese groups ($P > 0.05$, ANOVA). Altogether, the relatively much larger contraction induced increase in K_f^* than K_{fs} (in both groups) indicates that the contraction-induced increase in FA clearance is almost exclusively diverted into oxidation.

Parameters reflecting *in vivo* metabolic fluxes of plasma FA in individual tissues are given in Tables 4 and 5. R_f^* indexes the total rate of plasma FA utilization into both oxidative and non-oxidative metabolism. R_{fs} is an estimate of the

TABLE 4: Indices of FA utilization (R_f^*) and FA incorporation into storage (R_{fs}) in individual tissues of lean and obese Zucker rats *in vivo*.

	R_f^* (μmol 100 g^{-1} min^{-1})		R_{fs} (μmol 100 g^{-1} min^{-1})	
	Lean	Obese	Lean	Obese
Cerebellum	0.14 ± 0.01	$0.32 \pm 0.04^{***}$	0.26 ± 0.03	$0.54 \pm 0.07^{***}$
Liver	11.5 ± 1.4	$21.8 \pm 3.1^{**}$	13.7 ± 1.1	$26.3 \pm 3.5^{**}$
WAT	0.21 ± 0.02	$0.47 \pm 0.07^{**}$	0.47 ± 0.03	$0.93 \pm 0.16^{*}$
BAT	2.02 ± 0.14	2.37 ± 0.44	2.62 ± 0.26	5.28 ± 1.25
Heart	9.1 ± 1.3	10.2 ± 1.0	4.7 ± 0.4	$10.1 \pm 1.2^{***}$

R_f^* is an index of FA utilization rate (for both oxidation and storage) in the tissue. R_{fs} indexes flux of plasma FA into storage. Data are mean $\pm$ SE ($n = 6$ per group).
$^{*}P < 0.05$, $^{**}P < 0.01$, $^{***}P < 0.001$ versus Lean.

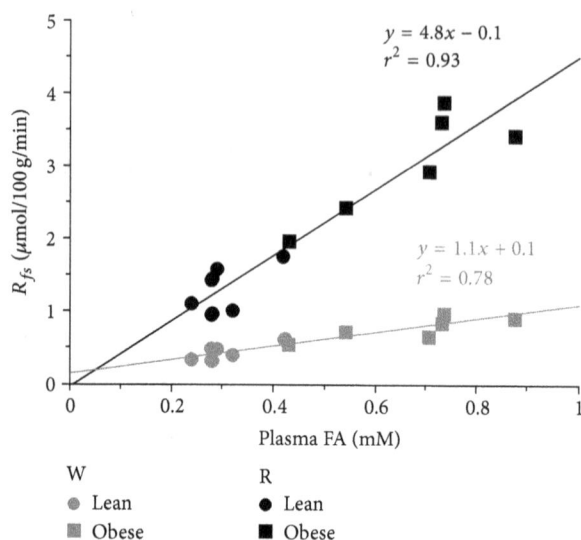

FIGURE 2: Relationship between *in vivo* flux of FA into storage in skeletal muscle and plasma FA level. Circles represent results from Lean, while squares represent results from Obese animals. Results for white quadriceps muscle (gray symbols, W) and red quadriceps muscle (black symbols, R) are plotted for individual animals ($n = 6$ rats per group). Straight lines represent linear regression equations.

rate of plasma FA incorporation into storage (non-oxidative metabolism) only. Obese animals had substantially higher R_f^* and R_{fs} values compared with Lean in the majority of tissues including cerebellum, liver, and WAT (Table 4), as well as all skeletal muscles examined (Table 5). This was caused by the higher FA levels in the obese compared to the lean animals resulting from the higher rate of entry of FA into plasma described above. Only in heart and BAT, was R_f^* similar in both Lean and Obese groups (Table 4).

Contraction increased R_f^* and R_{fs}, in the lower leg muscles, an expected consequence of the increases in K_f^* and K_{fs}, respectively (referred to above). The extent of the increase in R_f^* was greater in the Obese compared to Lean group: ($P < 0.05$, group effect, ANOVA).

Figure 2 shows the relationship between R_{fs} for WQ and RQ (Con-Leg only) and plasma FA level in individual animals. First it is apparent that in these quiescent muscles there is a simple linear dependence of plasma FA flux into storage on plasma FA concentration and that the same relationship seems to hold for both obese and lean groups, suggesting that the increased flux of FA into non-oxidative disposal in resting muscle in obese Zuckers is a direct result of the higher plasma FA levels. Second, the red oxidative muscle (RQ) has a much greater ability to store available FA than the glycolytic muscle (WQ) as evidenced by the greater slope in RQ compared to WQ ($P < 0.0001$). Linear relationships also apply for the other muscles (data not shown).

Tissue specific rates of FA oxidation cannot be calculated directly as the difference between R_f^* and R_{fs} because of the slower kinetics of ^{3}H-R-BrP compared with native FA. Plasma FA oxidation ($R_{f_{ox}}$) can however be estimated indirectly from R_{fs} and R_f^* in some tissues including muscles (see Table 6) as described in the Materials and Methods section. $R_{f_{ox}}$ in quiescent hindlimb muscles were low in comparison to the flux of FA into non-oxidative disposal (R_{fs}) and generally similar in obese compared to lean Zuckers. The exception was in the RQ muscle where the levels were higher in Obese versus Lean. Contraction (only occurring in WG, EDL and RG) induced substantial increases in $R_{f_{ox}}$ to levels similar in magnitude to the levels of R_{fs} in the corresponding muscles. Averaged across the three contracting

muscles, the contraction induced increase in FA oxidation was similar in obese versus lean rats, 2.0 ± 0.2 versus $1.6 \pm 0.3\,\mu$mol/100 g/min ($P > 0.05$) respectively. Obese rats apparently also had similar rates of plasma FA oxidation compared to lean rats in two other contracting muscle tissues, diaphragm and heart. Thus diaphragm $R_{f_{ox}}$ for Lean was 4.3 ± 0.7 versus Obese $5.9 \pm 1.8\,\mu$mol 100 g^{-1} min^{-1} while $R_{f_{ox}}$ for the heart was for Lean 43.4 ± 6.1 versus Obese $42.2 \pm 4.5\,\mu$mol 100 g^{-1} min^{-1}.

3.2. Series 2: Whole-Body FA Oxidation Rate and Its Dependence on Plasma FA Level. In Series 1, obese Zucker rats were observed to have a general elevation in plasma FA level and utilization at the whole body, as well as in muscle, fat, and liver compared to lean controls. To examine the dependence of FA oxidation on plasma FA availability, the antilipolytic agent nicotinic acid was used in several doses to suppress R_a in order to generate a range of FA levels that overlapped in the lean and obese animals. Figure 3 shows the relationships between R_a and rate of FA oxidation (R_{ox}) and plasma FA level, respectively. The nicotinic acid infusions applied succeeded in dose dependently reducing FA availability in both lean and obese Zucker rats. In both groups there was a tight linear dependence of R_{ox} on R_a (Figure 3(a)), with regression line intercepts virtually coinciding with the origin. The slope of this relationship in the obese Zucker was however 44% less than that in the lean animals ($P < 0.01$), indicating that under conditions of an equal rate of systemic FA supply the obese Zuckers would exhibit a reduced rate of FA oxidation compared to the lean Zuckers. Strong linear relationships were also apparent between R_{ox} and plasma FA level in both groups of Zucker rats (Figure 3(b)) with the

TABLE 5: R_f^* and R_{fs} in individual hind leg muscles in lean and obese Zucker rats.

| | R_f^* (μmol 100 g^{-1} min^{-1}) | | | | R_{fs} (μmol 100 g^{-1} min^{-1}) | | | |
| | Lean | | Obese | | Lean | | Obese | |
	Con-Leg	Stim-Leg	Con-Leg	Stim-Leg	Con-Leg	Stim-Leg	Con-Leg	Stim-Leg
WG	0.23 ± 0.03	$0.63 \pm 0.11^{\dagger}$	$0.47 \pm 0.06^{*}$	$1.22 \pm 0.11^{\dagger *}$	0.57 ± 0.06	$0.77 \pm 0.23^{\dagger}$	$1.32 \pm 0.20^{*}$	$1.59 \pm 0.23^{*}$
EDL	0.33 ± 0.04	$1.04 \pm 0.12^{\dagger}$	$0.72 \pm 0.10^{*}$	$1.67 \pm 0.16^{\dagger *}$	1.01 ± 0.11	$1.29 \pm 0.10^{\dagger}$	$2.84 \pm 0.42^{*}$	$3.08 \pm 0.28^{*}$
RG	0.36 ± 0.04	$0.81 \pm 0.17^{\dagger}$	$0.94 \pm 0.13^{*}$	$1.39 \pm 0.15^{\dagger *}$	1.14 ± 0.10	$1.56 \pm 0.20^{\dagger}$	$3.01 \pm 0.43^{*}$	$4.24 \pm 0.53^{\dagger *}$
WQ	0.19 ± 0.02	0.20 ± 0.02	$0.34 \pm 0.04^{*}$	$0.35 \pm 0.04^{*}$	0.42 ± 0.05	0.43 ± 0.04	$0.77 \pm 0.07^{*}$	$0.77 \pm 0.07^{*}$
RQ	0.50 ± 0.06	0.47 ± 0.06	$1.24 \pm 0.14^{*}$	$1.19 \pm 0.15^{*}$	1.29 ± 0.14	1.24 ± 0.13	$3.04 \pm 0.30^{*}$	$2.79 \pm 0.34^{*}$

Results represent mean $\pm$ SEM ($n = 6$ per group). $^{\dagger}P < 0.05$ versus Con-Leg (paired t-test, effect of sciatic nerve stimulation, within group), $^{*}P < 0.05$ versus Lean (unpaired t-test, effect of obesity status, between group).

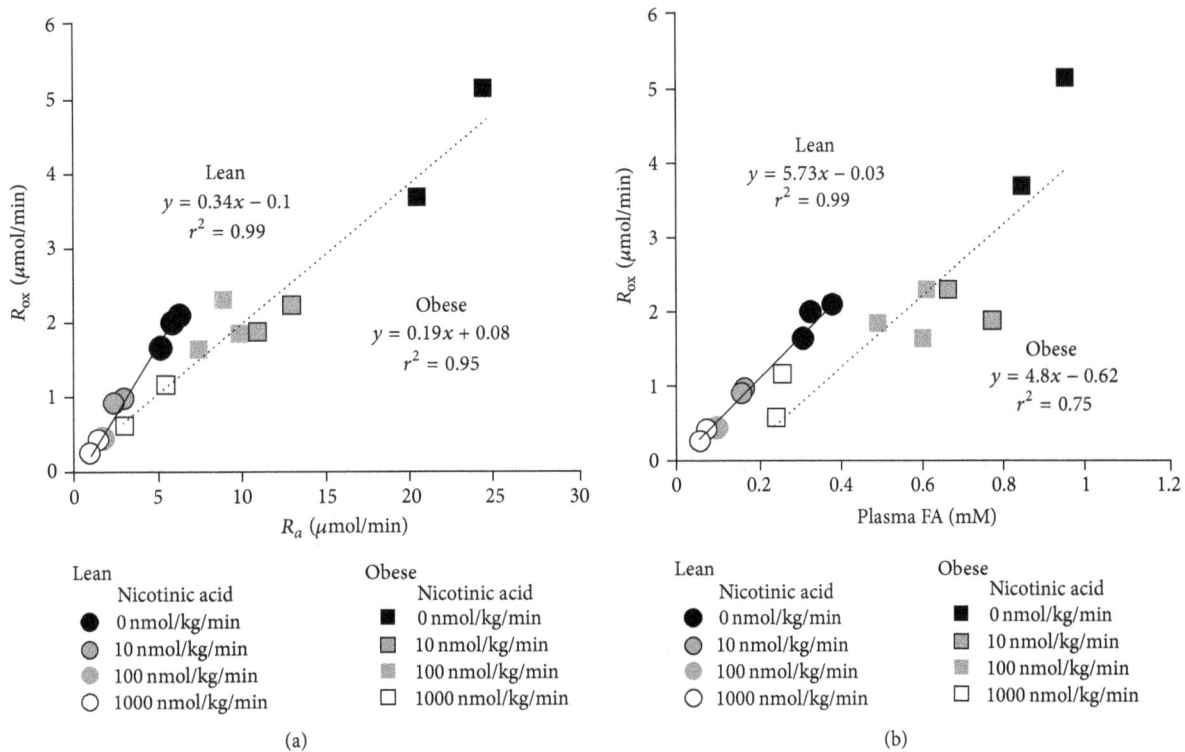

FIGURE 3: Dependence of FA oxidation (R_{ox}) on both the rate of FA appearance, R_a (a), and FA level (b), in lean and obese Zucker rats. FA availability has been pharmacologically manipulated using the antilipolytic agent nicotinic acid to generate a range of FA levels ($n = 2$-3 rats per dose).

regression line for the obese Zuckers having a lower intercept ($P < 0.05$) and tending to have a lower slope than the line for the lean Zuckers. Thus, under conditions of comparable FA levels, the obese Zuckers would manifest a reduced FA oxidation rate compared to lean Zuckers.

3.3. Series 3: Ex Vivo Fatty Acid Oxidation. In addition to assessing tissue specific FA metabolism *in vivo*, we also examined the tissues capacity to oxidize FA by measuring palmitate oxidation rates in tissue homogenates. Independent of obesity status, the results demonstrate a range of 2 orders of magnitude in the capacity to oxidize FA across different tissues of the body, lowest in WAT and highest in the heart (Figure 4). In the obese animals there was only one

tissue, BAT, where the fatty acid oxidation capacity was actually lower (-67%) than that observed in the lean animals, again consistent with the previously mentioned phenotypic difference. All other tissues of the obese animals examined showed either a similar capacity (heart, liver, cerebellum, and WAT) or a moderately enhanced ($+35\%$) capacity (skeletal muscles) to oxidize FA compared with the tissues of the lean animals.

To gain insight into the metabolic functionality of individual tissues, *in vivo* FA flux data are compared with the *ex vivo* estimates of FA oxidation capacity for individual tissues in Figure 5. Genuine rates of FA uptake (y-axis) were estimated from R_f^* values (Table 3) using procedures described in the Materials and Methods section.

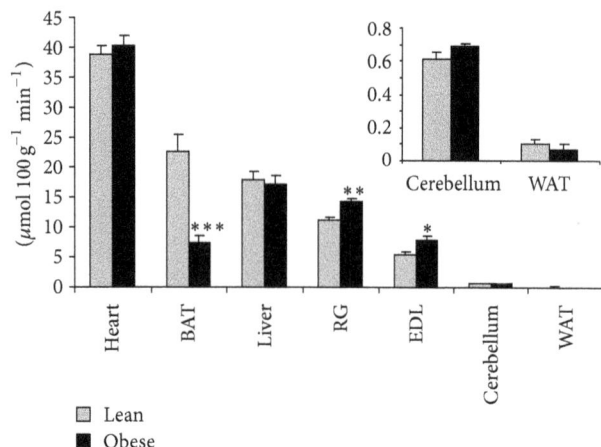

FIGURE 4: FA oxidation capacity of individual tissues, determined *ex vivo*, based on the ability of tissue homogenates to oxidize palmitate. Results are expressed as mean ± SE ($n = 5$ per group). $^{*}P < 0.05$, $^{**}P < 0.01$, $^{***}P < 0.001$ versus Lean.

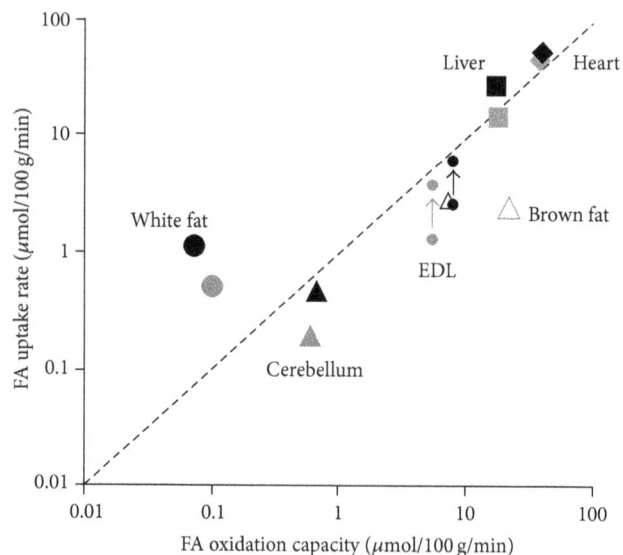

FIGURE 5: Relationship between FA uptake and FA oxidation capacity by individual tissues in lean (grey symbols) and obese (black symbols) Zucker rats. FA uptake has been calculated from R_f^* data (see Results section). The broken line indicates equality of uptake and oxidation capacity. FA uptake in the skeletal muscle, EDL (extensor digitorum longus), was assessed in the quiescent state and during sustainable twitch contractions: arrows indicate the effect of contraction.

4. Discussion

In association with marked insulin resistance, nondiabetic obese Zucker rats exhibit substantial accumulation of triglycerides and lipid intermediates in non-adipose tissues, including liver and muscle [26–28]. This lipid accumulation could result from disturbances in tissue fatty acid uptake or metabolic fate including systemic oversupply, a locally enhanced ability to take up plasma FA or impairment of their oxidation. Which of these factors predominates in obese animals *in vivo* has not been resolved. In this study *in vivo* FA metabolism, both at the whole-body and individual tissue levels, was characterized in obese Zucker rats and compared to lean Zucker rats. Completely novel information concerning the flux of FA from plasma and its metabolic fate was obtained by applying a method based on the combined use of the non-β-oxidizable FA analogue tracer, ^{3}H-R-BrP and ^{14}C-palmitate [21].

The results demonstrate that oversupply of plasma FA is a major factor in the fatty acid overload of non-adipose tissues of the obese Zucker rat. We have shown in the current study and previously [29] that the rate at which FA enters the plasma (R_a) in the fasting state in the obese animals is more than twice the rate of lean animals of the same age. The increased flux of fatty acid into the plasma has to be matched by a corresponding increase in flux of fatty acids into the tissues. Even after subtracting the fraction of R_a that disappears into the body fat, which is much greater in the obese animals, the remainder that supplies the non-adipose tissues of the body is still more than doubled in the obese compared to the lean animals, 14 versus 6 μmol min^{-1}, respectively, based on an average adipose tissue disposal equal to R_{fs} for epididymal fat (Table 4) and a body fat mass of 38% of body weight in obese versus 14% in lean animals (Table 1).

The consequence of systemic FA oversupply is a general elevation of FA flux into non-oxidative metabolism in the

tissues. Thus, in the obese animals the flux of plasma FA into storage metabolism was substantially increased in liver, skeletal muscle, WAT, and the heart compared to the corresponding fluxes in the lean animals (Tables 4 and 5). The group independent, linear relation between R_{fs} in quiescent muscles and plasma FA concentration (Figure 2) shows the importance of systemic FA in generating lipid overload in this metabolically important tissue, consistent with the old idea that FA metabolism is supply driven [30]. Our *in vivo* observations agree well with the results of previous *ex vivo* studies of skeletal muscle FA metabolism. Thus in both perfused hindlimbs [31] and incubated muscles [32] from normal rats, TG synthesis was found to be a linear function of the perfusate/media albumin-bound FA concentration, and synthesis rates were substantially higher in oxidative than glycolytic muscles.

Tissue FA uptake is determined both by plasma and extracellular FA availability and the ability of the tissue to take up FA. An enhanced ability of skeletal muscle to take up FA has been documented to occur *in vivo* in skeletal muscle of high fat fed rats, which like the obese Zucker rat, exhibit insulin resistance associated with lipid accumulation [33]. There is also evidence suggesting that this could be the case in the obese Zucker rat. *Ex vivo* studies in perfused hind limbs from lean and obese Zucker rats where perfusate FA concentration had been equalized showed that in noncontracting muscle in the absence of insulin, FA uptake is augmented in obese compared to lean Zuckers [34, 35]. However this difference, which appears to be causally associated with the degree of translocation of

TABLE 6: $R^*_{f_{ox}}$ (μmol/100 g/min) in individual hind leg muscles in lean and obese Zucker rats.

	Lean		Obese	
	Con-Leg	Stim-Leg	Con-Leg	Stim-Leg
WG	0.30 ± 0.06	$1.58 \pm 0.34^{\dagger}$	0.39 ± 0.23	$2.88 \pm 0.20^{\dagger*}$
EDL	0.22 ± 0.07	$2.57 \pm 0.31^{\dagger}$	-0.23 ± 0.36	$3.02 \pm 0.44^{\dagger}$
RG	0.20 ± 0.06	$1.43 \pm 0.44^{\dagger}$	0.43 ± 0.37	0.83 ± 0.40
WQ	0.27 ± 0.03	0.32 ± 0.04	0.47 ± 0.11	0.50 ± 0.12
RQ	0.55 ± 0.08	0.52 ± 0.11	$1.51 \pm 0.31^{*}$	$1.55 \pm 0.30^{*}$

Results represent mean $\pm$ SEM ($n = 6$ per group). $^{\dagger}P < 0.05$ versus Con-Leg (paired t-test, effect of sciatic nerve stimulation, within group), $^{*}P < 0.05$ versus Lean (unpaired t-test, effect of obesity status, between group).

putative FA transporter proteins [36], is condition dependent and can be abolished by insulin stimulation or contraction [37]. Moreover we found no evidence that the *in vivo* ability of muscle tissue to take up FA, assessed by the parameter K^*_f, is enhanced in the obese Zucker rats. While this is in contrast to the increased FA uptake reported in muscle of high fat fed rats [33], it should be noted that a comparable situation apparently exists in humans where fractional extraction of FA and perfusion of the leg were both determined to be similar in obese and lean volunteers in a classic study of Kelley and colleagues [38]. To summarize, our data suggests that at least under the conditions of this study, skeletal muscle FA oversupply in obese Zucker rats can be exclusively attributed to increased FA availability.

Another potential mechanism of muscle lipid overload is a reduction in local FA oxidation. However, our data provide no evidence that absolute rates of muscle FA oxidation *in vivo* were reduced in the obese compared with the lean Zucker rats. Information about the rate at which plasma FA enters the tissue and is immediately oxidized, rather than stored, was obtained by combining the tissue data derived from ^{3}H-R-BrP, which reflects oxidative and non-oxidative fate, with that derived from ^{14}C-palmitate incorporation into storage, reflecting non-oxidative metabolism only. Using this approach in the anesthetized preparation employed here, substantial levels of direct plasma FA oxidation were only in evidence in contracting muscles, including the beating heart and contracting diaphragm, as well as in contracting hindlimb muscles (Table 6). Most significantly, similar rates of direct oxidation of plasma FA were apparent in the working muscles of obese and lean animals.

While the absolute rate of muscle FA oxidation is not reduced in the obese animals, this does not preclude an abnormality in the control of FA oxidation. One potential abnormality relates to the primary genetic defect in the obese Zucker, the loss-of-function mutation in the leptin receptor [39]. Leptin has been shown to acutely increase FA oxidation in skeletal muscle [20] and in the isolated heart [19]. On the basis of these effects, one would expect an inappropriately low rate of FA oxidation in these animals, relative to the prevailing plasma FA concentration. Indirect support for this mechanism in the liver has been provided in diabetic obese Zucker rats where transgenic overexpression of functional

leptin receptor resulted in a remarkable reduction of the hepatic lipid content [39].

A confounding factor preventing direct comparison of oxidation rates in the obese compared with the lean animals is the higher prevailing plasma FA levels in the obese rats, which on its own would tend to drive higher rates of FA oxidation by simple mass action. To circumvent this issue we therefore studied the relationship between whole-body FA oxidation and FA availability by infusing the antilipolytic agent nicotinic acid in various doses to separate subgroups of animals (Figure 3). This revealed that indeed at comparable levels of plasma FA availability there was a lower rate of FA oxidation in the obese animals. Our *ex vivo* studies of palmitate oxidation seem to exclude the possibility that this results from a limitation in the capacity of tissues to oxidize FA. A much more likely explanation relates to the known elevation of tissue malonyl-CoA in the obese Zucker rats [40] which may be responsible by suppressing CPT I activity and diverting FA into storage as reviewed by [41]. In support of this, studies in our laboratory using pharmacological ACC inhibition in obese Zucker rats have shown normalization, relative to lean Zuckers, of both hepatic malonyl-CoA levels as well as ability to oxidize FA at the whole-body level (Oakes et al., unpublished observation).

To investigate whether a defect in fatty acid oxidation was apparent at the subcellular level we assessed the capacity of a comprehensive range of tissues to oxidize palmitate. Skeletal muscle homogenates from the obese animals had a modest enhancement in the rate of palmitate oxidation (Figure 4) compared with the lean animal, in qualitative agreement with the recently reported increases in the activities of several important enzymes involved in mitochondrial oxidation in the muscle of obese compared with lean Zucker rats [36, 42]. All other tissues of the obese Zucker rat, with the exception of BAT, possessed a similar capacity to oxidize FA compared to the lean rats. Our data are consistent with the findings of Noland et al. [28] obtained using similar methods in liver, muscle, and heart of lean and obese Zucker rats.

Overall the present data provide evidence that the excess lipid accumulation in non-adipose tissues, including skeletal muscle, of the obese Zucker rat is primarily due to an increased FA availability rather than a major intrinsic impairment in the ability to oxidize fatty acids. This deduction fits well with an *in vivo* study indicating equivalent mitochondrial oxidation capacity in skeletal muscle of diabetic Zucker rats compared to control rats [43]. Holloway et al. [36] also recently concluded, based on *ex vivo* studies, that intramyocellular lipid accumulation results from increased delivery of FA to the muscle cytosol rather than a defect in fatty acid oxidation. However, our *in vivo* findings diverge from those of Holloway et al. [36] in the major cause of the increased delivery (supply of FA to the myocyte versus enhanced plasma membrane transport) and these discrepancies likely relate to methodological differences in assessing fatty acid metabolism in the *in vivo* setting versus isolated *ex vivo* assessments.

Comparing the estimates of *in vivo* FA uptake versus *ex vivo* fatty acid oxidation capacity provides information about the metabolic functionality of the individual tissues

(Figure 5). Thus, the unique storage function of WAT is made apparent as it is the only tissue where FA uptake substantially exceeds oxidization capacity. By contrast, FA uptake in BAT lies well below its high capacity to oxidize, probably reflecting a rather low level of sympathetic activation under the physiological conditions of this study. Skeletal muscle also takes up much less FA than it is capable of oxidizing in the quiescent state, but sustainable twitch contractions causes uptake to approach oxidation capacity. In the beating heart, uptake and oxidation capacity were approximately matched, indicating that cardiac fatty acid oxidation is virtually maximal under the physiological conditions of this study. It is indeed remarkable that the intact heart seems to be able to extract FA from the plasma and oxidize it at approximately the maximal rate attainable *in vitro* where membrane barriers are absent and concentration gradients are presumably negligible. The points for the liver are also located on the line of equality of uptake and oxidation. However, unlike the heart, the liver exports a substantial fraction of the fatty acid taken up as plasma FA in the form of VLDL triglyceride and it also has a much greater capacity to transiently store FA as endogenous TG.

Based on ethical and practical considerations, whenever possible we use anesthetized preparations rather than conscious chronically catheterized animals for acute experiments. However one limitation is the fact that the skeletal musculature, of major importance for whole-body metabolism [44], in this situation is essentially inactive. To circumvent this problem we performed unilateral nerve stimulation to induce repetitive twitch contractions in a defined set of muscles in the lower hindlimb. This allows a comparison of substrate metabolism in specific contracting oxidative and glycolytic muscles of the stimulated leg with the corresponding contralateral resting muscles in the same animal. Our data show that contraction increases muscle FA uptake and oxidation without lowering the flux of FA into non-oxidative disposal. This implies that the contraction-induced increase in FA uptake is not simply the result of an elevated "pull" of fatty acids into mitochondrial β-oxidation. The latter is an important mechanism in the heart where increased respiration results in a fall in mitochondrial acetyl-CoA, in turn leading to a fall in cytosolic malonyl-CoA, relieving inhibition of carnitine palmitoyl transferase 1 (CPT I) and thereby increasing transport of fatty acyl-CoA into the mitochondria for oxidation [45]. A contraction-associated increase in skeletal muscle FA uptake without a reduction in FA storage is consistent with the involvement of a "push" mechanism and may be the result of enhanced membrane transport due to translocation of the fatty acid translocase (FAT/CD36) [46, 47].

While the methodology employed in this study offers a unique opportunity to study *in vivo* metabolism of plasma FA in a diverse range of tissues, an important limitation should be acknowledged: the method does not provide a complete assessment of overall tissue specific FA metabolism. Information is generated about the direct contribution of circulating FA to tissue specific FA metabolism but there are two significant alternative sources of FA: circulating TG and endogenous TG. The contribution of circulating TG

to tissue specific FA metabolism is potentially large. Thus, using results from a previous study [29], rates of VLDL esterified FA secretion in the fasting state corresponded to 44% and 71% of the FA R_a for the lean and obese Zucker rats respectively. Since TG secretion must be matched by tissue TG uptake, the substantially elevated (~4-fold higher) VLDL TG secretion in the obese Zucker compared to lean age-matched Zuckers [29] is almost certainly also a major contributor to tissue lipid overload. However, little is known about the true *in vivo* tissue fate of this TG-FA and methods that can provide quantitative information about the contribution of both plasma TG-FA and FA are needed. A perhaps even more challenging issue is experimentally determining the contribution of the intracellular TG pool. While pulse labeling wash-out studies might be achievable for highly oxidatively active tissues like the heart [48], assessing TG turnover in tissues with modest rates of turnover is likely to be difficult.

5. Conclusions

In summary, our assessments of *in vivo* FA fluxes demonstrate that the major factor responsible for non-adipose tissue lipid overload in the insulin resistant obese Zucker rat is systemic FA oversupply. There was no evidence for a widespread impairment in the capacity of tissues from the obese animals to oxidize FA. At the whole-body level the absolute rate of FA oxidation was not reduced in obese Zucker rats under physiological conditions, but the marked increase in FA availability was not being matched by an equivalent elevation in FA oxidation, indicating some disturbances in fatty acid utilization. When conditions of equivalent FA availability were achieved *in vivo*, the obese animals exhibited a mild defect in FA oxidation compared to Lean Zucker rats. We showed that in skeletal muscle, fatty acid uptake associated with contraction was channeled into oxidation. By contrast, non-oxidative disposal of FA in skeletal muscle was heavily influenced by availability and remarkably did not appear to be diminished by a contraction mediated increase in local oxidative metabolism.

Conflict of Interests

The authors declare they have no conflict of interests.

Acknowledgments

The authors are grateful to Anders Elmgrens group (AstraZeneca R&D Mölndal) for performing the clinical chemistry measurements. N. Turner is supported by a Future Fellowship from the Australian Research Council.

References

[1] E. W. Kraegen and G. J. Cooney, "Free fatty acids and skeletal muscle insulin resistance," *Current Opinion in Lipidology*, vol. 19, no. 3, pp. 235–241, 2008.

[2] V. T. Samuel and G. I. Shulman, "Mechanisms for insulin resistance: common threads and missing links," *Cell*, vol. 148, pp. 852–871, 2012.

[3] T. K. T. Lam, A. Carpentier, G. F. Lewis, G. Van de Werve, I. G. Fantus, and A. Giacca, "Mechanisms of the free fatty acid-induced increase in hepatic glucose production," *American Journal of Physiology*, vol. 284, no. 5, pp. E863–E873, 2003.

[4] F. Assimacopoulos-Jeannet, "Fat storage in pancreas and in insulin-sensitive tissues in pathogenesis of type 2 diabetes," *International Journal of Obesity*, vol. 28, supplement 4, no. 4, pp. S53–S57, 2004.

[5] H. N. Ginsberg, Y. L. Zhang, and A. Hernandez-Ono, "Regulation of plasma triglycerides in insulin resistance and diabetes," *Archives of Medical Research*, vol. 36, no. 3, pp. 232–240, 2005.

[6] J. Y. Kim, R. C. Hickner, R. L. Cortright, G. L. Dohm, and J. A. Houmard, "Lipid oxidation is reduced in obese human skeletal muscle," *American Journal of Physiology*, vol. 279, no. 5, pp. E1039–E1044, 2000.

[7] I. Dahlman, M. Forsgren, A. Sjögren et al., "Downregulation of electron transport chain genes in visceral adipose tissue in type 2 diabetes independent of obesity and possibly involving tumor necrosis factor-α," *Diabetes*, vol. 55, no. 6, pp. 1792–1799, 2006.

[8] M. Kaaman, L. M. Sparks, V. Van Harmelen et al., "Strong association between mitochondrial DNA copy number and lipogenesis in human white adipose tissue," *Diabetologia*, vol. 50, no. 12, pp. 2526–2533, 2007.

[9] H. Hwang, B. P. Bowen, N. Lefort et al., "Proteomics analysis of human skeletal muscle reveals novel abnormalities in obesity and type 2 diabetes," *Diabetes*, vol. 59, no. 1, pp. 33–42, 2010.

[10] K. Morino, K. F. Petersen, S. Dufour et al., "Reduced mitochondrial density and increased IRS-1 serine phosphorylation in muscle of insulin-resistant offspring of type 2 diabetic parents," *Journal of Clinical Investigation*, vol. 115, no. 12, pp. 3587–3593, 2005.

[11] V. B. Ritov, E. V. Menshikova, J. He, R. E. Ferrell, B. H. Goodpaster, and D. E. Kelley, "Deficiency of subsarcolemmal mitochondria in obesity and type 2 diabetes," *Diabetes*, vol. 54, no. 1, pp. 8–14, 2005.

[12] H. J. Choo, J. H. Kim, O. B. Kwon et al., "Mitochondria are impaired in the adipocytes of type 2 diabetic mice," *Diabetologia*, vol. 49, no. 4, pp. 784–791, 2006.

[13] J. A. Houmard, "Intramuscular lipid oxidation and obesity," *American Journal of Physiology*, vol. 294, no. 4, pp. R1111–R1116, 2008.

[14] G. P. Holloway, A. Bonen, and L. L. Spriet, "Regulation of skeletal muscle mitochondrial fatty acid metabolism in lean and obese individuals," *American Journal of Clinical Nutrition*, vol. 89, no. 1, pp. 455S–462S, 2009.

[15] D. E. Kelley and L. J. Mandarino, "Fuel selection in human skeletal muscle in insulin resistance: a reexamination," *Diabetes*, vol. 49, no. 5, pp. 677–683, 2000.

[16] K. Koyama, G. Chen, Y. Lee, and R. H. Unger, "Tissue triglycerides, insulin resistance, and insulin production: implications for hyperinsulinemia of obesity," *American Journal of Physiology*, vol. 273, no. 4, pp. E708–E713, 1997.

[17] J. S. Lally, L. A. Snook, X. X. Han, A. Chabowski, A. Bonen, and G. P. Holloway, "Subcellular lipid droplet distribution in red and white muscles in the obese Zucker rat," *Diabetologia*, vol. 55, pp. 479–488, 2012.

[18] R. H. Unger, "Minireview: weapons of lean body mass destruction: the role of ectopic lipids in the metabolic syndrome," *Endocrinology*, vol. 144, no. 12, pp. 5159–5165, 2003.

[19] L. L. Atkinson, M. A. Fischer, and G. D. Lopaschuk, "Leptin activates cardiac fatty acid oxidation independent of changes in the AMP-activated protein kinase-acetyl-CoA carboxylase-malonyl-CoA axis," *Journal of Biological Chemistry*, vol. 277, no. 33, pp. 29424–29430, 2002.

[20] Y. Minokoshi, Y. B. Kim, O. D. Peroni et al., "Leptin stimulates fatty-acid oxidation by activating AMP-activated protein kinase," *Nature*, vol. 415, no. 6869, pp. 339–343, 2002.

[21] N. D. Oakes and S. M. Furler, "Evaluation of free fatty acid metabolism *in vivo*," *Annals of the New York Academy of Sciences*, vol. 967, pp. 158–175, 2002.

[22] N. D. Oakes, A. Kjellstedt, G. B. Forsberg et al., "Development and initial evaluation of a novel method for assessing tissue-specific plasma free fatty acid utilization *in vivo* using (R)-2-bromopalmitate tracer," *Journal of Lipid Research*, vol. 40, no. 6, pp. 1155–1169, 1999.

[23] N. D. Oakes, P. Thalén, E. Aasum et al., "Cardiac metabolism in mice: tracer method developments and *in vivo* application revealing profound metabolic inflexibility in diabetes," *American Journal of Physiology*, vol. 290, no. 5, pp. E870–E881, 2006.

[24] L. E. Wright, A. E. Brandon, A. J. Hoy et al., "Amelioration of lipid-induced insulin resistance in rat skeletal muscle by overexpression of Pgc-1beta involves reductions in long-chain acyl-CoA levels and oxidative stress," *Diabetologia*, vol. 54, pp. 1417–1426, 2011.

[25] N. D. Oakes, P. Thalén, T. Hultstrand et al., "Tesaglitazar, a dual PPARα/γ agonist, ameliorates glucose and lipid intolerance in obese Zucker rats," *American Journal of Physiology*, vol. 289, no. 4, pp. R938–R946, 2005.

[26] J. Franch, J. Knudsen, B. A. Ellis, P. K. Pedersen, G. J. Cooney, and J. Jensen, "Acyl-CoA binding protein expression is fiber type-specific and elevated in muscles from the obese insulin-resistant Zucker rat," *Diabetes*, vol. 51, no. 2, pp. 449–454, 2002.

[27] P. D. Hockings, K. K. Changani, N. Saeed et al., "Rapid reversal of hepatic steatosis, and reduction of muscle triglyceride, by rosiglitazone: MRI/S studies in Zucker fatty rats," *Diabetes, Obesity and Metabolism*, vol. 5, no. 4, pp. 234–243, 2003.

[28] R. C. Noland, T. L. Woodlief, B. R. Whitfield et al., "Peroxisomal-mitochondrial oxidation in a rodent model of obesity-associated insulin resistance," *American Journal of Physiology*, vol. 293, no. 4, pp. E986–E1001, 2007.

[29] N. D. Oakes, P. G. Thalén, S. M. Jacinto, and B. Ljung, "Thiazolidinediones increase plasma-adipose tissue FFA exchange capacity and enhance insulin-mediated control of systemic FFA availability," *Diabetes*, vol. 50, no. 5, pp. 1158–1165, 2001.

[30] B. Issekutz Jr., W. M. Bortz, H. I. Miller, and P. Paul, "Turnover rate of plasma FFA in humans and in dogs," *Metabolism*, vol. 16, no. 11, pp. 1001–1009, 1967.

[31] L. Budohoski, J. Gorski, K. Nazar, H. Kaciuba-Uscilko, and R. L. Terjung, "Triacylglycerol synthesis in the different skeletal muscle fiber sections of the rat," *American Journal of Physiology*, vol. 271, no. 3, pp. E574–E581, 1996.

[32] D. J. Dyck, S. J. Peters, J. Glatz et al., "Functional differences in lipid metabolism in resting skeletal muscle of various fiber types," *American Journal of Physiology*, vol. 272, no. 3, pp. E340–E351, 1997.

[33] B. D. Hegarty, G. J. Cooney, E. W. Kraegen, and S. M. Furler, "Increased efficiency of fatty acid uptake contributes to lipid accumulation in skeletal muscle of high fat-fed insulin-resistant rats," *Diabetes*, vol. 51, no. 5, pp. 1477–1484, 2002.

[34] J. J. F. P. Luiken, Y. Arumugam, D. J. Dyck et al., "Increased rates of fatty acid uptake and plasmalemmal fatty acid transporters

in obese Zucker rats," *Journal of Biological Chemistry*, vol. 276, no. 44, pp. 40567–40573, 2001.

[35] L. P. Turcotte, J. R. Swenberger, M. Z. Tucker, and A. J. Yee, "Increased fatty acid uptake and altered fatty acid metabolism in insulin-resistant muscle of obese Zucker rats," *Diabetes*, vol. 50, no. 6, pp. 1389–1396, 2001.

[36] G. P. Holloway, C. R. Benton, K. L. Mullen et al., "In obese rat muscle transport of palmitate is increased and is channeled to triacylglycerol storage despite an increase in mitochondrial palmitate oxidation," *American Journal of Physiology*, vol. 296, no. 4, pp. E738–E747, 2009.

[37] X. X. Han, A. Chabowski, N. N. Tandon et al., "Metabolic challenges reveal impaired fatty acid metabolism and translocation of FAT/CD36 but not FABPpm in obese Zucker rat muscle," *American Journal of Physiology*, vol. 293, no. 2, pp. E566–E575, 2007.

[38] D. E. Kelley, B. Goodpaster, R. R. Wing, and J. A. Simoneau, "Skeletal muscle fatty acid metabolism in association with insulin resistance, obesity, and weight loss," *American Journal of Physiology*, vol. 277, no. 6, pp. E1130–E1141, 1999.

[39] Y. Lee, M. Y. Wang, T. Kakuma et al., "Liporegulation in diet-induced obesity: the antisteatotic role of hyperleptinemia," *Journal of Biological Chemistry*, vol. 276, no. 8, pp. 5629–5635, 2001.

[40] M. J. Azain and J. A. Ontko, "An explanation for decreased ketogenesis in the liver of the obese Zucker rat," *American Journal of Physiology*, vol. 257, no. 4, pp. R822–R828, 1989.

[41] D. Saggerson, "Malonyl-CoA, a key signaling molecule in mammalian cells," *Annual Review of Nutrition*, vol. 28, pp. 253–272, 2008.

[42] N. Turner, C. R. Bruce, S. M. Beale et al., "Excess lipid availability increases mitochondrial fatty acid oxidative capacity in muscle: evidence against a role for reduced fatty acid oxidation in lipid-induced insulin resistance in rodents," *Diabetes*, vol. 56, no. 8, pp. 2085–2092, 2007.

[43] H. M. De Feyter, E. Lenaers, S. M. Houten et al., "Increased intramyocellular lipid content but normal skeletal muscle mitochondrial oxidative capacity throughout the pathogenesis of type 2 diabetes," *The FASEB Journal*, vol. 22, no. 11, pp. 3947–3955, 2008.

[44] G. I. Shulman, D. L. Rothman, T. Jue, P. Stein, R. A. DeFronzo, and R. G. Shulman, "Quantitation of muscle glycogen synthesis in normal subjects and subjects with non-insulin-dependent diabetes by 13C nuclear magnetic resonance spectroscopy," *The New England Journal of Medicine*, vol. 322, no. 4, pp. 223–228, 1990.

[45] V. Saks, P. Dzeja, U. Schlattner, M. Vendelin, A. Terzic, and T. Wallimann, "Cardiac system bioenergetics: metabolic basis of the frank-starling law," *Journal of Physiology*, vol. 571, no. 2, pp. 253–273, 2006.

[46] A. Bonen, J. J. F. P. Luiken, Y. Arumugam, J. F. C. Glatz, and N. N. Tandon, "Acute regulation of fatty acid uptake involves the cellular redistribution of fatty acid translocase," *Journal of Biological Chemistry*, vol. 275, no. 19, pp. 14501–14508, 2000.

[47] Y. Yoshida, S. S. Jain, J. T. McFarlan, L. A. Snook, A. Chabowski, and A. Bonen, "Exercise-, and training-induced upregulation of skeletal muscle fatty acid oxidation are not solely dependent on mitochondrial machinery and biogenesis," *The Journal of Physiology*, 2012.

[48] M. Saddik and G. D. Lopaschuk, "Myocardial triglyceride turnover and contribution to energy substrate utilization in isolated working rat hearts," *Journal of Biological Chemistry*, vol. 266, no. 13, pp. 8162–8170, 1991.

Molecular Characterization of Lipopolysaccharide Binding to Human α-1-Acid Glycoprotein

Johnny X. Huang,[1] Mohammad A. K. Azad,[2] Elizabeth Yuriev,[3] Mark A. Baker,[4] Roger L. Nation,[2] Jian Li,[2] Matthew A. Cooper,[1] and Tony Velkov[2]

[1] Institute for Molecular Bioscience, The University of Queensland, 306 Carmody Road, St. Lucia, QLD 4072, Australia
[2] Drug Development and Innovation, Drug Delivery, Disposition and Dynamics, Monash Institute of Pharmaceutical Sciences, Monash University, 381 Royal Parade, Parkville, VIC 3052, Australia
[3] Medicinal Chemistry, Monash Institute of Pharmaceutical Sciences, Monash University, 381 Royal Parade, Parkville, VIC 3052, Australia
[4] Priority Research Centre in Reproductive Science, School of Environmental and Life Sciences, University of Newcastle, Callaghan, NSW 2308, Australia

Correspondence should be addressed to Tony Velkov, tony.velkov@monash.edu.au

Academic Editor: Igor C. Almeida

The ability of AGP to bind circulating lipopolysaccharide (LPS) in plasma is believed to help reduce the proinflammatory effect of bacterial lipid A molecules. Here, for the first time we have characterized human AGP binding characteristics of the LPS from a number of pathogenic Gram-negative bacteria: *Escherichia coli*, *Salmonella typhimurium*, *Klebsiella pneumonia*, *Pseudomonas aeruginosa*, and *Serratia marcescens*. The binding affinity and structure activity relationships (SAR) of the AGP-LPS interactions were characterized by surface plasma resonance (SPR). In order to dissect the contribution of the lipid A, core oligosaccharide and O-antigen polysaccharide components of LPS, the AGP binding affinity of LPS from smooth strains, were compared to lipid A, Kdo2-lipid A, R_a, R_d, and R_e rough LPS mutants. The SAR analysis enabled by the binding data suggested that, in addition to the important role played by the lipid A and core components of LPS, it is predominately the unique species- and strain-specific carbohydrate structure of the O-antigen polysaccharide that largely determines the binding affinity for AGP. Together, these data are consistent with the role of AGP in the binding and transport of LPS in plasma during acute-phase inflammatory responses to invading Gram-negative bacteria.

1. Introduction

The human body is continuously challenged by infectious microorganisms. Accordingly, it has evolved numerous mechanisms for the early recognition and efficient elimination of viable microbes and their remnants [1, 2]. For defense against invading Gram-negative bacteria, the recognition of bacterial cellular components such as LPS by the innate immune system is an important event for induction of the inflammatory immune response, which is responsible for targeting the invading microorganisms and for elimination and clearance of highly endotoxic LPS [1–4].

LPS is present only in the outer leaflet of the outer membrane (OM) in Gram-negative bacteria [5–7]. Structurally,

LPS of enterobacteria consists of three components: (1) lipid A, a disaccharide acylated with fatty acid chains which is the toxic component of LPS; (2) the core region, a nonrepetitive oligosaccharide (~9 sugars in length) which can be subdivided into the inner and outer parts; (3) O-antigen, a serogroup-specific polysaccharide of repetitive oligosaccharide units (Figure 1(c); Table 1) [5–7]. LPS mediates a range of pathophysiological processes, more specifically, it is the lipid A component that is responsible for inducing the immunopathogenic processes that can lead to endotoxemia-associated high mortality [3, 8, 9]. Lipid A is bound by the toll-like receptor 4 (TLR4) expressed on the membrane of macrophages and neutrophils [2–4]. Activation of TLR4 by LPS is also dependent on interactions with an additional cell

TABLE 1: The structures of the core oligosaccharide and O-antigen polysaccharides of the LPS samples used in this study. The general structure of the *Salmonella* LPS is shown in the first row. The stages at which the biosynthetic enzyme defects disrupt biosynthesis resulting in the production of the truncated "rough" LPS chemotypes $R_a \rightarrow R_e$ are indicated by the segmented arrows.

General structure of the *Salmonella* LPS

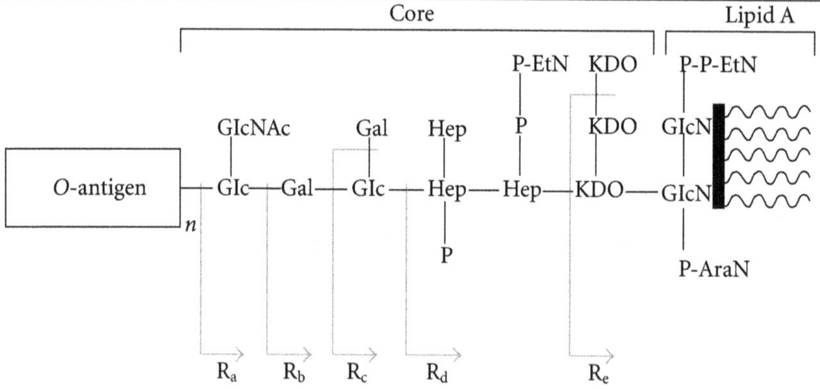

O-antigen structures

E. coli serotype O111:B4 [24]

α-Col-(1→6)
|
→4)-α-D-Glc-(1→4)-α-D-Gal-(1→3)-β-D-GlcNAc-(1→
|
α-Col-(1→3)

Neutral

E. coli serotype O127:B8 [24]

→2)-α-L-Fuc-(1→2)-β-D-Gal-(1→3)-α-D-GalNAc-(1→3)-α-D-GalNAc-(1→

Neutral

E. coli serotype O55:B5 [24]

→6)-β-D-GlcNAc-(1→3)-α-D-Gal-(1→3)-β-D-GalNAc-(1→
|
α-Col-(1→2)-β-D-Gal-(1→3)

Neutral

Klebsiella pneumoniae serotype O2a,c [25, 26]

→3)-α-D-Gal*p*-(1→3)-β-D-Gal*f*-(1→

Neutral

Pseudomonas aeruginosa serotype O10 [27–29]

Ac-(1→2)
|
→3)-α-L-Rha-(1→4)-α-L-GalNAcA-(1→3)-α-D-QuiNAc-(1→

Acidic

Salmonella typhimurium O-antigen 4,5,12 [30, 31]

Ac-(1→2)
|
α-Abe-(1→3) α-D-Glc-(1→6)
| |
→4)-α-D-Man-(1→4)-β-L-Rha-(1→3)-β-D-Gal-(1→

Neutral

TABLE 1: Continued.

LPS core region structures

E. coli R2 core Serotype O127:B8 [7, 30, 31]

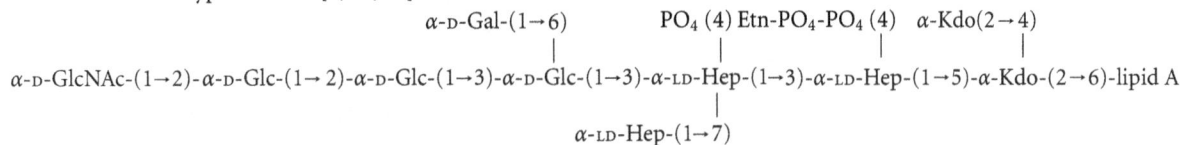

α-D-Gal-(1→6)   PO$_4$ (4) Etn-PO$_4$-PO$_4$ (4) α-Kdo(2→4)
| | |
α-D-GlcNAc-(1→2)-α-D-Glc-(1→2)-α-D-Glc-(1→3)-α-D-Glc-(1→3)-α-LD-Hep-(1→3)-α-LD-Hep-(1→5)-α-Kdo-(2→6)-lipid A
|
α-LD-Hep-(1→7)

Acidic

E. coli R3 core Serotype O111:B4 and O55:B5 [7, 30, 31]

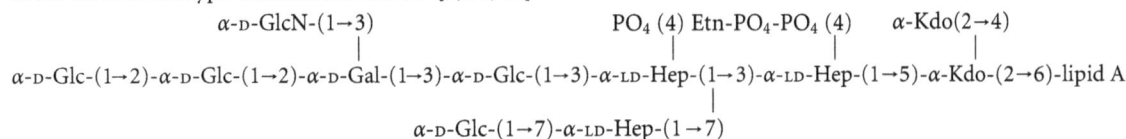

α-D-GlcN-(1→3)   PO$_4$ (4) Etn-PO$_4$-PO$_4$ (4) α-Kdo(2→4)
| | |
α-D-Glc-(1→2)-α-D-Glc-(1→2)-α-D-Gal-(1→3)-α-D-Glc-(1→3)-α-LD-Hep-(1→3)-α-LD-Hep-(1→5)-α-Kdo-(2→6)-lipid A
|
α-D-Glc-(1→7)-α-LD-Hep-(1→7)

Acidic

E. coli EH100 (Rough, R$_a$) [7, 30, 31]

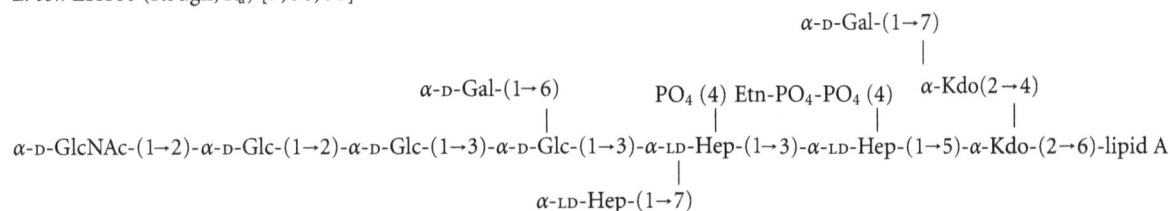

α-D-Gal-(1→7)
|
α-D-Gal-(1→6)   PO$_4$ (4) Etn-PO$_4$-PO$_4$ (4) α-Kdo(2→4)
| | |
α-D-GlcNAc-(1→2)-α-D-Glc-(1→2)-α-D-Glc-(1→3)-α-D-Glc-(1→3)-α-LD-Hep-(1→3)-α-LD-Hep-(1→5)-α-Kdo-(2→6)-lipid A
|
α-LD-Hep-(1→7)

Acidic

E. coli F583 (Rough, R$_d$) [7, 30–32]

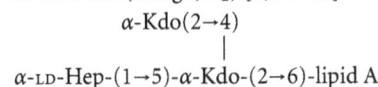

α-Kdo(2→4)
|
α-LD-Hep-(1→5)-α-Kdo-(2→6)-lipid A

Acidic

Klebsiella pneumoniae serotype O2 [7, 25, 26]

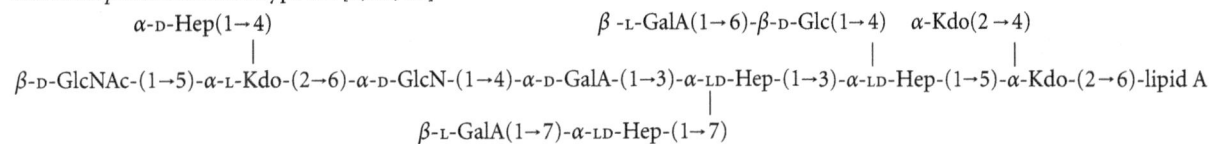

α-D-Hep(1→4) β-L-GalA(1→6)-β-D-Glc(1→4) α-Kdo(2→4)
| | |
β-D-GlcNAc-(1→5)-α-L-Kdo-(2→6)-α-D-GlcN-(1→4)-α-D-GalA-(1→3)-α-LD-Hep-(1→3)-α-LD-Hep-(1→5)-α-Kdo-(2→6)-lipid A
|
β-L-GalA(1→7)-α-LD-Hep-(1→7)

Acidic

Pseudomonas aeruginosa serotype O10 [27–29]

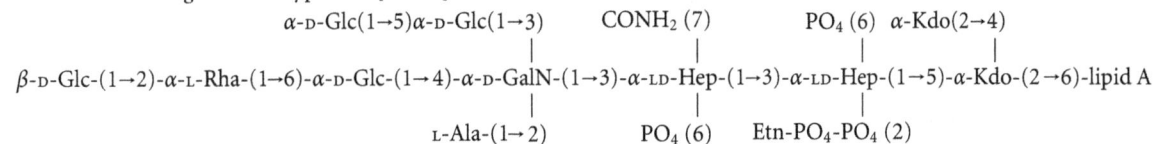

α-D-Glc(1→5)α-D-Glc(1→3)   CONH$_2$ (7)   PO$_4$ (6) α-Kdo(2→4)
| | | |
β-D-Glc-(1→2)-α-L-Rha-(1→6)-α-D-Glc-(1→4)-α-D-GalN-(1→3)-α-LD-Hep-(1→3)-α-LD-Hep-(1→5)-α-Kdo-(2→6)-lipid A
| | |
L-Ala-(1→2)   PO$_4$ (6)   Etn-PO$_4$-PO$_4$ (2)

Acidic

Salmonella typhimurium (smooth) [30, 31]

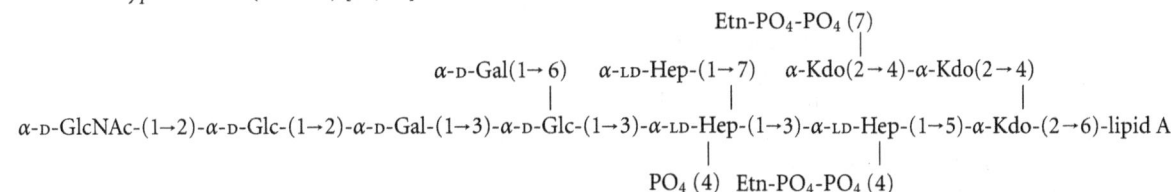

Etn-PO$_4$-PO$_4$ (7)
|
α-D-Gal(1→6) α-LD-Hep-(1→7) α-Kdo(2→4)-α-Kdo(2→4)
| | | |
α-D-GlcNAc-(1→2)-α-D-Glc-(1→2)-α-D-Gal-(1→3)-α-D-Glc-(1→3)-α-LD-Hep-(1→3)-α-LD-Hep-(1→5)-α-Kdo-(2→6)-lipid A
| |
PO$_4$ (4) Etn-PO$_4$-PO$_4$ (4)

Acidic

TABLE 1: Continued.

LPS core region structures

Salmonella typhimurium TV119 (Rough, R_a) [30, 31]

$$Etn\text{-}PO_4\text{-}PO_4 \ (7)$$
$$|$$
$$\alpha\text{-}D\text{-}Gal(1{\to}6) \quad \alpha\text{-}LD\text{-}Hep\text{-}(1{\to}7) \quad \alpha\text{-}Kdo(2{\to}4)\text{-}\alpha\text{-}Kdo(2{\to}4)$$
$$| \qquad\qquad\qquad\qquad | \qquad\qquad\qquad\qquad |$$
$$\alpha\text{-}D\text{-}GlcNAc\text{-}(1{\to}2)\text{-}\alpha\text{-}D\text{-}Glc\text{-}(1{\to}2)\text{-}\alpha\text{-}D\text{-}Gal\text{-}(1{\to}3)\text{-}\alpha\text{-}D\text{-}Glc\text{-}(1{\to}3)\text{-}\alpha\text{-}LD\text{-}Hep\text{-}(1{\to}3)\text{-}\alpha\text{-}LD\text{-}Hep\text{-}(1{\to}5)\text{-}\alpha\text{-}Kdo\text{-}(2{\to}6)\text{-}lipid\ A$$
$$| \qquad\qquad\qquad\qquad |$$
$$PO_4\ (4) \quad Etn\text{-}PO_4\text{-}PO_4\ (4)$$

Acidic

Salmonella typhimurium SL1181 (Rough, R_e) [30, 31]

$$Etn\text{-}PO_4\text{-}PO_4\ (7)$$
$$|$$
$$\alpha\text{-}Kdo(2{\to}4)\text{-}\alpha\text{-}Kdo(2{\to}4)$$
$$|$$
$$\alpha\text{-}Kdo\text{-}(2{\to}6)\text{-}lipid\ A$$

Acidic

Serratia marcescens [33, 34]

$$\beta\text{-}D\text{-}Glc\text{-}(1{\to}4) \quad \alpha\text{-}Kdo(2{\to}4)$$
$$| \qquad\qquad\qquad\qquad |$$
$$\alpha\text{-}L,D\text{-}Hep\text{-}(1{\to}2)\text{-}\alpha\text{-}D,D\text{-}Hep\text{-}(1{\to}2\ )\text{-}\alpha\text{-}D\text{-}GalA\text{-}(1{\to}3)\text{-}\alpha\text{-}L,D\text{-}Hep\text{-}(1{\to}3)\text{-}\alpha\text{-}L,D\text{-}Hep\text{-}(1{\to}5)\text{-}\alpha\text{-}Kdo\text{-}(2{\to}6)\text{-}lipid\ A$$
$$| \qquad\qquad\qquad\qquad |$$
$$\alpha\text{-}LD\text{-}Hep\text{-}(1{\to}7) \qquad\qquad L\text{-}Ara4N(1{\to}8)$$

Acidic

surface co-receptor MD-2 [2–4]. Moreover, CD14 and LPS-binding protein (LBP) are known to facilitate the presentation of LPS to MD-2 [2–4]. Once activated, the LPS-TLR4 complex stimulates signal transduction pathways that initiate the production of inflammatory cytokines, chemokines and, after hepatocyte activation, acute-phase proteins such as α_1-acid glycoprotein (AGP) that are central components of the inflammatory immune response to the invading microbe [3, 4]. The LPS-induced overstimulation of the immune system can lead to the excessive release of these endogenous inflammatory mediators, resulting in multi-organ failure, septic shock syndrome and even death [2, 3, 8].

Human AGP exists as three genetic variants, the A variant, the F1 and S variants [10–15]. The expression of human AGP is under the control of two adjacent genes ORM1 (*syn.* AAG-A) and ORM2 (*syn.* AAG-B/B′), situated on chromosome 9 [11, 12]. The more active of the two, ORM1, that is induced during acute phase reactions, encodes the F1 and S variants, and ORM2 encodes the A variant [10–15]. The precursor product of the ORM1 gene is a 201 amino acid polypeptide with an 18 residue N-terminal secretory peptide that is cleaved [10–15]. The F1 and S variants, encoded by two alleles of the ORM1 gene differ only in a single amino acid codon (Gln20 → Arg), and hereon in shall be referred to collectively as the F1*S variant. The ORM2 gene displays 22 base substitutions, which translates into 21 amino acid substitutions between the F1*S and A protein variants [10–12]. The resolution of the crystallographic structures of the human F1*S and A AGP variants revealed that the binding cavity of each variant is sub-divided into lobes [16, 17]. This intricate cavity organization suggests the AGP-drug binding site consists of partially overlapping sub-sites as opposed to the existence of distinct binding sites for acidic, basic and non-polar ligands (Figure 1(d)) [16–18]. On the primary level, AGP is composed of a single polypeptide chain of 183 amino acids [15, 19, 20]. The polypeptide component only contributes about a half of its total molecular mass of approximately 41 kDa, the rest of its mass derives from the five N-linked sialyl-glycans which confer AGP with a net negative charge at physiological pH [21–23]. These features also render AGP very soluble and acidic (pI ∼ 2.8–3.8) [15, 19, 20].

In healthy individuals, the basal plasma concentration of AGP is approximately 0.7–1.0 g/L making it one of the predominant plasma proteins. The plasma AGP concentrations can fluctuate widely between health and disease, in diseased states such as sepsis, AGP levels can increase up to 5-fold [15, 35–37]. Therefore, the effect of AGP binding on the activity of highly bound substance scan be significant during acute-phase reactions. Although AGP is an abundant plasma protein, its true physiological significance remains enigmatic. However, the time course of AGP production during acute-phase responses together with its high avidity for both exogenous and endogenous inflammatory mediators supports an immune-modulatory and/or transporter activity [19, 20, 38, 39]. AGP has been implicated in being part of the physiological response to a variety of insults such as major trauma, tissue necrosis, microbial infection, and exacerbations of

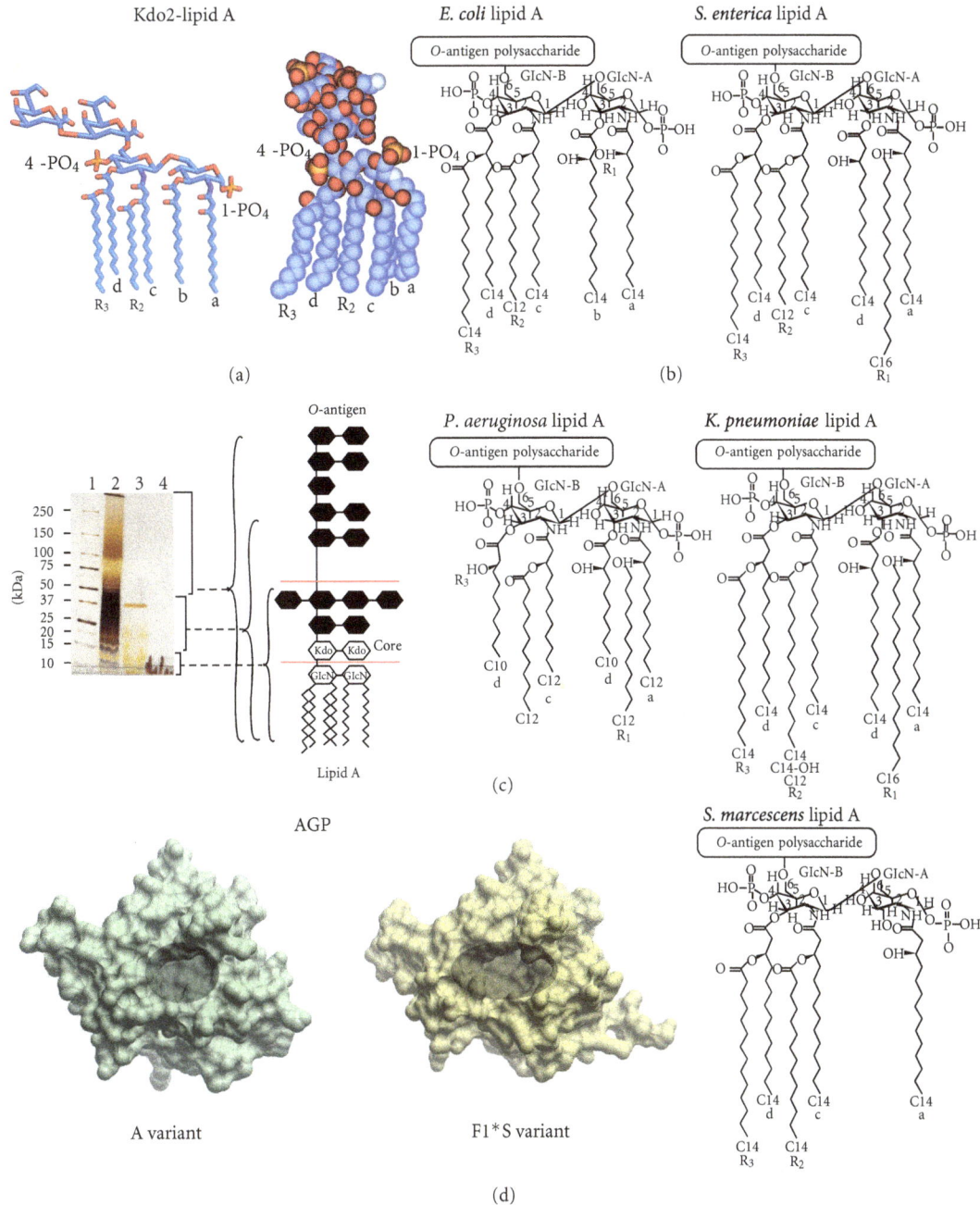

FIGURE 1: (a) Structure of Kdo2-lipid A shown in stick (left-hand panel) and CPK (right-hand panel) representation. (b) Chemical structures of the lipid A component of the LPS samples used in this study [41–43]. (c) 4–20% SDS-PAGE separation of the LPS samples used in this study. The well samples contained 30 μg of each LPS sample. The gel was stained by the silver staining technique used for LPS [44]. The structural organization of LPS is shown schematically in the right-hand panel. Lane 1: molecular weight; Lane 2: *E. coli* O111:B4 LPS; Lane 3: *E. coli* EH100 (R$_a$) LPS: Lane 4. *E. coli* F583 (R$_d$) LPS. (d) Surface representation of the A (PDB ID: 3APU) and F1*S (PDB ID: 3KQ0) variants of human AGP. The ligand binding pocket of each variant is highlighted by the shaded area.

inflammatory diseases [19]. Several lines of evidence suggest that one of the potential physiological functions of AGP is to bind LPS and exert immune-modulatory and/or transporter functions in relation to LPS during the acute-phase inflammatory response to Gram-negative bacterial infection: (1) AGP was demonstrated to bind directly with *E. coli* O111:B4 LPS using dynamic light scattering particle sizing, particle mobility and flow microcalorimetry techniques [39]. (2) AGP agglutinated LPS impregnated rabbit red blood cells [39]. (3) In a meningococcal endotoxemia mouse model, the intraperitoneal administration of 8 mg of bovine AGP 2 h prior to LPS challenge was shown to protect against sepsis [39]. (4) Intraperitoneal administration of 10 mg AGP, 2 h prior to a lethal challenge of *K. pneumoniae* increased the survival of mice [40]. (5) Transgenic over expression of rat AGP protected mice from a lethal challenge of *K. pneumonia* [40].

Notwithstanding the noted *in vitro-in vivo* correlations, these previous studies have not addressed any structural details of the LPS-AGP interaction. A better understanding of the molecular mechanisms that drive the interaction between LPS and central components of the host inflammatory response to infection (such as LPS-AGP complexation) is needed to enable the development of new treatment strategies for severe sepsis. This study is the first to utilize SPR and fluorometric binding assays together with molecular modeling to characterize the SAR that drive the binding of LPS to human AGP.

2. Materials and Methods

2.1. Materials. Human AGP, fluorescein isothiocyanate-(FITC) labeled *Escherichia coli* O111:B4LPS, and LPS from *E. coli* O111:B4, *E. coli* O127:B8 (ATCC 12740), *E. coli* EH-100 (R_a mutant) LPS, *E. coli* F-583 (R_d mutant) LPS, *Klebsiella pneumonia* (ATCC 15380) LPS, *Pseudomonas aeruginosa* serotype 10 LPS, *Salmonella typhimurium* (ATCC 7823) LPS, *Salmonella typhimurium* strain TV119 (R_a mutant) LPS, *Salmonella typhimurium* strain SL1181 (R_e mutant) LPS, *Serratia marcescens* (ATCC 21639) LPS, and diphosphoryl lipid A from *E. coli* were obtained from Sigma-Aldrich (Sydney, NSW, Australia). Kdo2-lipid A was obtained from Avanti Polar Lipids (Alabama, USA). All other reagents were of the highest grade commercially available.

2.2. Fluorometric Assay of FITC-E. Coli O111:B4 LPS Binding to AGP. The binding of FITC-LPS to AGP was measured by titrating a 1 mL solution of FITC-LPS (1.5 μM) in a quartz cuvette with aliquots of AGP (20–50 μM). Fluorescence was measured using a Cary Eclipse Fluorescence spectrophotometer (Varian, Mulgrave, VIC, Australia) set at an excitation wavelength specific for the FITC fluorophore (Exλ 494 nm). Slit widths were set to 5 nm for both the excitation and emission monochromators. The emission spectrum was collected in the 500 $\rightarrow$ 650 nm range.

2.3. Surface Plasmon Resonance (SPR) Assay of LPS Binding to Human AGP. SPR experiments were performed using a Biacore T200 instrument. AGP was immobilized on a CM5 sensor chip surface using the surface thiol coupling method following manufacturer's instructions (GE Health Care, Melbourne, Australia). In brief, 0.5 mg AGP was dissolved in 0.5 mL of 0.1 M morpholino-ethane sulfonic acid (MES) buffer pH 5.0. Then 250 μL of 15 mg/mL 2-(2-Pyridinyldithil)ethaneamine hydrochloride (PDEA) and 25 μL of 0.4 M 1-ethyl-3-(3-dimethylaminopropyl)carbodiimide hydrochloride (EDC) were added and incubated at room temperature for 10 min. After the reaction, excess reagents were removed using a PD10 desalting column. The PDEA-modified AGP was then used for thiol coupling. After the introduction and reduction of disulfide group on the CM5 chip, PDEA-modified AGP was injected across the surface at a flow rate of 10 μL/min for 7 min. The

immobilization level of AGP was 5000 RU. Solutions of LPS samples were prepared in either HBS-EP (10 mM HEPES pH 7.4, 150 mM NaCl, 3 mM EDTA, and 0.005% (v/v) surfactant P20) or HBS-P (10 mM HEPES pH 7.4, 150 mM NaCl, and 0.005% (v/v) surfactant P20) running buffer. The LPS concentration range used in SPR experiment was 10 to 3000 μg/mL, in 3-fold dilutions. LPS samples were injected across two flow cells on a CM5 chip at a flow rate of 30 μL/min. Flow cell 2 was immobilized with AGP; whereas flow cell 1 was deactivated and served as the blank reference surface. All SPR assays were repeated at least 3 times at 25°C. Association period was set up for 30 sec followed by 60 sec dissociation. Binding responses were recorded 10 sec before the end of association period. The reference flow cell response was subtracted from all data used in the analysis.

In the case of lipid A and Kdo2-lipid A binding, a L1 sensor chip was used (GE Health Care, Melbourne Australia). Small unilamellar vesicles (SUVs) were prepared in PBS by sonication and extrusion. Lipids (DMPC, lipid A, and Kdo2-lipid A) were dissolved in ethanol-free chloroform in 25 mL round-bottom flasks. 10% (mol/mol) of lipid A or Kdo2-lipid A was added into DMPC solution to make 10% lipid A/Kdo2-lipid A-DMPC mixtures, which were then deposited as a thin film by removal of the solvent (chloroform) under reduced pressure on a rotary evaporator and dried under high vacuum for at least 2 hours. PBS was then added into each flask to give a 1 mM suspension, which was sonicated 5 min for 5 times. The suspension was passed 17 times through a 50 nm polycarbonate filter in an Avestin Lipofast Basic extrusion apparatus to give a translucent solution of vesicles, which should possess a mean diameter of 50 nm. The SUVs were then injected into the flow cells of the L1 sensor chip for 2000 sec at a low flow rate of 2 μL/min to form a bilayer membrane model on the chip surface. Then, a series of AGP solutions was injected across the flow cells at a flow rate of 30 μL/min, having an injection phase of 180 s and a dissociation phase of 300 s. A regeneration step was added before and after every cycle using 40 mM octyl-glucoside, which cleans the chip surface for a new cycle. Pure DMPC bilayer was applied as a reference and all data used in analysis were reference-subtracted.

2.4. 3-Deoxy-D-manno-oct-2-ulopyranosonic Acid (Kdo) Assay. The molar concentration of Kdo in the LPS samples was measured following the purpald assay as previously described in detail [45].

2.5. Cytotoxicity Assay. HEK293 cell line was purchased from American Type Culture Collection (ATCC). The cells were cultured in DMEM (Invitrogen, Australia) containing 10% fetal bovine serum at 37°C, 5% CO_2. Serum-free media were used in cytotoxicity assays. Cytotoxicity of LPS to HEK293 cells was determined using Alamar Blue cell viability reagent (Invitrogen, Australia). In brief, HEK293 cells were seeded as 2×10^4 cells per well in a clear 96-well plate and incubated for 24 hr at 37°C, 5% CO_2. Then the media were replaced with serum-free media, which contained AGP and LPS (100 μg/mL). After a 24 hr incubation, 10 μL of Alamar Blue reagent were added per well and incubated at 37°C for

1 hr. Then the fluorescence intensity was read using Polar star Omega with excitation/emission wavelengths of 560/590.

2.6. Molecular Modeling of the Kdo2 Lipid A-AGP Complex.

The crystallographic coordinates of the F1*S variant of *apo*-human AGP were retrieved from the protein data bank (PDB ID: 3KQ0) [17]. The tetraantennary *N*-glycans characteristic for human AGP [19, 22, 46] were added to the crystallographic AGP structure using the GlyProt server (http://www.glycosciences.de/modeling/glyprot/php/main.php) [47, 48]. A docking model of Kdo2-lipid A in complex with the F1*S variant of human AGP was constructed using the Accelrys Discovery Studio V2.1 CDOCKER algorithm as per the standard protocol in the manufacturer's instructions (Accelrys, San Diego, CA, USA).

3. Results

3.1. Fluorometric Assay of the Binding Interaction between FITC-Labeled E. coli O111:B4 LPS with Human AGP.

FITC-LPS displays a fluorescence emission maximum as a wavelength of ~515 nm (Figure 2). Fluorescence intensity was noted to increase upon titration with AGP (Figure 2). EDTA has a well-documented effect of sequestering the divalent cations that help bridge adjacent LPS molecules when they are arranged in a leaflet or aggregate structure [9]. The titration of FITC-LPS with AGP in the presence of EDTA (1 mM final concentration) produced a higher level of fluorescence emission compared to an identical titration in the absence of EDTA (Figure 2). The addition of chlorpromazine (20 μM final concentration) to the FITC-LPS:AGP complex produced a decrease in fluorescence emission (Figure 2).

3.2. Surface Plasmon Resonance Assay of LPS Binding to Human AGP.

The highly variable length of the *O*-antigen polysaccharide means that LPS from different bacterial strains have different molecular weights which complicates quantitative comparisons of binding affinity. Accordingly, the molar concentration of the LPS samples was determined using the purpald Kdo assay and we standardized the concentrations of LPS in the SPR experiments. The SPR binding measurements were performed using AGP immobilized to a CM5 sensor chip surface and titrated with LPS (see Figure S1 available online at doi:10.1155/2012/475153). In order to investigate the impact of EDTA on binding, the experiments were replicated in HBS-P and HBS-EP buffers (HBS-EP buffer contains 3 mM EDTA). The binding of the LPS samples to AGP was dose-dependent (Figures 3(a) and 3(b)). The binding affinity varied between LPS isolates from different genera and between LPS isolates from different strains (Figure 3). Ranking of LPS binding to AGP was investigated at single concentrations of 40 μM of LPS, which gave a rank order of affinity *P. aeruginosa* > *K. pneumonia* > *E. coli* O127:B8 > *E. coli* O111:B4 > *S. enteric* > *E. coli* EH100 (R_a) ≥ *S. enteric* TV119 (R_a) > *S. enteric* SL1181 (R_e) > *S. marcescens* > *E. coli* F583 (R_d) (Figure 3(b)). The binding of *E. coli* O111:B4, *E. coli* O127:B8, *E. coli* EH100 (R_a), *S. enterica* TV119 (R_a), and *S. enteric* SL1181 (R_e) was higher in HBS-EP than in HBS-P buffer. Whereas, in the

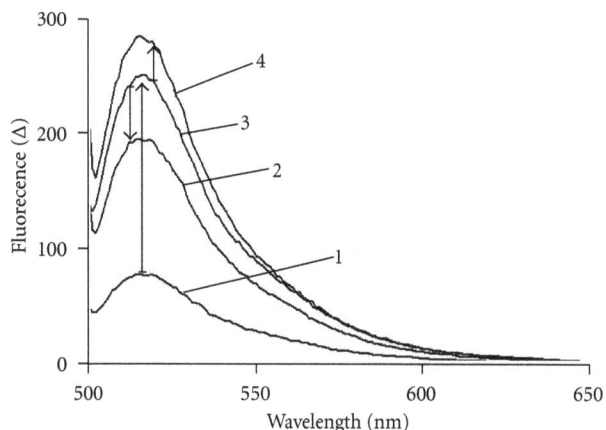

FIGURE 2: Fluorescence emission spectra of (1) FITC-labeled *E. coli* O111:B4 LPS (5 μM). (2) Decrease in fluorescence observed upon addition of chlorpromazine (20 μM) to the FITC-LPS:AGP complex. (3) FITC-LPS (5 μM) in complex with human AGP (25 μM). (4) FITC-LPS (5 μM) in complex with human AGP (25 μM) in the presence of EDTA (1 mM).

case of *K. pneumonia*, *Pseudomonas aeruginosa*, *S. enterica*, and *S. marcescens*, the presence of EDTA did not markedly affect the AGP binding levels. Because of its insolubility in the flow buffer, the binding of diphosphoryl lipid A from *E. coli* was investigated using hybrid SUVs. There was no specific AGP binding detected for diphosphoryl lipid A (data not shown). Hybrid SUV experiments were also performed with fully synthetic Kdo2-lipid A and *E. coli* F583 (R_d) LPS (Figure 3(d)). There was no specific AGP binding detected for *E. coli* F583 (R_d) lipid A, while as the concentrations of AGP were higher than 7 μM, a dose-dependent response to Kdo2-lipid A was observed (Figure 3(d)).

3.3. Molecular Modeling of the Kdo2-Lipid A in Complex with F1*S AGP.

In an attempt to provide a structurale rational for how LPS binds to AGP we have constructed a molecular docking model of the complex using the crystallographic structure of the F1*S variant of human AGP [17] as the receptor and Kdo2-lipid A as the ligand (Figure 4(a)). The model suggests that the LPS-AGP complex is in part stabilized through contacts between the fatty acyl chains of lipid A and a set of polar and nonpolar sidechains within the ligand binding cavity of AGP (Figure 4(b)). The major contact points with the lipid A fatty acyl chains involve the AGP side chains of Phe32, Glu36, Tyr37, Val41, Ile44, Thr47, Leu62, Tyr65, Glu64, Gln66, Asn75, Thr76, Thr77, Leu79, Val92, His97, Phe98, Leu112, Phe114, Val116, Asn117, Asn121, Trp122, and the 3-carbon aliphatic segment of the Arg90 side chain (Figure 4(b)). AGP displays five *N*-linked sialyl-oligosaccharides attached to the side chain of Asn residues found at positions 15, 38, 54, 75, and 85 in its amino acid sequence [21–23]. The Asn residues at positions 15, 54, and 85 are situated at the closed end of the β-barrel on the opposite side of the molecule from the entrance of the ligand binding cavity and, therefore, are unlikely to interfere with ligand entry [17]. While Asn38 and 75-line the entrance

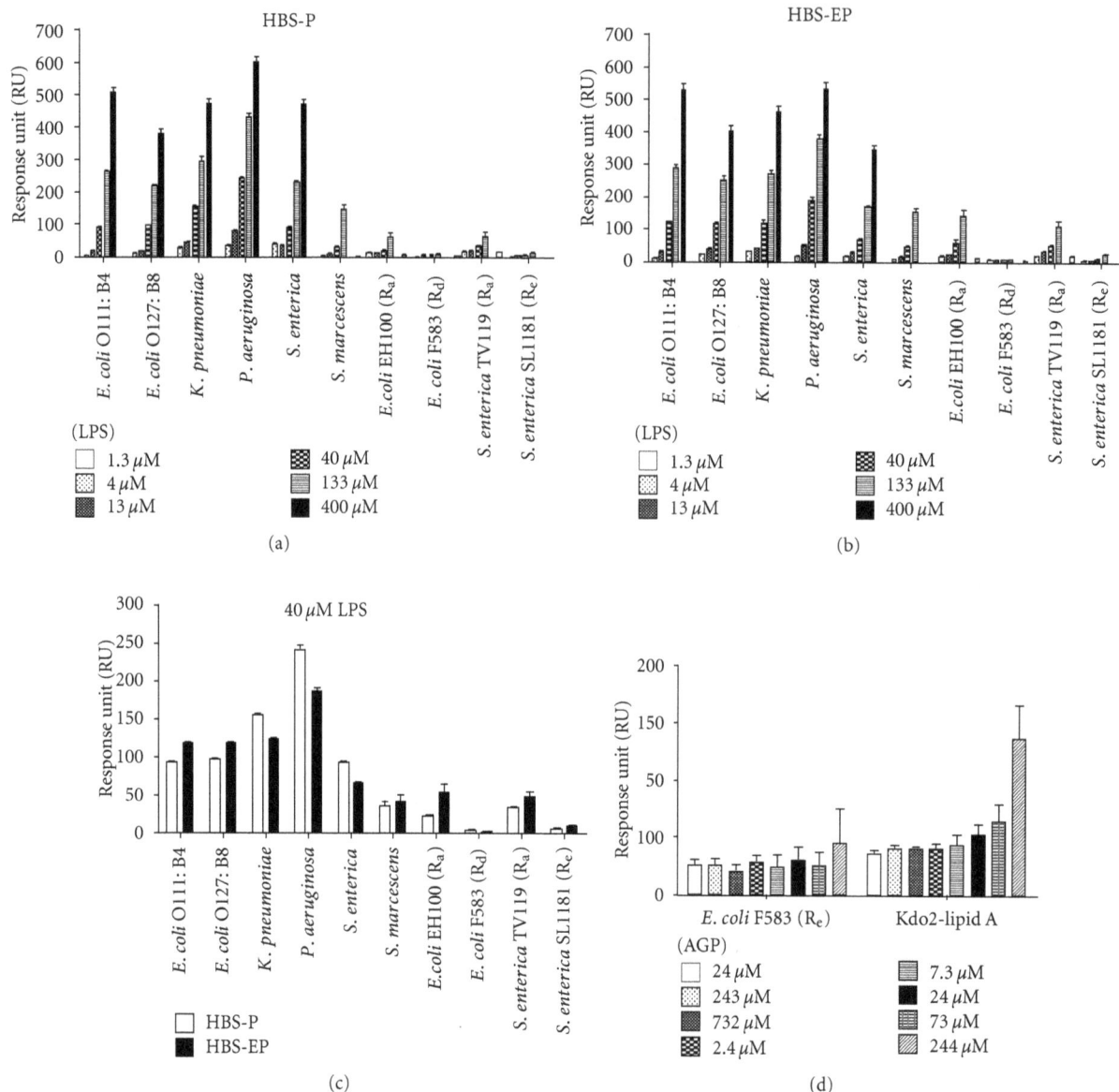

FIGURE 3: SPR assay of the binding of LPS or lipid A to AGP. (a) Binding responses over a concentration range of LPS in EDTA free (HBS-P) and (b) EDTA buffer (HBS-EP). (c) Binding responses at 40 μMLPS in EDTA free (HBS-P) and EDTA buffer (HBS-EP). (d) Binding responses of AGP to DMPC bilayer incorporated 10 (mol/mol) % lipid A.

to the ligand binding pocket [17, 21–23]. The five N-glycan sites were modeled using the reported tetra-antennary glycan structures for human AGP [19, 22, 46]. The docking model shows that the Kdo2 sugars of lipid A make polar contacts with the N-linked sialyl-oligosaccharides attached to Asn38 and 75 that decorate the entrance of the AGP cavity. This would suggest that the AGP-LPS complex is further stabilized through polar contacts between the core oligosaccharide and O-antigen with the N-glycan structures that decorate the surface of AGP (Figure 4(a)).

3.4. Protection against LPS Cytotoxicity by AGP.
In order to investigate the protective effect of AGP on mammalian cells in the present of LPS, we performed a cell viability assay using

HEK293 cells. The exposure of HEK293 cells to 100 μg/mL $E.$ $coli$ O111:B4 LPS induced more than 80% cell death with an IC50 value of 89.7 μg/mL (Figure 5). Similar results were observed with $E.$ $coli$ O127:B8 LPS which elicited an IC50 value of 71.4 μg/mL (Figure 5). The cytotoxic effect of LPS could be ameliorated in a concentration-dependent manner by AGP (Figure 5). In the presence of 1 mg/mL AGP, which is approximately equivalent to the human plasma concentration of AGP in healthy individuals (20 μM) [14, 49, 50], 80% of the HEK293 cells retained viability after a 24 hr incubation with 100 μg/mL $E.$ $coli$ O111:B4 LPS (Figure 5). The IC50 values for both LPS samples decreased in the presence of 1 mg/mL AGP, indicative of a protective effect (Figure 5, insets).

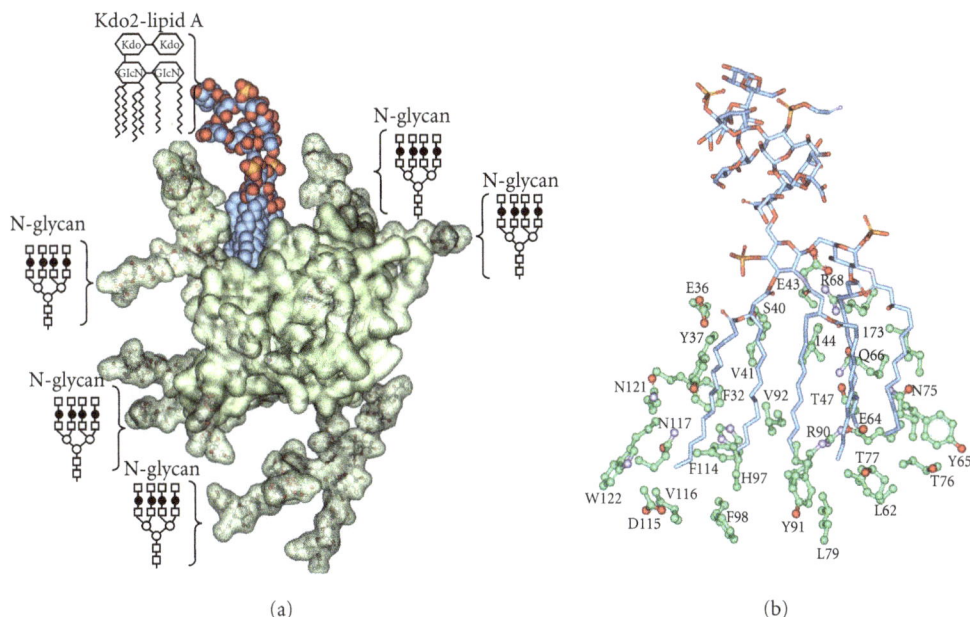

(a)   (b)

FIGURE 4: Molecular model of the Kdo2-lipid A F1*S AGP complex. (a) The AGP F1*S variant crystal structure (PDB code: 3KQ0) is shown in semitransparent surface representation, the bound Kdo2-lipid A is shown in CPK, carbon is colored blue. The Kdo2-lipid A and quaternary N-glycans on the AGP structure are indicated schematically (○ Man; □ GlcNAc; ● D-Gal) and are shown in ball- and stick-representation on the model. (b) Interactions between the fatty acyl chains of lipid A and the side chains of the ligand binding cavity of AGP.

4. Discussion

4.1. Structure-Activity Relationships for the LPS-AGP Interaction.
In order to investigate the SAR of the LPS-AGP interaction we examined the contribution of the three main components of LPS, namely, (1) lipid A; (2) the core oligosaccharide; (3) the O-antigen polysaccharide.

Bacteria that express LPS with a complete core and O-polysaccharide are referred to as "smooth" strains, a term originally coined by Griffith based on his observation of their colony morphology [51]. In contrast, mutant bacteria that have defects in the LPS biosynthetic pathway and produce LPS without O-polysaccharide and/or contain truncated core oligosaccharides are referred to as "rough" strains [6, 51]. The structural analysis of LPS from *Salmonella* rough mutants led to the differentiation of the R_a to R_e chemotypes [6]. R_a mutants have defects in the O-antigen biosynthetic pathway and display a LPS chemotype with a complete core structure minus the O-polysaccharide [6]. Whereas at the other extreme, R_e mutants display the smallest viable core structure required for growth, consisting of the Kdo2-lipid A "backbone" which lacks all of the core sugars except for the Kdo disaccharide (Figure 1(a); Table 1) [6]. The other chemotypes R_b, R_c, and R_d display core lengths intermediate to the R_a and R_e extremes [6] (Table 1). To determine the contribution of the O-antigen and core carbohydrate structures towards AGP binding we examined the binding of AGP with LPS isolated from R_a, R_d, and R_e mutants of *S. enterica* and *E. coli*, and in addition, fully synthetic Kdo2-lipid A. The contribution of the lipid A component for AGP binding was measured using diphosphoryl lipid A prepared from *E. coli* LPS (Figure 1(b); Table 1).

4.1.1. Lipid A.
The AGP binding affinity of the lipid A component of LPS was measured using diphosphoryl *E. coli* lipid A (Figure 1(b)) embedded in SUVs. The binding of lipid A could not be measured directly due to the insolubility of lipid A in the SPR flow buffer. The SPR data indicated that AGP did not bind lipid A embedded in SUVs, suggesting that the core and O-antigen carbohydrate structures are indispensable for the binding of LPS to AGP. Alternatively, it is also possible that the key structural components of lipid A required for the interaction with AGP are inaccessible in the SUV arrangement.

4.1.2. Core Oligosaccharide.
The core polysaccharide consists of a short chain of sugars that is invariably attached to the lipid A component via a Kdo residue through a ketosidic linkage [7, 30, 31]. The core is usually conserved across members of a bacterial genus, but can be quite variable between genera of Gram-negative bacteria [7, 30, 31]. Nevertheless, this structural variability is minor in comparison with that of the O-antigen polysaccharide. The inner core polysaccharide of Enterobacteriaceae is composed of L-glycero-D-manno heptose (heptose) and Kdo [7, 30, 31], usually carrying charged moieties such as phosphate and phosphorylethanolamine (Table 1). The heptoses sugar occurs as either L,D-Hep or in addition to D,D-Hep as per the core of *S. marcescens* (Table 1) [7, 30, 31, 33, 34]. The outer core is composed predominantly of common hexose sugars, glucose (Glu), galactose (Gal), and N-acetylglucosamine (GlcNAc) (Table 1). These basal sugars are often decorated with groups such as phosphate, sulfate, ethanolamine, acetyl, amino acids, or phosphodiester-linked derivatives [7, 30, 31]. The

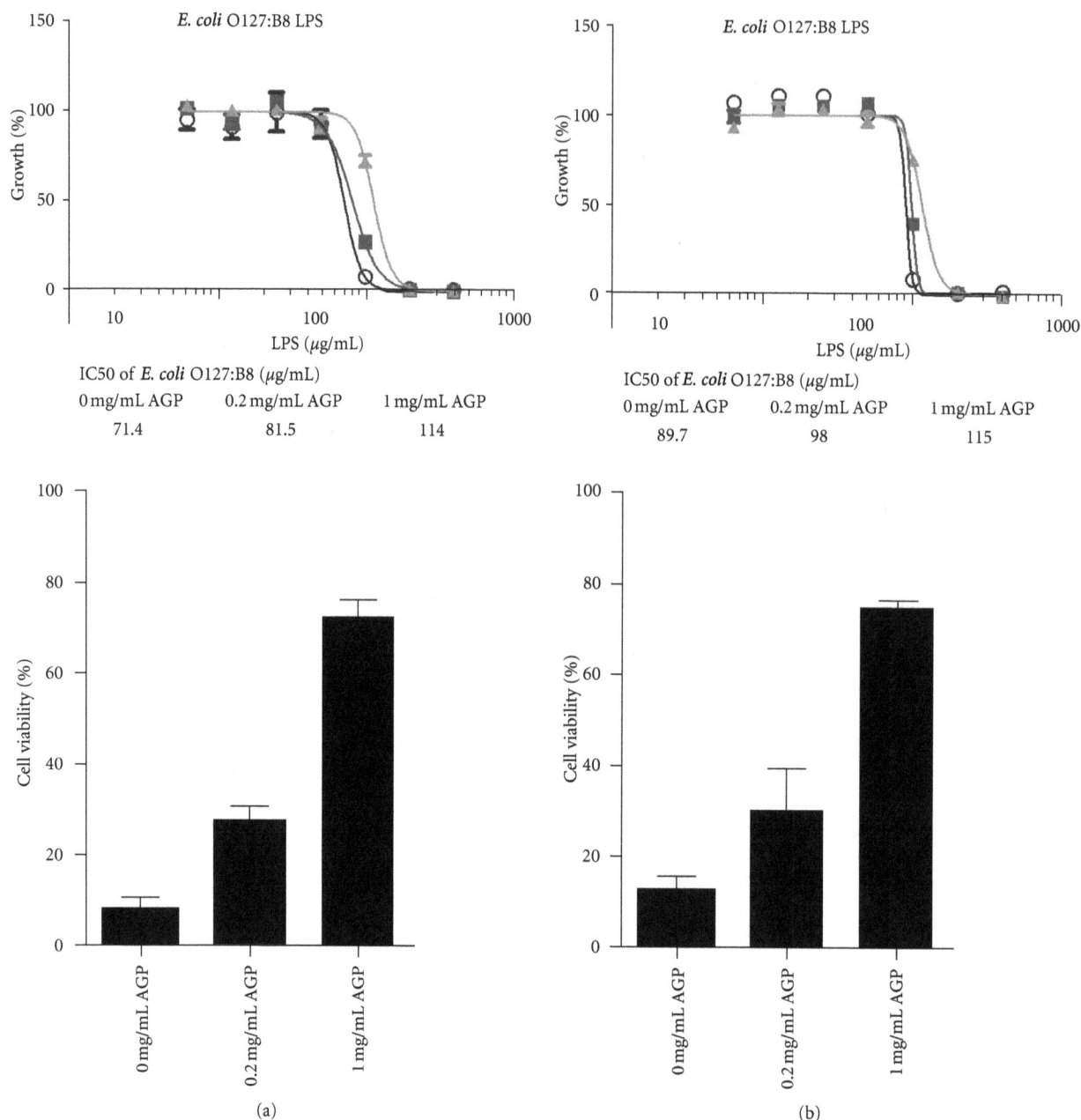

FIGURE 5: The protective effect of AGP against LPS-induced cytotoxicity in HEK293 cell culture. (a) *Top panel.* The percentage of cell growth in the presence of increasing concentrations of *E. coli* O127:B8 LPS. (○) 0 mg/mL AGP; (■) 0.2 mg/mL AGP; (▲) 1 mg/mL AGP. *Bottom panel.* The percentage of cell viability upon exposure to 100 μg/mL *E. coli* O127:B8 LPS, in the presence and absence of AGP. (b) *Top panel.* The percentage of cell growth in the presence of increasing concentrations of *E. coli* O111:B4 LPS. (○) 0 mg/mL AGP; (■) 0.2 mg/mL AGP; (▲) 1 mg/mL AGP. *Bottom panel.* The percentage of cell viability upon exposure to 100 μg/mL *E. coli* O111:B4 LPS, in the presence and absence of AGP. The inset table documents the IC50 values for LPS in the presence of increasing AGP levels.

chemical structures of core oligosaccharides from the LPS samples used in this study are been summarized in Table 1. While each of these core structures is structurally similar, there are a number of strain- and genera-specific differences in sugar composition and linkage configuration within the respective core regions. In *E. coli* LPS five distinct core structures have been characterized, termed K-12 and R1, R2, R3, R4 [7, 30, 31]. The serotype O127:B8 and EH100

R_a LPS possess the R2 core type, whereas the serotype O111:B4 strain possesses the R3 core type (Table 1). The SPR results did not show any marked differences in the AGP binding between O111:B4 and O127:B8 LPS suggesting the contribution of the core structure is secondary to the O-antigen polysaccharide (Figure 3). Two core types have been described for *S. enterica*, the test LPS from the smooth strain and from the TV119 R_a rough mutant both display the same core structure (Table 1) [30, 31]. The cores of the *E. coli* and *S. enterica* test strains are composed of the common hexoses, glucose (Glu), galactose (Gal), and *N*-acetylglucosamine (GlcNAc). The core regions of test LPS samples from *E. coli*, *S. enterica*, and *P. aeruginosa* are all phosphorylated at the two inner heptoses [27–31, 52]. Furthermore, the phosphate on the first heptose in these core structures is further derivatized with ethanolamine phosphate (PEtN), forming ethanolamine diphosphate [27–31, 52]. In addition, the *P. aeruginosa* core displays the L-rhamnose (Rha) and D-galactosamine (GalN) "special" sugars; 7-*O*-carbamoylation of HepII and *N*-acylation of GalN residue with L-alanine [27–29, 52]. As per *S. enterica*, thus far, only two core types have been described for *K. pneumonia* [25, 26]. In addition to the common hexoses, the core of the LPS from the *K. pneumonia* serotype O2 test strain contains the Kdo and galacturonic acid (GalA) "special" sugars (Table 1) [25, 26]. Unlike the core oligosaccharides of *E. coli*, *S. enterica* and *P. aeruginosa* are rich in negative charges which arise from the phosphate substitution of the Hep sugars [27–31, 33, 52]; one noticeable feature of the *K. pneumonia* and *S. marcescens* scores is the absence of phosphate residues. Instead the carboxylic acid groups of the Kdo and GalA residues provide the negative charges in their core oligosaccharides [25, 26, 33, 34]. Despite these structural differences across the core structures of the LPS test samples, it is difficult to differentiate the contribution of the individual core oligosaccharides in the smooth LPS strains due to the presence of the highly variable *O*-antigen in these structures. However, the SPR results from the titrations with the *E. coli* and *S. enteric* R_a, R_d, and R_e rough mutant LPS samples (which lack the *O*-antigen component), suggest that the full-length core oligosaccharide structure is required for avid binding to AGP (Figure 3). The AGP binding affinity was much weaker with the R_d and R_e rough LPS samples (in which the outer core sugars are absent) compared to the R_a samples, which displays the full-length core oligosaccharide structure (Figure 3; Table 1). Similarly, the AGP binding of the fully synthetic Kdo2-lipid A structure (homologous with the R_e LPS structure) was found to be very weak (Figure 3).

4.1.3. O-Antigen Polysaccharide. The SPR experiments revealed that the binding of AGP to LPS is predominantly dependent on its interaction with the *O*-antigen polysaccharide region of LPS, which appears to be more important than the interactions with the core and lipid A regions. The *O*-antigen is a large polymer attached to the outer core that extends into the environment and represents the most heterogenous region of the LPS molecule [24]. The *O*-antigen is made up of repeats of the same oligosaccharide unit, generally consisting of two to six sugar residues (Table 1) [24]. The *O*-antigen polysaccharide chain length is highly variable among bacterial strains and can range up to 10–40 repeating units [24]. So it follows, the LPS from a given bacteria strain is a heterogeneous mixture of LPS molecules differing in the length of the *O*-antigen polysaccharide. The heterogeneity in the number and distribution of *O*-antigen repeats of LPS samples are seen as a ladder of high molecular weight bands following SDS-PAGE

and silver staining (Figure 1(c)). In addition to the highly variable chain length, there are variations in the sugar content, at least 20 different types of sugar are found in the *O*-antigens of Enterobacteriaceae [24]. This inherent variability gives rise to the diversity of antigenic types and forms the foundation of *O*-serotyping for species of Gram-negative bacteria [24]. For example, *Salmonella spp.* display over 1000 distinct *O*-antigens [24]. Moreover, the heterogeneity within the *O*-antigen structure also arises from the anomeric linkages between sugars and from nonstoichiometric tailoring modifications to the sugar moieties, such as the addition of phosphate, acetyl, or methyl groups [24]. The heterogeneity of the *O*-antigen is a strategy Enterobacteriaceae employ to escape the host immune system [24]. Based on the SPR experiments, we can conclude that AGP displays the strongest affinity for the *K. pneumonia* and *P. aeruginosa O*-antigen chemotypes (Figure 3). The *E.coli* O111:B4 and O127:B8 LPS chemotypes displayed comparable affinity, whereas the *S. enterica* and *S. marcescens* chemotypes displayed the lowest AGP affinity (Figure 3). In view of the variability of the *O*-antigen repeats across the test samples in terms of their sugar composition, acidic or neutral character, anomeric sugar linkage configuration, and branched structure (Table 1), we could not draw any correlations with these *O*-antigen properties and the AGP binding affinity of the LPS samples. Firstly, this finding would point out the importance of the unique composition of the *O*-antigen region for recognition by AGP, and secondly it would suggest that the ultrastructure and highly variable chain length of the *O*-antigen might play a larger role for the recognition of LPS by AGP.

4.2. The Significance of LPS-AGP Complexation In Vivo. The excessive release of inflammatory mediators in response to circulating LPS during Gram-negative bacterial infection can produce septic shock and multiple organ failure [2, 8]. When LPS circulates in the bloodstream and initiates endotoxemia, its first interactions are with the protein and cellular elements of blood. Plasma proteins that are known to interact with LPS include human serum albumin [53], immunoglobulin G [54], high-density lipoprotein (HDL) particles [55, 56], apolipoprotein A-II [57], LBP [58, 59], bactericidal/permeability-increasing protein (BPI) [58, 59], lactoferrin [60], hemoglobin [61], and AGP [39]. The early recognition of LPS by these factors is crucial for a number of protective host mechanisms to infection including catalyzing LPS transfer to TLR-4; restricting the spread of LPS from the site of infection via the bloodstream; directing LPS for clearance; binding and neutralization of the endotoxic lipid A component; and the eventual resolution of LPS-triggered inflammation [2, 3]. Our biophysical data underscore the observations that the endotoxic activity of LPS is subject to modification by binding to AGP [39].

The nanomolar LPS binding affinity of the LPS-specific plasma binding proteins, LBP and BPI, is significantly greater compared to the high micromolar affinity of AGP [58, 59]. Notwithstanding, the plasma concentrations of LBP (5–15 µg/L) and BPI (<0.5 ng/mL) are much lower compared to that of AGP (0.7–1.0 g/L) [58, 59]. During infection, the plasma levels of LBP, BPI and AGP can increase up to 5-fold [13, 15, 35–37]. This situation would still present much more AGP binding sites for the binding of LPS compared to LBP and BPI. Therefore, despite the low affinity nature of the LPS-AGP interaction, the abundance of AGP in plasma, particularly as a result of the elevated levels during sepsis, presents sufficient apo-AGP sites for binding to LPS. Moreover, AGP appears to be more selective for specific O-antigen chemotypes of LPS (Figure 3), which contrasts the relatively nonspecific LPS recognition mechanism employed by LBP and BPI that is afforded by their selectivity for the conserved lipid A region of LPS [58], as opposed to AGP which appears to predominately participate in interactions with the highly variable O-antigen polysaccharide.

4.3. Structure-Recognition Characteristics of the LPS-AGP Complex.

We have employed the available crystallographic structure of the F1*S variant of human AGP to construct a model of the complex with Kdo2-lipid A (Figure 4). The model revealed some resemblances between the LPS binding mechanism of AGP and that of the polymyxin antimicrobial peptides [62]. Biophysical studies have shown that polymyxins bind to the disphosphoryl lipid A component of LPS firstly through charge and polar attractive forces and then through hydrophobic forces to the fatty acyl chains of the lipid A, leading to disaggregation of LPS [62]. The model suggests that similar to the LPS-polymyxin complex, the LPS-AGP complex is stabilized by both hydrophobic and polar interactions. The hydrophobic interactions involve the fatty acyl chains of lipid A and side chains in the ligand binding cavity of AGP. AGP is about 48% carbohydrate by weight, which would mean a significant area of the AGP surface that is available for intermolecular interactions is covered by N-glycans. In line with the SPR data, the model also suggests that polar interactions between the core and O-antigen of LPS with the surface N-glycans of AGP are involved in stabilizing the LPS-AGP complex. It is tempting to speculate that the formation of the LPS-AGP complex involves a two-step mechanism wherein the association initially relies upon a polar attraction between the core and O-antigen carbohydrate structures with the N-glycans that are proximal to the entrance of the AGP binding cavity. Once the polar attraction has been established, the fatty acyl chains of lipid A are able to insert into the nonpolar AGP cavity. This putative mechanism would be coincident with the weak affinity of AGP for diphosphoryl lipid A embedded in SUVs where the fatty acyl chains are inaccessible.

The binding of AGP to LPS has previously been shown to be associated with the neutralization of the large negative electrophoretic mobility of AGP [39]. This would suggest that the surface charge distribution of AGP is altered upon LPS complexation, which is in agreement with the core/O-antigen-AGP N-glycan interactions inferred from our SAR data.

The reported cocrystallographic complex of chlorpromazine bound to the A variant of human AGP revealed that the chlorpromazine molecule largely occupies the central hydrophobic lobe I region of the AGP cavity [16]. The ability of chlorpromazine to displace FITC-LPS from AGP suggests that LPS competes for the same AGP binding site that is occupied by the chlorpromazine molecule. Coincidently, our model suggests that the d, R2, and c fatty acyl chains of lipid A occupy the lobe I region (Figure 4(b)).

To further our understanding of the structure-recognition relationships of the LPS-AGP complex, it is important to understand the ultra-structure and composition of the interacting components under the solution conditions used for this study. Owing to its amphipathic nature, the LPS molecule normally exists as aggregates, as opposed to the mono-molecular form [9]. Bivalent cations (Ca^{2+} or Mg^{2+}) are required for the dense packing of LPS within the outer membrane and when in an aggregated state [9]. The sequestration of these bivalent cations by chelating agents such as EDTA leads to the disruption of the intermolecular interactions between adjacent LPS molecules [9]. There were noticeable differences in the SPR interaction responses between AGP and some of the LPS samples in the presence and absence of EDTA (Figure 3), suggesting that the binding interaction is influenced by the ultrastructural organization of the LPS aggregates. In the case of the E. coli LPS samples, the binding responses were greater in the presence of EDTA, which decreases the aggregation state of LPS (Figure 3) [9]. In comparison, the binding responses with the LPS samples from the other strains was unaffected by the presence of EDTA in the SPR binding buffer. These differences would suggest that the binding of LPS by AGP may involve a complex series of interfacial molecular events in which the mono-molecular LPS is sequestered from the aggregate by AGP. Binding to LPS aggregates has been reported for other plasma proteins including lactoferrin [63], apolipoprotein A-II [64], HDL [65], hemoglobin [61], and the antimicrobial peptide polymxin B [62, 66].

The commercial AGP preparations derived from human plasma that we have employed throughout this study consist of proportions of the F1, S, and A variants in a nearly constant ratio of 40 : 30 : 30 (F1 : S : A) and are not desialylated [14, 49, 50]. In a previous study it was reported that asialo-AGP potentiated LPS-induced secretion of interleukin-1 β, interleukin-6, and tumor necrosis factor-α by human monocytes and macrophages to the same level as did native AGP [67]. This finding is inconsistent with the key role of the surface N-glycans on AGP for binding to LPS inferred from our biophysical data.

4.4. Conclusions and Therapeutic Potential.

The overtly promiscuous ligand binding cavity of AGP allows it to interact with generic structural components of many molecules which include exogenous inflammatory mediators such as LPS that are commonly circulating remnants of dead bacteria during infection. Based on available evidence, it has been proposed that AGP serves a protective role during infection by directly binding to LPS, neutralizing its direct toxicity and thereby downregulating the inflammatory response [39]. In this paper we have examined the interaction of LPS with AGP at a molecular level and studied the SAR. The O-antigen polysaccharide was found to be the key structure on the LPS molecule responsible for the recognition of LPS by AGP. The species and strain-specific variations in the *O*-antigen structure greatly affected the LPS binding affinity for AGP. Moreover, we present data that demonstrates that AGP protects against LPS-induced cytotoxicity *in vitro* (Figure 5). Our data is broadly consistent with the proposed role of AGP as an essential component in nonspecific resistance to Gram-negative infections [39]. If our findings are considered in terms of potential therapeutic applications, given that AGP is well tolerated [40], AGP could theoretically be administered prophylactically prior to large bowel surgery or during the acute stages of sepsis.

Abbreviations

AGP: Human α-1-acid glycoprotein
FITC: Fluorescein isothiocyanate
ITC: Isothermal titration calorimetry
LPS: Lipopolysaccharide
LBP: Lipopolysaccharide-binding protein
Kdo: 3-deoxy-D-manno-oct-2-ulopyranosonic acid
OM: Outer-membrane
SPR: Surface plasmon resonance
SAR: Structure activity relationships
SUV: Small unilamellar vesicles
TLR4: Toll-like receptor 4.

Acknowledgments

R. L. Nation and J. Li are supported by research grants from the National Institute of Allergy and Infectious Diseases of the National Institutes of Health (R01A1070896 and R01AI079330). T. Velkov, R. L. Nation, and J. Li are also supported by the Australian National Health and Medical Research Council (NHMRC). The content is solely the responsibility of the authors and does not necessarily represent the official views of the National Institute of Allergy and Infectious Diseases or the National Institutes of Health. J. Li is an Australian NHMRC Senior Research Fellow. T. Velkov is an Australian NHMRC Industry Career Development Level 1 Research Fellow. M. A. Cooper is a NHMRC Australia Fellow supported by AF511105.

References

[1] I. Lerouge and J. Vanderleyden, "O-antigen structural variation: mechanisms and possible roles in animal/plant-microbe interactions," *FEMS Microbiology Reviews*, vol. 26, no. 1, pp. 17–47, 2002.

[2] J. Cohen, "The immunopathogenesis of sepsis," *Nature*, vol. 420, no. 6917, pp. 885–891, 2002.

[3] J. Schletter, H. Heine, A. J. Ulmer, and E. T. Rietschel, "Molecular mechanisms of endotoxin activity," *Archives of Microbiology*, vol. 164, no. 6, pp. 383–389, 1995.

[4] H. Kumar, T. Kawai, and S. Akira, "Toll-like receptors and innate immunity," *Biochemical and Biophysical Research Communications*, vol. 388, no. 4, pp. 621–625, 2009.

[5] M. Caroff and D. Karibian, "Structure of bacterial lipopolysaccharides," *Carbohydrate Research*, vol. 338, no. 23, pp. 2431–2447, 2003.

[6] S. Gronow and H. Brade, "Lipopolysaccharide biosynthesis: which steps do bacteria need to survive?" *Journal of Endotoxin Research*, vol. 7, no. 1, pp. 3–23, 2001.

[7] S. Müller-Loennies, L. Brade, and H. Brade, "Neutralizing and cross-reactive antibodies against enterobacterial lipopolysaccharide," *International Journal of Medical Microbiology*, vol. 297, no. 5, pp. 321–340, 2007.

[8] K. F. Bayston and J. Cohen, "Bacterial endotoxin and current concepts in the diagnosis and treatment of endotoxaemia," *Journal of Medical Microbiology*, vol. 31, no. 2, pp. 73–83, 1990.

[9] K. A. Brogden and M. Phillips, "The ultrastructural morphology of endotoxins and lipopolysaccharides," *Electron Microscopy Reviews*, vol. 1, no. 2, pp. 261–278, 1988.

[10] L. Dente, G. Ciliberto, and R. Cortese, "Structure of the human α1acid glycoprotein gene: sequence homology with other human acute phase protein genes," *Nucleic Acids Research*, vol. 13, no. 11, pp. 3941–3952, 1985.

[11] L. Dente, M. G. Pizza, A. Metspalu, and R. Cortese, "Structure and expression of the genes coding for human alpha 1-acid glycoprotein," *The EMBO Journal*, vol. 6, no. 8, pp. 2289–2296, 1987.

[12] L. Dente, U. Rüther, M. Tripodi, E. F. Wagner, and R. Cortese, "Expression of human alpha 1-acid glycoprotein genes in cultured cells and in transgenic mice," *Genes & Development*, vol. 2, no. 2, pp. 259–266, 1988.

[13] C. B. Eap, J. F. Fischer, and P. Baumann, "Variations in relative concentrations of variants of human α1-acid glycoprotein after acute-phase conditions," *Clinica Chimica Acta*, vol. 203, no. 2-3, pp. 379–385, 1991.

[14] F. Herve, E. Gomas, J. C. Duche, and J. P. Tillement, "Fractionation of the genetic variants of human α1-acid glycoprotein in the native form by chromatography on an immobilized copper(II) affinity adsorbent. Heterogeneity of the separate variants by isoelectrofocusing and by concanavalin A affinity chromatography," *Journal of Chromatography—Biomedical Applications*, vol. 615, no. 1, pp. 47–57, 1993.

[15] J. M. H. Kremer, J. Wilting, and L. H. M. Janssen, "Drug binding to human α-1-acid glycoprotein in health and disease," *Pharmacological Reviews*, vol. 40, no. 1, pp. 1–47, 1988.

[16] K. Nishi, T. Ono, T. Nakamura et al., "Structural insights into differences in drug-binding selectivity between two forms of human α1-acid glycoprotein genetic variants, the A and F1*S forms," *The Journal of Biological Chemistry*, vol. 286, no. 16, pp. 14427–14434, 2011.

[17] D. L. Schönfeld, R. B. G. Ravelli, U. Mueller, and A. Skerra, "The 1.8-Å crystal structure of α1-acid glycoprotein (orosomucoid) solved by UV RIP reveals the broad drug-binding activity of this human plasma lipocalin," *Journal of Molecular Biology*, vol. 384, no. 2, pp. 393–405, 2008.

[18] K. Matsumoto, K. Sukimoto, K. Nishi, T. Maruyama, A. Suenaga, and M. Otagiri, "Characterization of ligand binding sites on the alpha1-acid glycoprotein in humans, bovines and dogs," *Drug Metabolism and Pharmacokinetics*, vol. 17, no. 4, pp. 300–306, 2002.

[19] T. Fournier, N. Medjoubi-N, and D. Porquet, "α-1-acid glycoprotein," *Biochimica et Biophysica Acta*, vol. 1482, no. 1-2, pp. 157–171, 2000.

[20] T. Hochepied, F. G. Berger, H. Baumann, and C. Libert, "α1-acid glycoprotein: an acute phase protein with inflammatory and immunomodulating properties," *Cytokine and Growth Factor Reviews*, vol. 14, no. 1, pp. 25–34, 2003.

[21] K. Schmid, R. B. Nimerg, A. Kimura, H. Yamaguchi, and J. P. Binette, "The carbohydrate units of human plasma α1-acid

glycoprotein," *Biochimica et Biophysica Acta*, vol. 492, no. 2, pp. 291–302, 1977.

[22] M. Nakano, K. Kakehi, M. H. Tsai, and Y. C. Lee, "Detailed structural features of glycan chains derived from α1-acid glycoproteins of several different animals: the presence of hypersialylated, O-acetylated sialic acids but not disialyl residues," *Glycobiology*, vol. 14, no. 5, pp. 431–441, 2004.

[23] M. J. Treuheit, C. E. Costello, and H. B. Halsall, "Analysis of the five glycosylation sites of human α1-acid glycoprotein," *Biochemical Journal*, vol. 283, no. 1, pp. 105–112, 1992.

[24] R. Stenutz, A. Weintraub, and G. Widmalm, "The structures of *Escherichia coli* O-polysaccharide antigens," *FEMS Microbiology Reviews*, vol. 30, no. 3, pp. 382–403, 2006.

[25] E. Vinogradov, E. Frirdich, L. L. MacLean et al., "Structures of lipopolysaccharides from *Klebsiella pneumoniae*: elucidation of the structure of the linkage region between core and polysaccharide O chain and identification of the residues at the non-reducing termini of the O chains," *The Journal of Biological Chemistry*, vol. 277, no. 28, pp. 25070–25081, 2002.

[26] E. Vinogradov and M. B. Perry, "Structural analysis of the core region of the lipopolysaccharides from eight serotypes of *Klebsiella pneumoniae*," *Carbohydrate Research*, vol. 335, no. 4, pp. 291–296, 2001.

[27] H. L. Rocchetta, L. L. Burrows, and J. S. Lam, "Genetics of O-antigen biosynthesis in *Pseudomonas aeruginosa*," *Microbiology and Molecular Biology Reviews*, vol. 63, no. 3, pp. 523–553, 1999.

[28] S. G. Wilkinson, "Composition and structure of lipopolysaccharides from *Pseudomonas aeruginosa*," *Reviews of Infectious Diseases*, vol. 5, pp. S941–S949, 1983.

[29] Y. A. Knirel, O. V. Bystrova, N. A. Kocharova, U. Zähringer, and G. B. Pier, "Conserved and variable structural features in the lipopolysaccharide of *Pseudomonas aeruginosa*," *Journal of Endotoxin Research*, vol. 12, no. 6, pp. 324–336, 2006.

[30] O. Holst, "The structures of core regions from enterobacterial lipopolysaccharides—an update," *FEMS Microbiology Letters*, vol. 271, no. 1, pp. 3–11, 2007.

[31] P. E. Jansson, A. A. Lindberg, B. Lindberg, and R. Wollin, "Structural studies on the hexose region of the core in lipopolysaccharides from enterobacteriaceae," *European Journal of Biochemistry*, vol. 115, no. 3, pp. 571–577, 1981.

[32] G. Schmidt, B. Jann, and K. Jann, "Immunochemistry of R lipopolysaccharides of *Escherichia coli*. Studies on R mutants with an incomplete core, derived from E. coli O8:K27," *European Journal of Biochemistry*, vol. 16, no. 2, pp. 382–392, 1970.

[33] H. M. Aucken and T. L. Pitt, "Different O and K serotype distributions among clinical and environmental strains of *Serratia marcescens*," *Journal of Medical Microbiology*, vol. 47, no. 12, pp. 1097–1104, 1998.

[34] E. Vinogradov, B. Lindner, G. Seltmann, J. Radziejewska-Lebrecht, and O. Holst, "Lipopolysaccharides from *Serratia marcescens* possess one or two 4-amino-4-deoxy-L-arabinopyranose 1-phosphate residues in the lipid a and D-glycero-D-talo-Oct-2-ulopyranosonic acid in the inner core region," *Chemistry*, vol. 12, no. 25, pp. 6692–6700, 2006.

[35] C. B. Kristensen, "Imipramine serum protein binding in healthy subjects," *Clinical Pharmacology and Therapeutics*, vol. 34, no. 5, pp. 689–694, 1983.

[36] P. J. Ojala, M. Hermansson, M. Tolvanen et al., "Identification of α-1 acid glycoprotein as a lysophospholipid binding protein: a complementary role to albumin in the scavenging of lysophosphatidylcholine," *Biochemistry*, vol. 45, no. 47, pp. 14021–14031, 2006.

[37] F. Voulgari, P. Cummins, T. I. M. Gardecki, N. J. Beeching, P. C. Stone, and J. Stuart, "Serum levels of acute phase and cardiac proteins after myocardial infarction, surgery, and infection," *British Heart Journal*, vol. 48, no. 4, pp. 352–356, 1982.

[38] S. Pirnes-Karhu, R. Sironen, L. Alhonen, and A. Uimari, "Lipopolysaccharide-induced anti-inflammatory acute phase response is enhanced in spermidine/spermine N1-acetyltransferase (SSAT) overexpressing mice," *Amino Acids*, vol. 42, no. 2-3, pp. 473–484, 2012.

[39] D. F. Moore, M. R. Rosenfeld, P. M. Gribbon, C. P. Winlove, and C. M. Tsai, "α-1-acid (AAG, Orosomucoid) glycoprotein: interaction with bacterial lipopolysaccharide and protection from sepsis," *Inflammation*, vol. 21, no. 1, pp. 69–82, 1997.

[40] T. Hochepied, W. van Molle, F. G. Berger, H. Baumann, and C. Libert, "Involvement of the acute phase protein α1-acid glycoprotein in nonspecific resistance to a lethal gram-negative infection," *The Journal of Biological Chemistry*, vol. 275, no. 20, pp. 14903–14909, 2000.

[41] Y. Makimura, Y. Asai, A. Sugiyama, and T. Ogawa, "Chemical structure and immunobiological activity of lipid A from *Serratia marcescens* LPS," *Journal of Medical Microbiology*, vol. 56, no. 11, pp. 1440–1446, 2007.

[42] T. E. Rietschel and H. Brade, "Lipopolysaccharides, endotoxins and O-antigens of gram-negative bacteria: chemical structure, biological effect and serological properties," *Infection*, vol. 15, supplement 2, pp. S76–S84, 1987.

[43] E. T. Rietschel, H. Brade, L. Brade et al., "Lipid A, the endotoxic center of bacterial lipopolysaccharides: relation of chemical structure to biological activity," *Progress in Clinical and Biological Research*, vol. 231, pp. 25–53, 1987.

[44] P. J. Hitchcock and T. M. Brown, "Morphological heterogeneity among Salmonella lipopolysaccharide chemotypes in silver-stained polyacrylamide gels," *Journal of Bacteriology*, vol. 154, no. 1, pp. 269–277, 1983.

[45] C. H. Lee and C. M. Tsai, "Quantification of bacterial lipopolysaccharides by the purpald assay: measuring formaldehyde generated from 2-keto-3-deoxyoctonate and heptose at the inner core by periodate oxidation," *Analytical Biochemistry*, vol. 267, no. 1, pp. 161–168, 1999.

[46] K. D. Smith, A. Pollacchi, M. Field, and J. Watson, "The heterogeneity of the glycosylation of α-1-acid glycoprotein between the sera and synovial fluid in rheumatoid arthritis," *Biomedical Chromatography*, vol. 16, no. 4, pp. 261–266, 2002.

[47] A. Bohne-Lang and C. W. von der Lieth, "GlyProt: in silico glycosylation of proteins," *Nucleic Acids Research*, vol. 33, no. 2, pp. W214–W219, 2005.

[48] T. Lütteke, A. Bohne-Lang, A. Loss, T. Goetz, M. Frank, and C. W. von der Lieth, "GLYCOSCIENCES.de: an internet portal to support glycomics and glycobiology research," *Glycobiology*, vol. 16, no. 5, pp. 71R–81R, 2006.

[49] F. Hervé, G. Caron, J. C. Duché et al., "Ligand specificity of the genetic variants of human α1-acid glycoprotein: generation of a three-dimensional quantitative structure-activity relationship model for drug binding to the a variant," *Molecular Pharmacology*, vol. 54, no. 1, pp. 129–138, 1998.

[50] F. Hervé, J. C. Duché, P. D'Athis, C. Marché, J. Barré, and J. P. Tillement, "Binding of disopyramide, methadone, dipyridamole, chlorpromazine, lignocaine and progesterone to the two main genetic variants of human α1-acid glycoprotein: evidence for drug-binding differences between the variants and for the presence of two separate drug-binding sites on α1-acid glycoprotein," *Pharmacogenetics*, vol. 6, no. 5, pp. 403–415, 1996.

[51] F. Griffith, "The significance of pneumococcal types," *The Journal of Hygiene*, vol. 27, no. 2, pp. 113–159, 1928.

[52] Y. A. Knirel, E. V. Vinogradov, N. A. Kocharova et al., "The structure of O-specific polysaccharides and serological classification of *Pseudomonas aeruginosa*," *Acta Microbiologica Hungarica*, vol. 35, no. 1, pp. 3–24, 1988.

[53] G. Jürgens, M. Müller, P. Garidel et al., "Investigation into the interaction of recombinat human serum albumin with re-lipopolysaccharide and lipid A," *Journal of Endotoxin Research*, vol. 8, no. 2, pp. 115–126, 2002.

[54] M. H. Ginsberg and D. C. Morrison, "The selective binding of aggregated IgG to lipid A rich bacterial lipopolysaccharides," *Journal of Immunology*, vol. 120, no. 1, pp. 317–319, 1978.

[55] R. J. Ulevitch, A. R. Johnston, and D. B. Weinstein, "New function for high density lipoproteins. Their participation in intravascular reactions of bacterial lipopolysaccharides," *Journal of Clinical Investigation*, vol. 64, no. 5, pp. 1516–1524, 1979.

[56] R. J. Ulevitch, A. R. Johnston, and D. B. Weinstein, "New function for high density lipoproteins. Isolation and characterization of a bacterial lipopolysaccharide-high density lipoprotein complex formed in rabbit plasma," *Journal of Clinical Investigation*, vol. 67, no. 3, pp. 827–837, 1981.

[57] J. F. P. Berbée, L. M. Havekes, and P. C. N. Rensen, "Apolipoproteins modulate the inflammatory response to lipopolysaccharide," *Journal of Endotoxin Research*, vol. 11, no. 2, pp. 97–103, 2005.

[58] J. Weiss, "Bactericidal/permeability-increasing protein (BPI) and lipopolysaccharide-binding protein (LBP): structure, function and regulation in host defence against gram-negative bacteria," *Biochemical Society Transactions*, vol. 31, no. 4, pp. 785–790, 2003.

[59] J. Zweigner, H. J. Gramm, O. C. Singer, K. Wegscheider, and R. R. Schumann, "High concentrations of lipopolysaccharide-binding protein in serum of patients with severe sepsis or septic shock inhibit the lipopolysaccharide response in human monocytes," *Blood*, vol. 98, no. 13, pp. 3800–3808, 2001.

[60] B. J. Appelmelk, Y. Q. An, M. Geerts et al., "Lactoferrin is a lipid A-binding protein," *Infection and Immunity*, vol. 62, no. 6, pp. 2628–2632, 1994.

[61] W. Kaca, R. I. Roth, and J. Levin, "Hemoglobin, a newly recognized lipopolysaccharide (LPS)-binding protein that enhances LPS biological activity," *The Journal of Biological Chemistry*, vol. 269, no. 40, pp. 25078–25084, 1994.

[62] T. Velkov, P. E. Thompson, R. L. Nation, and J. Li, "Structure—activity relationships of polymyxin antibiotics," *Journal of Medicinal Chemistry*, vol. 53, no. 5, pp. 1898–1916, 2010.

[63] K. Brandenburg, G. Jürgens, M. Müller, S. Fukuoka, and M. H. J. Koch, "Biophysical characterization of lipopolysaccharide and lipid A inactivation by lactoferrin," *Biological Chemistry*, vol. 382, no. 8, pp. 1215–1225, 2001.

[64] P. A. Thompson, J. F. P. Berbée, P. C. N. Rensen, and R. L. Kitchens, "Apolipoprotein A-II augments monocyte responses to LPS by suppressing the inhibitory activity of LPS-binding protein," *Innate Immunity*, vol. 14, no. 6, pp. 365–374, 2008.

[65] K. Brandenburg, G. Jürgens, J. Andrä et al., "Biophysical characterization of the interaction of high-density lipoprotein (HDL) with endotoxins," *European Journal of Biochemistry*, vol. 269, no. 23, pp. 5972–5981, 2002.

[66] J. Lopes and W. E. Inniss, "Electron microscopy of effect of polymyxin on *Escherichia coli* lipopolysaccharide," *Journal of Bacteriology*, vol. 100, no. 2, pp. 1128–1129, 1969.

[67] A. Boutten, M. Dehoux, M. Deschenes, J. D. Rouzeau, P. N. Bories, and G. Durand, "α1-acid glycoprotein potentiates lipopolysaccharide-induced secretion of interleukin-1 β, interleukin-6 and tumor necrosis factor-α by human monocytes and alveolar and peritoneal macrophages," *European Journal of Immunology*, vol. 22, no. 10, pp. 2687–2695, 1992.

Additional Common Polymorphisms in the *PON* Gene Cluster Predict PON1 Activity but Not Vascular Disease

Daniel S. Kim,[1] **Amber A. Burt,**[2] **Jane E. Ranchalis,**[2]
Rebecca J. Richter,[2] **Julieann K. Marshall,**[2] **Jason F. Eintracht,**[3]
Elisabeth A. Rosenthal,[2] **Clement E. Furlong,**[1,2] **and Gail P. Jarvik**[1,2]

[1] *Department of Genome Sciences, University of Washington School of Medicine, Seattle, WA 98195, USA*
[2] *Department of Medicine, Division of Medical Genetics, University of Washington School of Medicine, P.O. Box 357720, Seattle, WA 98195, USA*
[3] *Department of General Medicine, Virginia Mason Medical Center, Seattle, WA 98101, USA*

Correspondence should be addressed to Gail P. Jarvik, gjarvik@medicine.washington.edu

Academic Editor: Bianca Fuhrman

Background. Paraoxonase 1 (PON1) enzymatic activity has been consistently predictive of cardiovascular disease, while the genotypes at the four functional polymorphisms at *PON1* have not. The goal of this study was to identify additional variation at the *PON* gene cluster that improved prediction of PON1 activity and determine if these variants predict carotid artery disease (CAAD). *Methods.* We considered 1,328 males in a CAAD cohort. 51 tagging single-nucleotide polymorphisms (tag SNPs) across the *PON* cluster were evaluated to determine their effects on PON1 activity and CAAD status. *Results.* Six SNPs (four in *PON1* and one each in *PON2/3*) predicted PON1 arylesterase (AREase) activity, in addition to the four previously known functional SNPs. In total, the 10 SNPs explained 30.1% of AREase activity, 5% of which was attributable to the six identified predictive SNPs. We replicate rs854567 prediction of 2.3% of AREase variance, the effects of rs3917510, and a *PON3* haplotype that includes rs2375005. While AREase activity strongly predicted CAAD, none of the 10 SNPs predicting AREase predicted CAAD. *Conclusions.* This study identifies new genetic variants that predict additional PON1 AREase activity. Identification of SNPs associated with PON1 activity is required when evaluating the many phenotypes associated with genetic variation near PON1.

1. Introduction

Paraoxonase 1 (PON1) is a liver-produced glycoprotein enzyme bound to the surface of high-density lipoprotein (HDL) whose activity is consistently correlated with atherosclerotic vascular disease and end-organ damage [1–3]. PON1 is at least partially responsible for the inhibitory effects of HDL on low-density lipoprotein (LDL) peroxidation [4–6] and also has been demonstrated to hydrolyze oxidized lipid or lipid hydroperoxides in LDL [7]. Accordingly, Watson et al. reported that inactivation of PON1 reduced the ability of HDL to inhibit both the oxidation of LDL and the interaction between macrophages and endothelium [6], both likely key factors in the inflammatory changes underlying atherogenesis. It has also been shown that PON1-deficient mice cannot neutralize the oxidized LDL lipids and

have an increased susceptibility to organophosphate toxicity and coronary heart disease (CHD) [8, 9]. Finally, PON1 activity appears to play a role in maintaining the endothelial-atheroprotective effects of HDL [10].

There are four currently established functional common *PON1* single-nucleotide polymorphisms (SNPs) amongst the nearly 200 SNPs in the gene [11]: two missense mutations ($PON1_{Q192R}$ [rs662] and $PON1_{M55L}$ [rs854560]) and two that alter promoter activity ($PON1_{-108C/T}$ [rs705379] and $PON1_{-162A/G}$ [rs705381]). $PON1_{-108C/T}$ has the largest effect, altering expression likely due to modification of an Sp1 binding site [12]. Rare functional variants have also been identified [13].

While PON1 activity is predictive of vascular disease, studies investigating the role of *PON1* SNPs in vascular disease have been contradictory [14–18]. A recent meta-analysis of 88 case-control studies by Wang et al. found that $PON1_{Q192}$ was correlated with CHD [19]. However, removal of smaller studies from the meta-analysis resulted in none of the functional PON1 SNPs having significant association with CHD, thereby replicating the results of past meta-analyses [20–22]. Similarly, our own past investigations have found that while PON1 enzyme levels are predictive of carotid artery disease (CAAD), the genotypes at the four common functional SNPs fail to predict CAAD status [2, 3]. However, studies of CAAD or ischemic strokes are generally more positive for associations with the PON1 functional SNPs [18, 23–26] than those for CHD. It should be noted that these studies generally have small sample size and several of such studies reported negative results [27, 28].

PON1 has broad substrate specificity and is protective against exposure to toxic organophosphorus insecticides [29]. For biological purposes, PON1 activity is generally measured with regard to the rate of hydrolysis of paraoxon, diazoxon, and phenylacetate (arylesterase activity) [30, 31]. These are termed POase, DZOase, and AREase activities, respectively. AREase enzymatic activity is unaffected by the functional $PON1_{Q192R}$ polymorphism, thus making it the best reflection of the levels of PON1 protein [32].

PON1 activity has also been linked to a number of other health-related phenotypes in addition to vascular disease and diabetes [33]. For example, PON1 also influences the metabolism of a variety of drugs, including statins, in addition to its aforementioned properties of reducing oxidized LDL and breaking down pesticides [34]. $PON1_{Q192R}$ is a reported determinant of clopidogrel efficacy [35], although this result has not been replicated [36] and remains controversial. PON1 has also been associated with diverse diseases [37]. The $PON1_{L55M}$ polymorphism has been repeatedly associated with Parkinson's disease [38–40], including a meta-analysis [41], but null results have also been reported [42]. Recent meta-analyses reported the association of PON1 coding SNPs and breast cancer [43, 44]. Both PON1 activity and genotypes have been associated with age-related macular degeneration [45–51]. PON1 activity is reportedly lower in subjects with systemic lupus erythematosus (SLE) [48, 52–54]. Finally, diabetes is associated with both reduced PON1 activity and PON1 genotypes [55].

PON1 is one of three paraoxonase gene family members, located in a gene cluster on chromosome 7q21.3-22.1. All of the paraoxonases have antioxidant activity [56]. PON1 and PON3 share similar functions in association with HDL as described previously; however, PON3 has lower expression levels [57]. In contrast, PON2 is ubiquitously expressed in human cells [58], particularly in endothelial and aortic smooth muscle cells [59]. PON2 polymorphisms have also been associated with CHD [58, 60]. In addition, all three PON gene products have been reported to hydrolyze the quorum sensing factor of Pseudomonas aeruginosa N-3-oxododecanoyl homoserine lactone (3OC12-HSL) [61], with PON1 and 2 enzymes specifically being shown in animal knock-out studies to be protective against p. aeruginosa infection [62, 63].

Carlson et al. previously performed a tagSNP analysis of the PON1, 2, and 3 gene cluster for association with AREase activity and CAAD status in an overlapping, but much smaller, cohort (n = 500 versus 1328) [27]. That study found evidence that additional functional SNPs likely exist in PON1, but that the majority of the genetic effect on AREase variation was explained by the four functional SNPs previously described. They did not find evidence for PON2 or PON3 SNPs predicting additional AREase activity.

However, the investigation by Carlson et al. still left a large portion of the variation in PON1 activity unexplained. Thus, the goals of this study are to followup on these previous results and utilize an enlarged cohort and denser tagSNP genotyping to attempt to identify novel common SNPs in the PON gene cluster that associate with PON1 activity and/or predict CAAD.

2. Methods

2.1. Sample. The study population for this analysis consisted of 1,328 samples from the previously described Carotid Lesion Epidemiology And Risk (CLEAR) study [2, 3, 64]. Only Caucasian males were analyzed due to underrepresentation of female and minority samples in this primarily Seattle-Veterans-based cohort. Current smoking status and reported ancestry were obtained by self-report. Ancestry was confirmed using STRUCTURE with three ancestral groups [65]. CAAD status was determined via ultrasound of the internal carotid arteries, with cases defined as having >50% stenosis in either artery or a relevant procedure on their carotid arteries in their medical history. Controls had <15% stenosis in both arteries. 88 subjects had intermediate stenosis (15–49%) and were not included for prediction of CAAD, though they were included for prediction of PON1 enzyme activity.

2.2. Genotyping and PON1 Phenotypes. The four known functional PON1 SNPs, $PON1_{Q192R}$, $PON1_{M55L}$, $PON1_{-108C/T}$, and $PON1_{-162A/G}$ and two SNPs identified as potentially predictive by Carlson et al. but not represented on the CVD chip, $PON1_{-909}$ (rs854572) and rs3917510 [27], were genotyped using previously described methods [12, 66]. An additional 86 SNPs in PON1, PON2, and PON3 cluster were genotyped using the Illumina HumanCVD BeadChip (http://www.illumina.com/products/humancvd_whole_genome_genotyping_kits.ilmn). Duplicate genotyping for 34 individuals showed 99.7% consistency in calls. The PON cluster genotypes were filtered with a minor allele frequency cutoff of 1% and did not show deviation from Hardy-Weinberg equilibrium at the $P < 10^{-4}$ level. Rs3917564 was also found to be predictive by Carlson et al. and was genotyped by the CVD chip but was not included in the full analysis due to low minor allele frequency (C/T, C allele frequency = 0.008).

The PON1 POase, DZOase, and AREase activities were measured by a continuous spectrophotometric assay with lithium heparin plasma, as previously described [66]. AREase activity was measured in duplicate and averaged.

AREase was utilized as the primary measured outcome of *PON* gene cluster variation, due to its closer correlation with protein levels. POase activity is largely determined by the $PON1_{Q192R}$ missense polymorphism, which predicts over 70% of its variance [2].

2.3. Analysis. LDselect was used to create tagSNPs from the 86 *PON1*, *PON2*, and *PON3* SNPs genotyped on the Illumina HumanCVD chip [67]. Functional annotation for these SNPs were taken from SNP-Nexus [68]. 51 bins were created, using a linkage disequilibrium (LD) r^2 threshold of 0.64. The first 13 of these bins, corresponding to the *PON1* gene, had multiple SNPs within them, while the remaining bins consisted of singletons. One SNP from each bin was randomly included in the regression analysis for a total of 51 SNPs in the *PON* gene cluster. These 51 SNPs did not include the four functional SNPs, which were included in the analysis separately.

We also made an effort to independently replicate SNPs identified as predictive of PON1 activity by Carlson et al. [27]. As our full sample overlaps with that smaller sample, these were tested in a nonoverlapping sample of 523 subjects with complete genotype and phenotype data which were not available at the time of that study.

Regression analysis was done in R (http://www.r-project .org/) using the standard regression tools available. Genotypes were coded using an additive model. Stepwise linear regression was performed, and model comparison was done using Akaike's Information Criterion (AIC) to examine the fit of each model, beginning with a base model that included current smoking status, age, and the genotypes for the four functional PON1 SNPs as covariates [2, 3, 27]. SNPs that are included in the final model increased the ability of the model to predict the dependent variable. Statin drug use can influence *PON1* expression, and this appears to be influenced by $PON1_{-108}$ genotype [69]. However, statin drug use could not be included as a covariate due to confounding with CAAD status; the preferential use of statins in cases can lead to an erroneous estimation of statin effects on PON1 activity.

3. Results

The sample included 1,328 males with a mean age of 67.8 years; 16.5% of participants reported being current smokers. The subjects included 596 cases and 644 controls considered in the prediction of case status as well as 88 subjects with carotid stenosis between 15–49% who were considered only for the genotype effects on PON1 activity. Cases had a mean censored (CAAD onset) age of 66.5 years and mean current age at enrollment age of 70.9; controls had a mean current age of 64.6 years. The rates of current smoking and statin use, respectively, were 25.8% and 64.7% for cases and 9.6% and 19.5% for controls. Descriptions of the 51 tag SNPs for the *PON* gene cluster are available in Table 1. The AREase activity showed an approximately normal distribution, with a mean of 134 U/I and standard deviation of 51.8.

A regression model containing functional *PON1* SNPs ($PON1_{Q192R}$, $PON1_{M55L}$, $PON1_{-108C/T}$, and $PON1_{-162A/G}$),

age, and current smoking status explained 25.2% of the variance in AREase activity. To explore the possibility of novel SNPs influencing AREase activity, we examined a best-fit model utilizing the stepwise regression including the aforementioned variables plus the 51 tagSNPs. AIC was used to assess whether the additional SNP provided a better fit to the prediction of AREase activity. Only SNPs that added to the predictive power of the best-fit model were kept; others that did not influence the model were discarded. In addition to the four functional SNPs, age, and current smoking, six SNPs were retained in the best-fit model. Together with the original 4 functional SNPs, these additional six SNPs in the *PON* gene cluster explained 30.1% of variance in AREase activity (see Table 2). Addition of these SNPs, rs854567, rs2299257, rs2237583, rs2375005, rs3917486, and rs11768074 serially explained an additional 2.34%, 0.85%, 0.5%, 0.34%, 0.58%, and 0.26% of total variance in PON1 activity. Amongst these six SNPs, four SNPs were in *PON1*, one was in *PON2* (rs2375005), and one was in *PON3* (rs11768074); all are intronic.

Five of the six SNPs found to predict PON1 activity were the only SNPs in their bin (singletons). The sixth SNP, rs854567, was binned with one other typed SNP, rs2299260, $r^2 = 0.80$. To observe whether it was superior at predicting PON1 AREase activity, we replaced rs854567 with rs2299260 in the complete model of 10 SNPs plus covariates. The model including rs2299260 did not predict additional AREase activity as compared to the model including rs854567, with a total of 30.1% of AREase variance explained in the full model. Therefore, either SNP or an untyped SNP in LD may be the functional SNP resulting in the association identified.

To address the potential that untyped SNPs are the functional SNPs that underlie the identified AREase associations, the 1000 Genomes database for European ancestry was consulted via SNP-Nexus [68] for these six SNPs. Five of the six SNPs we found to predict PON1 AREase activity were not in strong LD ($r^2 \geq 0.8$) with other regional SNPs, suggesting that they may be functional. Rs2375005, in contrast, is in strong LD with an additional five SNPs in *PON3* ($r^2 = 0.901$ with intronic rs978903 and synonymous A99A SNP rs1053275; $r^2 = 0.837$ with intronic rs10953146; $r^2 = 0.81$ for intronic rs11970910 and rs117154505) [10].

Prediction of POase activity utilizing these six SNPs that predicted AREase activity (including the base model with age, current smoking status, and the four functional *PON1* SNPs) resulted in 84.02% of POase enzymatic variance explained (see Table 3). This compared to 82.74% of variance explained with the base model with the four functional SNPs, age, and smoking status, with the high percentage of variation explained largely due to the effects of the PON_{Q192R} polymorphism on paraoxon catalytic efficiency. Five of the six SNPs (excluding rs2237583) showed the same directionality of their effects as seen in the AREase analysis, and three had significant effects on POase: rs854567, rs2299257, and rs3917486. When creating a best-fit model that allowed any of the 51 SNPs studied to enter regression in addition to the base model, 84.96% of POase variance in activity was explained.

Table 1: Characteristics of the 51 SNPs studied in the PON gene cluster.

SNP	Gene	Function[a]	Minor allele[b]	Major allele	MAF[c]
rs854549	PON1	3′-downstream	A	C	0.337
rs3735590	PON1	3′-UTR	A	G	0.060
rs3917577	PON1	3′-UTR	G	A	0.089
rs854552	PON1	3′-UTR	G	A	0.265
rs3917551	PON1	Intronic	A	G	0.051
rs3917550	PON1	Intronic	A	G	0.137
rs2269829	PON1	Intronic	G	A	0.278
rs3917542	PON1	Intronic	A	G	0.227
rs3917538	PON1	Intronic	A	G	0.236
rs2299257	PON1	Intronic	C	A	0.391
rs854560	PON1	Coding	T	A	0.360
rs3917498	PON1	Intronic	A	C	0.345
rs28699500	PON1	Intronic	G	A	0.289
rs854561	PON1	Intronic	A	G	0.357
rs854565	PON1	Intronic	A	G	0.294
rs2272365	PON1	Intronic	C	A	0.154
rs854567	PON1	Intronic	A	G	0.185
rs3917490	PON1	Intronic	A	G	0.490
rs2299261	PON1	Intronic	G	A	0.354
rs854568	PON1	Intronic	G	A	0.219
rs2299262	PON1	Intronic	A	G	0.399
rs854569	PON1	Intronic	A	C	0.216
rs2237583	PON1	Intronic	A	G	0.284
rs3917486	PON1	Intronic	A	G	0.054
rs3917481	PON1	Intronic	A	G	0.015
rs2237584	PON1	Intronic	A	G	0.058
rs3917478	PON1	Intronic	G	A	0.118
rs3917476	PON1	Intronic	A	C	0.031
rs854571	PON1	5′-upstream	A	G	0.289
rs13236941	PON1	5′-upstream	A	G	0.164
rs13228784	PON1	Intronic	G	A	0.255
rs17883513	PON1	Intronic	G	A	0.032
rs17886762	PON1	Intronic	A	G	0.072
rs17883952	PON1	Intronic	A	G	0.052
rs17884000	PON3	Intronic	G	A	0.202
rs9640632	PON3	3′-UTR	G	A	0.456
rs468	PON3	Intronic	G	A	0.066
rs11768074	PON3	Intronic	A	G	0.157
rs10487132	PON3	Intronic	G	A	0.390
rs740264	PON3	Intronic	C	A	0.254
rs17884563	Intergenic	Intergenic	T	A	0.109
rs17880030	Intergenic	Intergenic	A	G	0.199
rs17881071	Intergenic	Intergenic	A	G	0.198
rs2375005	PON2	Intronic	T	A	0.462
rs12026	PON2	Coding	C	G	0.240
rs2299264	PON2	Intronic	A	G	0.241
rs7803148	PON2	Intronic	A	G	0.405
rs2158806	PON2	Intronic	C	A	0.237
rs2286233	PON2	Intronic	A	T	0.131

TABLE 1: Continued.

SNP	Gene	Function[a]	Minor allele[b]	Major allele	MAF[c]
rs10259688	PON2	Intronic	G	A	0.179
rs730365	PON2	Intronic	A	G	0.132

Abbreviations: UTR = untranslated region, MAF = minor allele frequency, intergenic = located between two gene regions.
[a] SNP functional annotation from SNP-Nexus.
[b] Major and minor allele annotation from the Illumina HumanCVD Bead Chip.
[c] Minor allele frequencies calculated from the CLEAR study cohort.

TABLE 2: Best-fit model from stepwise linear regression predicting PON1 AREase activity.

Variable	Coefficient (± SE)	Gene[a]	MAF[b]	t-statistic[c]	AREase Variation %	P
(Intercept)	284.09 (±13.99)	—	—	20.304	—	$< 2.0 \times 10^{-16}$
$PON1_{C-108T}$	−24.82 (±2.61)	(PON1)	0.43	−9.498	14.10%	$< 2.0 \times 10^{-16}$
$PON1_{G-162A}$	4.61 (±4.60)	(PON1)	0.18	1.002	0.21%	0.317
$PON1_{Q192R}$	−22.09 (±4.20)	PON1	0.33	−5.258	1.17%	1.8×10^{-7}
$PON1_{M55L}$	−7.05 (±3.64)	PON1	0.42	−1.94	1.01%	0.053
Age	−1.33 (±0.15)	—	—	−9.014	4.29%	$< 2.0 \times 10^{-16}$
Current smoker	−28.25 (±3.63)	—	—	−7.776	4.42%	1.95×10^{-14}
rs854567	−8.19 (±4.77)	(PON1)	A = 0.185	−1.719	2.34%	0.086
rs2299257	12.66 (±3.57)	(PON1)	C = 0.391	3.546	0.85%	4.11×10^{-4}
rs2237583	11.36 (±3.12)	(PON1)	A = 0.284	3.645	0.5%	2.82×10^{-4}
rs2375005	−8.32 (±2.56)	(PON2)	T = 0.462	−3.25	0.34%	0.001
rs3917486	14.91 (±4.97)	(PON1)	A = 0.054	2.998	0.58%	0.003
rs11768074	8.42 (±4.48)	(PON3)	A = 0.157	1.878	0.26%	0.061

SE = standard error, MAF = minor allele frequency.
[a] Noncoding SNPs are presented in parentheses, for example, (PON1).
[b] Minor allele frequencies for the four functional SNPs reported through dbSNP. The remaining minor allele frequencies were calculated via the CLEAR cohort.
[c] t-statistics and P values were calculated from the coefficients (from all subjects) and standard errors within the best-fit multivariate model by the glm function in R.

TABLE 3: Application of best-fit model for PON1 AREase activity to predict PON1 POase activity.

Variable	Coefficient (± SE)	Gene[a]	MAF[b]	t-Statistic[c]	POase Variation %	P
(Intercept)	29.36 (±1.17)	—	—	24.986	—	$< 2.0 \times 10^{-16}$
$PON1_{C-108T}$	−1.91 (±0.22)	(PON1)	0.43	−8.762	11.78%	$< 2.0 \times 10^{-16}$
$PON1_{G-162A}$	0.78 (±0.39)	(PON1)	0.18	2.023	3.93%	0.043
$PON1_{Q192R}$	9.67 (±0.35)	PON1	0.33	27.27	65.61%	$< 2.0 \times 10^{-16}$
$PON1_{M55L}$	−1.59 (±0.30)	PON1	0.42	−5.133	0.35%	3.5×10^{-7}
Age	−0.09 (±0.01)	—	—	−7.475	0.78%	1.81×10^{-13}
Current smoker	−1.27 (±0.31)	—	—	−4.155	0.31%	3.56×10^{-5}
rs854567	−1.69 (±0.40)	(PON1)	A = 0.185	−4.246	0.54%	2.41×10^{-5}
rs2299257	0.92 (±0.30)	(PON1)	C = 0.391	3.085	0.15%	0.002
rs2237583	−0.35 (±0.26)	(PON1)	A = 0.284	−1.347	0.09%	0.179
rs2375005	−0.23 (±0.21)	(PON2)	T = 0.462	−1.081	0.00%	0.28
rs3917486	2.20 (±0.42)	(PON1)	A = 0.054	5.271	0.47%	1.7×10^{-7}
rs11768074	0.42 (±0.37)	(PON3)	A = 0.157	1.114	0.02%	0.266

SE = standard error, MAF = minor allele frequency.
[a] Non-coding SNPs are presented in parentheses, for example, (PON1).
[b] Minor allele frequencies for the four functional SNPs reported through dbSNP. The remaining minor allele frequencies were calculated via the CLEAR cohort.
[c] t-statistics and P values were calculated from the coefficients (from all subjects) and standard errors within the best-fit multivariate model by the glm function in R.

Similar application to the prediction of DZOase activity utilizing the six SNPs from the predictive AREase model plus the base model (age, current smoking status, and the four functional *PON1* SNPs) resulted in 54.85% of variance explained (see Table 4). Five of the six SNPs (excluding rs11768074) showed the same directionality of their effects, and 4 had significant effects (rs2299257, rs2237583, rs2375005, and rs3917486). When using the four functional SNPs, age, and sex alone, 50.99% of DZOase activity was explained. However, when allowing any of the 51 tagSNPs to enter the best-fit model, 55.60% of DZOase activity was accounted for, suggesting that different SNPs may affect DZOase.

We attempted to replicate SNPs previously identified by Carlson et al. as predicting PON1 activity in a nonoverlapping sample of 523 subjects (Table 5), because significance in an overlapping subset does not constitute replication. The SNPs identified by Carlson et al. were rs854549, rs3917564, rs2269829, rs854566, rs854572, and rs3917510. In our full analysis, rs854566 was tagged by rs854567 (r^2 = 0.93), which did enter the full model using the full sample and predicted 2.34% of AREase activity. Rs2269829 and rs854549 were not predictive of AREase in the full dataset. Rs3917564 was not included in the full model analysis due to minor allele frequency <0.01. Rs3917510 and rs854572 (*PON*$_{-909}$ promoter) were not tagged in the CVD chip analysis and were genotyped separately for the replication analysis. When we considered the independent sample to test the six Carlson SNPs in a linear model predicting AREase, which also included age, current smoking status, and the four functional *PON1* SNPs, two of the six Carlson et al. findings were replicated. Both rs854566 (Carlson P = 0.014, current rs854567 P = 1.64×10^{-5}) and rs3917510 (Carlson P = 0.016, current P = 0.028) were significant in predicting AREase. Moreover, the direction of effect for rs854566 (Carlson coefficient = −10.6, current coefficient = −20.4) and rs3917510 (Carlson coefficient = 16.6, current coefficient = 14.3) were the same in both analyses.

None of the 10 SNPs identified in our full analyses, including the four previously known and the six newly reported to predict AREase, predicted CAAD status, considering the covariates censored age and current smoking status. Moreover, none of the SNPs had a P value <0.10. However, AREase activity, adjusted by age and current smoking status, was highly associated with CAAD status (P = 3.62×10^{-6}), as previously reported in a smaller sample.

4. Discussion

Only four *PON1* SNPs are well established to affect PON1 activity. These mutations alone account for approximately only 25% of PON1 AREase activity, leaving a large amount of variation left unexplained. In this study, we utilized denser tagSNP genotyping and a 2.65-fold increased sample size than those previously used in the Carlson et al. study [27] to examine the effects of common variants, demonstrating the presence of additional functional genetic variance within the *PON* gene cluster. We identified six additional SNPs that

predicted AREase activity (rs854567, rs2299257, rs2237583, rs2375005, rs3917486, and rs11768074). All are intronic, with four in *PON1* and one each in *PON2* (rs2375005) and *PON3* (rs11768074). Of these, only rs2375005 was found to be in strong LD with other regional SNPs in the 1000 genomes data, which included a *PON3* synonymous SNP (rs1053275). This LD block SNP is also reported to be in weaker LD with a *PON1*$_{-1741GA}$ promoter region polymorphism (maximum r^2 = 0.47) [34]. The remaining 5 SNPs associated with AREase may be functional or in weaker LD with a functional site. Rs854567 alone predicted 2.3% of the additional variance in AREase; it lies in the first intron of *PON1*, a common regulatory area.

For the many phenotypes with genetic associations to the PON cluster, knowledge of which SNPs are associated with functional changes is helpful in determining true associations from spurious ones. As discussed above, rs2375005 is in strong LD with an additional five SNPs in *PON3* (rs978903, rs1053275, rs10953146, rs11970910, and rs117154505). These include SNPs that have a reported association with sporadic amyotrophic lateral sclerosis [10]. In addition, Riedmaier et al. have demonstrated that a haplotype block including rs2375005 was associated with atorvastatin lactose hydrolysis and increased *PON1* mRNA expression in liver tissue [34]. Our results validate the presence of a functional SNP in this haplotype block.

In comparing these results to the six SNPs identified by Carlson et al, we replicate the effects of two SNPs, rs85466 and rs3917510, while failing to replicate four (Table 5) in nonoverlapping data. Rs854566 was represented in our analyses by the tagSNP, rs854567 (r^2 = 0.93). In contrast, the effects of rs854549, rs854572, rs3917564, and rs2269829 are not replicated here. Rs854572 is 5′ SNP *PON1*$_{-909C/G}$; while it has been associated with AREase level, smaller studies suggested that all of its activity was attributable to LD with the four functional SNPs [66]. The Carlson et al. paper suggested that this site may have independent activity, but we find no additional effects of this site, in an independent sample of 523 subjects. In sum, our current study confirms both the effects of rs854566 or its bin-mate rs854567, predicting 2.3% of AREase activity and the effects of rs3917510, while also identifying five additional tagSNPs that accounted for approximately 2.7% of PON1 AREase activity that were not accounted for by Carlson et al.

The finding of *PON2* and *PON3* SNPs (rs2375005 and rs11768074, resp.) predicting PON1 AREase activity is intriguing. The *PON* genes are in a cluster and arranged in order from the centromere as *PON1, PON3*, and *PON2*. Each is transcribed in the same direction, toward the centromere. Therefore, variants in the *PON2* or *PON3* genes lie 5′ to *PON1*. Rs2375005 is in the sixth of eight *PON2* introns. Rs11768074 is in the last *PON3* intron. Neither PON2 nor PON3 has intrinsic AREase activity [70], suggesting that these SNPs tag effects on PON1. As noted above, SNPs in the *PON3* rs2375005 haplotype block have been described to affect *PON1* mRNA level [34], thus the effects of these SNPs, or SNPs in LD with them, may regulate *PON1* expression.

Recent research in a cohort investigating SLE has linked rs17884563 and rs740264 in the *PON3* region [53] and

TABLE 4: Application of best-fit model for PON1 AREase activity to predict PON1 DZOase activity.

Variable	Coefficient (± SE)	Gene[a]	MAF[b]	t-statistic[c]	DZOase Activity %	P
(Intercept)	154.26 (±4.24)	—	—	36.365	—	$< 2.0 \times 10^{-16}$
$PON1_{C-108T}$	−8.69 (±0.79)	(PON1)	0.43	−11.054	12.82%	$< 2.0 \times 10^{-16}$
$PON1_{G-162A}$	5.03 (±1.40)	(PON1)	0.18	3.597	5.10%	3.4×10^{-4}
$PON1_{Q192R}$	−20.41 (±1.28)	PON1	0.33	−15.944	23.71%	$< 2.0 \times 10^{-16}$
$PON1_{M55L}$	−5.03 (±1.12)	PON1	0.42	−4.498	3.39%	7.75×10^{-6}
Age	−0.44 (±0.44)	—	—	−9.797	4.21%	$< 2.0 \times 10^{-16}$
Current Smoker	−5.65 (±1.11)	—	—	−5.103	1.28%	4.08×10^{-7}
rs854567	−2.26 (±1.44)	(PON1)	A = 0.185	−1.566	1.74%	0.118
rs2299257	4.03 (±1.08)	(PON1)	C = 0.391	3.73	0.70%	2.03×10^{-4}
rs2237583	4.69 (±0.95)	(PON1)	A = 0.284	4.956	0.92%	8.57×10^{-7}
rs2375005	−1.73 (±0.76)	(PON2)	T = 0.462	−2.258	0.22%	0.024
rs3917486	3.42 (±1.51)	(PON1)	A = 0.054	2.275	0.27%	0.023
rs11768074	−0.64 (±1.36)	(PON3)	A = 0.157	−0.466	0.01%	0.641

SE = standard error, MAF = minor allele frequency.

[a]Noncoding SNPs are presented in parentheses, for example, (PON1).

[b]Minor allele frequencies for the four functional SNPs reported through dbSNP. The remaining minor allele frequencies were calculated via the CLEAR cohort.

[c]t-statistics and P values were calculated from the coefficients (from all subjects) and standard errors within the best-fit multivariate model by the glm function in R.

TABLE 5: Comparison of SNPs found significant in prior Carlson et al.[a] study with current, non-overlapping sample.

SNP	Seattle SNP annotation	Carlson coefficient (±SE)	Carlson t-Statistic[b]	Carlson P[c]	Current coefficient (±SE)[d]	Current t-Statistic[b]	Current P[c]
rs854566[e]	$PON1_{6842}$	−10.6 (±4.3)	−2.480	0.014	−20.4 (±4.68)	−4.353	$1.64 \times 10-5$
rs3917510	$PON1_{12471}$	16.6 (±6.9)	2.424	0.016	14.3 (±6.48)	2.208	0.028
rs2269829	$PON1_{19470}$	−16.5 (±10.8)	−1.520	0.129	13.6 (±21.82)	0.625	0.533
rs3917564	$PON1_{23887}$	−39.0 (±18.1)	−2.153	0.032	15.0 (±26.67)	0.564	0.573
rs854549	$PON1_{29021}$	9.2 (±4.5)	2.051	0.041	−1.3 (±4.90)	−0.260	0.795
rs854572	$PON1_{895}$	13.0 (±4.9)	2.677	0.008	−0.28 (±4.97)	−0.056	0.955

SE = standard error.

[a]Carlson et al. study $n = 500$ European male subjects [27].

[b]t-Statistics and P-values were calculated from the coefficients from each subgroup (Carlson $n = 500$, current study $n = 523$) and standard errors within the best-fit multivariate model by the glm function in R.

[c]Both Carlson and current study utilized a linear regression model adjusting for age, current smoking status, and the four functional PON1 SNPs.

[d]Current study subset of 523 European male subjects not considered by Carlson et al.

[e]Represented by proxy SNP, rs854457, with LD $r^2 = 0.93$ in the current study.

five PON2 SNPs [52] (rs6954345, rs13306702, rs987539, rs11982486, and rs4729189) with PON1 POase activity [52, 53]. These investigations utilized POase rather than AREase activity [71]; this is not optimal, as the $PON1_{Q192R}$ activity accounts for most POase activity. Of the PON3 SNPs found to predict POase activity [53], both rs17884563 (intergenic between PON2 and PON3 in our annotation) and rs740264 were directly genotyped and included in our regression model for PON1 AREase activity but were not predictive. When applying rs17884563 or rs740264 to POase activity, which the aforementioned investigators used as their PON1 phenotype, neither was predictive of POase activity. For the PON2 SNPs predictive of POase in the SLE cohort, all five were represented by tag SNPs ($r^2 > 0.6$), but only rs2375005 ($r^2 = 1$ with rs987539) was predictive of PON1 AREase

activity. Interestingly, none of these five PON2 SNPs predict POase in our data, including rs2375005 ($P = 0.28$). The differences in PON2 and PON3 SNP associations between our data and the SLE cohort may reflect differences in cohort selection criteria (older male vascular disease versus younger female SLE, cases and controls) or sizes (1,322 in our data versus 922 in the SLE data).

Application of the six SNPs from the AREase best-fit model to predicting POase and DZOase activity resulted in the prediction of 98.89% and 98.65% of enzymatic activity predicted by models, where all 51 SNPs were allowed to enter. Three of these six SNPs, all in PON1, also predict both PON1 POase and DZOase activities. While it is clear why coding SNPs would differentially influence the PON1 degradation of these three substrates, it is less clear why regulatory variants would. Further investigation is required to determine if and how these noncoding SNPs differentially influence PON1's multiple activities at the genomic, molecular, or cellular level.

None of the six new SNPs that predicted AREase activity were predictive of CAAD. In addition, none of the four functional *PON1* SNPs were predictive of CAAD, which is consistent with past findings with smaller sample sizes in this cohort [2, 3, 27]. Important sources of variance in AREase activity that are not captured by these genotypes or the covariates of age and current smoking must account for the strong association between this activity and CAAD. Possible sources of AREase variation include rare regional variants, regional gene regulation not captured by genotyping (such as methylation), variation in genes outside the *PON* cluster, nongenetic factors including statin drug use [72] and diet [73, 74], as recently reviewed [75], as well as interactions among these. Evidence of interactions includes the report of the association of *PON1* genotype and CHD only in subjects with diabetes [76]. These results emphasize the importance of researching the correlation of PON1 and cardiovascular disease more broadly by utilizing "PON status," taking into account both the genotype of *PON1* SNPs and the plasma activity [11, 77], as well as investigating factors which affect the specific activity of PON1. PON1 has been suggested as a drug target for vascular and other diseases, thus a clear understanding of its role in disease is crucial [78].

Some limitations of this study must be considered. First, the study was comprised entirely of males of European descent, thereby limiting the generalizations that can be drawn from these findings. Second, this investigation considered only SNPs from the *PON* gene cluster. Variation in other genes may influence PON1 activity [79]. For example, peroxisome proliferator-activated receptor gamma (PPARG) activates PON1 expression in hepatocytes [80], leading to the possibility that variation in the *PPARG* gene could alter levels of PON1 protein. However, the larger size of this study and the denser tagSNPing of the *PON* cluster, relative to the earlier Carlson et al. work [27], allowed us to detect novel genetic variation that predicts *PON1* AREase activity.

In conclusion, our analysis of the *PON* gene cluster identifies six additional common genetic variants that predict AREase activity: four are novel, predicting 2.4% of AREase activity and two replicate past findings. The replicated SNPs include rs854567, which tags 2.3% of AREase variance, rs3917510, and rs2375005, which tags 0.3% of AREase variance. We do not identify additional effects of the $PON1_{-909}$ polymorphism. Future studies to quantify the role of rare genetic variation and variation outside the *PON* cluster on PON1 activity will be important. Finally, the continued lack of an association between *PON1, 2,* or *3* genetic variants and CAAD, while PON1 activity is highly predictive, underscores the importance of utilizing PON status in future studies investigating the link between PON1 and vascular or other disease.

Abbreviations

AIC: Akaike's Information Criterion
CAAD: Carotid artery disease
CHD: Coronary heart disease
CLEAR: Carotid Lesion Epidemiology and Risk cohort
DZOase: Diazoxon enzymatic hydrolysis
POase: Paraoxon enzymatic hydrolysis
PON: Paraoxonase
tagSNP: Tagging single-nucleotide polymorphism.

Conflict of Interests

The authors declare that they have no conflict of interests.

Acknowledgments

The authors would like to thank the study participants. They would also like to thank the following people for their technical assistance: Tamara Bacus, Edward Boyko, Julieann Marshall, Laura McKinstry, Karen Nakayama, Jane Ranchalis, and Jeff Rodenbaugh. This paper was funded by NIH RO1 HL67406 and a State of Washington Life Sciences Discovery Award to the Northwest Institute of Genetic Medicine. D. S. Kim is supported by a Sarnoff Cardiovascular Research Fellowship for Medical Students Award. This work utilized resources of SeattleSNPs; NHLBI Program for Genomic Applications, SeattleSNPs, Seattle, WA (http://pga.gs.washington.edu/). Past work in this cohort was supported in part by resources from the VA Puget Sound Health Care System, Seattle, Washington, including the Veteran Affairs Epidemiology Research and Information Center Program (Award CSP 701S).

References

[1] M. I. Mackness, B. Mackness, and P. N. Durrington, "Paraoxonase and coronary heart disease," *Atherosclerosis Supplements*, vol. 3, no. 4, pp. 49–55, 2002.

[2] G. P. Jarvik, L. S. Rozek, V. H. Brophy et al., "Paraoxonase (PON1) phenotype is a better predictor of vascular disease than is PON1192 or PON155 genotype," *Arteriosclerosis, Thrombosis, and Vascular Biology*, vol. 20, no. 11, pp. 2441–2447, 2000.

[3] G. P. Jarvik, T. S. Hatsukami, C. Carlson et al., "Paraoxonase activity, but not haplotype utilizing the linkage disequilibrium structure, predicts vascular disease," *Arteriosclerosis, Thrombosis, and Vascular Biology*, vol. 23, no. 8, pp. 1465–1471, 2003.

[4] M. I. Mackness, S. Arrol, C. Abbott, and P. N. Durrington, "Protection of low-density lipoprotein against oxidative modification by high-density lipoprotein associated paraoxonase," *Atherosclerosis*, vol. 104, no. 1-2, pp. 129–135, 1993.

[5] A. Graham, D. G. Hassall, S. Rafique, and J. S. Owen, "Evidence for a paraoxonase-independent inhibition of low-density lipoprotein oxidation by high-density lipoprotein," *Atherosclerosis*, vol. 135, no. 2, pp. 193–204, 1997.

[6] A. D. Watson, J. A. Berliner, S. Y. Hama et al., "Protective effect of high density lipoprotein associated paraoxonase. Inhibition of the biological activity of minimally oxidized low density lipoprotein," *Journal of Clinical Investigation*, vol. 96, no. 6, pp. 2882–2891, 1995.

[7] H. Cao, A. Girard-Globa, F. Berthezene, and P. Moulin, "Paraoxonase protection of LDL against peroxidation is independent of its esterase activity towards paraoxon and is

unaffected by the Q → R genetic polymorphism," *Journal of Lipid Research*, vol. 40, no. 1, pp. 133–139, 1999.

[8] D. M. Shih, L. Gu, Y. R. Xia et al., "Mice lacking serum paraoxonase are susceptible to organophosphate toxicity and atherosclerosis," *Nature*, vol. 394, no. 6690, pp. 284–287, 1998.

[9] D. M. Shih, Y. R. Xia, X. P. Wang et al., "Combined serum paraoxonase knockout/apolipoprotein E knockout mice exhibit increased lipoprotein oxidation and atherosclerosis," *Journal of Biological Chemistry*, vol. 275, no. 23, pp. 17527–17535, 2000.

[10] C. Besler, K. Heinrich, L. Rohrer et al., "Mechanisms underlying adverse effects of HDL on eNOS-activating pathways in patients with coronary artery disease," *Journal of Clinical Investigation*, vol. 121, no. 7, pp. 2693–2708, 2011.

[11] R. J. Richter, G. P. Jarvik, and C. E. Furlong, "Paraoxonase 1 status as a risk factor for disease or exposure," in *Paraoxonases in Inflammation, Infection, and Toxicology*, S. T. Reddy, Ed., vol. 660, pp. 29–35, Humana Press, 2010.

[12] V. H. Brophy, M. D. Hastings, J. B. Clendenning, R. J. Richter, G. P. Jarvik, and C. E. Furlong, "Polymorphisms in the human paraoxonase (PON1) promoter," *Pharmacogenetics*, vol. 11, no. 1, pp. 77–84, 2001.

[13] G. P. Jarvik, R. Jampsa, R. J. Richter et al., "Novel paraoxonase (PON1) nonsense and missense mutations predicted by functional genomic assay of PON1 status," *Pharmacogenetics*, vol. 13, no. 5, pp. 291–295, 2003.

[14] T. Bhattacharyya, S. J. Nicholls, E. J. Topol et al., "Relationship of paraoxonase 1 (PON1) gene polymorphisms and functional activity with systemic oxidative stress and cardiovascular risk," *Journal of the American Medical Association*, vol. 299, no. 11, pp. 1265–1276, 2008.

[15] M. Serrato and A. J. Marian, "A variant of human paraoxonase/arylesterase (HUMPONA) gene is a risk factor for coronary artery disease," *Journal of Clinical Investigation*, vol. 96, no. 6, pp. 3005–3008, 1995.

[16] M. Antikainen, S. Murtomäki, M. Syvänne et al., "The Gln-Arg191 polymorphism of the human paraoxonase gene (HUMPONA) is not associated with the risk of coronary artery disease in Finns," *Journal of Clinical Investigation*, vol. 98, no. 4, pp. 883–885, 1996.

[17] S. M. Herrmann, H. Blanc, O. Poirier et al., "The Gln/Arg polymorphism of human paraoxonase (PON 192) is not related to myocardial infarction in the ECTIM Study," *Atherosclerosis*, vol. 126, no. 2, pp. 299–303, 1996.

[18] H. Schmidt, R. Schmidt, K. Niederkorn et al., "Paraoxonase PON1 polymorphism Leu-Met54 is associated with carotid atherosclerosis: results of The Austrian Stroke Prevention Study," *Stroke*, vol. 29, no. 10, pp. 2043–2048, 1998.

[19] M. Wang, X. Lang, L. Zou, S. Huang, and Z. Xu, "Four genetic polymorphisms of paraoxonase gene and risk of coronary heart disease: a meta-analysis based on 88 case-control studies," *Atherosclerosis*, vol. 214, no. 2, pp. 377–385, 2011.

[20] D. A. Lawlor, I. N. M. Day, T. R. Gaunt et al., "The association of the PON1 Q192R polymorphism with coronary heart disease: findings from the British Women's Heart and Health cohort study and a meta-analysis," *BMC Genetics*, vol. 5, no. 1, arricle 17, 2004.

[21] J. G. Wheeler, B. D. Keavney, H. Watkins, R. Collins, and J. Danesh, "Four paraoxonase gene polymorphisms in 11 212 cases of coronary heart disease and 12 786 controls: meta-analysis of 43 studies," *The Lancet*, vol. 363, no. 9410, pp. 689–695, 2004.

[22] K. E. Lohmueller, C. L. Pearce, M. Pike, E. S. Lander, and J. N. Hirschhorn, "Meta-analysis of genetic association studies supports a contribution of common variants to susceptibility to common disease," *Nature Genetics*, vol. 33, no. 2, pp. 177–182, 2003.

[23] B. Voetsch, K. S. Benke, B. P. Damasceno, L. H. Siqueira, and J. Loscalzo, "Paraoxonase 192 Gln → Arg polymorphism: an independent risk factor for nonfatal arterial ischemic stroke among young adults," *Stroke*, vol. 33, no. 6, pp. 1459–1464, 2002.

[24] B. Voetsch, K. S. Benke, C. I. Panhuysen, B. P. Damasceno, and J. Loscalzo, "The combined effect of paraoxonase promoter and coding region polymorphisms on the risk of arterial ischemic stroke among young adults," *Archives of Neurology*, vol. 61, no. 3, pp. 351–356, 2004.

[25] R. Schmidt, H. Schmidt, F. Fazekas et al., "MRI cerebral white matter lesions and paraoxonase PON1 polymorphisms: three-year follow-up of the Austrian stroke prevention study," *Arteriosclerosis, Thrombosis, and Vascular Biology*, vol. 20, no. 7, pp. 1811–1816, 2000.

[26] H. Markus, Z. Kapozsta, R. Ditrich et al., "Increased common carotid intima-media thickness in UK African Caribbeans and its relation to chronic inflammation and vascular candidate gene polymorphisms," *Stroke*, vol. 32, no. 11, pp. 2465–2471, 2001.

[27] C. S. Carlson, P. J. Heagerty, T. S. Hatsukami et al., "TagSNP analyses of the PON gene cluster: effects on PON1 activity, LDL oxidative susceptibility, and vascular disease," *Journal of Lipid Research*, vol. 47, no. 5, pp. 1014–1024, 2006.

[28] E. Topic, A. Timundic, M. Ttefanovic et al., "Polymorphism of Apoprotein E (APOE), methylenetetrahydrofolate reductase (MTHFR) and paraoxonase (PON1) genes in patients with cerebrovascular disease," *Clinical Chemistry and Laboratory Medicine*, vol. 39, pp. 346–350, 2001.

[29] L. G. Costa, B. E. McDonald, S. D. Murphy et al., "Serum paraoxonase and its influence on paraoxon and chlorpyrifos-oxon toxicity in rats," *Toxicology and Applied Pharmacology*, vol. 103, no. 1, pp. 66–76, 1990.

[30] R. J. Richter, G. P. Jarvik, and C. E. Furlong, "Determination of paraoxonase 1 status without the use of toxic organophosphate substrates," *Circulation*, vol. 1, no. 2, pp. 147–152, 2008.

[31] R. J. Richter, G. P. Jarvik, and C. E. Furlong, "Paraoxonase 1 (PON1) status and substrate hydrolysis," *Toxicology and Applied Pharmacology*, vol. 235, no. 1, pp. 1–9, 2009.

[32] C. E. Furlong, N. Holland, R. J. Richter, A. Bradman, A. Ho, and B. Eskenazi, "PON1 status of farmworker mothers and children as a predictor of organophosphate sensitivity," *Pharmacogenetics and Genomics*, vol. 16, no. 3, pp. 183–190, 2006.

[33] R. W. James, "A long and winding road: defining the biological role and clinical importance of paraoxonases," *Clinical Chemistry and Laboratory Medicine*, vol. 44, no. 9, pp. 1052–1059, 2006.

[34] S. Riedmaier, K. Klein, S. Winter, U. Hofmann, M. Schwab, and U. Zanger, "Paraoxonase (PON1 and PON3) polymorphisms: impact on liver expression and atorvastatin-lactone hydrolysis," *Frontiers in Pharmacology*, vol. 2, p. 41, 2011.

[35] H. J. Bouman, E. Schömig, J. W. Van Werkum et al., "Paraoxonase-1 is a major determinant of clopidogrel efficacy," *Nature Medicine*, vol. 17, no. 1, pp. 110–116, 2011.

[36] J. P. Lewis, A. S. Fisch, K. Ryan et al., "Paraoxonase 1 (PON1) gene variants are not associated with clopidogrel response," *Clinical Pharmacology and Therapeutics*, vol. 90, no. 4, pp. 568–574, 2011.

[37] B. Mackness, M. I. Mackness, M. Aviram, and G. Paragh, *Paraoxonases: Their Role in Disease Development and Xenobiotic Metabolism (Proteins and Cell Regulation)*, Springer, 2007.

[38] A. Carmine, S. Buervenich, O. Sydow, M. Anvret, and L. Olson, "Further evidence for an association of the paraoxonase 1 (PON1) met-54 allele with Parkinson's disease," *Movement Disorders*, vol. 17, no. 4, pp. 764–766, 2002.

[39] I. Kondo and M. Yamamoto, "Genetic polymorphism of paraoxonase 1 (PON1) and susceptibility to Parkinson's disease," *Brain Research*, vol. 806, no. 2, pp. 271–273, 1998.

[40] S. N. Akhmedova, A. K. Yakimovsky, and E. I. Schwartz, "Paraoxonase 1 Met-Leu 54 polymorphism is associated with Parkinson's disease," *Journal of the Neurological Sciences*, vol. 184, no. 2, pp. 179–182, 2001.

[41] E. Zintzaras and G. M. Hadjigeorgiou, "Association of paraoxonase 1 gene polymorphisms with risk of Parkinson's disease: a meta-analysis," *Journal of Human Genetics*, vol. 49, no. 9, pp. 474–481, 2004.

[42] S. N. Kelada, P. Costa-Mallen, H. Checkoway et al., "Paraoxonase 1 promoter and coding region polymorphisms in Parkinson's disease," *Journal of Neurology Neurosurgery and Psychiatry*, vol. 74, no. 4, pp. 546–547, 2003.

[43] S. Mostafa, "Paraoxonase 1 genetic polymorphisms and susceptibility to breast cancer: a meta-analysis," *Cancer Epidemiology*, vol. 36, no. 2, pp. e101–e103, 2012.

[44] C. Liu and L. Liu, "Polymorphisms in three obesity-related genes (LEP, LEPR, and PON1) and breast cancer risk: a meta-analysis," *Tumor Biology*, vol. 32, no. 6, pp. 1233–1240, 2011.

[45] A. Javadzadeh, A. Ghorbanihaghjo, E. Bahreini, N. Rashtchizadeh, H. Argani, and S. Alizadeh, "Serum paraoxonase phenotype distribution in exudative age-related macular degeneration and its relationship to homocysteine and oxidized low-density lipoprotein," *Retina*, vol. 32, no. 4, pp. 658–666, 2012.

[46] M. Brión, M. Sanchez-Salorio, M. Cortón et al., "Genetic association study of age-related macular degeneration in the Spanish population," *Acta Ophthalmologica*, vol. 89, no. 1, pp. e12–e22, 2011.

[47] G. J. T. Pauer, G. M. Sturgill, N. S. Peachey, and S. A. Hagstrom, "Protective effect of paraoxonase 1 gene variant Gln192Arg in age-related macular degeneration," *American Journal of Ophthalmology*, vol. 149, no. 3, pp. 513–522, 2010.

[48] O. Ates, S. Azzi, H. H. Alp et al., "Decreased serum paraoxonase 1 activity and increased serum homocysteine and malondialdehyde levels in Age-related macular degeneration," *Tohoku Journal of Experimental Medicine*, vol. 217, no. 1, pp. 17–22, 2009.

[49] H. Esfandiary, U. Chakravarthy, C. Patterson, I. Young, and A. E. Hughes, "Association study of detoxification genes in age related macular degeneration," *British Journal of Ophthalmology*, vol. 89, no. 4, pp. 470–474, 2005.

[50] G. Baskol, S. Karakucuk, A. O. Oner et al., "Serum paraoxonase 1 activity and lipid peroxidation levels in patients with age-related macular degeneration," *Ophthalmologica*, vol. 220, no. 1, pp. 12–16, 2005.

[51] P. N. Baird, D. Chu, E. Guida, H. T. V. Vu, and R. Guymer, "Association of the M55L and Q192R paraoxonase gene polymorphisms with age-related macular degeneration," *American Journal of Ophthalmology*, vol. 138, no. 4, pp. 665–666, 2004.

[52] S. Dasgupta, F. Y. Demirci, A. S. Dressen et al., "Association analysis of PON2 genetic variants with serum paraoxonase activity and systemic lupus erythematosus," *BMC Medical Genetics*, vol. 12, article 7, no. 1, 2011.

[53] D. K. Sanghera, S. Manzi, R. L. Minster et al., "Genetic variation in the paraoxonase-3 (PON3) gene is associated with serum PON1 activity," *Annals of Human Genetics*, vol. 72, no. 1, pp. 72–81, 2008.

[54] E. Kiss, I. Seres, T. Tarr, Z. Kocsis, G. Szegedi, and G. Paragh, "Reduced paraoxonase1 activity is a risk for atherosclerosis in patients with systemic lupus erythematosus," *Annals of the New York Academy of Sciences*, vol. 1108, pp. 83–91, 2007.

[55] B. MacKness, M. I. MacKness, S. Arrol et al., "Serum paraoxonase (PON1) 55 and 192 polymorphism and paraoxonase activity and concentration in non-insulin dependent diabetes mellitus," *Atherosclerosis*, vol. 139, no. 2, pp. 341–349, 1998.

[56] M. Rosenblat, D. Draganov, C. E. Watson, C. L. Bisgaier, B. N. La Du, and M. Aviram, "Mouse macrophage paraoxonase 2 activity is increased whereas cellular paraoxonase 3 activity is decreased under oxidative stress," *Arteriosclerosis, Thrombosis, and Vascular Biology*, vol. 23, no. 3, pp. 468–474, 2003.

[57] S. T. Reddy, D. J. Wadleigh, V. Grijalva et al., "Human paraoxonase-3 is an HDL-associated enzyme with biological activity similar to paraoxonase-1 protein but is not regulated by oxidized lipids," *Arteriosclerosis, Thrombosis, and Vascular Biology*, vol. 21, no. 4, pp. 542–547, 2001.

[58] C. J. Ng, D. J. Wadleigh, A. Gangopadhyay et al., "Paraoxonase-2 is a ubiquitously expressed protein with antioxidant properties and is capable of preventing cell-mediated oxidative modification of low density lipoprotein," *Journal of Biological Chemistry*, vol. 276, no. 48, pp. 44444–44449, 2001.

[59] H. Mochizuki, S. W. Scherer, T. Xi et al., "Human PON2 gene at 7q21.3: cloning, multiple mRNA forms, and missense polymorphisms in the coding sequence," *Gene*, vol. 213, no. 1-2, pp. 149–157, 1998.

[60] Q. Chen, S. E. Reis, C. M. Kammerer et al., "Association between the severity of angiographic coronary artery disease and paraoxonase gene polymorphisms in the National Heart, Lung, and Blood Institute-sponsored Women's Ischemia Syndrome Evaluation (WISE) study," *American Journal of Human Genetics*, vol. 72, no. 1, pp. 13–22, 2003.

[61] E. A. Ozer, A. Pezzulo, D. M. Shih et al., "Human and murine paraoxonase 1 are host modulators of Pseudomonas aeruginosa quorum-sensing," *FEMS Microbiology Letters*, vol. 253, no. 1, pp. 29–37, 2005.

[62] D. A. Stoltz, E. A. Ozer, C. J. Ng et al., "Paraoxonase-2 deficiency enhances Pseudomonas aeruginosa quorum sensing in murine tracheal epithelia," *American Journal of Physiology*, vol. 292, no. 4, pp. L852–L860, 2007.

[63] D. A. Stoltz, E. A. Ozer, P. J. Taft et al., "Drosophila are protected from *Pseudomonas aeruginosa* lethality by transgenic expression of paraoxonase-1," *Journal of Clinical Investigation*, vol. 118, no. 9, pp. 3123–3131, 2008.

[64] J. Ronald, R. Rajagopalan, J. E. Ranchalis et al., "Analysis of recently identified dyslipidemia alleles reveals two loci that contribute to risk for carotid artery disease," *Lipids in Health and Disease*, vol. 8, article 52, 2009.

[65] J. K. Pritchard, M. Stephens, and P. Donnelly, "Inference of population structure using multilocus genotype data," *Genetics*, vol. 155, no. 2, pp. 945–959, 2000.

[66] V. H. Brophy, R. L. Jampsa, J. B. Clendenning, L. A. McKinstry, G. P. Jarvik, and C. E. Furlong, "Effects of 5′ regulatory-region polymorphisms on paraoxonase-gene (PON1) expression," *American Journal of Human Genetics*, vol. 68, no. 6, pp. 1428–1436, 2001.

[67] C. S. Carlson, M. A. Eberle, M. J. Rieder, Q. Yi, L. Kruglyak, and D. A. Nickerson, "Selecting a Maximally Informative Set

of Single-Nucleotide Polymorphisms for Association Analyses Using Linkage Disequilibrium," *American Journal of Human Genetics*, vol. 74, no. 1, pp. 106–120, 2004.

[68] C. Chelala, A. Khan, and N. R. Lemoine, "SNPnexus: a web database for functional annotation of newly discovered and public domain single nucleotide polymorphisms," *Bioinformatics*, vol. 25, no. 5, pp. 655–661, 2009.

[69] S. Deakin, I. Leviev, S. Guernier, and R. W. James, "Simvastatin modulates expression of the PON1 gene and increases serum paraoxonase: a role for sterol regulatory element-binding protein-2," *Arteriosclerosis, Thrombosis, and Vascular Biology*, vol. 23, no. 11, pp. 2083–2089, 2003.

[70] D. I. Draganov, J. F. Teiber, A. Speelman, Y. Osawa, R. Sunahara, and B. N. La Du, "Human paraoxonases (PON1, PON2, and PON3) are lactonases with overlapping and distinct substrate specificities," *Journal of Lipid Research*, vol. 46, no. 6, pp. 1239–1247, 2005.

[71] L. M. Tripi, S. Manzi, Q. Chen et al., "Relationship of serum paraoxonase 1 activity and paraoxonase 1 genotype to risk of systemic lupus erythematosus," *Arthritis and Rheumatism*, vol. 54, no. 6, pp. 1928–1939, 2006.

[72] B. V. Kural, C. Örem, H. A. Uydu, A. Alver, and A. Örem, "The effects of lipid-lowering therapy on paraoxonase activities and their relationship with the oxidant-antioxidant system in patients with dyslipidemia," *Coronary Artery Disease*, vol. 15, no. 5, pp. 277–283, 2004.

[73] G. P. Jarvik, N. T. Tsai, L. A. McKinstry et al., "Vitamin C and E intake is associated with increased paraoxonase activity," *Arteriosclerosis, Thrombosis, and Vascular Biology*, vol. 22, no. 8, pp. 1329–1333, 2002.

[74] P. Koncsos, I. Seres, M. Harangi et al., "Favorable effect of short-term lifestyle intervention on human paraoxonase-1 activity and adipokine levels in childhood obesity," *Journal of the American College of Nutrition*, vol. 30, no. 5, pp. 333–339, 2011.

[75] L. G. Costa, G. Giordano, and C. E. Furlong, "Pharmacological and dietary modulators of paraoxonase 1 (PON1) activity and expression: the hunt goes on," *Biochemical Pharmacology*, vol. 81, no. 3, pp. 337–344, 2011.

[76] S. Bhaskar, M. Ganesan, G. R. Chandak et al., "Association of PON1 and APOA5 gene polymorphisms in a cohort of indian patients having coronary artery disease with and without type 2 diabetes," *Genetic Testing and Molecular Biomarkers*, vol. 15, no. 7-8, pp. 507–512, 2011.

[77] C. E. Furlong, S. M. Suzuki, R. C. Stevens et al., "Human PON1, a biomarker of risk of disease and exposure," *Chemico-Biological Interactions*, vol. 187, no. 1-3, pp. 355–361, 2010.

[78] Z. She, H. Chen, Y. Yan, H. Li, and D. Liu, "The human paraoxonase gene cluster as a target in the treatment of atherosclerosis," *Antioxidants & Redox Signaling*, vol. 16, no. 6, pp. 597–632, 2012.

[79] D. A. Winnier, D. L. Rainwater, S. A. Cole et al., "Multiple QTLs influence variation in paraoxonase 1 activity in Mexican Americans," *Human Biology*, vol. 78, no. 3, pp. 341–352, 2006.

[80] J. Khateeb, A. Gantman, A. J. Kreitenberg, M. Aviram, and B. Fuhrman, "Paraoxonase 1 (PON1) expression in hepatocytes is upregulated by pomegranate polyphenols: a role for PPAR-γ pathway," *Atherosclerosis*, vol. 208, no. 1, pp. 119–125, 2010.

Permissions

The contributors of this book come from diverse backgrounds, making this book a truly international effort. This book will bring forth new frontiers with its revolutionizing research information and detailed analysis of the nascent developments around the world.

We would like to thank all the contributing authors for lending their expertise to make the book truly unique. They have played a crucial role in the development of this book. Without their invaluable contributions this book wouldn't have been possible. They have made vital efforts to compile up to date information on the varied aspects of this subject to make this book a valuable addition to the collection of many professionals and students.

This book was conceptualized with the vision of imparting up-to-date information and advanced data in this field. To ensure the same, a matchless editorial board was set up. Every individual on the board went through rigorous rounds of assessment to prove their worth. After which they invested a large part of their time researching and compiling the most relevant data for our readers. Conferences and sessions were held from time to time between the editorial board and the contributing authors to present the data in the most comprehensible form. The editorial team has worked tirelessly to provide valuable and valid information to help people across the globe.

Every chapter published in this book has been scrutinized by our experts. Their significance has been extensively debated. The topics covered herein carry significant findings which will fuel the growth of the discipline. They may even be implemented as practical applications or may be referred to as a beginning point for another development. Chapters in this book were first published by Hindawi Publishing Corporation; hereby published with permission under the Creative Commons Attribution License or equivalent.

The editorial board has been involved in producing this book since its inception. They have spent rigorous hours researching and exploring the diverse topics which have resulted in the successful publishing of this book. They have passed on their knowledge of decades through this book. To expedite this challenging task, the publisher supported the team at every step. A small team of assistant editors was also appointed to further simplify the editing procedure and attain best results for the readers.

Our editorial team has been hand-picked from every corner of the world. Their multi-ethnicity adds dynamic inputs to the discussions which result in innovative outcomes. These outcomes are then further discussed with the researchers and contributors who give their valuable feedback and opinion regarding the same. The feedback is then collaborated with the researches and they are edited in a comprehensive manner to aid the understanding of the subject.

Apart from the editorial board, the designing team has also invested a significant amount of their time in understanding the subject and creating the most relevant covers. They scrutinized every image to scout for the most suitable representation of the subject and create an appropriate cover for the book.

The publishing team has been involved in this book since its early stages. They were actively engaged in every process, be it collecting the data, connecting with the contributors or procuring relevant information. The team has been an ardent support to the editorial, designing and production team. Their endless efforts to recruit the best for this project, has resulted in the accomplishment of this book. They are a veteran in the field of academics and their pool of knowledge is as vast as their experience in printing. Their expertise and guidance has proved useful at every step. Their uncompromising quality standards have made this book an exceptional effort. Their encouragement from time to time has been an inspiration for everyone.

The publisher and the editorial board hope that this book will prove to be a valuable piece of knowledge for researchers, students, practitioners and scholars across the globe.

List of Contributors

Jorge A. López-Velázquez, Luis D. Carrillo-Cordova, Norberto C. Chavez-Tapia, Misael Uribe and Nahum Mendez-Sanchez
Liver Research Unit, Medica Sur Clinic & Foundation, Puente de Piedra 150, Colonia Toriello Guerra, 14050 Tlalpan, Mexico City, Mexico

Raghavendra Pralhada Rao, Nanditha Vaidyanathan, Mathiyazhagan Rengasamy, Anup Mammen Oommen, Neeti Somaiya and M. R. Jagannath
Biology Group, Connexios Life Sciences Private Limited, No. 49 Shilpa Vidya, First Main Road, 3rd Phase JP Nagara, Bangalore 560078, India

Samar M. Hammad and Mohammed M. Al Gadban
Department of Regenerative Medicine and Cell Biology, Medical University of South Carolina, Charleston, SC 29425, USA

Andrea J. Semler
Division of Endocrinology, Metabolism, and Medical Genetics, Department of Medicine, Medical University of South Carolina, Charleston, SC 29425, USA

Richard L. Klein
Division of Endocrinology, Metabolism, and Medical Genetics, Department of Medicine, Medical University of South Carolina, Charleston, SC 29425, USA
Research Service, Ralph H. Johnson VA Medical Center, United States Department of Veterans Affairs, Charleston, SC 29401, USA

Ellen E. Quillen, David L. Rainwater, Thomas D. Dyer, Melanie A. Carless, Joanne E. Curran, Matthew P. Johnson, Harald H. H. Goring, Shelley A. Cole, Sue Rutherford, Jean W. MacCluer, Eric K. Moses, John Blangero, Laura Almasy and Michael C. Mahaney
Department of Genetics, Texas Biomedical Research Institute, P.O. Box 760549, San Antonio, TX 78245-0549, USA

Christopher Ngosong
Institute of Biology, Ecology Group, Humboldt-Universitat zu Berlin, Philippstraße 13, 10115 Berlin, Germany
Department of Natural Resource Sciences, MacCampus, McGill University, 21,111 Lakeshore Road, Ste Anne de Bellevue, QC, Canada H9X 3V9

Elke Gabriel
Institute of Vegetable and Ornamental Crops Großbeeren, Theodor-Echtermeyer-Weg 1, 14979 Großbeeren, Germany
Faculty of Food and Agriculture, UAE University, Jimi 1 Campus, Building 52, P.O. Box 17555, Al Ain, Abu Dhabi, UAE

Liliane Ruess
Institute of Biology, Ecology Group, Humboldt-Universitat zu Berlin, Philippstraße 13, 10115 Berlin, Germany

Tamotsu Tsukahara
Department of Integrative Physiology and Bio-System Control, Shinshu University School of Medicine, 3-1-1 Asahi, Matsumoto, Nagano 390-8621, Japan

J. A. Stephenson and B. Morgan
Department of Cancer Studies and Molecular Medicine, University of Leicester, Leicester Royal Infirmary, Leicester LE1 5WW, UK
Department of Imaging, Leicester Royal Infirmary, Leicester LE1 5WW, UK

O. Al-Taan and A. Arshad
Department of Cancer Studies and Molecular Medicine, University of Leicester, Leicester Royal Infirmary, Leicester LE1 5WW, UK
Department of Surgery, University Hospitals of Leicester, Leicester General Hospital, Leicester LE5 4PW, UK

M. S. Metcalfe and A. R. Dennison
Department of Surgery, University Hospitals of Leicester, Leicester General Hospital, Leicester LE5 4PW, UK

Maria C. de Beer
Saha Cardiovascular Research Center, University of Kentucky Medical Center, Lexington, KY 40536, USA
Department of Physiology, University of Kentucky Medical Center, Lexington, KY 40536, USA

Joanne M. Wroblewski, Victoria P. Noffsinger, Ailing Ji, Frederick C. de Beer and Nancy R. Webb
Saha Cardiovascular Research Center, University of Kentucky Medical Center, Lexington, KY 40536, USA
Department of Internal Medicine, University of Kentucky Medical Center, Lexington, KY 40536, USA

Jason M. Meyer
Saha Cardiovascular Research Center, University of Kentucky Medical Center, Lexington, KY 40536, USA
Department of Molecular and Cellular Biochemistry, University of Kentucky Medical Center, Lexington, KY 40536, USA

Deneys R. van der Westhuyzen
Saha Cardiovascular Research Center, University of Kentucky Medical Center, Lexington, KY 40536, USA
Department of Internal Medicine, University of Kentucky Medical Center, Lexington, KY 40536, USA
Department of Molecular and Cellular Biochemistry, University of Kentucky Medical Center, Lexington, KY 40536, USA
Department of Veterans Affairs Medical Center, Lexington, KY 40511, USA

Svetlana Dzitoyeva and Hari Manev
Department of Psychiatry, Psychiatric Institute, University of Illinois at Chicago, Chicago, IL 60612, USA

Ornella de Bari, Brent A. Neuschwander-Tetri and David Q.-H.Wang
Division of Gastroenterology and Hepatology, Department of Internal Medicine, Edward Doisy Research Center, Saint Louis University School of Medicine, 1100 S. Grand Boulevard, Room 205, St. Louis, MO 63104, USA

Min Liu
Department of Pathology and Laboratory Medicine, University of Cincinnati College of Medicine, Cincinnati, OH 45237, USA

Piero Portincasa
Department of Internal Medicine and Public Medicine, Clinica Medica "A. Murri", University of Bari Medical School, 70124 Bari, Italy

Gabriella Garruti
Section of Endocrinology, Department of Emergency and Organ Transplantations, University of Bari "Aldo Moro" Medical School, Piazza G. Cesare 11, 70124 Bari, Italy
Division of Gastroenterology and Hepatology, Department of Internal Medicine, Edward Doisy Research Center, Saint Louis University School of Medicine, 1100 S. Grand Boulevard, Room 205, St. Louis, MO 63104, USA

Helen H. Wang and David Q.-H.Wang
Division of Gastroenterology and Hepatology, Department of Internal Medicine, Edward Doisy Research Center, Saint Louis University School of Medicine, 1100 S. Grand Boulevard, Room 205, St. Louis, MO 63104, USA

Leonilde Bonfrate and Piero Portincasa
Department of Biomedical Sciences and Human Oncology, Clinica Medica "A. Murri", University of Bari Medical School, Piazza G. Cesare 11, 70124 Bari, Italy

Ornella de Bari
Division of Gastroenterology and Hepatology, Department of Internal Medicine, Edward Doisy Research Center, Saint Louis University School of Medicine, 1100 S. Grand Boulevard, Room 205, St. Louis, MO 63104, USA
Department of Biomedical Sciences and Human Oncology, Clinica Medica "A. Murri", University of Bari Medical School, Piazza G. Cesare 11, 70124 Bari, Italy

Ines Witte, Ulrich Foerstermann and Sven Horke
Department of Pharmacology, University Medical Center of the Johannes-Gutenberg University Mainz, Obere Zahlbacher Street 67, 55131 Mainz, Germany

Asokan Devarajan
Department of Medicine, University of California, Los Angeles, CA 90095, USA

Srinivasa T. Reddy
Department of Medicine, University of California, Los Angeles, CA 90095, USA
Department of Molecular and Medical Pharmacology, University of California, Los Angeles, CA 90095, USA

Eva-Maria Schweikert, Julianna Amort, Petra Wilgenbus, Ulrich Forstermann and Sven Horke
Institute of Pharmacology, University Medical Center of the Johannes Gutenberg-University Mainz, Obere Zahlbacher Straße 67, 55131 Mainz, Germany

John F. Teiber
Division of Epidemiology, Department of Internal Medicine, The University of Texas Southwestern Medical Center, 5323 Harry Hines Boulevard, Dallas, TX 75390, USA

Paulo J. Oliveira and Vilma A. Sardao
CNC—Center for Neuroscience and Cell Biology, University of Coimbra, 3004-517 Coimbra, Portugal

Rui A. Carvalho
CNC—Center for Neuroscience and Cell Biology, University of Coimbra, 3004-517 Coimbra, Portugal
Department of Life Sciences, University of Coimbra, 3004-517 Coimbra, Portugal

Piero Portincasa and Leonilde Bonfrate
Department of Internal Medicine and Public Medicine, Clinica Medica "A. Murri", University of Bari Medical School, 70124 Bari, Italy

Helena Van Overloop, Gerd Van der Hoeven and Paul P. Van Veldhoven
Department Cellular and Molecular Medicine, Katholieke Universiteit Leuven, Campus Gasthuisberg O&N1, LIPIT, Herestraat, Box 601, 3000 Leuven, Belgium

Nicholas D. Oakes, Ann Kjellstedt, Pia Thalén and Bengt Ljung
AstraZeneca R&D Molndal, 431 83 Molndal, Sweden

Nigel Turner
Diabetes and Obesity Program, Garvan Institute of Medical Research, Darlinghurst, NSW2010, Australia
School of Medical Sciences, University of New South Wales, Sydney, NSW2052, Australia

Johnny X. Huang and Matthew A. Cooper
Institute for Molecular Bioscience, The University of Queensland, 306 Carmody Road, St. Lucia, QLD 4072, Australia

Mohammad A. K. Azad, Roger L. Nation, Jian Li and Tony Velkov
Drug Development and Innovation, Drug Delivery, Disposition and Dynamics, Monash Institute of Pharmaceutical Sciences, Monash University, 381 Royal Parade, Parkville, VIC 3052, Australia

Elizabeth Yuriev
Medicinal Chemistry, Monash Institute of Pharmaceutical Sciences, Monash University, 381 Royal Parade, Parkville, VIC 3052, Australia

Mark A. Baker
Priority Research Centre in Reproductive Science, School of Environmental and Life Sciences, University of Newcastle, Callaghan, NSW 2308, Australia

Daniel S. Kim
Department of Genome Sciences, University of Washington School of Medicine, Seattle, WA 98195, USA

Amber A. Burt, Jane E. Ranchalis, Rebecca J. Richter, Julieann K. Marshall and Elisabeth A. Rosenthal
Department of Medicine, Division of Medical Genetics, University of Washington School of Medicine, P.O. Box 357720, Seattle, WA 98195, USA

Clement E. Furlong and Gail P. Jarvik
Department of Genome Sciences, University of Washington School of Medicine, Seattle, WA 98195, USA
Department of Medicine, Division of Medical Genetics, University of Washington School of Medicine, P.O. Box 357720, Seattle, WA 98195, USA

Jason F. Eintracht
Department of General Medicine, Virginia Mason Medical Center, Seattle, WA 98101, USA